Women in Bioorganic Chemistry

Women in Bioorganic Chemistry

Editors

**Francesca Cardona
Camilla Parmeggiani
Camilla Matassini**

MDPI • Basel • Beijing • Wuhan • Barcelona • Belgrade • Manchester • Tokyo • Cluj • Tianjin

Editors
Francesca Cardona
Università di Firenze
Sesto Fiorentino
Italy

Camilla Parmeggiani
Università di Firenze
Sesto Fiorentino
Italy

Camilla Matassini
Università di Firenze
Sesto Fiorentino
Italy

Editorial Office
MDPI
St. Alban-Anlage 66
4052 Basel, Switzerland

This is a reprint of articles from the Special Issue published online in the open access journal *Molecules* (ISSN 1420-3049) (available at: https://www.mdpi.com/journal/molecules/special_issues/women_bioorganic).

For citation purposes, cite each article independently as indicated on the article page online and as indicated below:

LastName, A.A.; LastName, B.B.; LastName, C.C. Article Title. *Journal Name* **Year**, *Volume Number*, Page Range.

ISBN 978-3-0365-4989-7 (Hbk)
ISBN 978-3-0365-4990-3 (PDF)

Cover image courtesy of Francesca Cardona, Camilla Parmeggiani, Camilla Matassini

© 2022 by the authors. Articles in this book are Open Access and distributed under the Creative Commons Attribution (CC BY) license, which allows users to download, copy and build upon published articles, as long as the author and publisher are properly credited, which ensures maximum dissemination and a wider impact of our publications.

The book as a whole is distributed by MDPI under the terms and conditions of the Creative Commons license CC BY-NC-ND.

Contents

About the Editors vii

Francesca Cardona, Camilla Parmeggiani and Camilla Matassini
Women in Bioorganic Chemistry
Reprinted from: *Molecules* **2022**, *27*, 4290, doi:10.3390/molecules27134290 1

David M. Campkin, Yuna Shimadate, Barbara Bartholomew, Paul V. Bernhardt, Robert J. Nash, Jennette A. Sakoff, Atsushi Kato and Michela I. Simone
Borylated 2,3,4,5-Tetrachlorophthalimide and Their 2,3,4,5-Tetrachlorobenzamide Analogues: Synthesis, Their Glycosidase Inhibition and Anticancer Properties in View to Boron Neutron Capture Therapy
Reprinted from: *Molecules* **2022**, *27*, 3447, doi:10.3390/molecules27113447 9

Crescenzo Coppa, Luca Sorrentino, Monica Civera, Marco Minneci, Francesca Vasile and Sara Sattin
New Chemotypes for the Inhibition of (p)ppGpp Synthesis in the Quest for New Antimicrobial Compounds
Reprinted from: *Molecules* **2022**, *27*, 3097, doi:10.3390/molecules27103097 37

Shaodan Chen, Bing Lin, Jiangyong Gu, Tianqiao Yong, Xiong Gao, Yizhen Xie, Chun Xiao, Janis Yaxian Zhan and Qingping Wu
Binding Interaction of Betulinic Acid to α-Glucosidase and Its Alleviation on Postprandial Hyperglycemia
Reprinted from: *Molecules* **2022**, *27*, 2517, doi:10.3390/molecules27082517 49

Michela Lupi, Martina Onori, Stefano Menichetti, Sergio Abbate, Giovanna Longhi and Caterina Viglianisi
Resolution of a Configurationally Stable Hetero[4]helicene
Reprinted from: *Molecules* **2022**, *27*, 1160, doi:10.3390/molecules27041160 61

Mariia Nesterkina, Serhii Smola, Nataliya Rusakova and Iryna Kravchenko
Terpenoid Hydrazones as Biomembrane Penetration Enhancers: FT-IR Spectroscopy and Fluorescence Probe Studies
Reprinted from: *Molecules* **2022**, *27*, 206, doi:10.3390/molecules27010206 77

Maksim Kukushkin, Vladimir Novotortsev, Vadim Filatov, Yan Ivanenkov, Dmitry Skvortsov, Mark Veselov, Radik Shafikov, Anna Moiseeva, Nikolay Zyk, Alexander Majouga and Elena Beloglazkina
Synthesis and Biological Evaluation of S-, O- and Se-Containing Dispirooxindoles
Reprinted from: *Molecules* **2021**, *26*, 7645, doi:10.3390/molecules26247645 91

Xi Chen, Yu-Cong Liu, Jing-Jing Cui, Fang-Ying Wu and Qiang Xiao
A Galactosidase-Activatable Fluorescent Probe for Detection of Bacteria Based on BODIPY
Reprinted from: *Molecules* **2021**, *26*, 6072, doi:10.3390/molecules26196072 115

Andrea Sartori, Kelly Bugatti, Elisabetta Portioli, Monica Baiula, Irene Casamassima, Agostino Bruno, Francesca Bianchini, Claudio Curti, Franca Zanardi and Lucia Battistini
New 4-Aminoproline-Based Small Molecule Cyclopeptidomimetics as Potential Modulators of $\alpha_4\beta_1$ Integrin
Reprinted from: *Molecules* **2021**, *26*, 6066, doi:10.3390/molecules26196066 125

Costanza Vanni, Anne Bodlenner, Marco Marradi, Jérémy P. Schneider, Maria de los Angeles Ramirez, Sergio Moya, Andrea Goti, Francesca Cardona, Philippe Compain and Camilla Matassini
Hybrid Multivalent Jack Bean α-Mannosidase Inhibitors: The First Example of Gold Nanoparticles Decorated with Deoxynojirimycin Inhitopes
Reprinted from: *Molecules* **2021**, *26*, 5864, doi:10.3390/molecules26195864 147

Aurore Dreneau, Fanny S. Krebs, Mathilde Munier, Chheng Ngov, Denis Tritsch, Didier Lièvremont, Michel Rohmer and Catherine Grosdemange-Billiard
α,α-Difluorophosphonohydroxamic Acid Derivatives among the Best Antibacterial Fosmidomycin Analogues
Reprinted from: *Molecules* **2021**, *26*, 5111, doi:10.3390/molecules26165111 161

Svetlana K. Vorontsova, Anton V. Yadykov, Alexander M. Scherbakov, Mikhail E. Minyaev, Igor V. Zavarzin, Ekaterina I. Mikhaevich, Yulia A. Volkova and Valerii Z. Shirinian
Novel D-Annulated Pentacyclic Steroids: Regioselective Synthesis and Biological Evaluation in Breast Cancer Cells
Reprinted from: *Molecules* **2020**, *25*, 3499, doi:10.3390/molecules25153499 177

Valentina Dell'Oste, Francesca Spyrakis and Cristina Prandi
Strigolactones, from Plants to Human Health: Achievements and Challenges
Reprinted from: *Molecules* **2021**, *26*, 4579, doi:10.3390/molecules26154579 195

Serena Silvestro, Giovanni Schepici, Placido Bramanti and Emanuela Mazzon
Molecular Targets of Cannabidiol in Experimental Models of Neurological Disease
Reprinted from: *Molecules* **2020**, *25*, 5186, doi:10.3390/molecules25215186 213

Shama S. M. Dissanayake, Manikandan Ekambaram, Kai Chun Li, Paul W. R. Harris and Margaret A. Brimble
Identification of Key Functional Motifs of Native Amelogenin Protein for Dental Enamel Remineralisation
Reprinted from: *Molecules* **2020**, *25*, 4214, doi:10.3390/molecules25184214 243

Camilla Matassini, Camilla Parmeggiani and Francesca Cardona
New Frontiers on Human Safe Insecticides and Fungicides: An Opinion on Trehalase Inhibitors
Reprinted from: *Molecules* **2020**, *25*, 3013, doi:10.3390/molecules25133013 261

About the Editors

Francesca Cardona

Francesca Cardona has been an Associate Professor since 2015. Born in 1971, she received her PhD in Chemical Sciences in Firenze in 1998 (Tutor Prof. A. Brandi). She was a post-doc fellow for 1 year at the University of Lausanne (CH) with Prof. P. Vogel. She was a permanent researcher (2002–2015) at the Department of Chemistry in Firenze. In 2006 she was awarded the "G. Ciamician" medal of the Italian Chemical Society for young organic chemists under 35. She is author of 107 papers, 12 chapters, 1 book (RSC) and 1 patent (h-index 36) on stereoselective syntheses of nitrogenated glycomimetics and new green oxidation methods.

Camilla Parmeggiani

Camilla Parmeggiani has been an Associate Professor since 2021. Born in 1981, she received her PhD in Chemical Science at the University of Florence with Prof. A. Goti. Since 2010, she has been an associate at LENS. In 2016, she was awarded the "Organic Chemistry for environment, energy and nanosciences" prize from the Organic Chemistry Division of the SCI and she was a finalist of the European Young Chemist Award. She has authored 55 papers, 1 book and 4 patents (h-index 29) on smart materials, stereoselective synthesis of iminosugars and new green oxidation methods, which have been cited over 3016 times.

Camilla Matassini

Camilla Matassini has been a Researcher Fellow since 2020. Born 1984, she received her PhD (*Doctor Europaeus*) in Chemical Science at the University of Firenze in 2014 with Prof. Goti, after spending one year at the Lab of GlycoNanotechnology (CICbiomaGUNE, San Sebastián, Spain) with Prof. Penadés. In 2016 she was awarded the Italian "Accademia dei Lincei" prize for her research in the field of Synthetic Organic Chemistry, and in 2020 the Organic Chemistry Junior Prize in Life Sciences from the Italian Chemistry Society. In 2019, by means of a MIUR grant (Leonardo da Vinci project), she spent a short-period placement in the group of Prof. Bols at the Department of Chemistry of the University of Copenhagen. Her interests focus on the synthesis of mono- and multivalent iminosugars and the preparation of gold nanoparticles for therapeutic applications. She is co-author of 2 patents, 2 chapters in scientific volumes and more than 40 papers in peer-reviewed journals (h-index 13).

Editorial

Women in Bioorganic Chemistry

Francesca Cardona [1,*], Camilla Parmeggiani [1,2] and Camilla Matassini [1]

1. Dipartimento di Chimica "Ugo Schiff" (DICUS), Università di Firenze, via della Lastruccia 3-13, 50019 Sesto Fiorentino, Italy; camilla.parmeggiani@unifi.it (C.P.); camilla.matassini@unifi.it (C.M.)
2. European Laboratory for Non Linear Spectroscopy (LENS), via Nello Carrara 1, 50019 Sesto Fiorentino, Italy
* Correspondence: francesca.cardona@unifi.it

Citation: Cardona, F.; Parmeggiani, C.; Matassini, C. Women in Bioorganic Chemistry. *Molecules* **2022**, *27*, 4290. https://doi.org/10.3390/molecules27134290

Received: 28 June 2022
Accepted: 30 June 2022
Published: 4 July 2022

Publisher's Note: MDPI stays neutral with regard to jurisdictional claims in published maps and institutional affiliations.

Copyright: © 2022 by the authors. Licensee MDPI, Basel, Switzerland. This article is an open access article distributed under the terms and conditions of the Creative Commons Attribution (CC BY) license (https://creativecommons.org/licenses/by/4.0/).

We are very happy to present this Special Issue, for which we acted as guest editors, and which includes scientific contributions from laboratories headed by women active in the field of bioorganic chemistry.

We made the decision to undertake this project since we deem that there are still gender biases that put women at a slight disadvantage when disseminating their research, preventing the science community from benefiting from a wider diversity of voices. The issues related to the gender scissor and the leaky pipeline that can be observed with career advancement in the academy, especially in the field of STEM disciplines, deserve our attention and the efforts of all of the scientific community to mitigate the gender gap. In order to embrace gender equality, recognize the career progression of women, and to celebrate the achievements of women in the field of bioorganic chemistry, we present in this Special Issue contributions both from highly renewed woman scientists and young woman researchers who are undertaking their early-stage careers.

This Special Issue includes fifteen manuscripts, among which eleven high-quality research articles and four comprehensive review articles in the area of bioorganic chemistry, published from mid-2020 to early 2022.

The scope of the Special Issue covers a wide range of topics at the organic chemistry-biology interface, including the synthesis and derivatization of natural compounds and their analogues, and the investigation of their biological activities in the human health field (for instance as antitumoral, antioxidants and antimicrobial agents) as well as their possible application in the crop protection field as agrochemicals. An example of nanoparticle-based biomaterial is also included. The techniques employed, besides organic synthesis, are in silico studies (docking procedures and molecular modeling), FT-IR spectroscopy, laser diffraction, PET, fluorescence, STD-NMR studies, enzymatic evaluation, experiments on cell lines, and in vivo studies on mice.

Cardona, Matassini and co-workers, from Sesto Fiorentino, Italy, reviewed the properties of carbohydrate-based natural compounds and other sugar mimics as trehalase inhibitors, in view of their potential use as non-toxic and therefore greener and safer pesticides [1].

Within the same field of agrochemicals, Dell'Oste, Spyrakis, Prandi and co-workers from Torino, Italy, described that strigolactones (SG), a class of sesquiterpenoid plant hormones, play a key role in the plants' response to biotic and abiotic stress. In addition, the authors highlighted the possibility that in the next future these compounds might have an application also in human health, and in particular in the control pathways related to apoptosis and inflammation (and therefore as anticancer and/or antimicrobial agents) [2]. Terpenes have a number of other different biological applications, as reported by the authors of this Special Issue. Zhan and co-workers, from Guangzhou, China, showed that the natural compound betulinic acid (BA), a pentacyclic triterpene widely distributed in nature, behave as a non-competitive inhibitor of α-glucosidase, showing a synergistic effect with acarbose, which is known for its use for alleviation of post-prandial hyperglycemia. The authors also performed molecular docking and molecular dynamics simulation and some preliminary in vivo experiments on mice [3]. Mazzon and co-workers, from Messina, Italy, reported on the numerous studies supporting the great properties of cannabidiol (CBD), a terpenophenol natural compound, for the management of neurological disorders (such as epilepsy, Alzheimer, multiple sclerosis and Parkinson), due to is antioxidant, anti-inflammatory, antidepressant, anxiolytic, anticonvulsant and antipsychotic properties. The biochemical and molecular mechanisms underlying the effects of CBD show that a multi-target mechanism of action takes place [4]. Nesterkina and co-workers from Odessa, Ukraina studied, with the aid of different techniques (FT-IR, laser diffraction, fluorescent measurements), the impact of terpenoids-based hydrazones on the molecular organization of lipid matrices using model liposomes based on lecithin or cardiolipin phospholipids, as well as lipids isolated from rat strata cornea [5].

Triterpenes are biosynthetic precursors of steroids, which are an important class of both natural and synthetic products. Volkova and co-workers from Moscow, Russia, described the synthesis of D-annulated pentacyclic steroids based on a regioselective interrupted Nazarov cyclization with trapping chloride ion, and evaluated the antiproliferative activity of the synthesized compounds against two breast cancer cell lines [6].

The interest in the design and synthesis of novel anticancer therapeutics is also present in the manuscript by Beloglazina and co-workers from Moscow, Russia, who reported the synthesis of a series of S-, O- and Se- containing dispirooxoindoles through 1,3-dipolar cycloaddition of azomethine ylides, assayed their cytotoxicity against different tumor cell lines and performed an in silico study to rationalize the results [7]. The group of Simone and co-workers from Callaghan, Australia, reported the synthesis, glycosidase inhibition and anticancer properties of highly chlorinated benzamide analogues bearing a boron-pinacolate ester group, with the perspective to use them in boron neutron capture therapy (BNCT) [8].

Sattin and co-workers from Milano, Italy, described, through virtual screening accompanied by STD-NMR studies, a structure-based approach to find new chemotypes able to target (p)ppGpp (guanosine tetra-or penta-phosphate) signaling, in view of overcoming antimicrobial resistance [9]. The issue of antimicrobial resistance was also addressed by the group of Grosdemange-Billiard and co-workers from Strasbourg, France, who synthesized fluorinated analogues of the natural compound fosmidomycin and tested them as E. coli 1-deoxy-D-xylulose 5-phosphate reductoisomerase (DXR) inhibitors as well as antimicrobial agents against E. coli on Petri dishes [10]. Pathogenic E. coli infection and food/water contamination by this pathogen was also the object of the article by Wu and co-workers from Nanchang, China, who designed and synthesized a β-galactosidase-activatable fluorescent probe (BOD-Gal) for the detection of this pathogen [11].

The process of microbial attack on dental enamel and the potential approaches for dental remineralization were described by Brimble and co-workers from Auckland, New Zealand, who highlighted the importance of the amelogenin protein and the efforts made by the researchers in the identification of the key structural motifs of this protein that enable dental remineralization, as well as the rational design of synthetic polypeptides for this aim [12].

Integrin $\alpha_4\beta_1$ belongs to the leukocyte integrin family and represents a target of relevant therapeutic interest due its role in mediating inflammation, autoimmune pathologies and cancer-related diseases. With the aim of discovering new compounds potentially able to recognize integrin $\alpha_4\beta_1$, Battistini and co-workers from Parma, Italy synthesized, through solid phase procedures followed by in-solution cyclization steps, seven new cyclic peptidomimetics bearing a 4-aminoproline core scaffold, and evaluated them in cell adhesion assays on Jurvat cells [13].

In the field of bionanomaterials, the Special Issue shows an example by Bodlenner from Strasbourg, France, and Matassini from Sesto Fiorentino, Italy, who reported the synthesis and biological evaluation as Jack Bean α-mannosidase inhibitors of hybrid multivalent glyco gold nanoparticles decorated with deoxynojirimycin inhitopes, among the best known glycomimetics in the field of glycosidases inhibition. The authors found a strong enhancement of the inhibitory activity consequent to the multivalent presentation of the inhitope [14].

Lastly, chirality is one of the most crucial aspects of nature and is of paramount importance in the area of bioorganic chemistry, and axial chirality represents an intriguing aspect of chirality itself. Viglianisi and co-workers from Sesto Fiorentino, Italy, developed an efficient chemical resolution of racemic aza[4]helicenes, interesting building blocks for the production of materials with chiroptical properties, using enantiopure camphanic acids as resolving agents [15].

We want to finish this Editorial by thanking again all of the authors who come from three different continents, namely Europe, Asia and Oceania, for having illustrated so well the importance of bioorganic chemistry in their contributions to this Special Issue.

A list of short biographical sketches of the authors, together with the description of the obstacles/challenges encountered during their career, or suggestions to a young woman keen to become a successful scientist in the field of bioorganic chemistry, follows in the Reference section.

Funding: MIUR-Italy ("Progetto Dipartimenti di Eccellenza 2018–2022") allocated to the Department of Chemistry "Ugo Schiff".

Acknowledgments: We thank the women researchers who kindly provided us their profiles and photos for the preparation of this manuscripts.

Conflicts of Interest: The authors declare no conflict of interest.

References

1. Matassini, C.; Parmeggiani, C.; Cardona, F. New Frontiers on Human Safe Insecticides and Fungicides: An Opinion on Trehalase Inhibitors. *Molecules* **2020**, *25*, 3013. [CrossRef] [PubMed]
2. Dell'Oste, V.; Spyrakis, F.; Prandi, C. Strigolactones, from Plants to Human Health: Achievements and Challenges. *Molecules* **2021**, *26*, 4579. [CrossRef]
3. Chen, S.; Lin, B.; Gu, J.; Yong, T.; Gao, X.; Xie, Y.; Xiao, C.; Zan, J.Y.; Wu, Q. Binding Interaction of Betulinic Acid to α-Glucosidase and Its Alleviation on Postprandial Hyperglycemia. *Molecules* **2022**, *27*, 2517. [CrossRef] [PubMed]
4. Silvestro, S.; Schepici, G.; Bramanti, P.; Mazzon, E. Molecular Targets of Cannabidiol in Experimental Models of Neurological Disease. *Molecules* **2020**, *25*, 5186. [CrossRef] [PubMed]
5. Nesterkina, M.; Smola, S.; Rusakova, N.; Kravchenko, I. Terpenoid Hydrazones as Biomembrane Penetration Enhancers: FT-IR Spectroscopy and Fluorescence Probe Studies. *Molecules* **2022**, *27*, 206. [CrossRef] [PubMed]
6. Vorontsova, S.K.; Yadykov, A.V.; Scherbakov, A.M.; Minyaev, M.E.; Zavarzin, I.V.; Mikhaevich, E.I.; Volkova, Y.A.; Shirinian, V.Z. Novel D-Annulated Pentacyclic Steroids: Regioselective Synthesis and Biological Evaluation in Breast Cancer Cells. *Molecules* **2020**, *25*, 3499. [CrossRef] [PubMed]
7. Kukushkin, M.; Novotortsev, V.; Filatov, V.; Ivanenkov, Y.; Skvortsov, D.; Veselov, M.; Shafikov, R.; Moiseeva, A.; Zyk, N.; Majouga, A.; et al. Synthesis and Biological Evaluation of S-, O- and Se-Containing Dispirooxindoles. *Molecules* **2021**, *26*, 7645. [CrossRef] [PubMed]
8. Campkin, D.M.; Shimadate, Y.; Bartholomew, B.; Bernhardt, P.V.; Nash, R.J.; Sakoff, J.A.; Kato, A.; Simone, M.I. Borylated 2,3,4,5-Tetrachlorophthalimide and Their 2,3,4,5-Tetrachlorobenzamide Analogues: Synthesis, Their Glycosidase Inhibition and Anticancer Properties in View to Boron Neutron Capture Therapy. *Molecules* **2022**, *27*, 3447. [CrossRef] [PubMed]
9. Coppa, C.; Sorrentino, L.; Civera, M.; Minneci, M.; Vasile, F.; Sattin, S. New Chemotypes for the Inhibition of (p)ppGpp Synthesis in the Quest for New Antimicrobial Compounds. *Molecules* **2022**, *27*, 3097. [CrossRef] [PubMed]
10. Dreneau, A.; Krebs, F.S.; Munier, M.; Ngov, C.; Tritsch, D.; Lièvremont, D.; Rohmer, M.; Grosdemange-Billiard, C. α, α-Difluorophosphonohydroxamic Acid Derivatives among the Best Antibacterial Fosmidomycin Analogues. *Molecules* **2021**, *26*, 5111. [CrossRef] [PubMed]
11. Chen, X.; Liu, Y.-C.; Cui, J.-J.; Wu, F.-Y.; Xiao, Q. A Galactosidase-Activatable Fluorescent Probe for Detection of Bacteria Based on BODIPY. *Molecules* **2021**, *26*, 6072. [CrossRef] [PubMed]
12. Dissanayake, S.S.M.; Ekambaram, M.; Li, C.K.; Harris, P.W.R.; Brimble, M.A. Identification of Key Functional Motifs of Native Amelogenin Protein for Dental Enamel Remineralisation. *Molecules* **2020**, *25*, 4214. [CrossRef] [PubMed]
13. Sartori, A.; Bugatti, K.; Portioli, E.; Baiula, M.; Casamassima, I.; Bruno, A.; Bianchini, F.; Curti, C.; Zanardi, F.; Battistini, L. New 4-Aminoproline-Based SmallMolecule Cyclopeptidomimetics as Potential Modulators of $\alpha_4\beta_1$ Integrin. *Molecules* **2021**, *26*, 6066. [CrossRef] [PubMed]

14. Vanni, C.; Bodlenner, A.; Marradi, M.; Schneider, J.P.; de los Angeles Ramirez, M.; Moya, S.; Goti, A.; Cardona, F.; Compain, P.; Matassini, C. Hybrid Multivalent Jack Bean α-Mannosidase Inhibitors: The First Example of Gold Nanoparticles Decorated with Deoxynojirimycin Inhitopes. *Molecules* **2021**, *26*, 5864. [CrossRef] [PubMed]
15. Lupi, M.; Onori, M.; Menichetti, S.; Abbate, S.; Longhi, G.; Viglianisi, C. Resolution of a Configurationally Stable Hetero[4]helicene. *Molecules* **2022**, *27*, 1160. [CrossRef] [PubMed]

Short Biography of Authors Who Contributed to the Special Issue

Lucia Battistini (Department of Food and Drug, University of Parma, Parco Area delle Scienze 27/A, 43124 Parma, Italy). Research interests: The main focus is on the development of new classes of peptidomimetic ligands for molecular recognition and their use for biomedical applications. *Molecules* **2021**, *26*, 6066; doi.org/10.3390/molecules26196066.

What is the most important challenge for a woman working in the field of bioorganic chemistry?

One of the most challenging issues for a bioorganic chemist (not just for a female scientist) working in a field that lies at the interface between different disciplines (chemistry, biology, medicine, etc.) is to create the best empathic and collaborative atmosphere in the working group, and fully recognize and value the originality and creativity within each contribution. I believe that women leading research groups owing to their empathy and sensitivity have a step ahead to contrast stereotypes, unconscious biases, and prejudices that sometimes spoil the teamwork.

Elena Beloglazkina (Department of Chemistry, Lomonosov Moscow State University, Leninskie gory 1-3, 119991 Moscow, Russia). Research interests: organic synthesis, biologically active organic compounds, organo-chalcogen compounds, metal complexes with organic ligands. *Molecules* **2021**, *26*, 7645; doi.org/10.3390/molecules26247645.

What are your suggestions for a young woman keen to become a successful scientist (in the field of bioorganic chemistry?

Do not be afraid to take on something completely new for your and do not give up in case of possible failures. Patience and interest in what you are doing sooner or later will yield results. It will not always be the same as you expected when starting your research, but that's the charm of our work.

Anne Bodlenner (Laboratoire d'Innovation Moléculaire et Applications UMR CNRS 7042-LIMA I ECPM), University of Strasbourg, 25 Rue Becquerel, 67087 Strasbourg, France). Research interests: bioorganic chemistry, interactions between small molecules and enzymes, multivalency, carbohydrates and glycomimetics. *Molecules* **2021**, *26*, 5864; doi.org/10.3390/molecules26195864.

What is the most important challenge for a woman working in the field of bioorganic chemistry?

As a woman researcher in chemistry, an important challenge was to find my own balance between research, teachings and personal life, as I wish to spend as much quality time with my son as possible.

What is the secret of being a successful female bioorganic chemist?

I found that being well-organized, defining clear objectives at work, and being in harmony with my priorities works well for me. My tasks being essentially intellectual, sport and nature also help me to find my physical and intellectual balance, which is necessary to raise enough energy to tackle all challenges. I am also very lucky to have a supportive partner who is actively involved in the daily running of things.

Margaret Brimble (School of Chemical Sciences, The University of Auckland, New Zealand). Research interests: Synthesis of bioactive natural products, antimicrobial peptides, antiviral peptides, lipopeptides and glycopeptides. Peptide based vaccines and adjuvants. *Molecules* **2020**, *25*, 4214; doi.org/10.3390/molecules25184214.

What are your suggestions for a young woman keen to become a successful scientist (in the field of bioorganic chemistry)?

I didn't realize that to be a scientist you have to be able to multi-task exceptionally well. There are so many things that need to be written—grants, reports, papers, patents, marketing material, references, reviews, outreach material, teaching material etc. These tasks are endless and I wish I could write quicker. I did organic chemistry since I liked doing things in the lab and not writing! The best thing to succeed is to remember you can't do it alone and you are only as successful as the people who work alongside you in your team. Take each day in your stride and seek out like-minded people as your team members and collaborators. Remember you only learn from setbacks and making mistakes and rise above the intimidating bravado that many of your colleagues are good at displaying. A lot of it is hype! Doing good science always takes time and a lot of hard work.

Francesca Cardona (Dipartimento di Chimica "Ugo Schiff" (DICUS), Università degli Studi di Firenze, Via della Lastruccia 3-13, 50019 Sesto Fiorentino, Italy). Research interests: Stereoselective syntheses of iminosugars as glycosidase inhibitors and/or pharmacological chaperones for lysosomal enzymes, green chemistry. *Molecules* **2020**, *25*, 3013; doi.org/10.3390/molecules25133013.

What are your suggestions for a young woman keen to become a successful scientist (in the field of bioorganic chemistry)?

The greatest challenge I had to face up during my career has been how to conciliate my passion in bioorganic chemistry with my private life. I love being a scientist and I also love being a mother. The secret? Not being too individualist! In my personal experience, the secret for being a quite good scientist and a quite good mother (as I hope to be) has been to create a good team, instead of just running alone.

Valentina Dell'Oste (Dipartimento di Scienze della Sanità Pubblica e Pediatriche), Università degli Studi di Torino, Via Santena 9, 10126 Torino, Italy). Research interests: Virus-host interactions, screening and characterization of new antiviral molecules, antiviral immunity. *Molecules* **2021**, *26*, 4579; doi.org/10.3390/molecules26154579.

What are your suggestions for a young woman keen to become a successful scientist (in the field of bioorganic chemistry)?

I strongly believe that to become a successful scientist, but more generally, to reach your professional goals, you need to apply three critical rules: perseverance in work and study, to be multitasking, and favor teamwork. Remember always that your colleagues and your family are your best allies! Then, try to be always open to new experiences, since just in this way you can improve your knowledge and transfer it to the work.

Catherine Grosdemange-Billiard (Chemistry), Université de Strasbourg/Institut de Chimie, 4, rue Blaise Pascal,67081 Strasbourg, France. Research interests: Development of novel and unexplored types of antibacterial drugs by synthesizing MEP pathway protein inhibitors as well as small molecules involved in the intra- and inter-species mechanisms of bacterial communication. *Molecules* **2021**, *26*, 5111; doi.org/10.3390/molecules26165111.

What are your suggestions for a young woman keen to become a successful scientist (in the field of bioorganic chemistry)?

Trust yourself and your passion for biomolecules and organic chemistry. Never give up but persevere the work you believe in and face obstacles with the right tools and by joining forces with scientist of other disciplines. Turn negative experiences into positives one. Share your knowledge and pass on your passion to the young scientists for keeping you inspired.

Camilla Matassini (Department of Chemistry "Ugo Schiff", DICUS), University of Florence, Via della Lastruccia 3-13, 50019, Sesto Fiorentino, Firenze (Italy). Research interests: nitrogen-containing glycomimetics; new oxidation methods, multivalency; gold glyconanoparticles; glycosidase inhibitors; lysosomal enzymes; pharmacological chaperones; Gaucher–Parkinson relationship. *Molecules* **2021**, *26*, 5864; doi.org/10.3390/molecules26195864.

As a women chemist, which obstacles did you encountered in your career, and how did you face them?

In my experience, most of friends and relatives knew very little (and were even wary), about an academic career, especially a scientific one. When I had my PhD fellowship my mother told me that I had been very good convincing people to pay me for studying ... However, I think that my passion for the laboratory life, my enthusiasm for chemistry and my curiosity for its application to biomedical issues, eventually convinced them that being a scientist was a real job! An added value? Being surrounded by inspiring and motivating mentors and colleagues.

Mariia Nesterkina (Drug Design and Optimisation, Helmholtz Institute for Pharmaceutical Research Saarland, Campus E8.1, 66123 Saarbrücken, Germany). Research interests: Chemistry of natural compounds, investigation of penetration enhancers, pharmacology. *Molecules* **2022**, *27*, 206; doi.org/10.3390/molecules27010206.

What is the secret of being a successful female bioorganic chemist?

As usually in science, success consists of 99% hard work and 1% luck. Bioorganic chemistry is no exception in this regard—you can spend several months in the laboratory synthesizing novel compounds that in the end were revealed as biologically inactive. Then a new path begins, new ideas and hypotheses are put forward needed to be experimentally confirmed. However, we understand the ultimate goal of our scientific research and its significance both for fundamental and applied investigations. Enthusiasm and inspiration for new discoveries and their potential importance to human society are the main incentives for success in bioorganic chemistry.

Cristina Prandi (Department of Chemistry), University of Turin, via P. Giuria 7, 10125 Torino). Research interests: organic synthesis in non-conventional media, organometallic chemistry, synthesis of natural compounds. *Molecules* **2021**, *26*, 4579; doi.org/10.3390/molecules26154579.

Have you ever felt disadvantaged in being a woman in your research field?

I have always felt that I would have to work much harder than my male colleagues to get the same recognition from the academic and scientific community, especially after the birth of my daughters. But, in the end, this turned out to be an advantage. I learned the value of time, to be more efficient and to combine family life with my role as a scientist.

Sara Sattin (Dipartimento di Chimica), Università degli Studi di Milano, via C. Golgi, 19, 20133 Milano, Italy). Research interests: design and synthesis of small molecules tailored to interact with specific protein targets (host and pathogen receptors and enzymes mediating pathogen adhesion, virulence and bacterial persistence). *Molecules* **2022**, *27*, 3097; doi.org/10.3390/molecules27103097.

What is the most important challenge for a woman working in the field of bioorganic chemistry?

I think the most important challenge I faced has been networking with peers and senior academic members of both organic chemistry and neighbouring fields. For instance, when I was already assistant or associate professor I often found myself in situations (e.g., conferences) where colleagues just assumed I was a student or a postdoc at most, rarely including me in relevant scientific or decision-making conversations. Women scientists should be more proactive in creating collaborative networks and advocating for female colleagues at all career levels.

Michela Simone Newcastle, CSIRO Energy Centre, CSIRO, 10 Murray Dwyer Court, Mayfield West, NSW2304, Australia. Previously in: Discipline of Chemistry, University of Newcastle, Callaghan, NSW 2308, Australia and Priority Research Centre for Drug Development, University of Newcastle, Callaghan, NSW 2308, Australia. Research interests: carbohydrate chemistry, synthetic organic chemistry, medicinal and bioorganic chemistry, nuclear magnetic resonance, carbohydrate active enzymes, heterocyclic chemistry, renewable energy technologies. *Molecules* **2022**, *27*, 3447; doi.org/10.3390/molecules27113447.

What is the secret of being a successful female bioorganic chemist?

I want to acknowledge the huge role my mother and my grandmother played in introducing me to scientific thinking and natural phenomena since I was a child. They also encouraged me to pursue excellence in everything I do. I owe them so much of what I have achieved, both in my private and public life. I found that the most important factors to being successful are a constant, ethical and honest approach to pursuing excellence, hard work, steadfast optimism, patience, good mentoring and cultivating fruitful relationships with a plethora of colleagues. Seek advice relentlessly, listen to all advice, read up as much as you can, but - at the end of the day - do your own thing, follow your instincts, be mightily proactive and do so fearlessly. It's difficult, but it's the only way forward. The other crucial message is: you don't have to choose between having a career and having a family. You can have both, if you want. There are no right or wrong times. It all comes down to how you manage what happens. Embrace anything that happens, be proactive, find ways of managing tough situations. Define your own success. Don't be afraid to restart from scratch (I have done so several times). Focus on the most important aspect of life: human relationships (especially with the children in your care).

Francesca Spyrakis (Department of Drug Science and Technology), University of Turin, Via Giuria 9, 10125, Turin, Italy. Research interests: Drug Design. *Molecules* **2021**, *26*, 4579; doi.org/10.3390/molecules26154579.

What is the most important challenge for a woman working in the field of bioorganic chemistry?

According to my opinion and personal experience, the most important challenge I found, and I still find, has been combining the professional life and commitments with the family ones. In other words, finding the right time to spend with my daughter and my partner. In the years I have learned to optimize time and to handle only fundamental commitments, while delegating the other ones. Also, I have found very smart collaborators, who help me in handling the research and teaching activities. Time is always short, when you are enjoying, but now I can get the best of it!

What are your suggestions for a young woman keen to become a successful scientist (in the field of bioorganic chemistry?

To young women willing to become scientists I would recommend to never give up! Even when things seem to be going wrong, there is always an opportunity around the corner. The important is to be ready to catch any occasion and not be scared to get in the game!

Caterina Viglianisi (Dipartimento di Chimica "Ugo Schiff" (DICUS), Università degli Studi di Firenze, Via della Lastruccia 3-13, 50019 Sesto Fiorentino, Italy). Research interests include redox chemistry with design and synthesis molecular, macromolecular and nano-supported antioxidants and the study of their potential applications in the medical field and new materials. A further research area is the synthesis and evaluation of the optoelectronic properties of condensed heterocyclic systems. *Molecules* **2022**, *27*, 1160; doi.org/10.3390/molecules27041160.

What is the secret of being a successful female bioorganic chemist? What are your suggestions for a young woman keen to become a successful scientist (in the field of bioorganic chemistry?

A young chemist must always remember that becoming a successful scientist requires countless hours of work, although we are in the exciting world of discovery, so work hard and enjoy your job!

Yulia Volkova (Laboratory of Steroid Compounds, N. D. Zelinsky Institute of Organic Chemistry, Russian Academy of Sciences, 47 Leninsky prosp., 119991 Moscow, Russia). Research interests: Design and synthesis of novel heterosteroids promising as anticancer agents against hormone-dependent cancers such as breast and prostate cancer. *Molecules* **2020**, *25*, 3499; doi.org/10.3390/molecules25153499.

What is the most important challenge for a woman working in the field of bioorganic chemistry?

Women were legally allowed to pursue careers in science just over 100 years ago. Having received new opportunities, women retained the old responsibilities. Today the main challenge women scientists' face is to find a work-life balance. The social burden of raising children and running a household in many countries is still regarded as predominantly female. Many women scientists are forced to take a break from their work or significantly reduce their work hours due to the above responsibilities. On this account, women in science are flexible, highly efficient, multitasking, and the able to quickly rearrange themselves.

Janis Ya-xian Zhan (School of Pharmaceutical Science, Guangzhou University of Chinese Medicine, Guangzhou University Town Waihuan East Road No. 232, 510006 Guangzhou, China). Research interests: Traditional Chinese medicine pharmacology, traditional Chinese medicine molecules, oxidative stress, immune inflammation, aging, environmental chemistry, pharmacokinetics. *Molecules* **2022**, *27*, 2517; doi.org/10.3390/molecules27082517.

Have you ever felt disadvantaged in being a woman in your research field?

As a female scientist, I am full of enthusiasm and interest in my research field. In the course of my research, I encountered many difficulties and setbacks. But fortunately, I have not been treated unfairly since I am a woman. Many colleagues and friends around have given selfless help and care. This makes me stick to my own research path and make continuous progress.

Article

Borylated 2,3,4,5-Tetrachlorophthalimide and Their 2,3,4,5-Tetrachlorobenzamide Analogues: Synthesis, Their Glycosidase Inhibition and Anticancer Properties in View to Boron Neutron Capture Therapy

David M. Campkin [1,2], Yuna Shimadate [3], Barbara Bartholomew [4], Paul V. Bernhardt [5], Robert J. Nash [4], Jennette A. Sakoff [2,6], Atsushi Kato [3] and Michela I. Simone [1,2,*]

1. Discipline of Chemistry, University of Newcastle, Callaghan, NSW 2308, Australia; campkindavid@gmail.com
2. Priority Research Centre for Drug Development, University of Newcastle, Callaghan, NSW 2308, Australia; jennette.sakoff@newcastle.edu.au
3. Department of Hospital Pharmacy, University of Toyama, 2630 Sugitani, Toyama 930-0194, Japan; m1961224@ems.u-toyama.ac.jp (Y.S.); kato@med.u-toyama.ac.jp (A.K.)
4. Phytoquest Ltd., Plas Gogerddan, Aberystwyth, Ceredigion SY23 3EB, UK; barbara.bartholomew@phytoquest.co.uk (B.B.); robert.nash@phytoquest.co.uk (R.J.N.)
5. School of Chemistry & Molecular Biosciences, University of Queensland, Brisbane, QLD 4072, Australia; p.bernhardt@uq.edu.au
6. Calvary Mater Newcastle Hospital, Edith Street, Waratah, NSW 2298, Australia
* Correspondence: michela_simone@yahoo.co.uk

Citation: Campkin, D.M.; Shimadate, Y.; Bartholomew, B.; Bernhardt, P.V.; Nash, R.J.; Sakoff, J.A.; Kato, A.; Simone, M.I. Borylated 2,3,4,5-Tetrachlorophthalimide and Their 2,3,4,5-Tetrachlorobenzamide Analogues: Synthesis, Their Glycosidase Inhibition and Anticancer Properties in View to Boron Neutron Capture Therapy. *Molecules* 2022, 27, 3447. https://doi.org/10.3390/molecules27113447

Academic Editor: René Csuk

Received: 19 February 2022
Accepted: 19 May 2022
Published: 26 May 2022

Publisher's Note: MDPI stays neutral with regard to jurisdictional claims in published maps and institutional affiliations.

Copyright: © 2022 by the authors. Licensee MDPI, Basel, Switzerland. This article is an open access article distributed under the terms and conditions of the Creative Commons Attribution (CC BY) license (https://creativecommons.org/licenses/by/4.0/).

Abstract: Tetrachlorinated phthalimide analogues bearing a boron-pinacolate ester group were synthesised via two synthetic routes and evaluated in their glycosidase modulating and anticancer properties, with a view to use them in boron neutron capture therapy (BNCT), a promising radiation type for cancer, as this therapy does little damage to biological tissue. An unexpected decarbonylation/decarboxylation to five 2,3,4,5-tetrachlorobenzamides was observed and confirmed by X-ray crystallography studies, thus, giving access to a family of borylated 2,3,4,5-tetrachlorobenzamides. Biological evaluation showed the benzamide drugs to possess good to weak potencies (74.7–870 µM) in the inhibition of glycosidases, and to have good to moderate selectivity in the inhibition of a panel of 18 glycosidases. Furthermore, in the inhibition of selected glycosidases, there is a core subset of three animal glycosidases, which is always inhibited (rat intestinal maltase α-glucosidase, bovine liver β-glucosidase and β-galactosidase). This could indicate the involvement of the boron atom in the binding. These glycosidases are targeted for the management of diabetes, viral infections (via a broad-spectrum approach) and lysosomal storage disorders. Assays against cancer cell lines revealed potency in growth inhibition for three molecules, and selectivity for one of these molecules, with the growth of the normal cell line MCF10A not being affected by this compound. One of these molecules showed both potency and selectivity; thus, it is a candidate for further study in this area. This paper provides numerous novel aspects, including expedited access to borylated 2,3,4,5-tetrachlorophthalimides and to 2,3,4,5-tetrachlorobenzamides. The latter constitutes a novel family of glycosidase modulating drugs. Furthermore, a greener synthetic access to such structures is described.

Keywords: boron; phthalimide; benzamide; glycosidase; cancer; boron neutron capture therapy

1. Introduction

The phthalimide scaffold appears in several drugs, including the fungicide *N*-(trichloromethylthio)phthalimide (Folpet®), thalidomide—now used for leprosy—and in the first line treatment of multiple myeloma [1], the antibacterial talmetoprim, the antifungal amphotalide, and the antiepileptic taltrimide (Figure 1) [2].

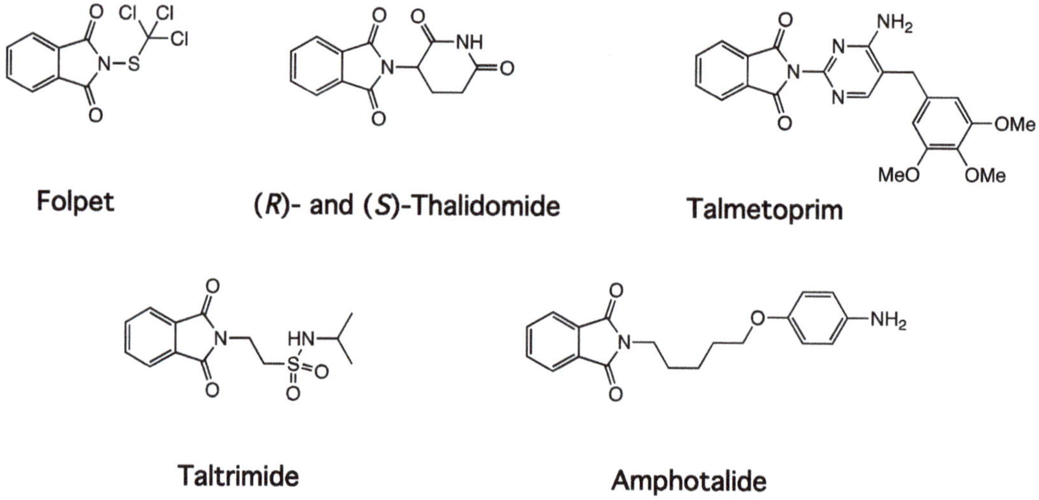

Folpet

(R)- and (S)-Thalidomide

Talmetoprim

Taltrimide

Amphotalide

Figure 1. Phthalimide-containing drugs.

Phthalimide analogues have also been shown to display glycosidase inhibition [3–5], with the 2,3,4,5-tetrachlorophthalimide scaffold deemed necessary for potent activity and the corresponding unsubstituted phthalimide derivatives showing reduced activity [6].

In our group, we are interested in the use of organic boron as a pharmacophoric group in its boronic acid (R-B(OH)$_2$) and boronate ester (R-B(OR')$_2$) [7,8] functional groups and in the development of synthetic methodologies for the installation of this pharmacophore on biologically active molecules to study and expand the palette of enzyme–drug interactions [9–13].

Glycosidase enzymes are involved in a number of disease states, ranging from diabetes and lysosomal storage disorders to viral infections, with their modulation being paramount in the management of these diseases [14–22]. The introduction of boron atoms to drug molecules also provides access to potential boron neutron capture therapy (BNCT) agents. BNCT provides the opportunity to utilise a type of radiotherapy that causes minimal damage to healthy tissue [23–25].

We report the synthesis of a family of novel drugs consisting of borylated 2,3,4,5-tetrachlorophthalimides and 2,3,4,5-tetrachlorobenzamides. The latter group arose from a decarbonylation side reaction, giving expedited access to them. Drugs have been characterized, including by X-ray crystallographic analysis in two instances. These confirmed the structural integrity, the outcome of the side reaction and the conformation of the boronate ester groups in the solid state. Furthermore, biological assays against glycosidase enzymes and cancer cell lines highlighted a good inhibitor for bovine liver β-galactosidase and three potent growth inhibitors and, of these, one selective growth inhibitor for cancer versus healthy cell lines in the cancer assay. These drugs represent an optimal set for further derivatisations.

2. Results and Discussion

2.1. Summary of Synthetic Work

Synthesis of the N-borylated 2,3,4,5-tetrachlorophthalimides was attempted via two synthetic strategies: the double acyl substitution route (reaction of **1** with **meta 2**, **para 2**, **ortho 4**, **meta 4** and **para 4**, Scheme 1) and the S$_N$2 route (reaction of **6** with **ortho 7** and **para 7**).

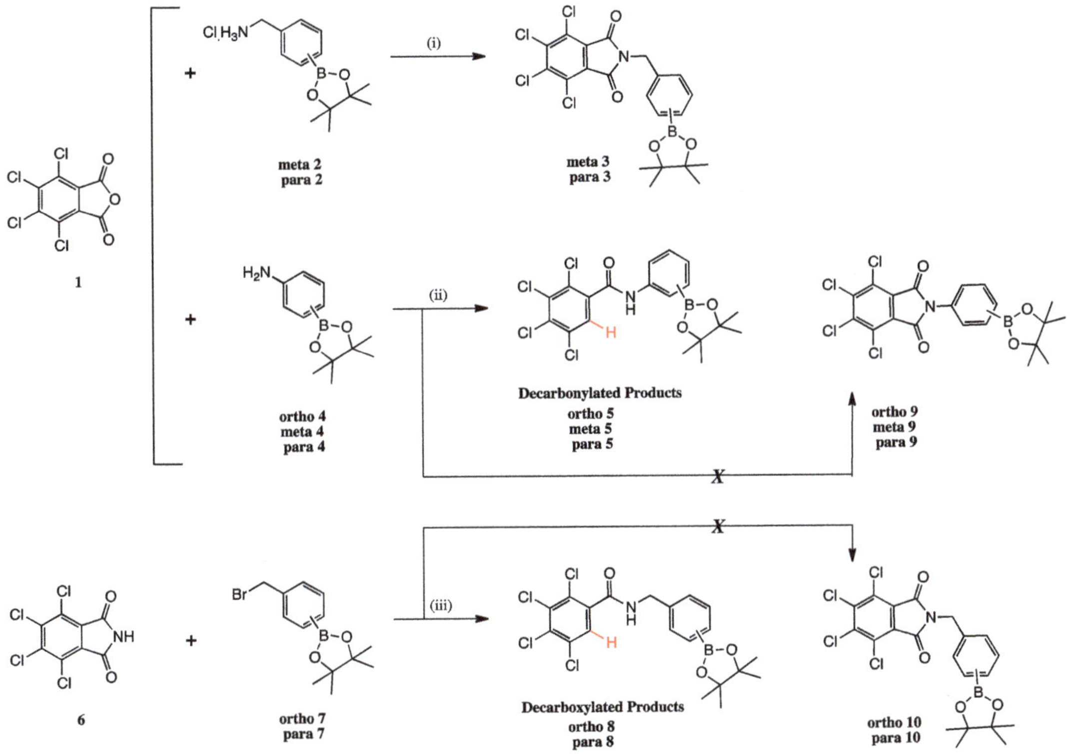

Scheme 1. Synthetic routes to borylated phthalimide analogues **meta 3**, **para 3**, **ortho 5**, **meta 5**, **para 5**, **ortho 8** and **para 8**. (i) Et$_3$N, DMF, 85 °C, 2d, 60% (**meta 3**), 44% (**para 3**); (ii) Et$_3$N, DMF, 85 °C, 2d, then 105 °C, 3d, 52% (**meta 5**), 37% (**ortho 5**); NaH, DMF, 100 °C, 3d, 46% (**para 5**); (iii) NaH, DMF, 100 °C, 2d, 60% (**ortho 8**), 56% (**para 8**).

Synthesis of the 2,3,4,5-tetrachlorophthalimides **meta 3** and **para 3** was achieved in moderate to good yields in one synthetic step via double acyl substitution from 2,3,4,5-tetrachlorophthalic anhydride **1**, which was reacted with 3- and 4-(aminomethyl)phenylboronic acid pinacol ester hydrochloride **meta 2** and **para 2** in DMF at 85 °C. Reagent 2-(aminomethyl)-phenylboronic acid pinacol ester hydrochloride was not commercially available for synthesis of **ortho 3** (not shown).

An unexpected decarbonylation reaction gave products **ortho 5**, **meta 5** and **para 5**, and decarboxylation reaction gave **ortho 8** and **para 8**.

2.2. Decarbonylation Reaction

To our knowledge, there are limited literature reports to the synthesis of 2,3,4,5-tetrachlorobenzyl scaffolds. Two main synthetic strategies were employed.

One method involves the synthesis of 2,3,4,5-tetrachlorophthalic acid diesters from direct esterification of 2,3,4,5-tetrachlorophthalic anhydride with primary alcohols at temperatures of 200 °C or above, which is accompanied by decarbonylation to give the expected tetrachlorophthalic acid diester and the decarbonylated 2,3,4,5-tetrachlorobenzoate ester in a ~2:1 ratio. [26] When the reaction is base catalysed (potassium carbonate, 3.63 mol%), the ratio of products is reversed (~1:2), with the decarbonylated product forming in greater amounts. The same authors also synthesised 2,3,4,5-tetrachlorobenzoic acid from 2,3,4,5-tetrachlorophthalic acid or anhydride by reaction in water, catalysed by sodium hydroxide (1 eq) at 200 °C for 7 h (93%). Then, synthesis of the corresponding secondary amides

(2,3,4,5-tetrachlorobenzanilide and 2,3,4,5-tetrachlorobenzamide) is achieved in two further steps by derivatising 2,3,4,5-tetrachlorobenzoic acid to the corresponding acid chloride, then by reaction with aniline and ammonium hydroxide, respectively, in excellent overall yield.

A second method, described by Harvey et al., achieves the synthesis of the same set of target molecules via a two-/three-step sequence from aromatic starting materials (e.g., toluene, ethylbenzene), firstly via perchlorination with chlorine bubbling through the aromatic starting material, iron powder and anhydrous ferric chloride boiling in carbon tetrachloride for 8 h, followed by reaction of the perchlorinated aromatics with sulfuric acid at 180–200 °C for heptachlorotoluene or 260–280 °C for nonachloroethylbenzene starting material used. The reaction with the nonachloroethylbenzene required a further step, namely the reaction with potassium permanganate in 2N sodium hydroxide at 80 °C. The yields for 2,3,4,5-tetrachlorobenzoic acid were 29% and 14%, respectively [27].

Earlier synthetic strategies include the reaction of tetrachlorophthalic acid in a sealed vessel at 300 °C in subcritical acetic acid [28], of tetrachloro(trichloromethyl)benzene (unknown isomer/s) in a sealed vessel at 280 °C in subcritical water [29]. A later publication, reporting on the phytotoxicity activity of benzoic acid derivatives, includes production of a small library of 2,3,4,5-tetrachlorobenzamides; however, the synthetic details to this sub-family are scarce [30].

Here, decarbonylation occurred during the reaction of boron-bearing amines with phthalic anhydride to produce the corresponding secondary amides (**ortho 5**, **meta 5** and **para 5**). The reaction mechanism is hypothesised to proceed through an acyl substitution reaction occurring at an anhydride carbonyl by the nitrogen atom of the amine reagent. This is followed by decarbonylation of the adjacent carboxylate and formation of an anionic intermediate, with the resulting electron pair held in the aromatic C-sp^2 orbital bearing the carboxylate group. This lone pair is thought to be stabilised by inductive and resonance effects via the nearby four electronegative chlorine atoms. This lone pair then strips off a proton intramolecularly from the quaternary nearby nitrogen atom, thus, providing the 2,3,4,5-tetrachlorobenzamide products.

2.3. Decarboxylation Reaction to Benzamides

Our methodology achieves the transformation of 2,3,4,5-tetrachlorophthalic anhydride and 2,3,4,5-tetrachlorophthalimide to the corresponding decarbonylated secondary amides in one synthetic step, avoiding the use of corrosive reagents and harsh conditions, and at lower temperature by heating the reaction mixture to 100 °C in DMF for 48 h to give the products **ortho 8** and **para 8** in moderate yields.

3. X-ray Crystallography Commentary

The structure of compound **meta 5** is shown in Figure 2. The asymmetric unit comprises two molecules, which exhibit essentially the same conformations and only one of these is shown. As expected, the trans-amide group is essentially planar (O1a-C-N1a-H 172.4°), while the two phenyl substituents are twisted (C_{Ar}-C_{Ar}-C-O1a 75.5°; C_{Ar}-C_{Ar}-N1a-C 33.1°) to minimise repulsion with the amide functional group.

In the structure of **ortho 8** (Figure 3), the substituents on the B-substituted ring are in ortho positions. In comparison with **meta 5**, the insertion of a methylene group between the amide and B-substituted phenyl ring relieves torsional strain (C_{Ar}-C_{Ar}-C(H$_2$)-N1 4.6°). The other structural features resemble those found in **meta 5**, defined by the dihedral angles O1-C-N1-H 173.8° and C_{Ar}-C_{Ar}-C-O1 76.6°, C_{Ar}-C-N1a-C 61.1°.

The boronic ester groups in both structures are close to coplanar with the adjacent phenyl ring (out of plane twist <10°) and the C-B bonds (**ortho 8** 1.561(6) Å; **meta 5** 1.558(6) and 1.557(6) Å) are reinforced due to π-bonding with the sp^2-hybridised B-atom. In twisted (purely σ-bonded) aromatic boronate esters (C_{Ar}-C_{Ar}-B-O ~90°), the C_{Ar}-B bond is typically in a range 1.57–1.59 Å [31,32].

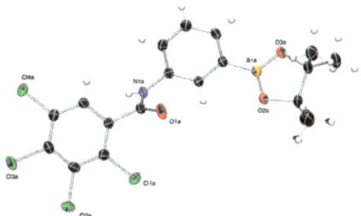

Figure 2. ORTEP plot of compound **meta 5** (30% ellipsoids).

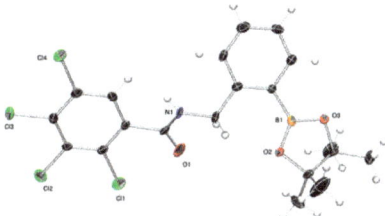

Figure 3. ORTEP plot of compound **ortho 8** (30% ellipsoids).

Intermolecular H-bonding in both structures comprises one-dimensional N-H ... O chains of the amide functional group (in its trans H-N-C=O conformation). There are differences in the symmetry of the chains in the two structures. When the boronate ester is *ortho* on the benzamide ring, adjacent molecules are related by the *c* glide place (Figure 4A), orthogonal to the place of the page and propagating right to left, leading to an alternating (zig-zag) array of H-bonds along the chain.

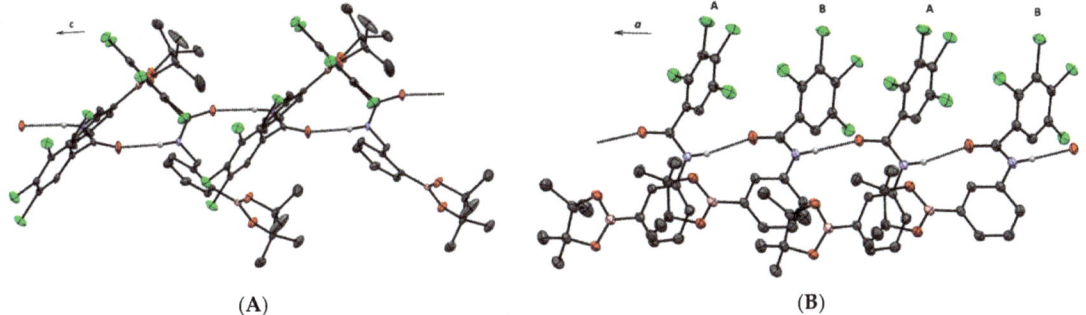

Figure 4. (**A**) H-bonding in **ortho 8** and (**B**) **meta 5** (molecules A and B comprise the asymmetric unit). The c and a axes respectively highlighted. Only amide H-atoms shown.

In the structure of the **meta 5** isomer (Figure 4B), the presence of two molecules in the asymmetric unit (molecules A and B) breaks any symmetry relationship between adjacent molecules within the H-bonded chain, and the orientations of the N-H ... O bonds are approximately the same along the chain, with rotation of the nearby aromatic rings facilitating closer packing.

4. Glycosidase Assay

In our laboratory we are interested in glycosidase modulation [13,18,19,21]. Screening for selectivity, as well as potency, is of paramount importance in carbohydrate-active enzyme research.

Our drugs and controls are, therefore, screened against two panels of glycosidases, respectively, in methanol (Table 1) and water (Table 2). This allows identification of the glycosidase-related disease area(s), selectivity profile and potency of biological action for each drug.

Following the glycosidase inhibition range recommendations [21], an IC_{50} value > 250 μm denotes weak inhibition, 100–249 μm denotes moderate inhibition, 10–99 μm good inhibition, 0.1–9 μm potent inhibition and <0.1 μm very potent inhibition. Our 2,3,4,5-tetrachlorobenzamides enter as a novel family of glycosidase inhibitors.

4.1. Glycosidases

The glycosidases screened are the following: rice α-glucosidase, yeast α-glucosidase, *Bacillus* α-glucosidase, rat intestinal maltase α-glucosidase, almond β-glucosidase, bovine liver β-glucosidase, coffee beans α-galactosidase, bovine liver β-galactosidase, Jack bean α-mannosidase, snail β-mannosidase, *Penicillium decumbens* α-L-rhamnosidase, bovine kidney α-L-fucosidase, *Eschierichia coli* β-glucuronidase, bovine liver β-glucuronidase, porcine kidney trehalase, *Aspergillus niger* amyloglucosidase and bovine kidney N-acetyl-β-glucosaminidase.

The disease areas for each glycosidase follow:

α-Glucosidase (EC 3.2.1.20) inhibition is linked to the management of diabetes, certain forms of hyperlipoproteinemia and obesity [33,34]. α-Glucosidase modulators also have potential as broad-spectrum anti-viral agents [35–38], for cancer [39] and lysosomal storage disorder Pompe disease [40,41].

β-Glucosidase (EC 3.2.1.21), α-Galactosidase (EC 3.2.1.22), and β-galactosidase (EC 3.2.1.23) are, respectively, linked to lysosomal storage disorders, such as Gaucher disease [42,43], Fabry disease [40], and GM1 gangliosidosis and Morquio syndrome B [40,44].

Other glycosidases involved in lysosomal storage disorders include α-mannosidase (EC 3.2.1.24), β-mannosidase (EC 3.2.1.25), α-L-fucosidase (EC 3.2.1.51), trehalase (EC 3.2.1.28), and β-glucuronidase (EC 3.2.1.31), whose malfunction cause, respectively, mannosidoses [45–47], fucosidosis [48,49], trehalase deficiency [50,51] and Sly disease [52]. Inhibition of β-glucuronidase may also help in controlling cancers and other diseases [53].

α-L-Rhamnosidase (EC 3.2.1.40) inhibition is linked to bacterial virulence [54,55]. Amyloglucosidase (EC 3.2.1.3) inhibition is important in diabetes management [56] and N-acetyl-β-D-glucosaminidase (EC 3.2.1.96) is involved in cancer progression and diabetic kidney disease [57].

4.2. Biological Activities for Phthalimides and Benzamides in the Literature

An overview of the literature in the field provides the following studies for phthalamides, 2,3,4,5-chlorophthalimides, and benzamides.

4.2.1. Phthalamides

A siastatin-derivatised phthalimide produced a very potent inhibition of bovine kidney α-L-fucosidase (IC_{50} 0.013 μM) [58].

SAR studies on phthalimide analogues with *Saccharomyces cerevisiae* (yeast) α-glucosidase highlight good inhibitions for N-phenylphthalimides derivatised at the *ortho*-position with non-polar groups [59]. This was not found by us with our phthalimide drugs, but with our benzamide drugs.

On the other side of the molecule, substitutions of the phthalimide scaffold H atoms with other groups, such as amine or hydroxyl, tend to largely abrogate potency, but the introduction of nitro or alkyl groups tend to produce good inhibitors.

Table 1. Glycosidase inhibition studies from laboratory 1. Percentage inhibition data and IC$_{50}$ values (on coloured background) for the most potent activities in a panel of 15 glycosidases.

Compound	Enzyme	α-Glucosidase			β-Glucosidase		α-Galactosidase	β-Galactosidase	α-Mannosidase	β-Mannosidase	α-L-Rhamnosidase	α-L-Fucosidase	β-Glucuronidase		Trehalase	Amyloglucosidase
		Rice	Yeast	Rat Intestinal Maltase	Almond	Bovine Liver	Coffee Beans	Bovine Liver	Jack Bean	Snail	Penicillium decumbens	Bovine Kidney	E. coli	Bovine Liver	Porcine Kidney	A. niger
BSH		^{a}NI^b (0%)	^{a}NI^b (6.9%)	^{a}NI^b (0%)	^{a}NI^b (0%)	^{a}NI^b (15%)	^{a}NI^b (12.3%)	^{a}NI^b (0%)	^{a}NI^b (7.9%)	^{a}NI^b (19.2%)	^{a}NI^b (0.2%)	^{a}NI^b (0%)	^{a}NI^b (6%)	^{a}NI^b (19.6%)	^{a}NI^b (4.2%)	^{a}NI^b (0%)
10B-BSH		^{a}NI^b (0%)	^{a}NI^b (5.6%)	^{a}NI^b (0%)	^{a}NI^b (0%)	^{a}NI^b (11.9%)	^{a}NI^b (3.9%)	^{a}NI^b (22.3%)	^{a}NI^b (7.7%)	^{a}NI^b (15%)	^{a}NI^b (0%)	^{a}NI^b (6.5%)	^{a}NI^b (3.2%)	^{a}NI^b (12.5%)	^{a}NI^b (2.3%)	^{a}NI^b (0%)
BPA		^{c}NI^d (0%)	^{c}NI^d (0%)	^{c}NI^d (0%)	^{c}NI^d (0%)	^{c}NI^d (14.2%)	^{c}NI^d (4.3%)	^{c}NI^d (10.9%)	^{c}NI^d (1.1%)	^{c}NI^d (2.1%)	^{c}NI^d (0%)	^{c}NI^d (0%)	^{c}NI^d (0.5%)	^{c}NI^d (7.4%)	^{c}NI^d (0%)	^{c}NI^d (0%)
10B-BPA		^{c}NI^d (0%)	^{c}NI^d (0%)	^{c}NI^d (0%)	^{c}NI^d (0%)	^{c}NI^d (0%)	^{c}NI^d (1.2%)	^{c}NI^d (0%)	^{c}NI^d (0%)	^{c}NI^d (0.3%)	^{c}NI^d (0%)	^{c}NI^d (0%)	^{c}NI^d (0%)	^{c}NI^d (0%)	^{c}NI^d (0%)	^{c}NI^d (0%)
para 3		^{a}NI^b (5.6%)	^{a}NI^b (0%)	^{a}NI^b (12.2%)	^{a}NI^b (0.7%)	^{a}NI^b (4.0%)	^{a}NI^b (0.2%)	^{a}NI^b (8.2%)	^{a}NI^b (0%)	^{a}NI^b (0%)	^{a}NI^b (0%)	^{a}NI^b (0%)	^{a}NI^b (2.2%)	^{a}NI^b (1.0%)	^{a}NI^b (0%)	^{a}NI^b (1.0%)
meta 3		^{c}NI^d (5.1%)	^{c}NI^d (0%)	^{c}NI^d (22.1%)	^{c}NI^d (5.0%)	^{c}NI^d (18.8%)	^{c}NI^d (3.5%)	^{c}NI^d (12.4%)	^{c}NI^d (2.7%)	^{c}NI^d (1.1%)	^{c}NI^d (0%)	^{c}NI^d (4.7%)	^{c}NI^d (6.1%)	^{c}NI^d (5.7%)	^{a}NI^b (0%)	^{c}NI^d (3.2%)
para 5		^{a}NI^b (0%)	^{a}NI^b (0%)	188	^{a}NI^b (17.1%)	543	^{a}NI^b (35.8%)	333	^{a}NI^b (19.3%)	^{a}NI^b (30.4%)	^{a}NI^b (14.4%)	^{a}NI^b (5.1%)	^{a}NI^b (45.5%)	^{a}NI^b (36.5%)	^{a}NI^b (0%)	^{a}NI^b (0%)
meta 5		^{a}NI^b (45.5%)	^{a}NI^b (12.8%)	274	^{a}NI^b (30.2%)	175	^{a}NI^b (30.2%)	213	^{a}NI^b (7.6%)	^{a}NI^b (39.1%)	^{a}NI^b (2.2%)	^{a}NI^b (3.3%)	932	^{a}NI^b (6.0%)	^{a}NI^b (0%)	^{a}NI^b (2.8%)
ortho 5		^{c}NI^d (0%)	^{c}NI^d (0%)	^{c}NI^d (3%)	^{c}NI^d (11.6%)	^{c}NI^d (21.2%)	^{c}NI^d (0%)	^{c}NI^d (0%)	^{c}NI^d (3.4%)	^{c}NI^d (0%)	^{c}NI^d (0%)	^{c}NI^d (0%)	^{c}NI^d (3.8%)	^{c}NI^d (0%)	^{c}NI^d (1.2%)	^{c}NI^d (0%)

Table 1. *Cont.*

Compound	Enzyme	α-Glucosidase			β-Glucosidase		α-Galactosidase	β-Galactosidase	α-Mannosidase	β-Mannosidase	α-L-Rhamnosidase	α-L-Fucosidase	β-Glucuronidase		Trehalase	Amyloglucosidase
		Rice	Yeast	Rat Intestinal Maltase	Almond	Bovine Liver	Coffee Beans	Bovine Liver	Jack Bean	Snail	*Penicillium decumbens*	Bovine Kidney	*E. coli*	Bovine Liver	Porcine Kidney	*A. niger*
para 8		ᵃNI ᵇ(2.1%)	ᵃNI ᵇ(0%)	207	ᵃNI ᵇ(29.6%)	702	ᵃNI ᵇ(18.0%)	74.7	ᵃNI ᵇ(4.5%)	ᵃNI ᵇ(34.7%)	ᵃNI ᵇ(3.3%)	ᵃNI ᵇ(7.2%)	ᵃNI ᵇ(34.7%)	ᵃNI ᵇ(18.8%)	ᵃNI ᵇ(4.1%)	ᵃNI ᵇ(0%)
ortho 8		ᵃNI ᵇ(0%)	ᵃNI ᵇ(0%)	305	254	922	ᵃNI ᵇ(114.4%)	278	ᵃNI ᵇ(12.5%)	ᵃNI ᵇ(39%)	ᵃNI ᵇ(6.2%)	ᵃNI ᵇ(19.9%)	ᵃNI ᵇ(41.7%)	ᵃNI ᵇ(16.3%)	ᵃNI ᵇ(16.8%)	ᵃNI ᵇ(0%)

ᵃNI: No inhibition (less than 50% inhibition at 1000 μM). ᵇ ( ): inhibition % at 1000 μM. ᶜNI: No inhibition (less than 50% inhibition at 100 μM). ᵈ ( ): inhibition % at 100 μM. For BSH and BPA. For our drugs: ᵃNI: No inhibition (less than 50% inhibition at 100 μM). ᵇ ( ): inhibition % at 100 μM. ᶜNI: No inhibition (less than 50% inhibition at 10 μM). ᵈ ( ): inhibition % at 10 μM.

Table 2. Glycosidase inhibition studies from laboratory 2. Results expressed as % inhibition at ~400 µM.

Compound	Enzyme	Appearance of the 1 mg/mL Aqueous Solution Tested	α-Glucosidase		β-Glucosidase	α-Mannosidase	N-Acetyl-β-Glucosaminidase	β-Glucuronidase
			Yeast	Bacillus	Almond	JACK BEAN	Bovine Kidney	Bovine Liver
	BSH	in solution	59	48.1	47.3	−18.7	37.1	31.6
	10B-BSH	in solution	65.9	53	49.9	−16.8	40.9	44.1
	BPA	some in solution with undissolved sediment	2.9	19.9	3.9	0.6	6.7	−0.7
	10B-BPA	some in solution with undissolved sediment	3.4	19.5	3	−0.7	6.3	−1
	para 3	some in solution with undissolved sediment	29.3	17.9	15.4	16.5	8.3	2.1
	meta 3	opaque suspension	32.2	31.7	19.3	30.3	17.1	5.8
	para 5	some in solution with undissolved sediment	38.4	20	11.3	−8.2	20.6	−2.2
	meta 5	some in solution with undissolved sediment	82.8	54.1	18.7	2.2	30.6	−5.7

Table 2. *Cont.*

Compound	Enzyme	Appearance of the 1 mg/mL Aqueous Solution Tested	α-Glucosidase		β-Glucosidase	α-Mannosidase	N-Acetyl-β-Glucosaminidase	β-Glucuronidase
			Yeast	Bacillus	Almond	JACK BEAN	Bovine Kidney	Bovine Liver
ortho 5		some in solution with undissolved sediment	17.3	14.4	21	0	28.8	−3.5
para 8		some in solution with undissolved sediment	99.5	62.2	19.1	−3.3	15	−1.1
ortho 8		some in solution with undissolved sediment	68.9	23.7	83.4	−6.1	28.3	−6.3

Another study on a library of N-phenylphthalimide derivatives showed the strongest potency against the three α-glucosidases screened was displayed by N-(2,4-dinitrophenyl)-phthalimide, having two nitro groups in *ortho* and *para* positions. This is a moderate inhibitor of yeast α-glucosidase (IC$_{50}$ 158 μM) and a good inhibitor of maltase (IC$_{50}$ 51 μM), displaying no inhibition of sucrase [60].

N-Phenylphthalimide derivatives substituted with non-polar groups departing from the *ortho* position of the phenyl group keep providing inhibition of *Saccharomyces cerevisiae* α-glucosidase as low as 16 μM [61].

Other examples include the Hashimoto papers [62–64].

Several members of a phthalimide moiety connected by an alkyl chain to variously substituted phenoxy rings were screened against α-glucosidase. The inhibition potency appeared to be governed by the chain length of the substrate. Substrates possessing 10 carbons afforded the highest levels of activity, which were one to two orders of magnitude more potent than the known inhibitor 1-DNJ [4,65].

Bian and coworkers screened, against α-glucosidase, a series of N-substituted-(p-toluenesulfonylamino)phthalimides. Many analogues provide good inhibitions of the enzymes, with aromatic pendants and tethers containing 1–3 atoms generally producing the most potent inhibitions [3].

Another study of N-phenoxy-substituted phthalimides showed that the presence of a thiazolidine-2,4-dione or a rhodanine group, located at the 4-position of the phenyl ring, resulted in the best activity, with IC$_{50}$ values as low as 5 μM against *Saccharomyces cerevisiae* α-glucosidase [66].

A series of phthalimide-benzamide-1,2,3-triazole hybrids showed good/moderate inhibitory activity against *Saccharomyces cerevisiae* α-glucosidase. The most potent compound displayed an IC$_{50}$ of 40 μm [67].

4.2.2. 2,3,4,5-Tetrachlorophthalimides

A 2,3,4,5-tetrachlorophthalimide, derivatised at the N-atom with a doubly *ortho*-substituted phenyl group, produced potent inhibitions. The use of linear alkyl chains departing from the phthalimide N atom produces potent and very potent inhibitions as the chain lengthens [59].

2,3,4,5-Tetrachlorophthalimides N-derivatised with a phenyl group attached directly to the N or through a linear alkyl tether (1–6 CH$_2$ units) all displayed potent IC$_{50}$ (3–11 μM) towards one α-glucosidase screened and more potent than the \1-DNJ control. The replacement of the four chlorine atoms with hydrogen atoms and the replacement of hydrogen atoms with other groups (e.g., nitro, amine) partially or completely abrogated the inhibitory activity of the drugs [6].

The same group also investigated other groups bonded to the phthalimide N, namely branched and cyclical alkyl groups and a dodecaborane group. All drugs displayed comparable or more potent activity (1–49 μM) than 1-DNJ. Cyclical alkyl groups and the borane group produced the most potent inhibitions [5].

The pendant groups attached to the phthalimide unit can clearly interact effectively with a number of sites in the vicinity of the active site, which is presumably occupied by the 2,3,4,5-tetrachlorophthalmide scaffold. This is highlighted by the variety and length of pendant groups. Hydrophobicity seems to be the common motif.

N-Phenyl-2,3,4,5-tetrachlorophthalimide derivatives substituted with non-polar groups departing from the *ortho*, *meta* and *para* positions of the phenyl group keep providing inhibition of α-glucosidase as low as 13 μM against *Saccharomyces cerevisiae* [61].

4.2.3. Benzamides

A series of N-substituted 1-aminomethyl-β-D-glucopyranoside derivatives was screened against *Saccharomyces cerevisiae* α-glucosidase, rat intestinal maltase α-glucosidase and sucrase. The most potent inhibitions were produced when the benzamide aromatic ring displayed groups in the *para* position to the amide. The three most potent compounds comprised *O*-acetyl groups in the 3-, 4- and 5-positions (IC$_{50}$ 7.7 and 15.6 μM against rat intestinal maltase α-glucosidase and sucrase), a nitro group in the 4-position (IC$_{50}$ 36.2 μM against *Saccharomyces cerevisiae* α-glucosidase) and an *O*-acetyl group in the 4-position (IC$_{50}$ 96.5 μM against *Saccharomyces cerevisiae* α-glucosidase) [68].

1-DNJ derivatised with benzamides also produced potent to very potent inhibitions of sweet almond and *A. faecalis* β-glucosidases (IC$_{50}$ 0.15–21 μM) [69].

Further, 24 Metronidazole-tethered benzamide triazoles demonstrated weak to good activity against β-glucuronidase, with no toxicity against 3T3 mouse fibroblast cell lines [70]. The most active compound has an IC$_{50}$ 12.4 μM with an activity ~four-times higher than the standard inhibitor, D-saccharic acid-1,4-lactone (IC$_{50}$ 45.8 μM) [53].

Iminosugar-benzamide derivatives displayed good to moderate inhibitions of *Aspergillus niger* amyloglucosidase, *Saccharomyces cerevisiae* α-glucosidase and human lymphocytes lysosomal α-glucosidase [71].

4.3. Biological Activities for Our Drug Library Screened in Methanol

Biological activities for controls, and our 2,3,4,5-tetrachloro phthalimides and benzamides, follow in Tables 1 and 2.

For Table 1:

4.3.1. Controls

Borocaptate sodium (BSH) and 4-borono-L-phenylalanine (BPA), and their ^{10}B-enriched congeners ^{10}B-BSH and ^{10}B-BPA, were the controls. To our knowledge, these drugs have never been reported in a glycosidase assay. BSH and BPA are the drugs currently clinically used in BNCT. It is possible to see that none of them significantly inhibit any of the glycosidases in the panel at 100 or 1000 μM. In the panel, percent inhibitions range from a minimum value of 0 to a maximum value of 19.6.

4.3.2. 2,3,4,5-Tetrachlorophthalimides

The 2,3,4,5-tetrachlorophthalimide drugs presented in this work do not provide any appreciable degree of inhibition, most likely because the spatial geometry and length of tether extending from the phthalimide scaffold are probably not able to reach the sites of interaction.

para 3 and **meta 3** possess the CH$_2$ spacer between the phthalimide and the aromatic boron group, with **para 3** displaying the boronate ester group in para position and **meta 3** in meta position to the phthalimide.

Of the benzamides, **para 5**, with no CH$_2$ spacer between the aromatic boron group and the benzamide, shows moderate inhibition towards maltase α-glucosidase, with an IC$_{50}$ 188 μM. Weak inhibitions are displayed towards bovine liver β-glucosidase (IC$_{50}$ 543 μM) and bovine liver β-galactosidase (IC$_{50}$ 333 μM).

meta 5, with no CH$_2$ spacer between the aromatic boron group and the benzamide, again, shows moderate inhibition towards bovine liver β-glucosidase (IC$_{50}$ 175 μM) and bovine liver β-galactosidase (IC$_{50}$ 213 μM). Weak inhibition is observed towards maltase α-glucosidase, with an IC$_{50}$ of 274 μM, and *E. coli* β-glucuronidase (IC$_{50}$ 932 μM).

para 5 and **meta 5** display the boronate ester group, respectively, in *para* and *meta* position from the benzamide. Both display partially selective inhibition profiles, inhibiting only 3–4 within the panel of 16 enzymes. Interestingly, their inhibitory profiles show a swap in potency, with **para 5** preferentially selecting one α-glucosidase and **meta 5** selecting β-glycosidases.

ortho 5, the *ortho* congener to **para 5** and **meta 5**, displays no significant inhibition of any of the enzymes. Hence, the location of the boronate ester group has a negative effect on drug–enzyme interactions, presumably either by preventing the drug from sitting in the active site as **para 5** and **meta 5** and/or abrogating any further interactions the aromatic boronate group may have with the enzyme.

4.3.3. 2,3,4,5-Tetrachlorobenzamides

Benzamides **para 8** and **ortho 8**, possessing the CH_2 spacer between the aromatic boron group and the benzamide, show partially selective inhibition of the same enzymes. **para 8** has the boronate ester group para to the benzamide, whereas **ortho 8** displays the boronate ester group ortho to the benzamide. Presumably, the lack of one of the carbonyl groups allows the pendant group to reach sites of favourable interactions with the enzymes. Since the enzymes inhibited are the same, it is surmised that the drug SAR profiles are similar.

para 8 elicits moderate inhibition of maltase α-glucosidase, with an IC_{50} of 207 μM. **ortho 8** inhibits the same enzyme weakly, with a similar value of 305 μM. Weak inhibitions are displayed towards bovine liver β-glucosidase (IC_{50} = 702 and 922 μM, respectively).

The two main differences are seen:

(a) In the inhibition of bovine liver β-galactosidase, which is inhibited weakly by **ortho 8** (IC_{50} 278 μM), but **para 8** inhibits the same enzyme with a good IC_{50} 74.7 μM. This is the most potent drug in the small libraries reported in this communication.

(b) In the inhibition of almond β-glucosidase by **ortho 8**, which is on the edge of moderate/weak, with an IC_{50} of 254 μM.

The drugs that show appreciable inhibition all inhibit the same glycosidase enzymes (maltase α-glucosidase, bovine liver β-glucosidase and bovine liver β-galactosidase). Furthermore, all drugs selectively inhibit glycosidases of animal origin vs. glycosidases of plant or bacterial origin within the same glycosidase class. This is a positive result for applications in a human disease medicinal chemistry context.

4.4. Biological Activities for Our Drug Library Screened in Water

For Table 2:

Some differences were noted in the results obtained from this second laboratory that used just water to suspend or dissolve the compounds. The compounds did not fully go into solution but nonetheless, activities were observed and so they are included here for comparison.

No appreciable inhibition was detected for any of the controls and compounds against *Bacillus* α-glucosidase, Jack bean α-glucosidase, bovine kidney A-acetyl-β-glucosaminidase and bovine liver β-glucuronidase.

Three potent % inhibitions at a concentration of ~400 μM were seen for **meta 5** (82.8%) and **para 8** (99.5%) against yeast α-glucosidase, and for **ortho 8** (83.4%) against almond β-glucosidase.

5. Cancer Assay and Structure Activity Relationships

In our laboratory, we are interested in BNCT as a potentially broad-spectrum approach to cancer management. It would be advantageous upon irradiation, if the boron-containing drugs accumulate more selectively in cancer cells vs. healthy cells [72,73]. In case the drugs do not accumulate selectively in cancer vs. healthy cells, the delivery of radiation is required with greater precision.

BNCT is essentially a non-invasive radiation technique and the least destructive currently available [23–25]. Use of a borylated drug in BNCT would ideally require that it is non-toxic in the absence of radiation. Following a first study of synthesis, purification and toxicity of organic-boron-containing drugs for BNCT applications [13], we report here, two further families of potential BNCT agents.

BNCT agents that contain organic boron groups are preferable to ones containing inorganic boron. The currently utilised sodium borocaptate, BSH, with its inorganic boron atoms, raises several toxicity concerns [74,75]. Boronophenylalanine, BPA, which contains the organic boronic acid moiety, has long been known to show no discernible toxicity [76]. Similarly, in this area, candidate BNCT agents containing an organic boron group should be more likely to reach the clinic.

It has long been known that organic boron is an essential element for plants [77,78] and is likely to be essential for human and animal health [79].

When comparing toxicological data for organic boron-containing molecules with their non-borylated congeners, the trend is that the presence of organic boron lowers toxicity profiles. For example, benzene has an LD_{50} (lethal dose) of 125 mg/kg (human, oral) [80] and an LCLO (lethal concentration) of 20,000 ppm (human, 5 min); it is carcinogenic, and also possibly mutagenic. The NIOSH Permissible Exposure Limit for benzene is 1 ppm, the Recommended Exposure Limit is 0.1 ppm, and the Immediately Dangerous to Life and Health concentration is at 500 ppm [81].

On the other hand, phenylboronic acid has an LD_{50} 740 mg/kg (rat, oral) [82], with no entry for RTECS, ACGIH, IARC, or NTP.

If a BNCT agent also has growth inhibition capability against cancer cells, then it is important to screen them in more complex biological systems, such as spheroids, as we did recently [72,73].

Table 3 shows, on a blue background, the percentage cell growth inhibition in response to 25 µM of the drug. In this case, a higher value correlates with a greater growth inhibition. Inhibition value ranges have been colour coded, according to potency, with 80–100% inhibition in red, 60–79% in orange, 35–59% in green, 10–34% in blue and 0–9% in black.

On the green background, the GI_{50} values are provided for the most potent drugs. The GI_{50} value provides the concentration in µM that induces a 50% cell growth inhibition. In this case, a lower value correlates with greater growth inhibition.

5.1. Analysis of the Percent Cell Growth Inhibition Data

BSH, BPA and their ^{10}B-enriched congeners ^{10}B-BSH and ^{10}B-BPA are the controls. BSH and BPA are the drugs currently clinically used in BNCT. It is possible to see that none of them significantly inhibit cell growth at 25 µM. Percent cell growth inhibitions range from <0% to 19% in our panel of 10 cancer cell lines (HT29, U87, MCF-7, A2780, H460, A431, Du145, BE2-C, SJ-G2 and MIA-Pa-Ca2) and a normal cell line (MCF10A). Specifically, only 10 out of 44 entries for the controls were double-digit (10% or greater) percent inhibitions. The other 34 entries contained inhibition values between 0.01% and 9.99%, and nine entries with values of 0% or <0%.

It is evident that more efficacious BNCT agents are required.

When analysing the data for the borylated drugs, a number of Structure–Activity Relationship considerations can be evinced.

A general overarching consideration is that the vast majority of percent cell growth inhibitions for the borylated drugs are significantly greater than the percent cell growth inhibitions for the BSH and BPA controls.

There are three drugs that possess potent inhibitions, namely tetrachlorophthalimides **para 3** and **meta 3**, and tetrachlorobenzamide **ortho 5**.

para 3 is the only tetrachlorophthalimide that displays potent inhibition. It displays the boronate ester group in *para* position to the phthalimide and it has the CH_2 spacer between the phthalimide and the aromatic boron group. Inhibitions range from 98% to >100% for 9 out of 10 cancer cell lines and for the normal cell line. Only the A431 cancer cell line displays a lower inhibition (85%).

Tetrachlorophthalimide **meta 3** possesses a similar structure to **para 3**, with the boronate ester group in *meta* position to the phthalimide. The installation of the boronate ester group in the *meta* position reduces potency significantly for all cancer lines and the normal cell line, though to varying extents. The smallest reduction in inhibition is seen in cell lines U87 (70%), MCF-7 (67%), A2780 (70%) and MCF10A (69%). A further loss in growth inhibition is seen in HT29 (53%), H460 (42%), A431 (48%), Du145 (43%) and BE2-C (45%). The greatest loss of growth inhibition is displayed in cell lines SJ-G2 (35%) and MIA-Pa-Ca2 (29%). Hence, the location of the boron ester group in *para* position greatly favours cell growth inhibition. These two drugs likely interact in similar ways with the cells, with the boronate ester group likely trying to interact with the same site/s (designated Site A for discussion purposes), but not managing quite as effectively when it is in the *meta* position.

Table 3. Cancer Screening. On blue background DOSE SCREEN: Percentage (%) Cell Growth Inhibition in response to 25 μM of Drug (the higher the value the greater the growth inhibition) and inhibition value ranges: 80–100% (red), 60–79% (orange), 35–59% (green), 10–34% (blue) and 0–9% (black). On green background DOSE RESPONSE: GI50 = Concentration (μM) that inhibits cell growth by 50% (the lower the value the greater the growth inhibition). In bold are highlighted the values of the three most potent drugs.

Compound	HT29 Colon Carcinoma	U87 Glioblastoma	MCF-7 Breast Carcinoma	A2780 Ovarian Carcinoma	H460 Lung Carcinoma	A431 Skin Carcinoma	Du145 Prostate Carcinoma	BE2-C Neuroblastoma	SJ-G2 Glioblastoma	MIA-Pa-Ca2 Pancreatic Carcinoma	MCF10A Breast (Normal)	Mode of Action	Selective
BSH	3 ± 2	<0	15 ± 3	2 ± 5	8 ± 2	<0	0 ± 8	10 ± 6	3 ± 8	2 ± 6	8 ± 3	NA	
10B-BSH	5 ± 1	0 ± 2	5 ± 3	5 ± 4	4 ± 2	<0	7 ± 7	8 ± 7	1 ± 9	2 ± 4	13 ± 4	NA	
BPA	14 ± 0	<0	<0	4 ± 1	7 ± 8	4 ± 6	19 ± 10	13 ± 10	5 ± 8	3 ± 3	4 ± 1	NA	
10B-BPA	15 ± 4	<0	1 ± 3	8 ± 4	8 ± 5	4 ± 4	15 ± 9	10 ± 6	5 ± 10	11 ± 3	<0	NA	
para 3	99 ± 2	>100	>100	>100	100 ± 2	85 ± 8	>100	>100	>100	98 ± 3	>100	A	
	5.3 ± 1.3	4.2 ± 1.0	3.0 ± 1.3	11 ± 1.1	13 ± 0.33	18 ± 2.3	5.5 ± 0.23	8.1 ± 1.7	6.6 ± 2.4	9.2 ± 1.9	12 ± 0.58		
meta 3	53 ± 3	70 ± 8	67 ± 7	70 ± 2	42 ± 6	48 ± 3	43 ± 13	45 ± 2	35 ± 4	29 ± 2	69 ± 7	A	
	26 ± 2.6	19 ± 1.5	23 ± 3.7	19 ± 0.33	33 ± 2.3	28 ± 0.85	30 ± 0.41	29 ± 0.58	35 ± 1.7	38 ± 2.9	18 ± 1.0		
para 5	69 ± 7	21 ± 16	42 ± 5	50 ± 1	54 ± 2	64 ± 4	1 ± 13	54 ± 1	28 ± 6	52 ± 1	2 ± 11	A	Y
meta 5	10 ± 7	<0	16 ± 4	49 ± 4	18 ± 6	9 ± 9	<0	3 ± 4	<0	8 ± 5	<0	A	Y
	98 ± 2	58 ± 5	>100	97 ± 2	>100	100 ± 2	>100	>100	92 ± 5	100 ± 3	58 ± 9	B	Y
ortho 5	14 ± 1.7	25 ± 2.6	12 ± 1.2	16 ± 1.2	13 ± 1.0	11 ± 1.8	16 ± 1.9	16 ± 0.58	16 ± 0.91	14 ± 0.25	27 ± 2.8		

Table 3. Cont.

Compound	HT29	U87	MCF-7	A2780	H460	A431	Du145	BE-2-C	SJ-G2	MIA-Pa-Ca2	MCF10A	Mode of Action	Selective
	Colon Carcinoma	Glioblastoma	Breast Carcinoma	Ovarian Carcinoma	Lung Carcinoma	Skin Carcinoma	Prostate Carcinoma	Neuroblastoma	Glioblastoma	Pancreatic Carcinoma	Breast (Normal)		
para 8	36 ± 9	14 ± 9	18 ± 5	58 ± 2	21 ± 3	<0	<0	15 ± 5	8 ± 10	23 ± 6	2 ± 6	A	y
ortho 8	5 ± 10	<0	<0	19 ± 10	0 ± 8	<0	<0	2 ± 5	<0	6 ± 10	<0	A	

Percentage (%) cell growth inhibition in response to 25 µM of drug after 72 h exposure using the MTT cell growth assay. The experiment was carried out in duplicate and replicated on 3 separate occasions.

Tetrachlorobenzamide **ortho 5**, displaying the boronate ester group in *ortho* position and having no CH_2 spacer between the benzamide and the aromatic boron group, also displays potent cell growth inhibition and a significant level of cell selectivity. This capability makes this drug the most interesting from a medicinal chemistry perspective. Cell selectivity (in particular, cancer versus healthy cell selectivity) is an area of active research in our group [72]. The removal of one of the carbonyl groups allows for greater conformational flexibility to this molecule, which may allow the boronate ester to interact with a different site than **para 3** and **meta 3** (designated Site B for discussion purposes). The *ortho*-phthalimide congener has not been synthetically achievable so far, and so it was not tested. Percent cell growth inhibitions range from 97 to >100 for most cancer cell lines (HT29, MCF-7, A2780, H460, A431, Du145, BE2-C and MIA-Pa-Ca2). It is 92% for SJ-G2 and drops dramatically for U87 (58%) and for the normal cell line MCF10A (58%). This selectivity between cancer versus healthy cells is a highly desirable drug capability.

The comparison in percent inhibition between benzamide **ortho 5** and its congener **ortho 8** is particularly interesting. The only structural difference is the CH_2 spacer between the benzamide and the aromatic boron group. However, there is a complete abrogation of cell growth inhibition produced by **meta 3**. It can be evinced that **ortho 8** likely interacts with the same sites **para 3** and **meta 3** interact with, whereas **ortho 5** has a different mode of action, interacting with another site that seems to be overexpressed in all cancer cell lines, apart from U87 and the normal cancer cell line MCF10A.

Of the drugs that inhibited cell growth to a lesser extent, **para 8** is the benzamide analogue of **para 3**. **para 8** is thought to also interact with Site A due to structural similarities with **para 3**; however, it shows a significantly reduced growth inhibition, probably due to the boronate ester group interacting not as efficiently on Site A. In this case as well, cell selectivity is displayed in inhibition. Normal cells MCF10A (2%) and SJ-G2 (8%), Du145 (<0%) and A431 (<0%) were not inhibited, whereas BE2-C (15%), MCF-7 (18%) and U87 (14%) were minimally inhibited, MIA-Pa-Ca2 (23%), H460 (21%) and HT29 (36%) were somewhat more inhibited, and finally, A2780 was significantly inhibited (58%).

Benzamide **para 5**, not possessing the CH_2 spacer between the benzamide and the *para*-aromatic boron group, is thought to interact with Site A, due to structural similarities with its phthalimide congeners; however, it does not efficiently interact with Site A. This may be due to the boronate ester not reaching Site A due to the lack of the CH_2 spacer and the greater degree of conformational flexibility deriving from the removal of the carbonyl group from the phthalimide scaffold. This results in an overall reduction in cell growth inhibition. In this case as well, cell selectivity is displayed in inhibition. Normal cells MCF10A (2%) and Du145 (1%) were not inhibited, whereas U87 (21%) and SJ-G2 (28%) were somewhat more inhibited, MCF-7 (42%), A2780 (50%), H460 (54%), BE2-C (54%) and MIA-Pa-Ca2 (52%) were significantly inhibited, and finally, HT29 (69%) and A431 (64%) were inhibited the most.

Benzamide **meta 5**, not possessing the CH_2 spacer between the benzamide and the aromatic boron group, shows selective and significant inhibition for A2780 (49%). The benzamide structure provides a greater degree of conformational flexibility, likely placing the *meta*-positioned boronate ester somewhere in between Site A and Site B, and preventing it from interacting efficiently with either.

Benzamide **ortho 8**, possessing the CH_2 spacer between the benzamide and the aromatic boron group, is thought to interact with Site A, due to structural similarities with its phthalimide congeners; however, it does not efficiently interact with Site A, thus, showing almost complete abrogation of cell growth inhibition.

Based on Structure–Activity Relationship data obtained, two modes of cell growth inhibition are put forward, Mode of Action A, which arises from drugs interacting at Site A, and Mode of Action B, arising from drug **ortho 5** interacting at Site B. Both Modes of Action can be elicited in selective ways by drugs **para 8** (for Mode of Action A) and **ortho 5** (for Mode of Action B) and displaying minimal or zero inhibition on the normal cell line.

5.2. Analysis of the GI$_{50}$ Data

The GI$_{50}$ was measured for the three most potent drugs, tetrachlorophthalimides **para 3** and **meta 3**, and tetrachlorobenzamide **ortho 5**. **para 3** shows consistent potency via GI$_{50}$ values between 3 and 18 µM for all cancer cell lines and the normal cell line. Similarly, **meta 3** displays consistently potent GI$_{50}$ values between 18 and 38 µM for all cancer cell lines and the normal cell line. **ortho 5** displays consistently potent GI$_{50}$ values between 11 and 27 µM for all cancer cell lines and the normal cell line.

6. Experimental

6.1. Glycosidase Inhibition Experimental from Laboratory 1

In Table 1

The enzymes α-glucosidase (from yeast), β-glucosidases (from almond and bovine liver), α-galactosidase (from coffee beans), β-galactosidase (from bovine liver), α-mannosidase (from Jack bean), β-mannosidase (from snail), α-L-rhamnosidase (from *Penicillium decumbens*), α-L-fucosidase (from bovine kidney), trehalase (from porcine kidney), β-glucuronidases (from *E. coli* and bovine liver), amyloglucosidase (from *A. niger*), para-nitrophenyl glycosides, and various disaccharides were purchased from Sigma-Aldrich Co (St Louis, MO, USA).

Brush border membranes were prepared from the rat small intestine according to the method of Kessler et al. [83] and were assayed at pH 6.8 for rat intestinal maltase using maltose. For rat intestinal maltase, porcine kidney trehalase, and *A. niger* amyloglucosidase activities, the reaction mixture contained 25 mM maltose and the appropriate amount of enzyme, and the incubations were performed for 10–30 min at 37 °C. The reaction was stopped by heating at 100 °C for 3 min. After centrifugation (600× *g*; 10 min), the resulting reaction mixture was added to the Glucose CII-test Wako (Wako Pure Chemical Ind., Osaka, Japan). The absorbance at 505 nm was measured to determine the amount of the released D-glucose. Other glycosidase activities were determined using an appropriate para-nitrophenyl glycoside as substrate at the optimum pH of each enzyme. The reaction mixture contained 2 mM of the substrate and the appropriate amount of enzyme. The reaction was stopped by addition of 400 mM Na$_2$CO$_3$. The released para-nitrophenol was measured spectrometrically at 400 nm. All reactions run in methanol.

6.2. Glycosidase Inhibition Experimental from Laboratory 2

In Table 2

All enzymes and para-nitrophenyl substrates were purchased from Sigma. Enzymes were assayed at 27 °C in 0.1 M citric acid/0.2 M disodium hydrogen phosphate buffers at the optimum pH for the enzyme. The incubation mixture consisted of 10 µL enzyme solution, 10 µL of 1 mg/mL aqueous solution of extract and 50 µL of the appropriate 5 mM para-nitrophenyl substrate made up in buffer at the optimum pH for the enzyme. The reactions were stopped by addition of 70 µL 0.4 M glycine (pH 10.4) during the exponential phase of the reaction, which had been determined at the beginning using uninhibited assays in which water replaced inhibitor. Final absorbances were read at 405 nm using a Versamax microplate reader (Molecular Devices). Assays were carried out in triplicate, and the values given are means of the three replicates per assay. All reactions run in water.

6.3. Cancer Screening Experimental

All test agents were prepared as stock solutions (20 mM) in dimethyl sulfoxide (DMSO) and stored at −20 °C. Cell lines used in the study included HT29 (colorectal carcinoma); U87, SJ-G2, (glioblastoma); MCF-7, (breast carcinoma); A2780 (ovarian carcinoma); H460 (lung carcinoma); A431 (skin carcinoma); Du145 (prostate carcinoma); BE2-C (neuroblastoma); MiaPaCa-2 (pancreatic carcinoma); and SMA560 (spontaneous murine astrocytoma), together with the one non-tumour-derived normal breast cell line (MCF10A). All cell lines were incubated in a humidified atmosphere 5% CO$_2$ at 37 °C. The cancer cell lines were maintained in Dulbecco's modified Eagle's medium (DMEM; Sigma, Australia) supplemented with foetal bovine serum (10%), sodium pyruvate (10 mM), penicillin (100 IUmL^{-1}), streptomycin (100 µg mL^{-1}) and L-glutamine (2 mM).

The non-cancer MCF10A cell line was maintained in DMEM:F12 (1:1) cell culture media, 5% heat-inactivated horse serum, supplemented with penicillin (50 IUmL^{-1}), streptomycin (50 µg mL^{-1}), HEPES (20 mM), L-glutamine (2 mM), epidermal growth factor (20 ng mL^{-1}), hydrocortisone (500 ng mL^{-1}), cholera toxin (100 ng mL^{-1}) and insulin (10 mg mL^{-1}).

Growth inhibition was determined by plating cells in duplicate in medium (100 µL) at a density of 2500–4000 cells per well in 96-well plates. On day 0 (24 h after plating), when the cells were in logarithmic growth, medium (100 µL) with or without the test agent was added to each well. After 72 h drug exposure, growth inhibitory effects were evaluated using the MTT (3-(4,5-dimethyltiazol-2-yl)-2,5-diphenyltetrazolium bromide) assay and absorbance read at 540 nm. The percentage growth inhibition was calculated at a fixed concentration of 25 µM, based on the difference between the optical density values on day 0 and those at the end of drug exposure. Each data point is the mean ± the standard error of the mean (SEM) calculated from three replicates which were performed on separate occasions and separate cell line passages.

6.4. Chemistry Experimental

6.4.1. General Experimental

Reaction solvents were purchased from the Aldrich Chemical Company (St Louis, MO, USA) in sure-seal™ reagent bottles. All other solvents (analytical or HPLC grade) were used as supplied without further purification. Deuterated chloroform (CDCl$_3$) and water (D$_2$O) were used as NMR solvent. Triethylamine, sodium hydride (60% dispersion in mineral oil), tetrachlorophthalimide and tetrachlorophthalic anhydride were purchased from Sigma Aldrich. All boron-containing reagents were purchased from Boron Molecular, apart from BSH (>97%), ^{10}B-BSH (>97%), BPA (>98%) and ^{10}B-BPA (>98%) which came from Katchem spol. s r. o. The reagents were used as provided without further purification, with NMR analysis confirming an acceptable degree of purity and correct structural identity.

Purification via silica gel column chromatography was performed on Davisil 40–63-micron silica gel.

Thin layer chromatography (t.l.c.) was performed on aluminium sheets coated with 60 F254 silica by Merck and visualised using UVG-11 Compact UV lamp (254 nm) or stained with the cerium molybdate stain (12.0 g ammonium molybdate, 0.5 g ceric ammonium molybdate in 15 mL concentrated sulfuric acid and 235 mL distilled water).

Nuclear Magnetic Resonance (NMR) spectra were recorded on Bruker Ascend™ 400 in deuterated chloroform (CDCl$_3$). Chemical shifts (δ) are quoted in ppm and coupling constants (J) in Hz. Residual signals from the CDCl$_3$ (7.26 ppm for ^{1}H-NMR and 77.16 ppm for ^{13}C-NMR) were used as an internal reference [84].

Infrared spectroscopy (IR) spectra were obtained on a PerkinElmer Spectrum Two Spectrometer and on a PerkinElmer Spectrum 2 with UATR. Only characteristic peaks are quoted and in units of cm^{-1}.

High-resolution mass spectrometry (HRMS) spectra were obtained from samples suspended in acetonitrile (1 mL with 0.1% formic acid at a concentration of ~1 mg/mL, before being further diluted to ~10 ng/µL in 50% acetonitrile/water containing 0.1% formic acid). Samples were infused directly into the HESI source of a Thermo Scientific Q Exactive™ Plus Hybrid Quadrupole-Orbitrap™ Mass Spectrometer using an on-board syringe pump at 5 µL/min. Data were acquired on the QE+ in both positive and negative ion mode at a target resolution of 70,000 at 200 *m/z*. The predominant ions were manually selected for MS/MS fragmentation (collision energies were altered for each compound to obtain sufficient fragmentation). Data analysis of each sample was performed manually using Thermo Qualbrowser whilst the Isotopic Patterns of predicted chemical formula were modelled using Bruker Compass Isotope Pattern.

Crystallographic data were collected on an Oxford Diffraction Gemini CCD diffractometer employing either graphite-monochromated Mo-Kα radiation (0.71073 Å) or Cu-Kα (1.54184 Å). The sample was cooled to 190 K with and Oxford Cryosystems Desktop Cooler. Data reduction and empirical absorption corrections were performed with Oxford Diffraction CrysAlisPro software. Structures were solved by direct methods and refined with SHELXL [85]. All non-H atoms were refined with anisotropic thermal parameters. The crystal of **ortho 8** was a non-merohedral twin which was refined using the HKLF 5 mode in SHELX. Molecular structure diagrams were produced with Mercury [86]. The data in CIF format were deposited at the Cambridge Crystallographic Data Centre (CCDC 215189 and 2151899).

Melting points were taken on a Dynalon SMP100 Digital Melting Point Device and are uncorrected.

6.4.2. Experimental

From Synthetic Strategy 1

4,5,6,7-Tetrachloro-2-(3-(4,4,5,5-tetramethyl-1,3,2-dioxaborolan-2-yl)benzyl)isoindoline-1,3-dione meta 3

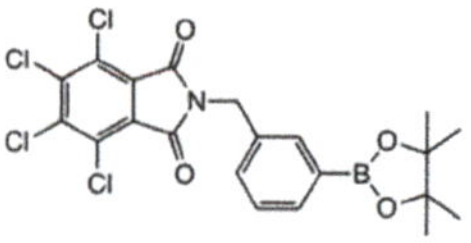

Tetrachlorophthalic anhydride **1** (445 mg, 1.558 mmol, 1.2 equiv.) and 3-(aminomethyl)-phenylboronic acid pinacol ester hydrochloride **meta 2** (353 mg, 1.309 mmol, 1.0 equiv.) were stirred in N,N dimethylformamide (8 mL) until dissolved. Triethylamine (328 mg, 0.45 mL, 3.245 mmol, 2.5 equiv.) was then added dropwise to reaction mixture. A white precipitate crashed out upon addition of triethylamine. The stirring reaction mixture was then heated to 85 °C. After 48 h the reaction mixture was gravity filtered to remove the precipitate (triethylamine salt and excess tetrachlorophthalic anhydride by NMR analysis), which was collected in a sample vial and retained for analysis. The filtrate was then evaporated and the solid recrystallized from methanol (20 mL) to afford the product **meta 3** as a pale creamy yellow solid (389 mg, 60.0%). M.p. 316–320 °C. m/z (HRMS ES$^+$): Relative intensities for $C_{21}H_{19}BCl_4NO_4$, $[M + H]^+$, found 499.01957 (22%), 500.01581 (82%), 501.01824 (40%), 502.01293 (100%), 503.04453 (34%), 504.01001 (45%), 505.04176 (16%), 506.00706 (9%); calculated 499.01921 (18%), 500.01596 (76%), 501.01748 (40%), 502.01323 (100%), 503.01528 (34%), 504.01064 (50%), 505.01292 (13%), 506.00829 (12%). ν_{max} (thin film, cm^{-1}): 2979 (w, alkyl CHs), 1776 (w, C=O), 1716 (s, C=ON, amide I), 1606, 1486 (w, ArC=C), 1432 (m, C-B), 1389, 1361, 1330, 1075 (s, sp^2 B-O). δ_H (CDCl$_3$, 400 MHz): 7.85 (1 H, s, H^a), 7.73 (1 H, d, $J_{Hb,Hc}$ 7.6 Hz, H^b), 7.52 (1 H, d, $J_{Hd,Hc}$ 7.6 Hz, H^d), 7.33 (1 H, t, $J_{Hc,Hb/Hd}$ 7.6 Hz, H^c), 4.85 (2 H, s, CH$_2$), 1.34 (12 H, s, 4 × CH$_3$). δ_C (CDCl$_3$, 100 MHz): 163.2 (2 × C=O), 140.1 (ArC$_q$-CH$_2$), 135.00, 134.6 (2 × ArCH), 134.5 (2 × ArC$_q$-CO), 131.7 (ArCH), 129.7, 128.2 (ArCH), 127.60 (4 × ArC$_q$-Cl), 83.9 (2 × pinacol C$_q$), 42.4 (N-CH$_2$) and 24.8 (4 × pinacol CH$_3$). ArC$_q$-B is not visible. δ_B (dissolved in CDCl$_3$, 96 MHz): 31.5 ppm (Figures S1 and S2, in the Supplementary Materials).

4,5,6,7-Tetrachloro-2-(4-(4,4,5,5-tetramethyl-1,3,2-dioxaborolan-2-yl)benzyl)isoindoline-1,3-dione para 3

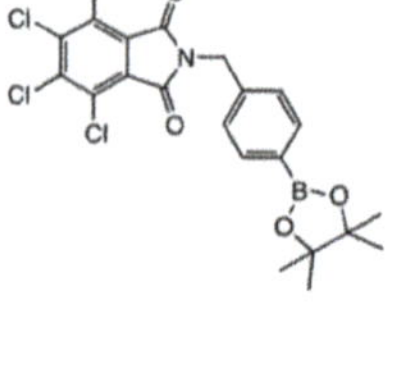

Tetrachlorophthalic anhydride **1** (453 mg, 1.558 mmol, 1.2 equiv.) and 4-(aminomethyl)-phenylboronic acid pinacol ester hydrochloride **para 2** (356 mg, 1.321 mmol, 1.0 equiv.) were stirred in N,N-dimethylformamide (8 mL) until dissolved. Triethylamine (328 mg, 0.45 mL, 3.245 mmol, 2.5 equiv.) was then added dropwise to reaction mixture. A white precipitate crashed out upon addition of triethylamine. The stirring reaction mixture was then heated to 85 °C. After 48 h the reaction mixture was gravity filtered. The filtrate was then evaporated and the solid recrystallised from petroleum ether (60:80)/ethanol (30/80 mL) to afford the product **para 3** as a pale creamy yellow solid (289 mg, 44.4%). M.p. 242–246 °C. m/z (HRMS ES$^+$): Relative intensities for $C_{21}H_{19}BCl_4NO_4$, [M + H]$^+$, found 499.01957 (22%), 500.01581 (82%), 501.01824 (40%), 502.01293 (100%), 503.04453 (34%), 504.01001 (45%), 505.04176 (16%), 506.00706 (9%); calculated 499.01921 (18%), 500.01596 (76%), 501.01748 (40%), 502.01323 (100%), 503.01528 (34%), 504.01064 (50%), 505.01292 (13%), 506.00829 (12%). ν_{max} (thin film, cm^{-1}): 2978 (m, ArCH$_2$), 2938, 2893 (w, alkyl C-H), 1778 (w, C=O), 1715 (s, C=ON, amide I), 1512 (s, C=ON, amide II), 1433 (m, C-B), 1373, 1360, 1345, 1087 (s, sp^2 B-O), 662 (s, C-B(-O)$_2$, out of plane bending). δ_H (CDCl$_3$, 400 MHz): 7.77 (2 H, d, J 8.0 Hz, 2 × ArH), 7.42 (2 H, d, J 8.0 Hz, 2 × ArH), 4.85 (2 H, s, CH$_2$), 1.32 (12 H, s, 4 × CH$_3$). δ_C (CDCl$_3$, 100 MHz): 163.4 (C=O), 140.3 (ArC$_q$-CH$_2$), 138.3 (2 × ArC$_q$-CO), 135.4 (2 × ArCH), 129.9, 127.7 (4 × ArC$_q$-Cl), 128.3 (2 × ArCH), 84.0 (2 × pinacol C$_q$), 42.6 (N-CH$_2$) and 25.0 (4 × pinacol CH$_3$). ArC$_q$-B is not visible. δ_B (dissolved in CDCl$_3$, 96 MHz): 31.1 ppm.

2,3,4,5-Tetrachloro-N-(2-(4,4,5,5-tetramethyl-1,3,2-dioxaborolan-2-yl)phenyl)benzamide **ortho 5**

Tetrachlorophthalic anhydride (560 mg, 1.959 mmol, 1.2 equiv.) and 2-amino phenylboronic acid pinacol ester (352 mg, 1.607 mmol, 1.0 equiv.) were stirred in N,N-dimethylformamide (9 mL) to give a cloudy orange solution. Triethylamine (404 mg, 0.56 mL, 3.994 mmol, 2.5 equiv.) was added dropwise and stirred at 85 °C for 72 h, and at 105 °C for a further 72 h. Evaporation gave a dark brown oily residue. Recrystallisation attempts using ethanol (20 mL) and petroleum ether (60/80) (20 mL)/chloroform (5 mL) did not yield enough product. The residue was purified by flash column chromatography (hexane:acetone, 1:1). After evaporation a recrystallization using chloroform (20 mL) gave a filtrate which was left to stand for 3–4 days. Crystallisation gave product **ortho 5** (284 mg, 37.4%). M.p. 106–108 °C. m/z (HRMS ES$^+$): Relative intensities for $C_{19}H_{19}BCl_4NO_3$, [M + H]$^+$, found 459.02226 (26%), 460.02068 (77%), 461.02147 (35%), 462.01769 (100%), 463.02069 (28%), 464.01455 (46%), 465.01838 (11%), 466.01142 (10%); calculated 459.02429 (18%), 460.02101 (77%), 461.02249 (39%), 462.01824 (100%), 463.02026 (32%), 464.01558 (50%), 465.01788 (12%), 466.01314 (11%). ν_{max} (thin film, cm^{-1}): 3357 (m, sh, NH, hydrogen-bonded), 2980, 2933 (m, alkyl CHs), 1690 (s, C=ONH, amide I), 1612 (w, ArC=C), 1580 (m, C=ONH, amide II), 1536 (w, C=ONH, bending), 1450 (m, C-N), 1407 (m, C-B), 1351, 1323, 1303, 1140 (s, sp^2 B-O), 758 (s, C-Cl), 652 (s, C-B(-O)$_2$, out of plane bending). δ_H (CDCl$_3$, 400 MHz), major:minor conformer 2:1. Major conformer: 9.90–9.81 (1 H, broad s, NH), 8.59 (1 H, d, $J_{Ha,Hb}$ 8.4 Hz, H^a), 7.82 (1 H, dd, $J_{Hd,Hc}$ 7.2 Hz, $J_{Hd,Hb}$ 1.6 Hz, H^d), 7.71 (1 H, s, H^e), 7.53 (1 H, td, $J_{Hb,Ha/Hc}$ 7.6 Hz, $J_{Hb,Hd}$ 1.5 Hz, H^b), 7.16 (1 H, td, $J_{Hc,Hd/Hb}$ 7.5 Hz, Hz, $J_{Hc,Ha}$ 0.9 Hz, H^c), 1.33 (12 H, s, 4 × CH$_3$). δ_C (CDCl$_3$, 100 MHz): 162.3 (C=O), 144.0 (NHCq), 136.8 (Cq), 136.5 (C^d), 134.9 (Cq), 133.2 (C^b), 132.7, 130.2 (2 × Cq), 127.8 (C^e), 124.1 (C^c), 120.0 (C^a), 84.8 (2 × pinacol Cq) and 25.0 (4 × pinacol CH$_3$); note, ArC$_q$-B is not visible. δ_B (dissolved in CDCl$_3$, 96 MHz): 31.1 ppm.

2,3,4,5-tetrachloro-N-(3-(4,4,5,5-tetramethyl-1,3,2-dioxaborolan-2-yl)phenyl)benzamide meta 5

Tetrachlorophthalic anhydride (580 mg, 2.029 mmol, 1.2 equiv.) and 3-aminophenylboronic acid pinacol ester (354 mg, 1.616 mmol, 1.0 equiv.) were stirred in N,N dimethylformamide (9 mL) to give a golden-yellow solution. Triethylamine (404 mg, 0.56 mL, 3.994 mmol, 2.5 equiv.) was added dropwise and the reaction stirred at 85 °C for 48 h and then at 105 °C for a further 72 h. The reaction mixture was evaporated and recrystallised with ethanol (20 mL) and the reaction solution left to crystallise in the freezer for three days at −25 °C to produce **meta 5** as light-brown crystals (395 mg, 52%). M.p. 36–38 °C. m/z (HRMS ES$^+$): Relative intensities for $C_{19}H_{18}BCl_4NO_3$, [M + H]$^+$, found 459.02469 (19%), 460.02087 (80%), 461.02264 (35%), 462.01773 (100%), 463.02026 (30%), 464.01450 (47%), 465.01779 (11%), 466.01114 (9%); calculated 459.02429 (18%), 460.02101 (77%), 461.02249 (39%), 462.01824 (100%), 463.02026 (32%), 464.01558 (50%), 465.01788 (12%), 466.01314 (11%). δ_H (CDCl$_3$, 400 MHz): 7.95 (1 H, dd, $J_{Hb,Hc}$ 7.7 Hz, $J_{Hb,Ha}$ 1.2 Hz, H^b), 7.80 (1 H, s, NH), 7.76 (1 H, d, $J_{Ha,Hb}$ 1.6 Hz, H^a), 7.72 (1 H, s, H^e), 7.63 (1 H, d, $J_{Hd,Hc}$ 7.2 Hz, H^d), 7.41 (1 H, t, $J_{Hc,Hb/Hd}$ 7.6 Hz, H^c) and 1.34 (12 H, s). δ_B (dissolved in CDCl$_3$, 96 MHz): 31.7. X-ray Crystallographic analysis.

2,3,4,5-tetrachloro-N-(4-(4,4,5,5-tetramethyl-1,3,2-dioxaborolan-2-yl)phenyl)benzamide para 5

Tetrachlorophthalic anhydride (560 mg, 1.958 mmol, 1.2 equiv.) and 4-amino phenylboronic acid pinacol ester (353 mg, 1.611 mmol, 1.0 equiv.) were stirred in N,N-dimethylformamide (9 mL) into a cloudy orange solution. Sodium hydride (168 mg, 17.51 mmol, 2.5 equiv.) was then carefully added as hydrogen gas was given off. The reaction mixture was then heated to 100 °C for 72 h. After cooling, two drops of R.O. water were added to quench the reaction and evaporated. The residue was eluted in ethanol (10 mL) and a black solid precipitated. The filtrate contained product **para 5**, which was coevaporated with DCM (354 mg, 46%). M.p. 218–220 °C. m/z (HRMS ES$^+$): Relative intensities for $C_{19}H_{19}BCl_4NO_3$, [M + H]$^+$, found 459.02445 (20%), 460.02049 (77%), 461.02242 (38%), 462.01751 (100%), 463.02008 (30%), 464.01432 (47%), 465.01768 (12%), 466.01123 (10%); calculated 459.02429 (18%), 460.02101 (77%), 461.02249 (39%), 462.01824 (100%), 463.02026 (32%), 464.01558 (50%), 465.01788 (12%), 466.01314 (11%). ν_{max} (thin film, cm^{-1}): 3266 (w, br, NH), 2976, 2926, 2855 (m, alkyl CHs), 1661 (s, C=ONH, amide I), 1597 (w, ArC=C), 1529 (m, C=ONH, amide II), 1505 (w, C=ONH, bending), 1444 (m, C-N), 1399 (m, C-B), 1389, 1356, 1316, 1084 (s, sp^2 B-O), 833 (s, C-Cl), 653 (s, C-B(-O)$_2$, out of plane bending). δ_H (CDCl$_3$, 400 MHz): 7.83 (2 H, d, J 8.0 Hz, 2 × ArH), 7.75 (1 H, s, H^e), 7.65 (1 H, s, NH), 7.61 (2 H, d, J 7.6 Hz, 2 × ArH), 1.35 (12 H, s, 4 × CH$_3$). δ_C (CDCl$_3$, 100 MHz): 136.2 (2 × ArCH), 128.7 (C^e), 119.1 (2 × ArCH), 84.0 (2 × pinacol C$_q$) and 25.0 (4 × pinacol CH$_3$); note, other Cq are not visible. δ_B (dissolved in CDCl$_3$, 96 MHz): 31.8 ppm.

From Synthetic Strategy 2

2,3,4,5-tetrachloro-N-(2-(4,4,5,5-tetramethyl-1,3,2-dioxaborolan-2-yl)benzyl)benzamide ortho 8

Tetrachlorophthalimide **6** (352 mg, 1.235 mmol, 1.0 equiv.) and 2-(bromomethyl)phenylboronic acid pinacol ester **ortho 7** (455 mg, 1.838 mmol, 1.2 equiv.) were stirred in N,N-dimethylformamide (9 mL). Sodium hydride (123 mg, 12.80 mmol, 2.5 equiv.) was then carefully added as hydrogen gas was given off to give a pale creamy yellow reaction mixture and heated to 100 °C. After 48 h the reaction mixture was cooled to room temperature and a yellow precipitate formed. Two drops of R.O. water were added to quench. Evaporation afforded a crude residue, which was then dissolved in chloroform (30 mL) and gravity filtered. The filtrate was dried affording **ortho 8** (371 mg, 60%). M.p. 120–122 °C. m/z (HRMS ES$^+$): Relative intensities for $C_{21}H_{19}BCl_4NO_4$, [M + H]$^+$, 473.04023 (20%), 474.03660 (80%), 475.03831 (39%), 476.03341 (100%), 477.03592 (32%), 478.03016 (46%), 479.03318 (12%), 480.02677 (9%); calculated 473.03994 (18%), 474.03668 (76%), 475.03818 (40%), 476.03392 (100%), 477.03596 (33%), 478.03129 (50%), 479.03358 (13%), 480.02889 (11%). δ_H (CDCl$_3$, 400 MHz), major:minor conformers 5:1–3:1. Major conformer: 7.87 (1 H, dd, $J_{Ha,Hb}$ 7.6, $J_{Ha,Hc}$ 1.2 Hz, H^a), 7.61 (1 H, s, H^e), 7.49 (1 H, d, $J_{Hd,Hc}$ 7.5 Hz, H^d), 7.45 (1 H, td, $J_{Hc,Hd}$ 7.5, $J_{Hc,Ha}$ 1.3 Hz, H^c), 7.32 (2 H, td, J 7.5, 1.2 Hz, H^b and NH), 4.71 (2 H, d, J 6.4 Hz, CH$_2$), 1.35 (12 H, s, 4 × CH$_3$). Minor conformer: Only selected signals visible. 7.81 (1 H, d, J 6.7), 7.77 (1 H, d, J 7.4), 4.70 (2 H, s, CH$_2$), 1.35 (12 H, s, 4 × CH$_3$). δ_C (CDCl$_3$, 100 MHz): 163.5 (C=O), 143.6 (CH$_2$Cq), 136.9 (C^a), 136.1, 134.6, 133.6, 132.9 (4 × Cq), 132.0 (C^c), 130.3 (C^d), 128.7 (C^e), 127.4 (C^b), 125.7 (Cq), 84.4 (2 × pinacol Cq), 44.8 (CH$_2$) and 25.1 (4 × pinacol CH$_3$); note, ArC$_q$-B is not visible. δ_B (dissolved in CDCl$_3$, 96 MHz): 31.8 ppm. X-ray Crystallographic analysis.

4,5,6,7-Tetrachloro-2-(4-(4,4,5,5-tetramethyl-1,3,2-dioxaborolan-2-yl)benzyl)isoindoline-1,3-dione para 8

Tetrachlorophthalimide **6** (352 mg, 1.235 mmol, 1.0 equiv.) and 4-(bromomethyl)phenylboronic acid pinacol ester **para 7** (440 mg, 1.474 mmol, 1.2 equiv.) were stirred in N,N-dimethylformamide (8 mL). Sodium hydride (123 mg, 12.80 mmol, 2.5 equiv.) was then carefully added as hydrogen gas was given off to give a pale creamy yellow reaction mixture and heated to 100 °C. After 48 h the reaction mixture was cooled to room temperature and a golden honey colour solution with a golden yellow precipitate was visible. Two drops of R.O. water were added to quench and evaporated. The residue was dissolved in chloroform (30 mL) and gravity filtered. The filtrate was evaporated to give **para 8** (339 mg, 56%). M.p. 110–112 °C. m/z (HRMS ES$^+$): Relative intensities for $C_{21}H_{19}BCl_4NO_4$, [M + H]$^+$, 473.04016 (20%), 474.03632 (74%), 475.03815 (39%), 476.03317 (100%), 477.03586 (31%), 478.02997 (47%), 479.03320 (13%), 480.02664 (10%); calculated 473.03994 (18%), 474.03667 (78%), 475.03816 (41%), 476.03372 (100%), 477.03587 (33%), 478.03042 (46%), 479.03320 (13%), 480.02747 (10%). ν_{max} (thin film, cm^{-1}): 3265 (m, br, NH), 2976, 2926, 2853 (m, alkyl CHs), 1652 (s, C=ONH, amide I), 1575 (w, ArC=C), 1517 (m, C=ONH, amide II), 1459 (m, C-N), 1407 (m, C-B), 1358, 1340, 1088 (s, sp^2 B-O), 751 (m, C-Cl), 657 (s, C-B(-O)$_2$, out of plane bending). δ_H (CDCl$_3$, 400 MHz): 7.81 (2 H, d, J 8.0, 2 × ArH), 7.65 (1 H, s, H^e), 7.36 (2 H, d, J 8.0 Hz, 2 × ArH), 6.32–6.40 (1 H, br s, NH), 4.65 (2 H, d, J 5.6 Hz), 1.35 (12 H, s, 4 × CH₃). δc (CDCl₃, 400 MHz): 135.5 (2 × Ar-CH), 128.6 (C^e), 127.4 (2 × Ar-CH), 84.1 (2 × pinacol Cq) and 25.0 (4 × pinacol CH₃); note, Cq are not visible. δ_B (dissolved in CDCl₃, 96 MHz): 31.6 ppm.

7. Conclusions

We reported an expedited synthesis to a small library of novel borylated 2,3,4,5-tetrachlorophthalimides and 2,3,4,5-tetrachlorobenzamides. Biological assays against glycosidase enzymes and cancer cell lines highlighted a good inhibitor for bovine liver β-galactosidase and three potent growth inhibitors and, of these, one selective growth inhibitor for cancer versus healthy cell lines in the cancer assay. These drugs are set for further derivatisations and utilisation in BNCT.

Supplementary Materials: The following supporting information can be downloaded at: https://www.mdpi.com/article/10.3390/molecules27113447/s1, Figure S1: ^{1}H NMR spectrum of **para 3**; Figure S2: ^{13}C NMR spectrum of **para 3**; Figure S3: ^{1}H NMR spec-trum of **mata 3**; Figure S4: ^{13}C NMR spectrum of **mata 3**; Figure S5: ^{1}H NMR spectrum of **ortho 5**; Figure S6: ^{13}C NMR spectrum of **ortho 5**; Figure S7: ^{1}H NMR spectrum of **meta 5**; Figure S8: ^{1}H NMR spectrum of **para 5**; Figure S9: ^{13}C NMR spectrum of **para 5**; Figure S10: ^{1}H NMR spectrum of **ortho 8**; Figure S11: ^{13}C NMR spectrum of **ortho 8**; Figure S12: ^{1}H NMR spectrum of **para 8**; Figure S13: ^{13}C NMR spectrum of **para 8**; Table S1: Crystal data and structure refinement for **meta 5**; Table S2: Bond lengths [Å] and angles [°] for **meta 5**; Table S3: Crystal data and structure re-finement for **ortho 8**; Table S4: Bond lengths [Å] and angles [°] for **ortho 8**.

Author Contributions: Conceptualization, M.I.S.; Data curation, D.M.C., Robert Nash, J.A.S., A.K. and M.I.S.; Formal analysis, D.M.C., Y.S., B.B., P.V.B., J.A.S. and A.K.; Funding acquisition, M.I.S.; Investigation, D.M.C.; Methodology, R.J.N. and J.A.S.; Project administration, M.I.S.; Resources, M.I.S.; Supervision, R.J.N., A.K. and M.I.S.; Writing—original draft, M.I.S.; Writing—review and editing, M.I.S. All authors have read and agreed to the published version of the manuscript.

Funding: B18 Project and, at the University of Newcastle, the Faculty of Science and the Priority Research Centre for Drug Development for financial support.

Institutional Review Board Statement: Not applicable.

Informed Consent Statement: Not applicable.

Data Availability Statement: Not applicable.

Acknowledgments: MIS would like to thank the B18 Project and, at the University of Newcastle, the Faculty of Science and the Priority Research Centre for Drug Development for financial support, and the undergraduate student DC for dedicating his time gratis to this project on a voluntary basis.

Conflicts of Interest: The authors declare no conflict of interest.

References

1. Hashimoto, Y. Thalidomide as a Multi-Template for Development of Biologically Active Compounds. *Arch. Pharm. Chem. Life Sci.* **2008**, *341*, 536–547. [CrossRef] [PubMed]
2. Qiu, S.; Zhai, S.; Wang, H.; Tao, C.; Zhao, H.; Zhai, H. Efficient Synthesis of Phthalimides via Cobalt-Catalyzed C(sp^2)—H Carbonylation of Benzoyl Hydrazides with Carbon Monoxide. *Adv. Synth. Cat.* **2018**, *360*, 3271–3276. [CrossRef]
3. Bian, X.; Wang, Q.; Ke, C.; Zhao, G.; Li, Y. A new series of N2-substituted-5-(*p*-toluenesulfonylamino)phthalimide analogues as α-glucosidase inhibitors. *Bioorg. Med. Chem. Lett.* **2013**, *23*, 2022–2026. [CrossRef] [PubMed]
4. Pascale, R.; Carocci, A.; Catalano, A.; Lentini, G.; Spagnoletta, A.; Cavalluzzi, M.M.; Santis, F.D.; Palma, A.D.; Scalera, V.; Franchini, C. New *N*-(phenoxydecyl)phthalimide derivatives displaying potent inhibition activity towards α-glucosidase. *Bioorg. Med. Chem.* **2010**, *18*, 5903–5914. [CrossRef]
5. Sou, S.; Takahashi, H.; Yamasaki, R.; Kagechika, H.; Endo, Y.; Hashimoto, Y. α-Glucosidase Inhibitors with a 4,5,6,7-Tetrachlorophthalimide Skeleton Pendanted with a Cycloalkyl or Dicarba-*closo*-dodecaborane Group. *Chem. Pharm. Bull.* **2001**, *49*, 791–793. [CrossRef]
6. Sou, S.; Mayumi, S.; Takahashi, H.; Yamasaki, R.; Kadoya, S.; Sodeoka, M.; Hashimoto, Y. Novel α-Glucosidase Inhibitors with a Tetrachlorophthalimide Skeleton. *Bioorg. Med. Chem. Lett.* **2000**, *10*, 1081–1084. [CrossRef]
7. Baker, S.J.; Ding, C.Z.; Akama, T.; Zhang, Y.K.; Hernandez, V.; Xia, Y. Therapeutic potential of boron-containing compounds. *Future Med. Chem.* **2009**, *1*, 1275–1288. [CrossRef]

8. Leśnikowski, Z.J. Recent developments with boron as a platform for novel drug design. *Expert Opin. Drug Discov.* **2016**, *11*, 569–578. [CrossRef]
9. Jenkinson, S.F.; Thompson, A.L.; Simone, M.I. Methyl 2-(5,5-dimethyl-1,3,2-dioxaborinan-2-yl)-4-nitrobenzoate. *Acta Cryst.* **2012**, *E68*, o2699–o2700. [CrossRef]
10. Simone, M.I.; Houston, T.A. A brief overview of some latest advances in the applications of boronic acids. *J. Glycom. Lipidom.* **2014**, *4*, e124–e129. [CrossRef]
11. Pappin, B.B.; Levonis, S.M.; Healy, P.C.; Kiefel, M.J.; Simone, M.I.; Houston, T.A. Crystallization Induced Amide Bond Formation Creates a Boron-Centred Spirocyclic System. *Heterocycl. Commun.* **2017**, *23*, 167–169. [CrossRef]
12. Pappin, B.B.; Garget, T.A.; Healy, P.C.; Simone, M.I.; Kiefel, M.J.; Houston, T.A. Facile amidinations of 2-aminophenylboronic acid promoted by boronate ester formation. *Org. Biomol. Chem.* **2019**, *17*, 803–806. [CrossRef]
13. Legge, W.J.; Shimadate, Y.; Sakoff, J.; Houston, T.A.; Kato, A.; Bernhardt, P.V.; Simone, M.I. Borylated methyl cinnamates: Green synthesis, characterization, crystallographic analysis and biological activities in glycosidase inhibition and in cancer cells lines—Of (E)-methyl 3-(2/3/4-(4,4,5,5-tetramethyl-1,3,2-dioxaborolan-2-yl)phenyl)acrylates. *Beil. Arch.* **2021**, *2021*, 4.
14. Sarazin, H.; Prudent, S.; Defoin, A.; Tarnus, C. Evaluation of 6-Deoxy–amino–sugars as Potent Glycosidase Inhibitors. Importance of the $CH_2OH(6)$ Group for Enzyme-Substrate Interaction. *ChemistrySelect* **2017**, *2*, 1484–1490. [CrossRef]
15. Bonduelle, C.; Huang, J.; Mena-Barraga, T.; Mellet, C.O.; Decroocq, C.; Etame, E.; Heise, A.; Compain, P.; Lecommandoux, S. Iminosugar-based glycopolypeptides: Glycosidase inhibition with bioinspired glycoprotein analogue micellar self-assemblies. *Chem. Commun.* **2014**, *50*, 3350–3352. [CrossRef] [PubMed]
16. Glawar, A.F.G.; Martínez, R.F.; Ayers, B.J.; Hollas, M.A.; Ngo, N.; Nakagawa, S.; Kato, A.; Butters, T.D.; Fleet, G.W.J.; Jenkinson, S.F. Structural essentials for β-N-acetylhexosaminidase inhibition by amides of prolines, pipecolic and azetidine carboxylic acids. *Org. Biomol. Chem.* **2016**, *14*, 10371–10385. [CrossRef] [PubMed]
17. Cruz, F.P.D.; Newberry, S.; Jenkinson, S.F.; Wormald, M.R.; Butters, T.D.; Alonzi, D.S.; Nakagawa, S.; Becq, F.; Norez, C.; Nash, R.J.; et al. 4-C-Methyl-DAB and 4-C-Me-LAB—Enantiomeric alkyl branched pyrrolidine iminosugars—Are specific and potent α-glucosidase inhibitors; acetone as the sole protecting group. *Tetrahedron Lett.* **2011**, *52*, 219–223. [CrossRef] [PubMed]
18. Simone, M.I.; Soengas, R.G.; Jenkinson, S.F.; Evinson, E.L.; Nash, R.J.; Fleet, G.W.J. Synthesis of three branched iminosugars [(3R,4R,5S)-3-(hydroxymethyl)piperidine-3,4,5-triol, (3R,4R,5R)-3-(hydroxymethyl)piperidine-3,4,5-triol and (3S,4R,5R)-3-(hydroxymethyl)piperidine-3,4,5-triol] and a branched trihydroxynipecotic acid [(3R,4R,5R)-3,4,5-trihydroxypiperidine-3-carboxylic acid] from sugar lactones with a carbon substituent at C-2. *Tetrahedron Asymm.* **2012**, *23*, 401–408.
19. Soengas, R.G.; Simone, M.I.; Hunter, S.; Nash, R.J.; Fleet, G.W.J. Hydroxymethyl-branched piperidines from hydroxymethyl-branched lactones: Synthesis and biological evaluation of 1,5-dideoxy-2-C-hydroxymethyl-1,5-imino-D-mannitol, 1,5-dideoxy-2-hydroxymethyl-1,5-imino-L-gulitol and 1,5-dideoxy-2-hydroxymethyl-1,5-imino-D-talitol. *Eur. J. Org. Chem.* **2012**, *12*, 2394–2402.
20. Simone, M.I.; Mares, L.; Eveleens, C.; McCluskey, A.; Pappin, B.; Kiefel, M.; Houston, T.A. Back to (non)-Basic: An Update on Neutral and Charge-Balanced Glycosidase Inhibitors. *Mini-Rev. Med. Chem.* **2017**, *18*, 812–827. [CrossRef]
21. Prichard, K.; Campkin, D.; O'Brien, N.; Kato, A.; Fleet, G.W.J.; Simone, M.I. Biological Activities of 3,4,5-Trihydroxypiperidines and their O- and N-Alkylated Derivatives. *Chem. Biol. Drug Des.* **2018**, *2018*, 1171–1197. [CrossRef]
22. Brás, N.F.; Cerqueira, N.M.F.S.A.; Ramos, M.J.; Fernandes, P.A. Glycosidase inhibitors: A patent review (2008–2013). *Expert Opin. Ther. Pat.* **2014**, *24*, 857–874. [CrossRef] [PubMed]
23. Yamamoto, T.; Nakai, K.; Matsumura, A. Boron Neutron Capture Therapy for Glioblastoma. *Cancer Lett.* **2008**, *262*, 143–152. [CrossRef] [PubMed]
24. Zavjalov, E.; Zaboronok, A.; Kanygin, V.; Kasatova, A.; Kichigin, A.; Mukhamadiyarov, R.; Razumov, I.; Sycheva, T.; Mathis, B.J.; Maezono, S.E.B.; et al. Accelerator-based Boron Neutron Capture Therapy for Malignant Glioma: A Pilot Neutron Irradiation Study using Boron Phenylalanine, Sodium Borocaptate and Liposomal Borocaptate with a Heterotopic U87 Glioblastoma Model in SCID Mice. *Int. J. Radiat. Biol.* **2020**, *96*, 868–878. [CrossRef] [PubMed]
25. Nakagawa, Y.; Pooh, K.; Kobayashi, T.; Kageji, T.; Uyama, S.; Matsumura, A.; Kumada, H. Clinical Review of the Japanese Experience with Boron Neutron Capture Therapy and a Proposed Strategy using Epithermal Neutron Beams. *J. Neuro-Oncol.* **2003**, *62*, 87–99. [CrossRef]
26. Nordlander, B.W.; Cass, W.E. The Esterification of tetrachlorophthalic Anhydride. *J. Am. Chem. Soc.* **1947**, *69*, 2679. [CrossRef]
27. Harvey, P.G.; Smith, F.; Stacey, M.; Tatlow, J.C. Studies on the Nuclear Chlorination of Aromatic Compounds. *J. Appl. Chem.* **1954**, *4*, 325. [CrossRef]
28. Tust, P. Ueber Tetrachlorobenzoësäure und einige Derivate derselben. *Ber. Dtsch. Chem. Ges. Berl.* **1887**, *20*, 2439–2442. [CrossRef]
29. Beilstein, F.; Kuhlberg, A. Untersuchungen über Isomerie in der Benzoëreihe. *Justus Liebigs Ann. Chem.* **1869**, *152*, 224–246. [CrossRef]
30. Pagani, G.; Baruffini, A.; Mazza, M.; Vicarini, L.; Caccialanza, G. Sull'attività fitotossica selettiva di N,N-di,sec.butilamidi di acidi benzoici variamente alogenati. *Il Farm.-Ed. Sc.* **1973**, *28*, 570–589.
31. Medina, I.-M.R.Y.; Rohdenburg, M.; Mostaghimi, F.; Grabowsky, S.; Swiderek, P.; Beckmann, J.; Hoffmann, J.; Dorcet, V.; Hissler, M.; Staubitz, A. Tuning the Optoelectronic Properties of Stannoles by the Judicious Choice of the Organic Substituents. *Inorg. Chem.* **2018**, *57*, 12562–12575. [CrossRef] [PubMed]
32. Fan, W.; Winands, T.; Doltsinis, N.L.; Li, Y.; Wang, Z. A Decatwistacene with an Overall 170° Torsion. *Angew. Chem. Int. Ed.* **2017**, *56*, 15373–15377. [CrossRef] [PubMed]

33. Hanefeld, M.; Schaper, F. The role of alpha-glucosidase inhibitors (Acarbose). In *Pharmacotherapy of Diabetes: New Developments*; Mogesen, C.E., Ed.; Springer: Boston, MA, USA, 2007; pp. 143–152.
34. Laar, F.A.V.D.; Lucassen, P.L.; Akkermans, R.P.; Lisdonk, E.H.V.D.; Rutten, G.E.; Weel, C.V. α-Glucosidase Inhibitors for Patients with Type 2 Diabetes. *Diabetes Care* **2005**, *28*, 154–163. [CrossRef] [PubMed]
35. Ma, J.; Zhang, X.; Soloveva, V.; Warren, T.; Guo, F.; Wu, S.; Lu, H.; Guo, J.; Su, Q.; Shen, H.; et al. Enhancing the antiviral potency of ER α-glucosidase inhibitor IHVR-19029 against hemorrhagic fever viruses in vitro and in vivo. *Antivir. Res.* **2018**, *150*, 112–122. [CrossRef]
36. Chang, J.; Block, T.M.; Guo, J.-T. Antiviral therapies targeting host ER α-glucosidases: Current status and future directions. *Antivir. Res.* **2013**, *99*, 251–260. [CrossRef] [PubMed]
37. Karpas, A.; Fleet, G.W.J.; Dwek, R.A.; Petursson, S.; Namgoong, S.K.; Ramsden, N.G.; Jacob, G.S.; Rademacher, T.W. Aminosugar derivatives as potential anti-human immunodeficiency virus agents. *Proc. Natl. Acad. Sci. USA* **1988**, *85*, 9229–9233. [CrossRef]
38. Sunkara, P.S.; Bowlin, T.L.; Liu, P.S.; Sjoerdsma, A. Antiretroviral activity of castanospermine and deoxynojirimycin, specific inhibitors of glycoprotein processing. *Biochem. Biophys. Res. Commun.* **1987**, *148*, 206–210. [CrossRef]
39. Bernacki, R.; Niedbala, M.; Korytnyk, W. Glycosidases in cancer and invasion. *Cancer Metastasis* **1985**, *4*, 81–102. [CrossRef]
40. Martín-Banderas, L.; Holgado, M.A.; Durán-Lobato, M.; Infante, J.J.; Álvarez-Fuentes, J.; Fernández-Arévalo, M. Role of Nanotechnology for Enzyme Replacement Therapy in Lysosomal Diseases. A Focus on Gaucher's Disease. *Curr. Med. Chem.* **2016**, *23*, 929–952. [CrossRef]
41. Raben, N.; Fukuda, T.; Gilbert, A.L.; Jong, D.D.; Thurberg, B.L.; Mattaliano, R.J.; Meikle, P.; Hopwood, J.J.; Nagashima, K.; Nagaraju, K.; et al. Replacing Acid α-Glucosidase in Pompe Disease: Recombinant and Transgenic Enzymes are Equipotent, but Neither Completely Clears Glycogen from Type II Muscle Fibers. *Mol. Ther.* **2005**, *11*, 48–56. [CrossRef]
42. Cox, T.M. *Gaucher's Disease*; Academic Press: Cambridge, MA, USA, 2001; pp. 753–757.
43. Mena-Barragán, T.; García-Moreno, M.I.; Sevšek, A.; Okazaki, T.; Nanba, E.; Higaki, K.; Martin, N.I.; Pieters, R.J.; Fernández, J.M.G.; Mellet, C.O. Probing the Inhibitor versus Chaperone Properties of sp²-Iminosugars towards Human β-Glucocerebrosidase: A Picomolar Chaperone for Gaucher Disease. *Molecules* **2018**, *23*, 927. [CrossRef] [PubMed]
44. Hinek, A.; Zhang, S.; Smith, A.C.; Callahan, J.W. Impaired Elastic-Fiber Assembly by Fibroblasts from Patients with Either Morquio B Disease or Infantile GM1-Gangliosidosis Is Linked to Deficiency in the 67-kD Spliced Variant of β-Galactosidase. *Am. J. Hum. Genet.* **2000**, *67*, 23–26. [CrossRef] [PubMed]
45. Johnson, W.G. Chapter 33—Disorders of Glycoprotein Degradation: Sialidosis, Fucosidosis, α-Mannosidosis, β-Mannosidosis, and Aspartylglycosaminuria. In *Rosenberg's Molecular and Genetic Basis of Neurological and Psychiatric Disease*, 5th ed.; Academic Press: Cambridge, MA, USA, 2015; pp. 369–383.
46. Stütz, A.E.; Wrodnigg, T.M. Chapter Four—Carbohydrate-Processing Enzymes of the Lysosome: Diseases Caused by Misfolded Mutants and Sugar Mimetics as Correcting Pharmacological Chaperones. In *Advances in Carbohydrate Chemistry and Biochemistry*; Academic Press: Cambridge, MA, USA, 2016; Volume 73, pp. 225–302.
47. Malm, D.; Nilssen, Ø. Alpha-mannosidosis. *Orphanet J. Rare Dis.* **2008**, *3*, 21. [CrossRef] [PubMed]
48. Moreno-Clavijo, E.; Carmona, A.T.; Moreno-Vargas, A.J.; Molina, L.; Robina, I. Syntheses and Biological Activities of Iminosugars as α-L-Fucosidase Inhibitors. *Curr. Org. Synth.* **2011**, *8*, 102–133. [CrossRef]
49. Mohamed, F.E.; Al-Gazali, L.; Al-Jasmi, F.; Ali, B.R. Pharmaceutical Chaperones and Proteostasis Regulators in the Therapy of Lysosomal Storage Disorders: Current Perspective and Future Promises. *Front. Pharmacol.* **2017**, *8*, 448. [CrossRef]
50. National Centre for Advancing Translational Sciences, Genetic and Rare Diseases Information Centre. Available online: https://rarediseases.info.nih.gov/diseases/10372/trehalase-deficiency (accessed on 21 May 2022).
51. Gibson, R.P.; Gloster, T.M.; Roberts, S.; Warren, R.A.J.; Gracia, I.S.D.; García, Á.; Chiara, J.L.; Davies, G.J. Molecular Basis for Trehalase Inhibition Revealed by the Structure of Trehalase in Complex with Potent Inhibitors. *Angew. Chem. Int. Ed.* **2007**, *46*, 4115–4119. [CrossRef]
52. Matalon, R.; Matalon, K.M. Mucopolysacharidoses. In *Encyclopedia of the Neurological Sciences*; Academic Press: Cambridge, MA, USA, 2003; pp. 237–241.
53. Batool, F.; Khan, M.A.; Shaikh, N.N.; Iqbal, S.; Akbar, S.; Fazal-ur-rehman, S.; Choudhary, M.I.; Basha, F.Z. New Benzamide Analogues of Metronidazole-tethered Triazoles as Non-sugar Based Inhibitors of β-Glucuronidase. *ChemistrySelect* **2019**, *4*, 8634–8637. [CrossRef]
54. Guillotin, L.; Kim, H.; Traore, Y.; Moreau, P.; Lafite, P.; Coquoin, V.; Nuccio, S.; Vaumas, R.D.; Daniellou, R. Biochemical Characterization of the α-L-Rhamnosidase DtRha from *Dictyoglomus thermophilum*: Application to the Selective Derhamnosylation of Natural Flavonoids. *ACS Omega* **2019**, *4*, 1916–1922. [CrossRef]
55. Yadav, V.; Yadav, P.K.; Yadav, S.; Yadav, K.D.S. α-L-Rhamnosidase: A review. *Proc. Biochem.* **2010**, *45*, 1226–1235. [CrossRef]
56. Souza, P.M.D.; Sales, P.M.D.; Simeoni, L.A.; Silva, E.C.; Silveira, D.; Magalhães, P.D.O. Inhibitory Activity of α-Amylase and α-Glucosidase by Plant Extracts from the Brazilian Cerrado. *Planta Med.* **2012**, *78*, 393–399. [CrossRef]
57. Aoyama, T.; Kojima, F.; Imada, C.; Muraoka, Y.; Naganawa, H.; Okami, Y.; Takeuchi, T.; Aoyagi, T. Pyrostatins A and B, New inhibitors of N-Acetyl-β-D-Glucosaminidase, produced by *Streptomyces* sp. SA-3501. *J. Enz. Inhib.* **1995**, *8*, 223–232. [CrossRef] [PubMed]
58. Nishimura, Y.; Shitara, E.; Takeuchi, T. Enantioselective synthesis of a new family of α-L-fucosidase inhibitors. *Tetrahedron Lett.* **1999**, *40*, 2351–2354. [CrossRef]

59. Takahashi, H.; Sou, S.; Yamasaki, R.; Sodeoka, M.; Hashimoto, Y. α-Glucosidase inhibitors with a phthalimide skeleton: Structure-activity relationship study. *Chem. Pharm. Bull.* **2000**, *48*, 1494–1499. [CrossRef]
60. Pluempanupat, W.; Adisakwattana, S.; Yibchok-Anun, S.; Chavasiri, W. Synthesis of *N*-phenylphthalimide Derivatives as alpha-Glucosidase Inhibitors. *Arch. Pharm. Res.* **2007**, *30*, 1501–1506. [CrossRef] [PubMed]
61. Motoshima, K.; Noguchi-Yachide, T.; Sugita, K.; Hashimoto, Y.; Ishikawa, M. Separation of a-glucosidase-inhibitory and liver X receptor-antagonistic activities of phenethylphenyl phthalimide analogs and generation of LXRa-selective antagonists. *Bioorg. Med. Chem.* **2009**, *17*, 5001–5014. [CrossRef]
62. Dodo, K.; Aoyama, A.; Noguchi-Yachide, T.; Makishima, M.; Miyachi, H.; Hashimoto, Y. Co-existence of α-glucosidase-inhibitory and liver X receptor-regulatory activities and their separation by structural development. *Bioorg. Med. Chem.* **2008**, *16*, 4272–4285. [CrossRef]
63. Noguchi-Yachide, T.; Aoyama, A.; Makishima, M.; Miyachi, H.; Hashimoto, Y. Liver X receptor antagonists with a phthalimide skeleton derived from thalidomide-related glucosidase inhibitors. *Bioorg. Med. Chem. Lett.* **2007**, *17*, 3957–3961. [CrossRef]
64. Noguchi-Yachide, T.; Miyachi, H.; Aoyama, H.; Aoyama, A.; Makishima, M.; Hashimoto, Y. Structural Development of Liver X Receptor (LXR) Antagonists Derived from Thalidomide-Related Glucosidase Inhibitors. *Chem. Pharm. Bull.* **2007**, *55*, 1750–1754. [CrossRef]
65. Mbarki, S.; Hallaoui, M.E.; Dguigui, K. 3D-QSAR for α-glucosidase inhibitory activity of n-(phenoxyalkyl) phthalimide derivatives. *Int. J. Rec. Res. Appl. Stud.* **2012**, *11*, 395–401.
66. Wang, G.; Peng, Y.; Xie, Z.; Wang, J.; Chen, M. Synthesis, α-glucosidase inhibition and molecular docking studies of novel thiazolidine-2,4-dione or rhodanine derivatives. *MedChemComm* **2017**, *8*, 1477–1484. [CrossRef]
67. Sadat-Ebrahimi, S.E.; Rahmani, A.; Mohammadi-Khanaposhtani, M.; Jafari, N.; Mojtabavi, S.; Faramarzi, M.A.; Emadi, M.; Yahya-Meymandi, A.; Larijani, B.; Biglar, M.; et al. New phthalimide-benzamide-1,2,3-triazole hybrids; design, synthesis, α-glucosidase inhibition assay, and docking study. *Med. Chem. Res.* **2020**, *29*, 868–876. [CrossRef]
68. Bian, X.; Fan, X.; Ke, C.; Luan, Y.; Zhao, G.; Zeng, A. Synthesis and alpha-glucosidase inhibitory activity evaluation of N-substituted aminomethyl-b-D-glucopyranosides. *Bioorg. Med. Chem.* **2013**, *21*, 5442–5450. [CrossRef] [PubMed]
69. Hoos, R.; Naughton, A.B.; Thiel, W.; Vasella, A.; Weber, W.; Rupitz, K.; Withers, S.G. D-Gluconhydroximo-1,5-Lactam And Related *N*-Arylcarbamates Theoretical Calculations, Structure, Synthesis, and Inhibitory Effect on Beta-Glucosidases. *Helv. Chim. Acta* **1993**, *76*, 2666–2686. [CrossRef]
70. Awolade, P.; Cele, N.; Kerru, N.; Gummidi, L.; Oluwakemi, E.; Singh, P. Therapeutic significance of b-glucuronidase activity and its inhibitors: A review. *Eur. J. Med. Chem.* **2020**, *187*, 111921. [CrossRef] [PubMed]
71. Ferhati, X.; Matassini, C.; Fabbrini, M.G.; Goti, A.; Morrone, A.; Cardona, F.; Moreno-Vargas, A.J.; Paoli, P. Dual targeting of PTP1B and glucosidases with new bifunctional iminosugar inhibitors to address type 2 diabetes. *Bioorg. Chem.* **2019**, *87*, 534–549. [CrossRef] [PubMed]
72. Glenister, A.; Simone, M.I.; Hambley, T.W. A Warburg effect targeting vector designed to increase the uptake of compounds by cancer cells demonstrates glucose and hypoxia dependent uptake. *PLoS ONE* **2019**, *14*, e0217712. [CrossRef]
73. Glenister, A.; Chen, C.K.J.; Renfrew, A.K.; Simone, M.I.; Hambley, T.W. Warburg Effect Targeting Cobalt(III) Cytotoxin Chaperone Complexes. *J. Med. Chem.* **2021**, *64*, 2678–2690. [CrossRef]
74. Venhuizen, J.R. *Idaho National Engineering Laboratory, INEL BNCT Research Program Annual Report*; 1995. Available online: https://www.osti.gov/biblio/421334 (accessed on 20 February 2022).
75. Koo, M.-S.; Ozawa, T.; Santos, R.A.; Lamborn, K.R.; Bollen, A.W.; Deen, D.F.; Kahl, S.B. Synthesis and Comparative Toxicology of a Series of Polyhedral Borane Anion-Substituted Tetraphenyl Porphyrins. *J. Med. Chem.* **2007**, *50*, 820–827. [CrossRef]
76. Cano, W.G.; Solares, G.R.; Dipetrillo, T.A.; Meylaerts, S.A.G.; Lin, S.C.; Zamenhof, R.G.; Saris, S.C.; Duker, J.S.; Goad, E.; Madoc-Jones, H.; et al. Toxicity Associated with Boronophenylalanine and Cranial Neutron Irradiation. *Radiat. Oncol. Investig.* **1995**, *3*, 108–118. [CrossRef]
77. Pereira, G.L.; Siqueira, J.A.; Batista-Silva, W.; Cardoso, F.B.; Nunes-Nesi, A.; Araújo, W.L. Boron: More Than an Essential Element for Land Plants? *Front. Plant Sci.* **2021**, *11*, 610307. [CrossRef]
78. Warington, K. The effect of boric acid and borax on the broad bean and certain other plants. *Ann. Bot.* **1923**, *37*, 629–672. [CrossRef]
79. Uluisika, I.; Karakaya, H.C.; Koc, A. The Importance of Boron in Biological Systems. *J. Trace Elem. Med. Biol.* **2018**, *45*, 156–162. [CrossRef] [PubMed]
80. U.S. Department of Health and Human Services; Public Health Service Agency for Toxic Substances and Disease Registry. Division of Toxicology and Environmental Medicine/Applied Toxicology Branch 1600 Clifton Road NE, Mailstop F-32, Atlanta, Georgia 30333, *Toxicological Profile for Benzene*. 2007. Available online: https://www.atsdr.cdc.gov/ (accessed on 20 February 2022).
81. Centers for Disease Control and Prevention; The National Institute for Occupational Safety and Health (NIOSH). Benzene (Immediately Dangerous to Life or Health Concentrations (IDLH)). Available online: https://www.cdc.gov/niosh/idlh/71432.html (accessed on 20 February 2022).
82. Thermo Fisher Scientific, Safety Data Sheet, Phenylboronic Acid. Available online: https://www.fishersci.com/store/msds?partNumber=AC130360100&productDescription=PHENYLBORIC+ACID%2C+98%2B%25+10GR&vendorId=VN00032119&countryCode=US&language=en (accessed on 20 February 2022).

83. Kessler, M.; Acuto, O.; Storelli, C.; Murer, H.; Müller, M.; Semenza, G. A modified procedure for the rapid preparation of efficiently transporting vesicles from small intestinal brush border membranes. Their use in investigating some properties of D-glucose and choline transport systems. *Biochim. Biophys. Acta* **1978**, *506*, 136–154. [CrossRef]
84. Gottlieb, H.E.; Kolyar, V.; Nudelman, A. NMR chemical shifts of common laboratory solvents as trace impurities. *J. Org. Chem.* **1997**, *62*, 7512–7515. [CrossRef]
85. Sheldrick, G.M. A short history of SHELX. *Acta Crystallogr. Sect. A Found. Crystallogr.* **2008**, *64*, 112–122. [CrossRef]
86. Macrae, C.F.; Sovago, I.; Cottrell, S.J.; Galek, P.T.A.; McCabe, P.; Pidcock, E.; Platings, M.; Shields, G.P.; Stevens, J.S.; Towler, M.; et al. Mercury 4.0: From visualization to analysis, design and prediction. *J. Appl. Cryst.* **2020**, *53*, 226–235. [CrossRef]

Article

New Chemotypes for the Inhibition of (p)ppGpp Synthesis in the Quest for New Antimicrobial Compounds

Crescenzo Coppa [†], Luca Sorrentino [†], Monica Civera, Marco Minneci, Francesca Vasile and Sara Sattin *

Dipartimento di Chimica, Università degli Studi di Milano, Via C. Golgi, 19, 20133 Milano, Italy; crescenzo.coppa@unimi.it (C.C.); luca.sorrentino@unimi.it (L.S.); monica.civera@unimi.it (M.C.); marco.minneci@unimi.it (M.M.); francesca.vasile@unimi.it (F.V.)
* Correspondence: sara.sattin@unimi.it
† These authors contributed equally to this work.

Abstract: Antimicrobial resistance (AMR) poses a serious threat to our society from both the medical and economic point of view, while the antibiotic discovery pipeline has been dwindling over the last decades. Targeting non-essential bacterial pathways, such as those leading to antibiotic persistence, a bacterial bet-hedging strategy, will lead to new molecular entities displaying low selective pressure, thereby reducing the insurgence of AMR. Here, we describe a way to target (p)ppGpp (guanosine tetra- or penta-phosphate) signaling, a non-essential pathway involved in the formation of persisters, with a structure-based approach. A superfamily of enzymes called RSH (RelA/SpoT Homolog) regulates the intracellular levels of this alarmone. We virtually screened several fragment libraries against the (p)ppGpp synthetase domain of our RSH chosen model Rel$_{Seq}$, selected three main chemotypes, and measured their interaction with Rel$_{Seq}$ by thermal shift assay and STD-NMR. Most of the tested fragments are selective for the synthetase domain, allowing us to select the aminobenzoic acid scaffold as a hit for lead development.

Keywords: AMR; persisters; (p)ppGpp; fragment screening; thermal shift assay; STD-NMR

1. Introduction

The rising of bacteria resistant to the currently available antibiotic arsenal is posing a serious threat to the way of life we have become accustomed to over the past century since Fleming isolated and characterized penicillin [1,2]. If left unchecked, antimicrobial resistance (AMR), paired with the decreasing number of new antibiotics progressing through the clinical pipeline, will lead to millions of deaths each year in a few decades [3].

AMR manifests itself in stable, heritable genetic forms, as well as in lesser-known transient phenotypes that are more elusive and difficult to recognize and tackle [4]. On the other hand, antimicrobial compounds currently in the clinical phase consist mostly of derivatives of established classes, with an urgent need for drugs that address multidrug-resistant Gram-negative bacteria [3].

The search for new chemical entities, targeting non-essential pathways that play a role in both the infection process (e.g., bacterial adhesion, quorum sensing, virulence, and biofilm formation) and the insurgence of genetic resistance, is an attractive strategy for attaining new antimicrobial drugs exerting minimal selective pressure.

We posit (p)ppGpp-signalling molecules (guanosine tetra- or penta-phosphate) [5] as key players in both processes [6]. Indeed, (p)ppGpp is a ubiquitous alarmone that directs bacterial adaptation to environmental changes (e.g., nutrient starvation, oxidative stress, etc.) by binding various targets involved, e.g., in nucleotide metabolism, DNA replication and repair [7], transcription [8], and translation [9,10]. This alarmone thus has pleiotropic effects on bacterial cell physiology [11], regulating cell size, virulence, quorum sensing, and biofilm formation [12–15]. In particular, its ability to downregulate cell metabolism and cell growth hints at its role in the insurgence of the bacterial non-heritable dormant phenotype called persister [16]. Persisters are transiently tolerant to antibiotic treatment (i.e., they are a form of phenotypic AMR) until they revert to the awake state and resume growth, constituting an infection reservoir that sustains chronic and recurrent infections and paves the way to the acquisition of genetic resistance [17,18]. Over the last two decades, extensive research has revealed different molecular mechanisms leading to their formation [19], and one of them is the accumulation of (p)ppGpp together with its downstream effects.

Intracellular (p)ppGpp levels are regulated by enzymes belonging to the RSH (RelA/SpoT Homolog) superfamily [20]. These enzymes catalyze (p)ppGpp synthesis via a Mg^{2+}-dependent pyrophosphate transfer from ATP to the 3'-OH group of either GDP or GTP. They also catalyze (p)ppGpp hydrolysis leading to the release of pyrophosphate (PPi) using distinct active sites in different protein domains (Figure 1a). "Short" RSH proteins harbor either only the synthetase (SAS, small alarmone synthetases) or hydrolase (SAH, small alarmone hydrolases) domain, respectively. "Long" RSH proteins contain both catalytic domains and a C-terminal regulatory domain (CTD) that activates alarmone synthesis by promoting Rel oligomerization [21] and/or upon binding to stalled ribosomes (i.e., ribosomes bound to uncharged tRNAs during aminoacid starvation) [22,23] or favors alarmone hydrolysis by directly inhibiting the synthetase site [24]. In addition, reciprocal regulation of the two catalytic domains has been postulated with mechanisms that vary among species [25–28].

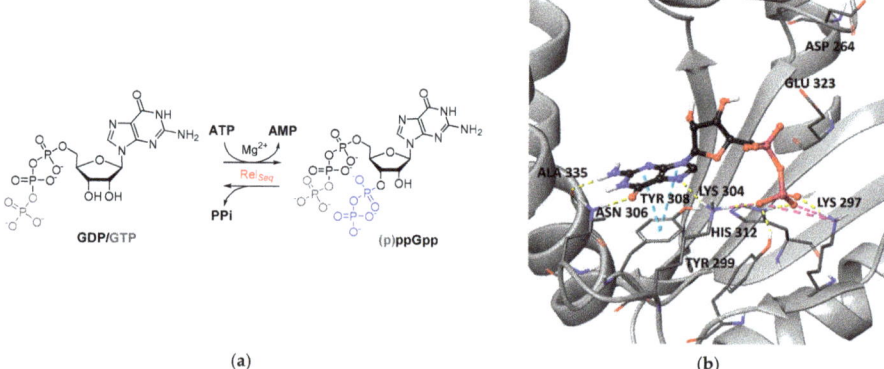

Figure 1. (a) Enzymatic reaction that regulates intracellular (p)ppGpp levels catalyzed by RSH enzymes such as Rel_{Seq}; (b) X-ray structure of GDP in the SYNTH site of Rel_{Seq} (1VJ7, chainA). GDP is represented in ball and stick, and the key interacting amino acids are labeled.

The search for inhibitors of Rel enzymes synthetase activity dates back to the early 2010s, when Relacin [29] and a few other (p)ppGpp analogs [30,31] were described by Wexselblatt and co-workers. In all cases, IC_{50} values measured against RelA (*E. coli*) and Rel from *D. radiodurans* were estimated to be between 1 and 5 mM, with relatively low ligand efficiency and no subsequent further optimization reported. More recently, extensive efforts, including high-throughput screening (Rel_{Bs}, *B. subtilis*) and an expanded library of (p)ppGpp analogs (RelA), have identified some low μM, non-specific Rel inhibitors [32,33], while screening of a large pharmaceutical library (GSK, >2 M compounds) for inhibitors of Rel_{Mtb} (*M. tuberculosis*) identified only one compound (X9) as a potential lead for a combination TB therapy with isoniazid [34].

In this framework, we optimized the synthesis of a fluorescent (p)ppGpp selective chemosensor [35], and here, we report the identification of three novel chemical scaffolds for the design of selective RSH inhibitors through fragments virtual screening campaigns followed by experimental validation of representative fragments on the synthetase site of the "long" RSH protein Rel_{Seq} (*S. equisimilis*).

2. Results and Discussion

2.1. Fragment Libraries Virtual Screening in Rel$_{Seq}$ Synthetase Site

We chose as a protein model the X-ray crystal structure reported for Rel$_{Seq}$ in a so-called synthetase-ON conformation, i.e., the chain A of the pdb structure 1VJ7 (residues 1–385), which carries the GDP substrate in the synthetase catalytic site [25]. This is a truncated form of the protein, lacking the C-terminal regulatory domain.

By analyzing the interactions of GDP within the catalytic site (Figure 1b), we could observe that it binds to the G-loop (Tyr299-Ser310), forming a π–π stacking interaction with the side chain of Tyr308 through its guanine ring. H-bonds with the side chains of Lys304 and Asn306 and with the backbone of Ala335 stabilize this core interaction. In addition, the GDP pyrophosphate moiety forms salt bridges with Lys304 and Lys297 side chains and H-bonds with Tyr299 and His312 side chains.

Our analysis revealed that the Rel$_{Seq}$ synthetase-ON catalytic site conformation could not be catalytically competent, as it lacks the space necessary to accommodate the pyrophosphate donor ATP, and the reported catalytic residues Asp264 and Glu323 are unfavorably oriented to promote the reaction (Figure 1b). On this basis, we constructed and reported a catalytically competent Rel$_{Seq}$ chimera model based on the structure of the SAS RelP from *S. aureus*, where the resulting catalytic site is considerably more extended [36]. However, we performed an initial virtual screening on the Rel$_{Seq}$ X-ray crystal structure in order to focus our binding site exploration on the region occupied by the enzyme substrate GDP.

During protein preparation (1VJ7, chain A, see Section 3), we took a closer look at the protonation state of His312, interacting with the β-phosphate group of GDP in the crystal. Although the neutral form should be more plausible, given the pH working conditions of the enzyme (activity usually tested at pH 7–9), we decided to generate a model with the residue in the protonated form as well (Hip312) in order to assess the validity of this assumption. We set up and validated a docking protocol within the synthetase binding site for both models by re-docking the co-crystallized GDP molecule using GLIDE v8.0 (Supplementary Figures S1 and S2). [37]

Several chemical vendors currently make available virtual structure datasets of fragment libraries, often organized according to specific experimental properties (e.g., solubility). In order to maximize the chemical space explored in our screening [38,39], we selected seven different libraries of commercially available fragments, *Maybridge* Rule of 3, *Asinex* Fragments, *Life Chemicals* Fragment Library with Experimental Solubility Data I and II, *OTAVA* Solubility fragment library, *Chembridge* Fragment library, and *SPECS* fragment library, amounting to a total of 58,321 2D entries (see Section 3 and Supplementary Table S1). We implemented the validated docking protocol for the virtual screening (VS) of the selected fragment libraries following the workflow shown in Figure 2. For each library, Ligprep [40] converted 2D entries into 3D structures considering stereoisomers, tautomers, and protonation states, leading to an increase in the total number of structures up to 114,966.

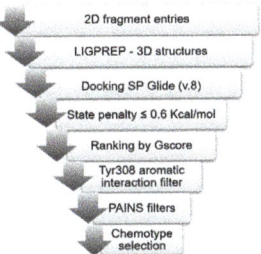

Figure 2. Virtual screening workflow. Ligprep generated 3D structure from 2D fragments; Glide docked the 3D structures into both Rel$_{Seq}$ models (His312 and Hip312). A state penalty filter excluded unfavorable states, and the resulting docking poses were ranked according to Gscore. Only fragments forming an aromatic interaction with Tyr308 were retained, and after removing PAINS, we evaluated the top 1% ranked poses identifying the most representative chemotypes.

2.2. Post-Docking Analysis and Chemotype Selection

We removed from the docking outputs (one pose saved for each fragment) the less stable tautomeric and ionization forms, as determined by Epik [41] (state penalty value ≤ 0.6 kcal/mol). We removed duplicates from the merged outputs and ranked them by Gscore. Finally, we applied an aromatic interaction filter with Tyr308 (i.e., an aromatic ligand atom must be within 5 Å from any heavy atom of the Tyr308 side chain), a key residue that, when mutated into asparagine or serine, inhibits the synthetic activity of the enzyme, and removed PAINS (pan-assay interference compounds) using the filters provided by Canvas [42,43] in order to exclude frequent hitters [44]. This work resulted in the selection of 30,126 and 30,960 fragments for the His312 and Hip312 grids, respectively. We visually inspected and assessed the top 1% of docked poses, identifying three recurrent chemotypes, i.e., benzimidazole, aminobenzoic acid, and indole (for the calculated enrichment factors, see Supplementary Table S2) and three singletons (the best pose of representative fragments is shown in Figure 3 and Supplementary Figure S3). Aminobenzoic acids emerged mainly from the Hip312 model and benzimidazoles mainly from the His312 model, while indoles emerged from both to a lesser extent.

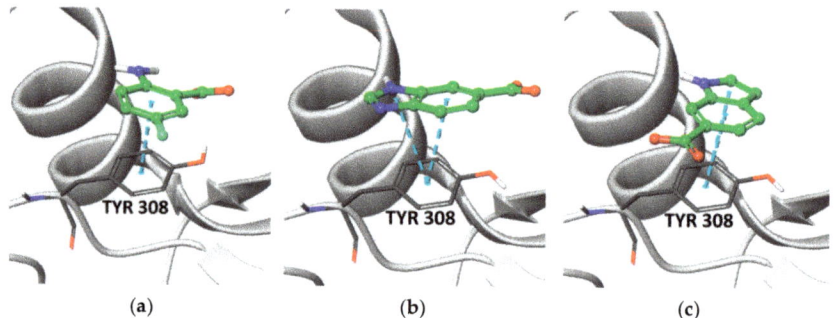

Figure 3. Best poses of representative structures of the three chemotypes selected for experimental validation. (**a**) aminobenzoic acid **A3** docked in the Hip312 grid; (**b**) Benzimidazole **B2** docked in the His312 grid; (**c**) indole **I1** docked in the His312 grid. Fragments are shown as balls and sticks with green carbon atoms. The protein is shown in grey, with the side chain of Tyr308 highlighted.

In order to maximize the chemical space explored, we performed a similarity search (Tanimoto index ≥ 90%) using the PubChem database [45] and the representative structures of each chemotype as input. We applied the same screening workflow (see Supplementary Information) to the expanded set of fragments leading to the final selection of eighteen fragments for experimental validation (Figure 4).

Prior to conducting biochemical assays, we assessed the docking poses stability by running molecular dynamics (MD) simulations. Starting from the best pose for each fragment we ran 100 ns simulations using Desmond [46] (NPT, T = 300 K, p = 1 atm, TIP3P [47], OPLS3e [48], dt = 2 fs). We considered the aromatic interaction with Tyr308 described above as the key feature to be maintained and monitored during the simulations. All the examined fragments form stable interactions with the residues involved in the binding of the GDP guanine ring in the X-ray structure, retaining, in particular, stable contact with the side chain of Tyr308, with the exception of fragments B1 and I2 that exit the binding pocket (Supplementary Figure S4).

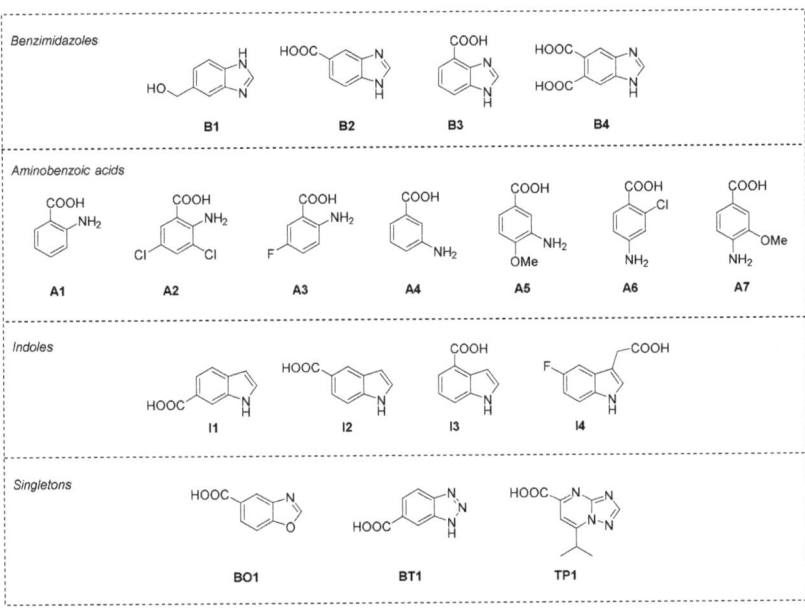

Figure 4. Fragments selected for experimental validation (i.e., thermal shift assay) grouped by chemotype: benzimidazoles (**B1–B4**), aminobenzoic acids (**A1–A7**), indoles (**I1–I4**), and the three singletons (**BO1, BT1,** and **TP1**).

2.3. Thermal Shift Assay on Selected Fragments vs. Rel$_{Seq}$ Protein Constructs

With the aim of studying protein–ligand interactions between Rel$_{Seq}$ constructs and selected fragments, we evaluated the use of different techniques, such as tryptophan assay, microscale thermophoresis (MST), isothermal calorimetry (ITC) or thermal shift assay (TSA). The tryptophan assay is based on ligand-induced conformational changes in the local environment surrounding tryptophan residues in the target protein. Irradiation at 280 nm is followed by the detection of tryptophan fluorescence emission at different wavelengths. We used this technique to measure the Kd values of the natural substrates GDP and ATP (see below), but it is not applicable in the case of fragments due to their high absorbance at 280 nm. On the other hand, the currently available protein-labeling reagents for MST were not compatible with our protein constructs. We chose TSA over ITC due to its higher potential throughput and lower amount of protein required.

Thermal shift is an experimental technique in which protein thermal denaturation is monitored following the increase in fluorescence reported by a protein-bound dye [49]. In particular, an environment-sensitive hydrophobic dye (e.g., SYPRO Orange) binds to hydrophobic regions that become progressively exposed during thermal denaturation, resulting in an increase in its fluorescence emission. Since the binding of small molecules (e.g., fragments) to the protein can cause conformational changes that affect its melting temperature, this technique allows the screening of several compounds in a range of concentrations without consuming sizeable amounts of protein.

We preliminary determined by TSA the dissociation constants (K_d) of complexes engaged by Rel$_{Seq}$ with its natural substrates, ATP (0.49 ± 0.09 mM) and GDP (0.26 ± 0.06 mM), finding values comparable to those obtained by tryptophan assay (K_d^{ATP} = 0.39 ± 0.04 mM, K_d^{GDP} = 0.15 ± 0.01 mM) and confirming the robustness of this technique (Table 1, entries 1 and 2).

Table 1. Thermal shift assay. K_d (mM) of protein–ligand complexes engaged by Rel$_{Seq}$ constructs with the tested fragments.

Entry	Compound	K_d Rel$_{Seq}$ 1–385	K_d Rel$_{Seq}$ SYNTH	K_d Rel$_{Seq}$ HYD
1	ATP	0.49 ± 0.09 (0.39 ± 0.04) [1]		
2	GDP	0.26 ± 0.06 (0.15 ± 0.01) [1]		
3	B1	No binding	No binding	No binding
4	B2	Biphasic	No binding	1.9 ± 0.7
5	B3	3.4 ± 0.3	4.3 ± 0.4	3.3 ± 0.9
6	B4	No binding	No binding	No binding
7	A1	1.2 ± 0.3	10.8 ± 2.2	No binding
8	A2	No binding	No binding	No binding
9	A3	1.5 ± 0.1	5.5 ± 0.9	No binding
10	A4	6.6 ± 0.1	9.8 ± 2.8	No binding
11	A5	1.1 ± 0.2	2.2 ± 0.4	No binding
12	A6	4.3 ± 1.1	6.5 ± 1.2	No binding
13	A7	4.0 ± 0.9	4.3 ± 0.5	No binding
14	I1	6.5 ± 1.1	9.6 ± 1.5	No binding
15	I2	2.5 ± 0.6	5.5 ± 0.9	No binding
16	I3	4.0 ± 0.5	9.9 ± 4.5	No binding
17	I4	3.2 ± 0.7	>15	No binding
18	BO1	2.2 ± 0.3	2.7 ± 0.6	No binding
19	BT1	No binding	No binding	No binding
20	TP1	3.4 ± 0.8	8.3 ± 1.6	No binding

[1] K_d values measured by tryptophan assay.

We therefore evaluated the affinity of the 18 fragments selected from our *in silico* screening for the bifunctional protein Rel$_{Seq}$ by titration of the protein in a thermal shift assay (Table 1). We initially used Rel$_{Seq}$ 1–385, a truncated construct lacking the abovementioned C-terminal regulatory domain with a catalytic activity intrinsically shifted towards (p)ppGpp synthesis (12-fold higher than the full-length protein) [50].

Interestingly, all but four of the tested fragments showed a dose-dependent interaction with the bifunctional protein with K_d values in the low millimolar range.

In particular, only one of the four tested benzimidazoles showed a measurable affinity for Rel$_{Seq}$ (1–385) (**B3**, Table 1, entry 5), whereas fragment **B2** yielded a biphasic curve that requires further investigation (Table 1, entry 4). All of the selected aminobenzoic acids showed good affinities for the protein, with the exception of **A2** (Table 1, entries 7–13), while among the four indoles tested, all interacted in a dose-dependent manner with Rel$_{Seq}$ (1–385) (Table 1, entries 14–17). Finally, two of the three singletons tested (**BO1** and **TP1**) showed a low mM affinity for the protein.

We assessed the fragments' selectivity for Rel$_{Seq}$ synthetase domain by performing TSA experiments on two mono-functional truncated protein constructs: Rel$_{Seq}$ SYNTH (residues 79–385) and Rel$_{Seq}$ HYD (residues 1–224), presenting only the synthetase or hydrolase protein domain, respectively. Both constructs retain part of the central 3-helix bundle to ensure proper folding, especially in the case of the less stable SYNTH domain [50].

The four fragments that failed to show binding to the bifunctional protein (**B1**, **B4**, **A2**, and **BT1**) also failed to show dose-dependent effects on the two truncated constructs, while the biphasic curve initially observed for **B2** was determined to be a selective interaction with the HYD domain, with no affinity for the SYNTH domain.

With the exception of **B3**, which shows comparable affinities for both catalytic domains, all the fragments binding to Rel$_{Seq}$ (1–385) showed a remarkable selectivity for the SYNTH domain, even with a generally increased absolute value for the measured K_d. This is probably due to the lower overall stability of the SYNTH domain compared to the full protein, as previously reported by Mechold et al. [50]. Indeed, Rel$_{Seq}$ SYNTH (79–385), despite being catalytically functional (data not shown), requires the use of a non-ionic surfactant in the purification steps in order to avoid precipitation.

2.4. STD-NMR Protein–Fragment Interaction Experiments

The relatively weak affinity of the fragments measured by TSA directed us towards NMR methods to assess the specificity of the fragments-protein interactions. Indeed, ligand-based NMR methods [51] can be applied to weak and transient protein–ligand complexes that are difficult to study with other structural techniques and do not require protein labeling (since only NMR signals of the small molecule are detected), and only a small amount of protein is required. In particular, STD (Saturation Transfer Difference) exploits NOE effects between the protein and the ligand to map target–ligand interactions and to characterize biologically relevant complexes [52,53].

Considering the promising selectivity profile of the aminobenzoic acids for the Rel$_{Seq}$ synthetase domain, we performed STD-NMR experiments with fragment **A1** as the representative chemotype. The results confirmed the binding event and showed a good interaction for the aromatic protons of **A1** with Rel$_{Seq}$ (1–385) (Figure 5b). Comparable overall STD intensities, suggesting a similar binding mode, were observed for Rel$_{Seq}$ SYNTH (79–385), confirming the specificity of the interaction (Figure 5c). On the other hand, the Rel$_{Seq}$ HYD (1–224) construct did not produce any magnetization transfer (Figure 5d). STD experiments performed on Rel$_{Seq}$ (1–385) with the two non-binding fragments **B1** and **A2** did not show any significant interaction, refuting artifacts or non-specific binding (Supplementary Figures S5 and S6, respectively).

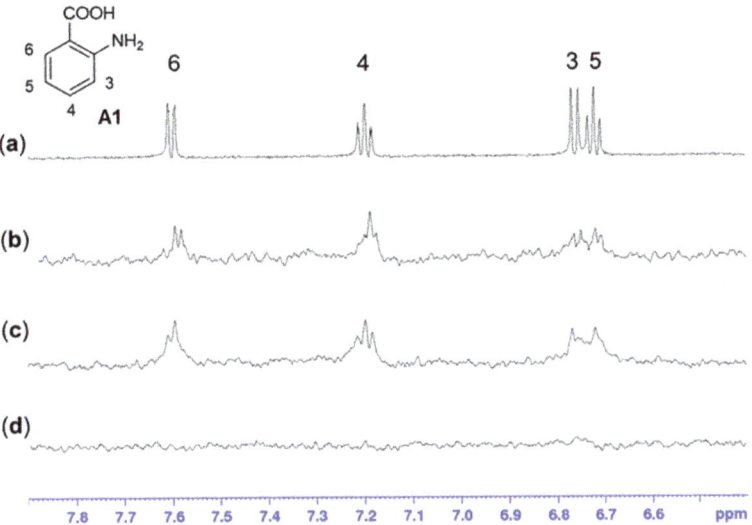

Figure 5. STD-NMR experiments (fragment:protein ratio 1000:1; fragment concentration 3 mM). (**a**) ^{1}H-NMR of fragment **A1** in phosphate buffer at 298 K. (**b**) STD-NMR experiment of **A1** with Rel$_{Seq}$ (1–385). (**c,d**) STD-NMR experiment of **A1** with Rel$_{Seq}$ SYNTH (79–385) and with Rel$_{Seq}$ HYD (1–224), respectively. The same binding mode can be observed for full-length Rel$_{Seq}$ and Rel$_{Seq}$ SYNTH, while no binding can be detected with Rel$_{Seq}$ HYD.

3. Materials and Methods

3.1. Computational Methods

Protein Preparation. Rel$_{Seq}$ three-dimensional structure (PDB 1VJ7, chain A, residues 5–341) was prepared for docking calculations using the 'Protein Preparation Wizard' of Schrödinger® suite [40] and OPLS_2005 force field [54]. All water molecules were deleted, and the gaps of the HD domain (K110-N123 and K153-D158) were built using Prime [55]. The residues' protonation states were determined according to PROPKA at pH 7. Two models were built considering the two possible protonation states of His312, i.e., the neutral (His312) and the protonated (Hip312) form. According to Epik [41] results at pH 7± 2, the GDP molecule is fully deprotonated (total formal charge of −3). Hydrogen bonds were optimized according to the exhaustive sampling option, and the entire complexes were optimized by using a restrained minimization (root-mean-square deviation on heavy atoms < 0.30 Å). The K110-N123 and K153-D158 gaps of the HD domain were built, and the former was further refined using the 'refine loops' tool of Prime (OPLS3e [48], VSGB [56]) with default parameters. Five structures were generated, and the model with the lowest Prime energy was selected for the docking calculation.

Docking protocol. Grid-Based Ligand Docking With Energetics (Glide) [37] v.8.0 software was used with the OPLS_2005 force field. Receptor grids for HIS312 and HIP312 systems were generated in a cubic region (24.5 Å) centered on GDP molecules with an inner cubic box of 10 Å. The receptor was considered a rigid body, while the ligands were considered flexible. The standard precision (SP) method was applied with default parameters. No Epik state penalty was added to the Glide score. The docking protocol was validated for the X-ray ligand by saving five poses after a post-minimization of the first 10 poses.

The top-ranked poses succeeded in reproducing the experimental binding mode of GDP (RMSD on heavy atoms of 0.78 Å and 1.23 Å in the HIS312 and HIP312 models, respectively) (Supplementary Figures S1 and S2).

Fragment libraries preparation. Seven fragment libraries were downloaded:

- Maybridge Ro3 Diversity Set (2500 fragments) (www.maybridge.com, accessed on 9 June 2015);
- Asinex-Fragments-21872 (21,872 fragments) (www.asinex.com, accessed on 18 January 2019);
- 'Fragment Libraries with Experimental Solubility Data', two datasets from Life Chemicals (11,667 and 2921 fragments, respectively) (www.lifechemicals.com, accessed on 1 February 2019);
- OTAVA Solubility Fragment Library (1021 fragments) (www.otavachemicals.com, accessed on 4 February 2019);
- FragmentLibrary_sdf_13808 (13,808 fragments) from CHEMBRIDGE (www.chembridge.com, accessed on 18 April 2019)
- Preplated Fragment-Based Library (4532 fragments) from SPECS (www.specs.net, accessed on 11 February 2019).

For each fragment, we generated 3D structures, tautomers, stereoisomers (at most 32 per ligand), and protonation states (Epik at pH = 7 ± 2) using the Ligprep tool [40]. Their energy was minimized using 'MacroModel' [40], implemented with truncated Newton conjugated gradient method [57], and the resulting structures were used as input for docking calculations (see Supplementary Table S1).

Molecular Dynamics simulations. Molecular dynamics (MD) simulations (100 ns, NPT, OPLS3e [48], T = 300 K, Langevin thermostat [58] relaxation time = 1.0 ps; p = 1 atm; barostat relaxation time = 2.0 ps [59]) were carried out using Desmond [46] starting from the best pose of the eighteen fragments selected for the TSA (Hip312 best pose for aminobenzoic acids, His312 best pose for all the other fragments). Atomic coordinates were saved every 100 ps for a total of about 1000 frames. The systems were solvated into a (10 Å side) cubic box of TIP3P [47] water molecules and neutralized by adding Cl$^-$ and Na$^+$ ions at a physiological concentration of 0.15 M NaCl. The systems were equilibrated by applying the 'desmond_npt_relax.msj' protocol available in Desmond with default parameters.

3.2. Experimental Methods

Cloning, Expression, and Purification of Rel$_{Seq}$ constructs. A pET21 expression vector containing the DNA sequence coding for Rel$_{Seq}$ 1–385 fused with a C-terminal His-tag was purchased from Giotto Biotech. Two truncated constructs of the bifunctional enzyme, Rel$_{Seq}$ 79–385 (Rel$_{Seq}$ SYNTH) and Rel$_{Seq}$ 1–224 (Rel$_{Seq}$ HYD) [50], presenting only the synthetase or hydrolase domain, respectively, were obtained with the Q5 Site-directed mutagenesis kit (New England Biolabs). Each protein construct was overproduced in BL21(DE3) *Escherichia coli* cells (Merck), grown in LB medium. Protein expression was induced by the addition of 0.5 mM IPTG and prolonged overnight at 25 °C for Rel$_{Seq}$ 1–385 and Rel$_{Seq}$ HYD and at 20 °C for Rel$_{Seq}$ SYNTH.

In a typical purification, bacterial cells harvested by centrifugation were resuspended in lysis buffer (50 mM Tris-HCl pH 8.0, 250 mM NaCl, 10 mM imidazole, 0.5 mM TCEP), supplemented with 1 mM phenylmethanesulfonylfluoride, 20 µg/mL DNAse I (Merck) and, only in the case of Rel$_{Seq}$ SYNTH, 0.1% Triton X-100. Cell disruption was performed by sonication, and, after high-speed centrifugation and microfiltration, the resulting bacterial soluble extract was loaded on two 1 mL Ni Sepharose HisTrap columns (GE Healthcare), connected in series and equilibrated with lysis buffer. Elution of Rel$_{Seq}$ constructs was achieved by applying a linear gradient of elution buffer (50 mM Tris-HCl pH 8.0, 250 mM NaCl, 500 mM imidazole, 0.5 mM TCEP) over 15 column volumes. After a size exclusion chromatography step on a HiPrep 16/60 Sephacryl S-200 HR (GE Healthcare), Rel$_{Seq}$ constructs were stored at −80 °C in 50 mM Tris-HCl pH 8, 200 mM NaCl, 5% glycerol.

Thermal shift assays on selected fragments vs. Rel$_{Seq}$ constructs. The binding of the selected fragments to Rel$_{Seq}$ constructs was assessed by titration of the protein in thermal shift assays, performed using a Step One Real-Time PCR system (Thermo Fisher Scientific, Waltham, MA, USA). Assays were performed in 20 mM Tris-HCl, pH 8, containing 150 mM NaCl. The final protein concentration was kept at 0.5 mg/mL for all Rel$_{Seq}$ constructs, and the fluorescent Protein Thermal Shift Dye (Thermo Fisher Scientific, Waltham, MA, USA) was used to monitor protein unfolding within the excitation/emission ranges 470–505/540–700 nm. Each fragment was dissolved in DMSO at a stock concentration of 200 mM, and two-fold dilution series were prepared to have final compound concentrations ranging from 0.3 mM to 10 mM; 2.5% DMSO was added in place of the fragments for control samples. Assays were performed in triplicate at a final volume of 15 µL in MicroAmp™ Fast Optical 48-well reaction plates (Thermo Fisher Scientific, Waltham, MA, USA) sealed with adhesive films. Plates were heated from 25 to 90 °C with a heating rate of 0.5 °C/min. The K_d of protein–ligand complexes engaged by Rel$_{Seq}$ constructs with the tested fragments was calculated from the plot of protein melting temperature variations as a function of fragment concentrations with the equation for Ligand Binding (1 site) provided with the software GraFit 5.0 (Erithacus Software, Staines, UK).

STD-NMR experiments on selected fragments vs. Rel$_{Seq}$ constructs. Experiments were performed on a 600 MHz Bruker Avance spectrometer. All experiments were acquired at 298 and 283 K on the free ligands in deuterated phosphate buffer pH 7.4. A DMSO-d6 percentage of about 5% was added to dissolve the fragments. In 1D spectra, water suppression was achieved by excitation sculpting sequence. STD NMR experiments were performed using WATERGATE 3-9-19 pulse sequence for water suppression. On-resonance irradiation of the protein was performed at a chemical shift of −0.05 ppm and 10 ppm; off-resonance irradiation was applied at 200 ppm, where no protein signals were visible. Selective pre-saturation of the protein was achieved by a train of Gauss-shaped pulses of 49 ms in length each. STD spectra were acquired with a saturation time of 2.94 s for all compounds. Blank experiments were conducted in the absence of protein in order to avoid artifacts. We tested several protein/fragment ratios (i.e., 1:100, 1:200, and 1:1000) and found that a 1:1000 ratio with a protein concentration of 3 µM (500 µL final volume) afforded the best signal-to-noise ratio.

4. Conclusions

In conclusion, we targeted the intracellular accumulation of (p)ppGpp, a bacterial stringent-response-signaling molecule involved in the insurgence of persisters, a form of phenotypic AMR, and in bacterial virulence. New chemical entities with antimicrobial activity targeting non-essential pathways, such as (p)ppGpp signaling, are urgently needed to fight and prevent antimicrobial resistance.

We performed an extensive structure-based *in silico* fragment screening on the synthetase site of the bifunctional enzyme Rel$_{Seq}$, selecting three main chemotypes. Protein–fragment interaction experiments evidenced several low mM affinity binders. In particular, the aminobenzoic acid scaffold showed a marked synthetase domain selectivity and was therefore selected for the rational design of enzyme inhibitors that will be described in due course. Potent and selective Rel inhibitors will enable to shed light on the role of (p)ppGpp signaling in persisters' formation and pave the way to low-selective-pressure antimicrobial therapeutic approaches.

Supplementary Materials: The following are available online at https://www.mdpi.com/article/10.3390/molecules27103097/s1, Figure S1: Superposition of X-ray vs. Hip312 best pose GDP, Figure S2: Superposition of X-ray vs. His312 best pose GDP, Figure S3: Best poses of the three singletons selected from the His312 grid, Figure S4: Tyr308-fragment centroids distances monitored during MD simulations. Figure S5: STD-NMR of **B1** with Rel$_{Seq}$ (1–385), Figure S6: STD-NMR of **A2** with Rel$_{Seq}$ (1–385), Table S1: Databases used for the virtual screening campaign, Table S2: 3D Datasets used in the VS workflow and enrichment factors calculated for each chemotype, Table S3: Fragment datasets obtained by using PubChem database, Table S4: Docking results for the fragment sets into HIP312 and HIS312 models.

Author Contributions: Conceptualization, S.S.; methodology, S.S., L.S. and M.C.; validation, S.S., M.C. and L.S.; investigation, C.C., L.S., M.M. and F.V.; writing—original draft preparation, S.S., C.C., M.C., L.S. and F.V.; writing—review and editing, S.S. and M.C.; visualization, C.C. and F.V.; supervision, S.S.; project administration, S.S. and M.C.; funding acquisition, S.S. All authors have read and agreed to the published version of the manuscript.

Funding: This project received funding from the European Research Council (ERC) under the European Union's Horizon 2020 Research and Innovation Programme (ERACHRON project, grant agreement no. 758108). The APC was funded by the ERC (grant no. 758108).

Institutional Review Board Statement: Not applicable.

Informed Consent Statement: Not applicable.

Data Availability Statement: The data presented in this study are available upon request to the corresponding author.

Acknowledgments: The authors would like to thank Stefania Cimbari for the administrative support.

Conflicts of Interest: The authors declare no conflict of interest. The funders had no role in the design of the study; in the collection, analyses, or interpretation of data; in the writing of the manuscript, or in the decision to publish the results.

References

1. Fleming, A. On the Antibacterial Action of Cultures of a Penicillium, with Special Reference to their Use in the Isolation of B. influenzæ. *Br. J. Exp. Pathol.* **1929**, *10*, 226–236. [CrossRef]
2. O'Neill, J. Tackling Drug-Resistant Infections Globally: Final Report and Recommendations in the Review on Antimicrobial Resistance. 2016, pp. 1–81. Available online: https://apo.org.au/node/63983 (accessed on 19 April 2022).
3. Theuretzbacher, U.; Gottwalt, S.; Beyer, P.; Butler, M.; Czaplewski, L.; Lienhardt, C.; Moja, L.; Paul, M.; Paulin, S.; Rex, J.H.; et al. Analysis of the clinical antibacterial and antituberculosis pipeline. *Lancet Infect. Dis.* **2019**, *19*, e40–e50. [CrossRef]
4. Schrader, S.M.; Vaubourgeix, J.; Nathan, C. Biology of antimicrobial resistance and approaches to combat it. *Sci. Transl. Med.* **2020**, *12*, eaaz6992. [CrossRef] [PubMed]
5. Cashel, M.; Gallant, J. Two Compounds implicated in the Function of the RC Gene of Escherichia coli. *Nature* **1969**, *221*, 838–841. [CrossRef]
6. Pulschen, A.A.; Fernandes, A.Z.N.; Cunha, A.F.; Sastre, D.E.; Matsuguma, B.E.; Gueiros-Filho, F.J. Many birds with one stone: Targeting the (p)ppGpp signaling pathway of bacteria to improve antimicrobial therapy. *Biophys. Rev.* **2021**, *13*, 1039–1051. [CrossRef]
7. Kamarthapu, V.; Epshtein, V.; Benjamin, B.; Proshkin, S.; Mironov, A.; Cashel, M.; Nudler, E. ppGpp couples transcription to DNA repair in E. coli. *Science* **2016**, *352*, 993–996. [CrossRef]
8. Molodtsov, V.; Sineva, E.; Zhang, L.; Huang, X.; Cashel, M.; Ades, S.E.; Murakami, K.S. Allosteric Effector ppGpp Potentiates the Inhibition of Transcript Initiation by DksA. *Mol. Cell* **2018**, *69*, 828–839.e5. [CrossRef]
9. Diez, S.; Ryu, J.; Caban, K.; Gonzalez, R.L., Jr.; Dworkin, J. The alarmones (p)ppGpp directly regulate translation initiation during entry into quiescence. *Proc. Natl. Acad. Sci. USA* **2020**, *117*, 15565–15572. [CrossRef]
10. Steinchen, W.; Zegarra, V.; Bange, G. (p)ppGpp: Magic Modulators of Bacterial Physiology and Metabolism. *Front. Microbiol.* **2020**, *11*, 2072. [CrossRef]

11. Irving, S.E.; Choudhury, N.R.; Corrigan, R.M. The stringent response and physiological roles of (pp)pGpp in bacteria. *Nat. Rev. Microbiol.* **2021**, *19*, 256–271. [CrossRef]
12. Buke, F.; Grilli, J.; Cosentino Lagomarsino, M.; Bokinsky, G.; Tans, S.J. ppGpp is a bacterial cell size regulator. *Curr. Biol.* **2022**, *32*, 870–877.e5. [CrossRef] [PubMed]
13. Kim, K.; Islam, M.; Jung, H.W.; Lim, D.; Kim, K.; Lee, S.G.; Park, C.; Lee, J.C.; Shin, M. ppGpp signaling plays a critical role in virulence of Acinetobacter baumannii. *Virulence* **2021**, *12*, 2122–2132. [CrossRef] [PubMed]
14. Kalia, D.; Merey, G.; Nakayama, S.; Zheng, Y.; Zhou, J.; Luo, Y.; Guo, M.; Roembke, B.T.; Sintim, H.O. Nucleotide, c-di-GMP, c-di-AMP, cGMP, cAMP, (p)ppGpp signaling in bacteria and implications in pathogenesis. *Chem. Soc. Rev.* **2013**, *42*, 305–341. [CrossRef] [PubMed]
15. Hauryliuk, V.; Atkinson, G.C.; Murakami, K.S.; Tenson, T.; Gerdes, K. Recent functional insights into the role of (p)ppGpp in bacterial physiology. *Nat. Rev. Microbiol.* **2015**, *13*, 298–309. [CrossRef]
16. Bigger, J.W. Treatment of staphylococcal infections with penicillin by intermittent sterilisation. *Lancet* **1944**, *244*, 497–500. [CrossRef]
17. Cohen, N.R.; Lobritz, M.A.; Collins, J.J. Microbial persistence and the road to drug resistance. *Cell Host Microbe* **2013**, *13*, 632–642. [CrossRef]
18. Huemer, M.; Mairpady Shambat, S.; Brugger, S.D.; Zinkernagel, A.S. Antibiotic resistance and persistence—Implications for human health and treatment perspectives. *EMBO Rep.* **2020**, *21*, e51034. [CrossRef]
19. Wilmaerts, D.; Windels, E.M.; Verstraeten, N.; Michiels, J. General Mechanisms Leading to Persister Formation and Awakening. *Trends Genet* **2019**, *35*, 401–411. [CrossRef]
20. Atkinson, G.C.; Tenson, T.; Hauryliuk, V. The RelA/SpoT homolog (RSH) superfamily: Distribution and functional evolution of ppGpp synthetases and hydrolases across the tree of life. *PLoS ONE* **2011**, *6*, e23479. [CrossRef]
21. Kaspy, I.; Glaser, G. Escherichia coli RelA Regulation via Its C-Terminal Domain. *Front. Microbiol.* **2020**, *11*, 572419. [CrossRef]
22. Winther, K.S.; Roghanian, M.; Gerdes, K. Activation of the Stringent Response by Loading of RelA-tRNA Complexes at the Ribosomal A-Site. *Mol. Cell.* **2018**, *70*, 95–105.e4. [CrossRef] [PubMed]
23. Takada, H.; Roghanian, M.; Caballero-Montes, J.; Van Nerom, K.; Jimmy, S.; Kudrin, P.; Trebini, F.; Murayama, R.; Akanuma, G.; Garcia-Pino, A.; et al. Ribosome association primes the stringent factor Rel for tRNA-dependent locking in the A-site and activation of (p)ppGpp synthesis. *Nucleic Acids Res.* **2021**, *49*, 444–457. [CrossRef] [PubMed]
24. Pausch, P.; Abdelshahid, M.; Steinchen, W.; Schafer, H.; Gratani, F.L.; Freibert, S.A.; Wolz, C.; Turgay, K.; Wilson, D.N.; Bange, G. Structural Basis for Regulation of the Opposing (p)ppGpp Synthetase and Hydrolase within the Stringent Response Orchestrator Rel. *Cell Rep.* **2020**, *32*, 108157. [CrossRef] [PubMed]
25. Hogg, T.; Mechold, U.; Malke, H.; Cashel, M.; Hilgenfeld, R. Conformational Antagonism between Opposing Active Sites in a Bifunctional RelA/SpoT Homolog Modulates (p)ppGpp Metabolism during the Stringent Response. *Cell* **2004**, *117*, 57–68. [CrossRef]
26. Avarbock, A.; Avarbock, D.; Teh, J.-S.; Buckstein, M.; Wang, Z.-m.; Rubin, H. Functional Regulation of the Opposing (p)ppGpp Synthetase/Hydrolase Activities of RelMtb from Mycobacterium tuberculosis †. *Biochemistry* **2005**, *44*, 9913–9923. [CrossRef] [PubMed]
27. Tamman, H.; Van Nerom, K.; Takada, H.; Vandenberk, N.; Scholl, D.; Polikanov, Y.; Hofkens, J.; Talavera, A.; Hauryliuk, V.; Hendrix, J.; et al. A nucleotide-switch mechanism mediates opposing catalytic activities of Rel enzymes. *Nat. Chem. Biol.* **2020**, *16*, 834–840. [CrossRef]
28. Sinha, A.K.; Winther, K.S. The RelA hydrolase domain acts as a molecular switch for (p)ppGpp synthesis. *Commun. Biol.* **2021**, *4*, 434. [CrossRef]
29. Wexselblatt, E.; Oppenheimer-Shaanan, Y.; Kaspy, I.; London, N.; Schueler-Furman, O.; Yavin, E.; Glaser, G.; Katzhendler, J.; Ben-Yehuda, S. Relacin, a novel antibacterial agent targeting the Stringent Response. *PLoS Pathog.* **2012**, *8*, e1002925. [CrossRef]
30. Wexselblatt, E.; Katzhendler, J.; Saleem-Batcha, R.; Hansen, G.; Hilgenfeld, R.; Glaser, G.; Vidavski, R.R. ppGpp analogues inhibit synthetase activity of Rel proteins from Gram-negative and Gram-positive bacteria. *Bioorg. Med. Chem.* **2010**, *18*, 4485–4497. [CrossRef]
31. Wexselblatt, E.; Kaspy, I.; Glaser, G.; Katzhendler, J.; Yavin, E. Design, synthesis and structure-activity relationship of novel Relacin analogs as inhibitors of Rel proteins. *Eur. J. Med. Chem.* **2013**, *70*, 497–504. [CrossRef]
32. Andresen, L.; Varik, V.; Tozawa, Y.; Jimmy, S.; Lindberg, S.; Tenson, T.; Hauryliuk, V. Auxotrophy-based High Throughput Screening assay for the identification of Bacillus subtilis stringent response inhibitors. *Sci. Rep.* **2016**, *6*, 35824. [CrossRef] [PubMed]
33. Beljantseva, J.; Kudrin, P.; Jimmy, S.; Ehn, M.; Pohl, R.; Varik, V.; Tozawa, Y.; Shingler, V.; Tenson, T.; Rejman, D.; et al. Molecular mutagenesis of ppGpp: Turning a RelA activator into an inhibitor. *Sci. Rep.* **2017**, *7*, 41839. [CrossRef] [PubMed]
34. Dutta, N.K.; Klinkenberg, L.G.; Vazquez, M.J.; Segura-Carro, D.; Colmenarejo, G.; Ramon, F.; Rodriguez-Miquel, B.; Mata-Cantero, L.; Porras-De Francisco, E.; Chuang, Y.M.; et al. Inhibiting the stringent response blocks Mycobacterium tuberculosis entry into quiescence and reduces persistence. *Sci. Adv.* **2019**, *5*, eaav2104. [CrossRef] [PubMed]
35. Conti, G.; Minneci, M.; Sattin, S. Optimised Synthesis of the Bacterial Magic Spot (p)ppGpp Chemosensor PyDPA. *ChemBioChem* **2019**, *20*, 1717–1721. [CrossRef]

36. Civera, M.; Sattin, S. Homology Model of a Catalytically Competent Bifunctional Rel Protein. *Front. Mol. Biosci.* **2021**, *8*, 628596. [CrossRef] [PubMed]
37. Friesner, R.A.; Banks, J.L.; Murphy, R.B.; Halgren, T.A.; Klicic, J.J.; Mainz, D.T.; Repasky, M.P.; Knoll, E.H.; Shelley, M.; Perry, J.K.; et al. Glide: A New Approach for Rapid, Accurate Docking and Scoring. 1. Method and Assessment of Docking Accuracy. *J. Med. Chem.* **2004**, *47*, 1739–1749. [CrossRef]
38. Erlanson, D.A.; Fesik, S.W.; Hubbard, R.E.; Jahnke, W.; Jhoti, H. Twenty years on: The impact of fragments on drug discovery. *Nat. Rev. Drug Discov.* **2016**, *15*, 605–619. [CrossRef]
39. Lamoree, B.; Hubbard, R.E. Using Fragment-Based Approaches to Discover New Antibiotics. *SLAS Discov.* **2018**, *23*, 495–510. [CrossRef]
40. Schrödinger. *Maestro, Schrödinger Release 2018-1*; Schrödinger, LLC: New York, NY, USA, 2018.
41. Shelley, J.C.; Cholleti, A.; Frye, L.L.; Greenwood, J.R.; Timlin, M.R.; Uchimaya, M. Epik: A software program for pK(a) prediction and protonation state generation for drug-like molecules. *J. Comput. Aided Mol. Des.* **2007**, *21*, 681–691. [CrossRef]
42. Duan, J.; Dixon, S.L.; Lowrie, J.F.; Sherman, W. Analysis and comparison of 2D fingerprints: Insights into database screening performance using eight fingerprint methods. *J. Mol. Graph. Model.* **2010**, *29*, 157–170. [CrossRef]
43. Sastry, M.; Lowrie, J.F.; Dixon, S.L.; Sherman, W. Large-Scale Systematic Analysis of 2D Fingerprint Methods and Parameters to Improve Virtual Screening Enrichments. *J. Chem. Inf. Model.* **2010**, *50*, 771–784. [CrossRef] [PubMed]
44. Baell, J.B.; Holloway, G.A. New substructure filters for removal of pan assay interference compounds (PAINS) from screening libraries and for their exclusion in bioassays. *J. Med. Chem.* **2010**, *53*, 2719–2740. [CrossRef] [PubMed]
45. Kim, S.; Chen, J.; Cheng, T.; Gindulyte, A.; He, J.; He, S.; Li, Q.; Shoemaker, B.A.; Thiessen, P.A.; Yu, B.; et al. PubChem in 2021: New data content and improved web interfaces. *Nucleic Acids Res.* **2020**, *49*, D1388–D1395. [CrossRef]
46. Bowers, K.J.; Chow, E.; Xu, H.; Dror, R.O.; Eastwood, M.P.; Gregersen, B.A.; Klepeis, J.L.; Kolossváry, I.; Moraes, M.A.; Sacerdoti, F.D.; et al. Scalable Algorithms for Molecular Dynamics Simulations on Commodity Clusters. In Proceedings of the ACM/IEEE Conference on Supercomputing (SC06), Tampa, FL, USA, 11–17 November 2006.
47. Jorgensen, W.L.; Chandrasekhar, J.; Madura, J.D.; Impey, R.W.; Klein, M.L. Comparison of simple potential functions for simulating liquid water. *J. Chem. Phys.* **1983**, *79*, 926–935. [CrossRef]
48. Roos, K.; Wu, C.; Damm, W.; Reboul, M.; Stevenson, J.M.; Lu, C.; Dahlgren, M.K.; Mondal, S.; Chen, W.; Wang, L.; et al. OPLS3e: Extending Force Field Coverage for Drug-Like Small Molecules. *J. Chem. Theory Comput.* **2019**, *15*, 1863–1874. [CrossRef] [PubMed]
49. Huynh, K.; Partch, C.L. Analysis of protein stability and ligand interactions by thermal shift assay. *Curr. Protoc. Protein. Sci.* **2015**, *79*, 28.9.1–28.9.14. [CrossRef]
50. Mechold, U.; Murphy, H.; Brown, L.; Cashel, M. Intramolecular Regulation of the Opposing (p)ppGpp Catalytic Activities of RelSeq, the Rel/Spo Enzyme from Streptococcus equisimilis. *J. Bacteriol.* **2002**, *184*, 2878–2888. [CrossRef]
51. Meyer, B.; Peters, T. NMR Spectroscopy Techniques for Screening and Identifying Ligand Binding to Protein Receptors. *Angew. Chem. Int. Ed.* **2003**, *42*, 864–890. [CrossRef]
52. Guzzetti, I.; Civera, M.; Vasile, F.; Araldi, E.M.; Belvisi, L.; Gennari, C.; Potenza, D.; Fanelli, R.; Piarulli, U. Determination of the binding epitope of RGD-peptidomimetics to αvβ3 and αIIbβ3 integrin-rich intact cells by NMR and computational studies. *Org. Biomol. Chem.* **2013**, *11*, 3886–3893. [CrossRef]
53. Sattin, S.; Panza, M.; Vasile, F.; Berni, F.; Goti, G.; Tao, J.H.; Moroni, E.; Agard, D.; Colombo, G.; Bernardi, A. Synthesis of Functionalized 2-(4-Hydroxyphenyl)-3-methylbenzofuran Allosteric Modulators of Hsp90 Activity. *Eur. J. Org. Chem.* **2016**, *2016*, 3349–3364. [CrossRef]
54. Jorgensen, W.L.; Maxwell, D.S.; Tirado-Rives, J. Development and Testing of the OPLS All-Atom Force Field on Conformational Energetics and Properties of Organic Liquids. *J. Am. Chem. Soc.* **1996**, *118*, 11225–11236. [CrossRef]
55. Jacobson, M.P.; Pincus, D.L.; Rapp, C.S.; Day, T.J.; Honig, B.; Shaw, D.E.; Friesner, R.A. A hierarchical approach to all-atom protein loop prediction. *Proteins* **2004**, *55*, 351–367. [CrossRef]
56. Li, J.; Abel, R.; Zhu, K.; Cao, Y.; Zhao, S.; Friesner, R.A. The VSGB 2.0 model: A next generation energy model for high resolution protein structure modeling. *Proteins* **2011**, *79*, 2794–2812. [CrossRef] [PubMed]
57. Ponder, J.W.; Richards, F.M. An efficient newton-like method for molecular mechanics energy minimization of large molecules. *J. Comput. Chem.* **1987**, *8*, 1016–1024. [CrossRef]
58. Grest, G.S.; Kremer, K. Molecular dynamics simulation for polymers in the presence of a heat bath. *Phys. Rev. A* **1986**, *33*, 3628–3631. [CrossRef]
59. Berendsen, H.J.C.; Postma, J.P.M.; Vangunsteren, W.F.; Dinola, A.; Haak, J.R. Molecular-Dynamics with Coupling to an External Bath. *J. Chem. Phys.* **1984**, *81*, 3684–3690. [CrossRef]

Article

Binding Interaction of Betulinic Acid to α-Glucosidase and Its Alleviation on Postprandial Hyperglycemia

Shaodan Chen [1,†], Bing Lin [2,†], Jiangyong Gu [3], Tianqiao Yong [1], Xiong Gao [1], Yizhen Xie [1], Chun Xiao [1], Janis Yaxian Zhan [2,*] and Qingping Wu [1,*]

1. Guangdong Provincial Key Laboratory of Microbial Safety and Health, Key Laboratory of Agricultural Microbiomics and Precision Application, Ministry of Agriculture and Rural Affairs, State Key Laboratory of Applied Microbiology Southern China, Institute of Microbiology, Guangdong Academy of Science, Guangzhou 510070, China; chensd@gdim.cn (S.C.); tianqiao@mail.ustc.edu.cn (T.Y.); gaoxiong@gdim.cn (X.G.); xieyz@gdim.cn (Y.X.); xiaochun960@hotmail.com (C.X.)
2. School of Pharmaceutical Science, Guangzhou University of Chinese Medicine, Guangzhou 510006, China; bing135388@163.com
3. Research Centre for Integrative Medicine, School of Basic Medical Science, Guangzhou University of Chinese Medicine, Guangzhou 510006, China; gujy@gzucm.edu.cn
* Correspondence: zyx@gzucm.edu.cn (J.Y.Z.); wuqp203@163.com (Q.W.)
† These authors contributed equally to this work.

Abstract: Inhibiting the intestinal α-glucosidase can effectively control postprandial hyperglycemia for type 2 diabetes mellitus (T2DM) treatment. In the present study, we reported the binding interaction of betulinic acid (BA), a pentacyclic triterpene widely distributed in nature, on α-glucosidase and its alleviation on postprandial hyperglycemia. BA was verified to exhibit a strong inhibitory effect against α-glucosidase with an IC_{50} value of 16.83 ± 1.16 μM. More importantly, it showed a synergistically inhibitory effect with acarbose. The underlying inhibitory mechanism was investigated by kinetics analysis, surface plasmon resonance (SPR) detection, molecular docking, molecular dynamics (MD) simulation and binding free energy calculation. BA showed a non-competitive inhibition on α-glucosidase. SPR revealed that it had a strong and fast affinity to α-glucosidase with an equilibrium dissociation constant (K_D) value of 5.529×10^{-5} M and a slow dissociation. Molecular docking and MD simulation revealed that BA bound to the active site of α-glucosidase mainly due to the van der Waals force and hydrogen bond, and then changed the micro-environment and secondary structure of α-glucosidase. Free energy decomposition indicated amino acid residues such as PHE155, PHE175, HIE277, PHE298, GLU302, TRY311 and ASP347 of α-glucosidase at the binding pocket had strong interactions with BA, while LYS153, ARG210, ARG310, ARG354 and ARG437 showed a negative contribution to binding affinity between BA and α-glucosidase. Significantly, oral administration of BA alleviated the postprandial blood glucose fluctuations in mice. This work may provide new insights into the utilization of BA as a functional food and natural medicine for the control of postprandial hyperglycemia.

Keywords: betulinic acid; α-glucosidase; inhibition mechanism; postprandial hyperglycemia; synergistic effect

1. Introduction

Type 2 diabetes mellitus (T2DM) has become one of the major chronic diseases. According to International Diabetes Federation (IDF) 2019 estimation, around 463.0 million adults are suffering from T2DM, and this number will increase to about 578.4 million by 2030 and further around 700.2 million by 2045 [1]. Specially, T2DM is major featured with hyperglycemia, which leads to the occurrence of serious diabetes complications, such as vascular damage and cardiovascular disease [2,3]. Epidemiological studies suggested that postprandial hyperglycemia might be an independent risk factor of cardiovascular disease beyond and more powerful than fasting hyperglycemia for T2DM sufferers [4]. Thus, controlling postprandial hyperglycemia became an effective therapy in T2DM treatment and diabetes complications prevention.

Carbohydrate, such as starch and dextrin in foods, is the main source of human blood sugar. The α-glucosidase presented in the gastrointestinal tract can catalyze the cleavage of α-1,4 glycosidic bonds of carbohydrate to form glucose, which is absorbed by small intestine and then enters into blood circulation. Inhibiting the activity of intestinal α-glucosidase can effectively control postprandial hyperglycemia [5–7]. Acarbose, an α-glucosidase inhibitor, is effective in managing postprandial blood glucose; however, the side-effects have restricted its long-term applications [8]. Hence, searching for more effective and safer α-glucosidase inhibitors from natural sources is greatly demanded.

Betulinic acid (BA, Figure 1A) is a triterpene widely presented in medicinal plants such as *Betula* (Betulaceae), *Ziziphus* (Rhamnaceae), *Syzygium* (Myrtaceae) and Chaga mushroom [9,10]. BA can be easily isolated from the source with different solvents or be synthesized by oxidation from betulin, a precursor with a yield of 20% dry weight in *Betula* species [9,11]. BA had a wide variety of bioactive effects including antitumor, anti-inflammatory, and hepatoprotective activities [12–14]. The regulation of BA in metabolic syndrome has gained more and more research in the past ten years and BA has been recorded as a potential antidiabetic agent [10,15–22]. Several studies have reported its regulatory effects on glucose absorption and uptake [23,24], insulin resistance [25], glucose production and glycogen synthesis [17,26–28]. BA was also found to inhibit α-glucosidase activity in vitro [7,29–31] and its inhibition mechanism was investigated by spectroscopic analysis and molecular docking [7]; however, the binding mechanism as well as the alleviated effect on postprandial hyperglycemia was still unclear.

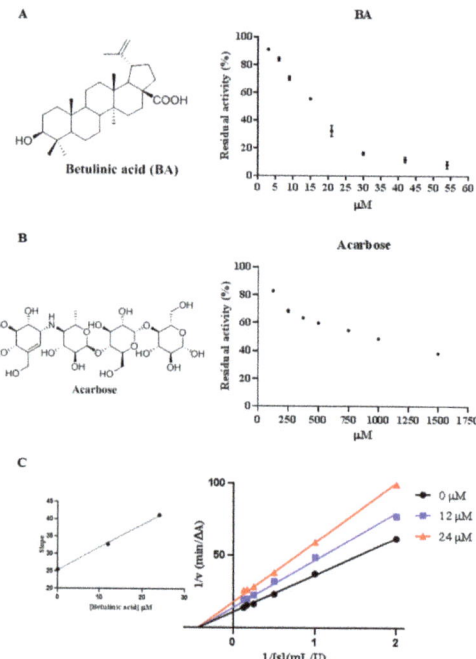

Figure 1. (**A**) Chemical structure of betulinic acid (**BA**) and its inhibitory effect of BA against α-glucosidase; (**B**) Chemical structure of acarbose and its inhibitory effect of acarbose against α-glucosidase; (**C**) Line-weaver-Burk diagrams (1/V vs. 1/[S]) for BA against α-glucosidase. The secondary plot of slope vs. BA was included.

In the present study, we firstly verified the inhibitory effect of BA and investigated its synergy with acarbose on α-glucosidase. The interaction mechanism was further explained by SPR analysis and molecular dynamics (MD) simulation in addition to molecular docking. Furthermore, the modulating effect of oral administration of BA on the postprandial blood glucose level was detected in vivo. This work provides novel insights into the use of BA as a functional food or a natural medicine for the control of hyperglycemia.

2. Results and Discussion

2.1. Inhibitory Effect of BA against α-Glucosidase

As shown in Figure 1A,B, BA and acarbose inhibited α-glucosidase activity in a dose-dependent manner. Compared to acarbose, BA displayed superior inhibition against α-glucosidase in very low concentrations, with IC_{50} value of 16.83 ± 1.16 μM, which was approximately 50-fold stronger than acarbose (841.3 ± 29.6 μM). Previous research had reported BA inhibited α-glucosidase in vitro with IC_{50} value from 10.6 to 83.6 μM, which was in accordance with our result [7,29,30]. Combined with previous papers [7,29,30,32–34], it was found that the different substitutions at C-3 and C-28 positions of BA could have influence on the inhibition activity. Additional polar interactions formed at C-3 and C-28 of BA would improve the inhibitory potency.

2.2. Inhibition of Acarbose Combined with BA against α-Glucosidase

Acarbose, a drug based on carbohydrate-related structures, is effective in contrasting postprandial hyperglycemia. However, it causes adverse gastrointestinal effects when administered in accompany with a high-carbohydrate diet [35]. It was found that the combination of two or more inhibitors could act synergistically, having an inhibitory effect against α-glucosidase; thus, the interaction of acarbose combined with BA on α-glucosidase inhibition was investigated. Based on the results summarized in Table 1, the combination of acarbose with BA showed a synergistic effect on inhibiting α-glucosidase activity as the values of $V_{ab} - V_c$ were below -0.1 at different concentrations. Acarbose was a competitive inhibitor, while BA was a non-competitive one; the two inhibitors may bind to the different sites of the active pocket of α-glucosidase. The presence of BA may enhance the binding affinity between α-glucosidase and acarbose, and lead to a strong joint inhibition. The result was similar to the combined interaction of corosolic acid or oleanolic acid (another two triterpenoids) and acarbose on α-amylase, and the combination of apigenin and acarbose on α-glucosidase [36,37]. Taking advantage of this synergistic effect, it would be possible to reduce the side effects of acarbose at a lower dose by combining with BA. A reduction in side effects and toxicity is considered as one of the rationales of drugs combinations [38]. BA exerted global antidiabetic effects mediated by specific mechanisms of action with low toxicity in addition to inhibiting intestinal α-glucosidase [21], while acarbose mainly acted on the gastrointestinal tract [39]. Combining compounds with different signaling pathways and non-overlapping side effects, BA and acarbose may allow for a more significant total efficacy with possibly fewer side effects, which resembles the "cocktail therapy". Moreover, the poor water solubility and low bioavailability of BA limits its clinical application [40]. It is possible that modification of BA by introduction of acarbose moiety to the parent molecule would increase the efficacy as well as the water solubility and bioavailability. It would be worthwhile to investigate the physiological mechanisms as well as bioavailability, pharmacokinetics, pharmacodynamics and toxicity of the combination of BA and acarbose in detail.

Table 1. Synergistic effect of acarbose with BA on the inhibitory activity of α-glucosidase*.

	Acarbose-600 μM			Acarbose-800 μM			Acarbose-1000 μM		
	Value		Interaction	Value		Interaction	Value		Interaction
	V_{ab}	V_c	$V_{ab} - V_c$	V_{ab}	V_c	$V_{ab} - V_c$	V_{ab}	V_c	$V_{ab} - V_c$
BA-5 μM	0.47	0.59	−0.12SY	0.42	0.56	−0.15SY	0.41	0.53	−0.13SY
BA-10 μM	0.39	0.57	−0.18SY	0.36	0.54	−0.18SY	0.34	0.51	−0.17SY
BA-15 μM	0.32	0.50	−0.18SY	0.29	0.48	−0.18SY	0.28	0.45	−0.16SY

V_{ab} and V_c are, respectively, defined as the observed and expected residual activity treated by an acarbose–BA mixture. SY means synergistic interaction.

2.3. Inhibition Types of BA on α-Glucosidase

The Lineweaver-Burk plot was employed to analyze the inhibition type of BA against α-glucosidase. As shown in Figure 1C, the horizontal axis intercept ($-1/K_m$) remained constant while the vertical axis intercept ($1/V_{max}$) ascended with the increasing concentration of BA, demonstrating that the inhibition type of BA was non-competitive, which was the same as another two triterpenes ursolic acid and oleanolic acid, but was different from acarbose [2]. However, Ding et al. had reported a different result on inhibition type of α-glucosidase by BA [7]. The difference may be attributed to the different source (i.e., yeast species) of α-glucosidase. Moreover, it had been reported that non-competitive inhibition is sometimes considered as a special case of mixed-type inhibition when the competitive constant and uncompetitive constant are exactly the same [41]. The Michaelis constant (K_m) was 2.272 and V_{max} was 0.0907 μM/min. Furthermore, the α-glucosidase inhibition kinetic constant (K_i) was calculated as 38.5 μM according to the secondary plot.

2.4. Surface Plasmon Resonance (SPR) Analysis

SPR biosensor is a powerful tool to analyze the biomolecular binding interactions, as it can measure the kinetics and affinity of bimolecular binding in a real-time and label-free fashion with low reagent consumption [42]. As shown in Figure 2, BA effectively bound to α-glucosidase in a concentration-dependent manner. The binding time and the dissociation time were 120 s and 120 s, respectively. The equilibrium dissociation constant (K_D) was 5.529×10^{-5} M, manifesting that BA had a high binding affinity to α-glucosidase and underwent a fast binding and slow dissociation reaction.

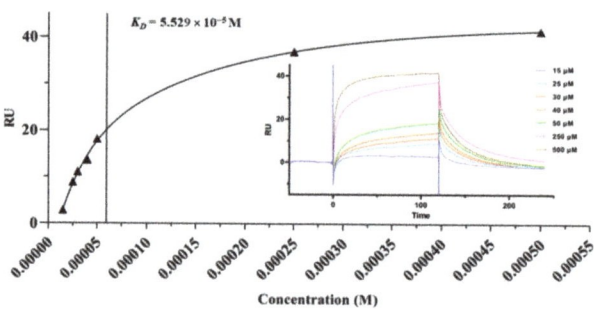

Figure 2. Binding kinetics and affinity of α-glucosidase to BA by SPR assay. Sensorgram showing the binding of BA at different concentrations to α-glucosidase was included.

2.5. Molecular Docking

Molecular docking is a widely used structure-based drug design technique. It can predict the conformation of a ligand in the target binding site and calculate the binding energy [43]. The binding interaction of BA to α-glucosidase was analyzed through molecular docking. BA sat at the active pocket of α-glucosidase with high overlay, mainly due to van de Walls interaction (Figure 3). Moreover, BA formed an H-bond to Arg310 of α-glucosidase, which could further stabilize the orientation of the interaction. Thus, molecular docking demonstrated the binding interaction of BA to the hydrophobic cavity of α-glucosidase and the formation of hydrogen bonds together changed the micro-environment and structure of α-glucosidase, and led to a decrease in enzyme activity, which was in accordance with the results reported previously [7].

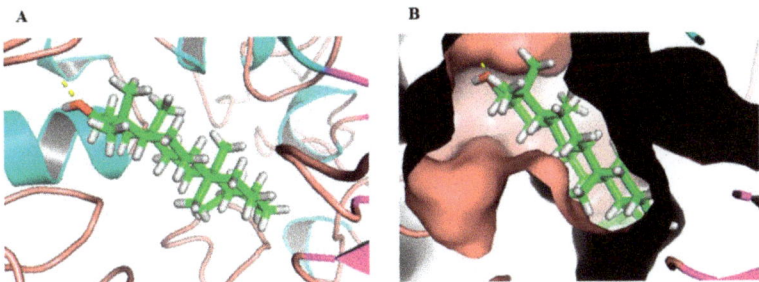

Figure 3. Molecular docking of BA with α-glucosidase. (**A**) Three-dimensional diagram of the interaction between BA and the binding pocket of α-glucosidase. (**B**) The receptor surface model of α-glucosidase with BA. BA was inserted into the hydrophobic cavity of α-glucosidase (light blue) in the surface structure; the yellow dashed line represents the hydrogen bond. The atoms of BA were color-coded as follows: C, green; H, white; O, red.

2.6. Molecular Dynamics (MD) Simulation and Calculation of Binding Free Energy

Molecular dynamics (MD) simulation is a prominent computational dynamics tool to analyze the molecular mechanism of an enzyme–ligand interaction. MD simulation can get the atomic trajectory at spatial and temporal scales, and provide detailed information about the conformational changes and fluctuations in protein [44,45].

The docked conformation based on molecular docking was used as initial conformation of MD simulation. The system was performed for 100 ns MD. The structural stability of enzyme-product complex was evaluated by calculation of the root mean square deviation (RMSD). RMSD measures the deviation of a set of coordinates of a protein to a reference set of coordinates. The RMSD for the protein backbones of α-glucosidase in complex with BA was shown in Figure 4A. Temporal RMSD was fluctuant initially and stable after 60 ns. RMSD for protein backbones converged to equilibrium during the last 5 ns. The average binding free energy for the system was calculated as −18.53 ± 3.44 kcal/mol based on the MD simulation by MM/GBSA method. For BA, the residues PHE175, ASP347, ARG310, PHE298, GLU302 and GLY278 presented van der Waals interactions to α-glucosidase. Meanwhile, HIE277, HIE237 and PHE155 contributed to Pi-alkyl interactions (Figure 4B–D). In order to further elucidate key amino acid residues, which had more contribution to the binding free energy, per-residue decomposition was used to generate the residue-product interaction spectra (Figure 4E). An amino acid residue may have a positive or negative contribution. The more negative the decomposed binding free energy of an amino acid residue means the more contributions to the binding affinity. Conversely, the more positive decomposed binding free energy of an amino acid residue, the stronger repulsive effect it will have. As shown in Figure 4E, conserved amino acid residues PHE155, PHE175, HIE277, PHE298, GLU302, TRY311 and ASP347 of α-glucosidase had strong interactions with the corresponding product in the complex, indicating that these amino acid residues would play important roles in the catalytic process. Consistent with the binding free energy analysis, most of the amino acid residues were hydrophobic, implying a hydrophobic–hydrophobic interaction. However, several amino acid residues, such as LYS153, ARG210, ARG310, ARG354 and ARG437, had repulsive effects on the binding of BA with α-glucosidase.

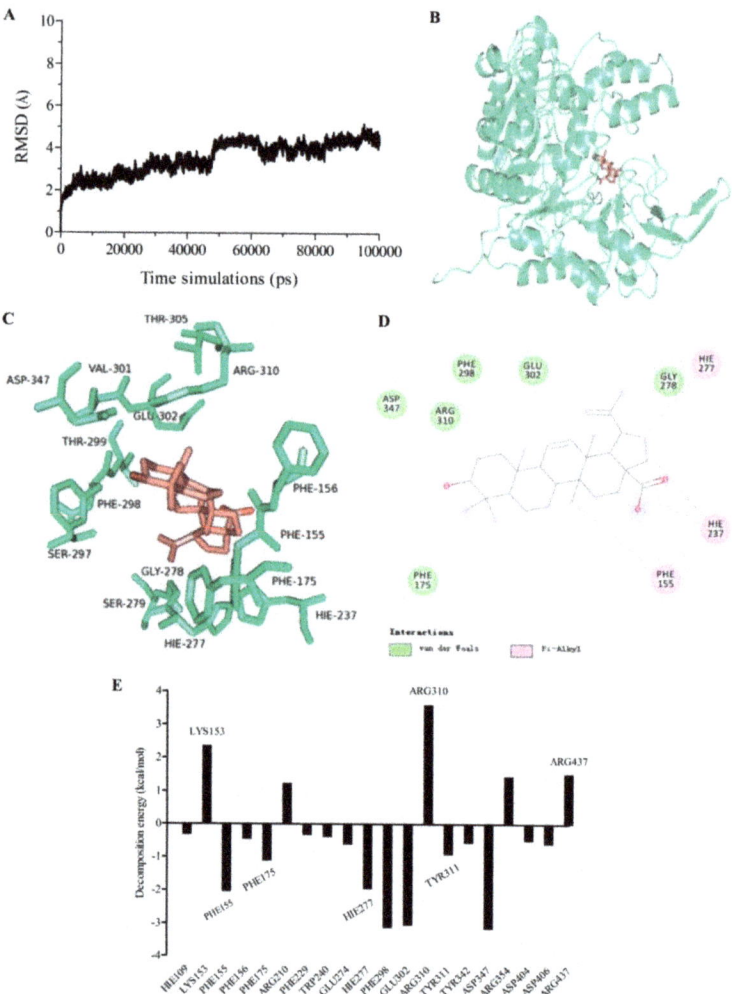

Figure 4. Molecular dynamics simulation. (**A**) Time dependence of root mean square deviation (RMSD) of the backbone of the protein in the complex; (**B**) Conformation of the binding of BA to α-glucosidase; (**C**) Interactions of BA to key residues of α-glucosidase; (**D**) 2D scheme of interactions of BA to key residues of α-glucosidase. Green and pink AA residues represent van der Waals interaction and Pi-alkyl interaction, respectively. (**E**) Decomposition of binding free energy of each residue for α-glucosidase.

2.7. Effects of BA on Oral Saccharides Tolerance in Normal Mice

Postprandial hyperglycemia is the most important symptom of T2DM. BA had high binding affinity to α-glucosidase and effectively inhibited its activity, suggesting BA may influence the postprandial blood glucose level through inhibiting α-glucosidase activity. Accordingly, an oral disaccharide tolerance test (ODTT) and an oral glucose tolerance test (OGTT) were carried out in mice, respectively. As shown in Figure 5, the blood glucose level reached the highest point at 30 min after administration of maltose-sucrose or glucose and dropped sharply in the next 30 min. Compared to the control group, the postprandial blood glucose levels in BA-treated group and acarbose-treated group after oral administration of mixed maltose-sucrose were significantly suppressed ($p < 0.01$) (Figure 5A). In detail, the blood glucose levels at 30 min in BA-treated group and acarbose-treated group were decreased from 12.58 ± 0.47 to 10.46 ± 0.31 mM ($p < 0.01$) and to 10.76 ± 0.39 mM ($p < 0.01$), respectively, and then gradually decreased to the initial level. However, the BA-treated group (20 mg/kg) and acarbose-treated group (7 mg/kg) had no significant effect on blood glucose level in mice after oral glucose administration (Figure 5B). The results suggested that BA may slow down the hydrolysis rate of maltose and sucrose into glucose by inhibiting the activity of α-glucosidase, thereby delaying the absorption of intestinal glucose and effectively controlling the acute postprandial blood glucose fluctuation.

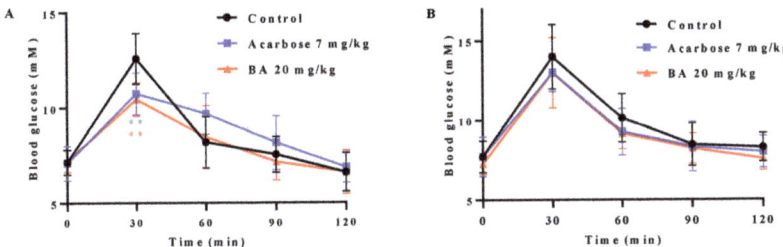

Figure 5. Effects of BA and acarbose on postprandial blood glucose levels in normal mice ($n = 8$). (**A**) Blood glucose after oral administration tests of mixed maltose and sucrose (each 1.0 g/kg); (**B**) Blood glucose after oral administration tests of glucose (2.0 g/kg). Data are expressed as means ± SD; ** $p < 0.01$ indicates a significant difference compared to control by t-test.

3. Materials and Methods

3.1. Materials

α-Glucosidase from yeast was purchased from Sigma-Aldrich (St. Louis, MO, USA). Acarbose (≥98%) and betulinic acid (≥98%) were obtained from Shanghai Yuanye Bio-Technology Co., Ltd. (Shanghai, China). p-Nitrophenyl-α-D-glucopyranoside (pNPG) were offered by Shanghai Aladdin Bio-Chem Technology Co., Ltd. (Shanghai, China). CM5 sensor chip, EDC/NHS and Amine Coupling Kit were purchased from GE Healthcare (Buckinghamshire, UK).

3.2. Inhibition Assay against α-Glucosidase

α-Glucosidase inhibition was assayed according to the methods described previously [46,47]. Briefly, 40 μL phosphate buffer (0.1 M, pH = 6.8), 10 μL BA solutions and 50 μL α-glucosidase solution (0.4 U/mL) were added into 96-well plates and mixed thoroughly. After being incubated at 37 °C for 10 min, 50 μL pNPG solution (2.5 mM) was added to initiate the reaction and stayed at 37 °C for 30 min. Then, 40 μL Na$_2$CO$_3$ solution (1.0 M) was added to terminate the reaction. A microplate reader was employed to read the absorbance at 405 nm. The inhibition rates (%) = [(OD$_{control}$ − OD$_{control\ blank}$) − (OD$_{test}$ − OD$_{test\ blank}$)]/(OD$_{control}$ − OD$_{control\ blank}$) × 100%.

3.3. Combination between BA and Acarbose

The interaction assay of acarbose with BA on α-glucosidase activity was performed as previously reported with minor modification [37,48]. The method was the same as the α-glucosidase activity assay. The OD value of the α-glucosidase reaction in the presence of acarbose divided by that obtained in its absence was named V_a, which represents the remnant activity fraction of α-glucosidase after addition of acarbose. The value for BA (V_b) was also defined. If the inhibitory effects on α-glucosidase of the two compounds are independent, the final remnant activity fraction of α-glucosidase (V_c) is equal to $V_a V_b$ when the reaction is sequentially treated by those two compounds. The result of the OD value of the α-glucosidase reaction in the presence of both acarbose and BA divided by that obtained in their absence was named V_{ab}. The types of interaction were determined by the relationship between V_{ab} and V_a, V_b, or V_c. $V_{ab} − V_c < −0.1$ is considered as the synergistic (SY) interaction between BA and acarbose, $-0.1 < V_{ab} - V_c < 0.1$ means additive (AD) and $V_{ab} - V_c > 0.1$ is regarded as subadditive (SU).

3.4. Enzyme Inhibition Kinetics

α-Glucosidase inhibition kinetics was studied as previously reported [47]. BA was set with the concentrations of 0, 12 and 24 µM for 0.2 U/mL α-glucosidase. A series of concentrations of *p*NPG were used as substrates. The inhibition type was determined by using the Lineweaver–Burk plots and its secondary equations. The kinetic parameters (V_{max} and K_i) were evaluated by the nonlinear regression of the Michaelis–Menton equation with GraphPad Prism 7.0. (https://www.graphpad.com/support/prism-7-updates/, accessed on 12 May 2021).

3.5. SPR Assay

SPR analysis was applied on a Biacore T200 system (GE Healthcare, Uppsala, Sweden). α-Glucosidase solution (50 µg/mL) was prepared and stayed at room temperature for 30 min. Then, the activated α-glucosidase was immobilized on a CM5 sensor chip pre-activated by EDC/NHS with an amine coupling kit at a flow rate of 10 µL/min. BA solution of different concentrations containing 5% DMSO was serially injected over α-glucosidase coated surface. PBS was used as the mobile phase with a flow rate of 30 µL/min. According to the 1:1 Langmuir model, the equilibrium dissociation constant (K_D) was calculated by the Biacore evaluation software. (https://www.cytivalifesciences.com/en/us/shop/protein-analysis/spr-label-free-analysis/software/biacore-insight-evaluation-software-p-23528, accessed on 26 October 2016).

3.6. Molecular Docking

CDOCKER was used to conduct the docking. Briefly, BA was minimized energetically with MMFF94 force field for 5000 iterations until the minimum RMS gradient was below 0.01. The structure of α-glucosidase was modeled as per our previous report [47]. Then, the molecular docking was performed in a graphical interface of CDOCKER in Discovery Studio. CDOCKER_Energy was used to evaluate the docking results.

3.7. Molecular Dynamic (MD) Simulation and Free Energy Calculation and Decomposition

MD simulation was performed with Amber 16. The structure of α-glucosidase docked with BA was utilized as the initial structure. Gaussian 16 was used to optimize the geometries and calculate the electrostatic potentials of compound at the DFT/6-31G* level. Antechamber module in Amber 16 was exploited to generate the topology files of the ligand with the charge distribution calculated at DFT/6-31G* level. The complex was added with the AMBER ff14SB force field and dissolved with TIP3P water at the size of 15Å. The system was neutralized with 0.15 M NaCl. Then, the system was minimized by the sequence of solvent molecules, side chains of proteins and then backbone. The temperature was then increased gradually to 300 K. Dynamics simulation production was performed for 100 ns and the last 5 ns were extracted for analysis. The Molecular Mechanics/Generalized Born Surface Area (MM/GBSA) method was employed to calculate the binding free energy. The MM/GBSA method includes the calculation of the van der Waals interaction energy, the electrostatic energy, the non-polar solvation free energy and the polar solvation. The contribution of binding energy of each residue was further decomposed into four parts: van der Waals energy, electrostatic energy, non-polar solvation energy and polar solvation energy. The key residues which are responsible for the binding affinity of the complex can be discovered by this method [43,45].

3.8. Oral Disaccharide Tolerance Test (ODTT) and Oral Glucose Tolerance Test (OGTT)

The dynamic characteristics of postprandial blood glucose level were evaluated by ODTT and OGTT, which were performed as described previously [49]. BALB/c mice (5 weeks old, male) were purchased from Guangdong Medical Laboratory Animal Center

(Guangzhou, China). The animal study was approved by the Committee on the Ethics of Animal Experiments of the Institute of Microbiology, Guangdong Academy of Science (GT-IACUC202012102). After one week of acclimation, the mice were randomly divided into a control group, BA-treated group and acarbose-treated group (n = 8), respectively. All the mice were fasted for 6 h, and then, the BA-treated and acarbose-treated groups were orally administered with BA (20 mg/kg) and acarbose (7 mg/kg), respectively, while the control group was given the same volume of distilled water. After 30 min, all the mice were intragastrically administered with the mixed maltose-sucrose solution (2 g/kg). In total, 10 μL of blood was collected by tail cut and glucose levels at 0, 30, 60, 90 and 120 min was measured by glucometer (Roche ACCU-CHEK Performa Nano), respectively. Two days later, all the mice were give the oral glucose tolerance test (OGTT), which was performed as the ODTT modus. The oral dose of glucose was also 2 g/kg.

3.9. Statistical Analysis

Data were performed by GraphPad Prism 7.0 software and the results were expressed as means ± SD. Comparison between groups was analyzed by Student's *t*-test. $p < 0.05$ was considered to be significant.

4. Conclusions

In summary, the strong inhibitory effect of BA against α-glucosidase with an IC$_{50}$ value of 16.83 ± 1.16 μM was verified and its non-competitive inhibition was investigated. Moreover, BA exerted a synergistically inhibitory effect combined with acarbose, which may allow reducing the side-effects of acarbose in T2DM treatment. More importantly, the underlying inhibitory mechanism of BA against α-glucosidase was explored by SPR analysis, molecular docking, molecular dynamics (MD) simulation and binding free energy calculation. SPR revealed that it had a strong and fast affinity to α-glucosidase with equilibrium dissociation constant (K_D) value of 5.529×10^{-5} M and a slow dissociation. Molecular docking revealed that BA bound to the active site of α-glucosidase mainly due to the van der Waals force and hydrogen bond, and then changed the micro-environment and secondary structure of α-glucosidase. MD simulation and MM/GBSA binding free energy calculations further strengthened the evidence. Free energy decomposition indicated conserved amino acid residues such as PHE155, PHE175, HIE277, PHE298, GLU302, TRY311 and ASP347 had strong interactions with BA and α-glucosidase, which were crucial for the interactions of BA and α-glucosidase. Chemical modification of key amino acids would be conducted to verify the credibility of the molecular docking and MD modeling results. Additionally, oral administration of BA significantly suppresses postprandial blood glucose levels in normal mice. These findings may provide implications to understand the inhibitory mechanism of BA on α-glucosidase, and contribute to the development of natural effective inhibitors for diabetes management in the future. Future work will focus on the role of BA in intestinal glucose absorption and glucose homeostasis.

Author Contributions: Conceptualization, S.C.; methodology, B.L., J.G. and T.Y.; software, B.L. and X.G.; validation, S.C. and B.L.; data curation, S.C.; writing—original draft, S.C. and B.L.; writing—review and editing, S.C. and T.Y.; supervision, C.X., Y.X., Y.Z. and Q.W.; funding acquisition, S.C., T.Y., Y.Z. and Q.W. All authors have read and agreed to the published version of the manuscript.

Funding: This research was funded by the National Natural Science Foundation of China (81803393, 31901696), Guangdong Basic and Applied Basic Research Foundation (2020A1515011337), Guangdong Province Science and Technology Project (2019A050520003), Science and Technology Planning Project of Guangzhou (202102021263), Guangdong Province Agriculture Research Project & Agricultural Technique Promotion Project (2021KJ103) and GDAS' Project of Science and Technology Development (2020GDASYL-20200104016).

Institutional Review Board Statement: The animal study was approved by the Committee on the Ethics of Animal Experiments of Institute of Microbiology, Guangdong Academy of Science (GT-IACUC202012102).

Informed Consent Statement: Not applicable.

Data Availability Statement: The data are available from the corresponding author on reasonable request.

Conflicts of Interest: The authors declare no conflict of interest.

Sample Availability: Not available.

References

1. Khursheed, R.; Singh, S.K.; Wadhwa, S.; Gulati, M.; Awasthi, A. Therapeutic potential of mushrooms in diabetes mellitus: Role of polysaccharides. *Int. J. Biol. Macromol.* **2020**, *164*, 1194–1205. [CrossRef] [PubMed]
2. Ding, H.; Hu, X.; Xu, X.; Zhang, G.; Gong, D. Inhibitory mechanism of two allosteric inhibitors, oleanolic acid and ursolic acid on α-glucosidase. *Int. J. Biol. Macromol.* **2018**, *107*, 1844–1855. [CrossRef] [PubMed]
3. Kato, A.; Hayashi, E.; Miyauchi, S.; Adachi, I.; Imahori, T.; Natori, Y.; Yoshimura, Y.; Nash, R.J.; Shimaoka, H.; Nakagome, I.; et al. α-1-C-butyl-1,4-dideoxy-1,4-imino-L-arabinitol as a second-generation iminosugar-based oral α-glucosidase inhibitor for improving postprandial hyperglycemia. *J. Med. Chem.* **2012**, *55*, 10347–10362. [CrossRef] [PubMed]
4. Esposito, K.; Giugliano, D.; Nappo, F.; Marfella, R.; Campanian Postprandial Hyperglycemia Study Group. Regression of carotid atherosclerosis by control of postprandial hyperglycemia in type 2 diabetes mellitus. *Circulation* **2004**, *110*, 214–219. [CrossRef] [PubMed]
5. Guo, L.; Yang, C.; Yang, R.; Zhao, W. Magnetically anchored antibody-coupled nanocomposite as α-Amylase inhibitor for long-time protection against glycemic variability. *Chem. Eng. J.* **2022**, *430*, 132984. [CrossRef]
6. Ma, Y.Y.; Zhao, D.G.; Zhang, R.Q.; He, X.; Li, B.Q.; Zhang, X.Z.; Wang, Z.J.; Zhang, K. Identification of bioactive compounds that contribute to the α-glucosidase inhibitory activity of rosemary. *Food Funct.* **2020**, *11*, 1692–1701. [CrossRef]
7. Ding, H.; Wu, X.; Pan, J.; Hu, X.; Gong, D.; Zhang, G. New insights into the inhibition mechanism of betulinic acid on α-glucosidase. *J. Agric. Food Chem.* **2018**, *66*, 7065–7075. [CrossRef]
8. Shah, M.A.; Khalil, R.; Ul-Haq, Z.; Panichayupakaranant, P. α-Glucosidase inhibitory effect of rhinacanthins-rich extract from *Rhinacanthus nasutus* leaf and synergistic effect in combination with acarbose. *J. Funct. Foods* **2017**, *36*, 325–331. [CrossRef]
9. Ali-Seyed, M.; Jantan, I.; Vijayaraghavan, K.; Bukhari, S.N. Betulinic Acid: Recent Advances in Chemical Modifications, Effective Delivery, and Molecular Mechanisms of a Promising Anticancer Therapy. *Chem. Biol. Drug Des.* **2016**, *87*, 517–536. [CrossRef]
10. Luis Rios, J.; Manez, S. New pharmacological opportunities for betulinic acid. *Planta Med.* **2018**, *84*, 8–19. [CrossRef]
11. Melnikova, N.; Burlova, I.; Kiseleva, T.; Klabukova, I.; Gulenova, M.; Kislitsin, C.A.; Vasin, V.; Tanaseichuk, B. A practical synthesis of betulonic acid using selective oxidation of betulin on aluminium solid support. *Molecules* **2012**, *17*, 11849–11863. [CrossRef] [PubMed]
12. Huang, L.; Zhu, L.; Ou, Z.; Ma, C.; Kong, L.; Huang, Y.; Chen, Y.; Zhao, H.; Wen, L.; Wu, J.; et al. Betulinic acid protects against renal damage by attenuation of oxidative stress and inflammation via Nrf2 signaling pathway in T-2 toxin-induced mice. *Int. Immunopharmacol.* **2021**, *101*, 108210. [CrossRef] [PubMed]
13. Harwansh, R.K.; Mukherjee, P.K.; Biswas, S. Nanoemulsion as a novel carrier system for improvement of betulinic acid oral bioavailability and hepatoprotective activity. *J. Mol. Liq.* **2017**, *237*, 361–371. [CrossRef]
14. Serain, A.F.; Morosi, L.; Ceruti, T.; Matteo, C.; Meroni, M.; Minatel, E.; Zucchetti, M.; Salvador, M.J. Betulinic acid and its spray dried microparticle formulation: In vitro PDT effect against ovarian carcinoma cell line and in vivo plasma and tumor disposition. *J. Photochem. Photobiol. B-Biol.* **2021**, *224*, 112328. [CrossRef] [PubMed]
15. Zeng, A.; Hua, H.; Liu, L.; Zhao, J. Betulinic acid induces apoptosis and inhibits metastasis of human colorectal cancer cells in vitro and in vivo. *Bioorg. Med. Chem.* **2019**, *27*, 2546–2552. [CrossRef] [PubMed]
16. Liao, L.; Liu, C.; Xie, X.; Zhou, J. Betulinic acid induces apoptosis and impairs migration and invasion in a mouse model of ovarian cancer. *J. Food Biochem.* **2020**, *44*, e13278. [CrossRef]
17. Kim, S.J.; Quan, H.Y.; Jeong, K.J.; Kim, D.Y.; Kim, G.W.; Jo, H.K.; Chung, S.H. Beneficial effect of betulinic acid on hyperglycemia via suppression of hepatic glucose production. *J. Agric. Food Chem.* **2014**, *62*, 434–442. [CrossRef]
18. Ou, Z.; Zhao, J.; Zhu, L.; Huang, L.; Ma, Y.; Ma, C.; Luo, C.; Zhu, Z.; Yuan, Z.; Wu, J.; et al. Anti-inflammatory effect and potential mechanism of betulinic acid on lambda-carrageenan-induced paw edema in mice. *Biomed. Pharmacother.* **2019**, *118*, 109347. [CrossRef]
19. Gautam, R.; Jachak, S.M. Recent developments in anti-inflammatory natural products. *Med. Res. Rev.* **2009**, *29*, 767–820. [CrossRef]
20. Kim, J.; Lee, Y.S.; Kim, C.S.; Kim, J.S. Betulinic acid has an inhibitory effect on pancreatic lipase and induces adipocyte lipolysis. *Phytother. Res.* **2012**, *26*, 1103–1106. [CrossRef]
21. Silva, F.S.G.; Oliveira, P.J.; Duarte, M.F. Oleanolic, ursolic, and betulinic acids as food supplements or pharmaceutical agents for type 2 diabetes: Promise or illusion? *J. Agric. Food Chem.* **2016**, *64*, 2991–3008. [CrossRef] [PubMed]
22. Vinayagam, R.; Xiao, J.; Xu, B. An insight into anti-diabetic properties of dietary phytochemicals. *Phytochem. Rev.* **2017**, *16*, 535–553. [CrossRef]
23. Kumar, S.; Kumar, V.; Prakash, O. Enzymes inhibition and antidiabetic effect of isolated constituents from *Dillenia indica*. *Biomed Res. Int.* **2013**, *2013*, 382063. [CrossRef] [PubMed]

24. de Melo, C.L.; Queiroz, M.G.R.; Arruda Filho, A.C.V.; Rodrigues, A.M.; de Sousa, D.F.; Almeida, J.G.L.; Pessoa, O.D.L.; Silveira, E.R.; Menezes, D.B.; Melo, T.S.; et al. Betulinic acid, a natural pentacyclic triterpenoid, prevents abdominal fat accumulation in mice fed a high-fat diet. *J. Agric. Food Chem.* **2009**, *57*, 8776–8781. [CrossRef]
25. Gomes Castro, A.J.; Silva Frederico, M.J.; Cazarolli, L.H.; Bretanha, L.C.; Tavares, L.d.C.; Buss, Z.d.S.; Dutra, M.F.; Pacheco de Souza, A.Z.; Pizzolatti, M.G.; Mena Barreto Silva, F.R. Betulinic acid and 1,25(OH)$_2$ vitamin D$_3$ share intracellular signal transduction in glucose homeostasis in soleus muscle. *Int. J. Biochem. Cell Biol.* **2014**, *48*, 18–27. [CrossRef]
26. Wen, X.; Sun, H.; Liu, J.; Cheng, K.; Zhang, P.; Zhang, L.; Hao, J.; Zhang, L.; Ni, P.; Zographos, S.E.; et al. Naturally occurring pentacyclic triterpenes as inhibitors of glycogen phosphorylase: Synthesis, structure-activity relationships, and X-ray crystallographic studies. *J. Med. Chem.* **2008**, *51*, 3540–3554. [CrossRef]
27. Ha, D.T.; Dao Trong, T.; Nguyen Bich, T.; Nhiem, N.X.; Ngoc, T.M.; Yim, N.; Bae, K. Palbinone and triterpenes from *Moutan Cortex* (*Paeonia suffruticosa*, Paeoniaceae) stimulate glucose uptake and glycogen synthesis via activation of AMPK in insulin-resistant human HepG2 Cells. *Bioorg. Med. Chem. Lett.* **2009**, *19*, 5556–5559. [CrossRef]
28. Heiss, E.H.; Kramer, M.P.; Atanasov, A.G.; Beres, H.; Schachner, D.; Dirsch, V.M. Glycolytic switch in response to betulinic acid in non-cancer cells. *PLoS ONE* **2014**, *9*, e115683. [CrossRef]
29. Choi, C.I.; Lee, S.R.; Kim, K.H. Antioxidant and α-glucosidase inhibitory activities of constituents from *Euonymus alatus* twigs. *Ind. Crops Prod.* **2015**, *76*, 1055–1060. [CrossRef]
30. Tri, M.D.; Phat, N.T.; Trung, N.T.; Phan, C.T.D.; Minh, P.N.; Chi, M.T.; Nguyen, T.P.; Dang, C.H.; Hong Truong, L.; Pham, N.K.T.; et al. A new 26-norlanostane from *Phlogacanthus turgidus* growing in Vietnam. *J. Asian Nat. Prod. Res.* **2021**, *24*, 196–202. [CrossRef]
31. Thengyai, S.; Thiantongin, P.; Sontimuang, C.; Ovatlarnporn, C.; Puttarak, P. α-Glucosidase and α-amylase inhibitory activities of medicinal plants in Thai antidiabetic recipes and bioactive compounds from *Vitex glabrata* R. Br. stem bark. *J. Herb. Med.* **2020**, *19*, 100302. [CrossRef]
32. Chukwujekwu, J.C.; Rengasamy, K.R.; de Kock, C.A.; Smith, P.J.; Slavetinska, L.P.; van Staden, J. α-glucosidase inhibitory and antiplasmodial properties of terpenoids from the leaves of *Buddleja saligna* Willd. *J. Enzym. Inhib. Med. Chem.* **2016**, *31*, 63–66. [CrossRef] [PubMed]
33. Kazakova, O.B.; Giniyatullina, G.V.; Mustafin, A.G.; Babkov, D.A.; Sokolova, E.V.; Spasov, A.A. Evaluation of Cytotoxicity and α-Glucosidase Inhibitory Activity of Amide and Polyamino-Derivatives of Lupane Triterpenoids. *Molecules* **2020**, *25*, 4833. [CrossRef] [PubMed]
34. Khusnutdinova, E.F.; Petrova, A.V.; Thu, H.N.T.; Tu, A.L.T.; Thanh, T.N.; Thi, C.B.; Babkov, D.A.; Kazakova, O.B. Structural modifications of 2,3-indolobetulinic acid: Design and synthesis of highly potent α-glucosidase inhibitors. *Bioorg. Chem.* **2019**, *88*, 102957. [CrossRef]
35. Cardullo, N.; Floresta, G.; Rescifina, A.; Muccilli, V.; Tringali, C. Synthesis and in vitro evaluation of chlorogenic acid amides as potential hypoglycemic agents and their synergistic effect with acarbose. *Bioorg. Chem.* **2021**, *117*, 105458. [CrossRef]
36. Zhang, B.; Xing, Y.; Wen, C.; Yu, X.; Sun, W.; Xiu, Z.; Dong, Y. Pentacyclic triterpenes as α-glucosidase and α-amylase inhibitors: Structure-activity relationships and the synergism with acarbose. *Bioorg. Med. Chem. Lett.* **2017**, *27*, 5065–5070. [CrossRef]
37. Yang, J.; Wang, X.; Zhang, C.; Ma, L.; Wei, T.; Zhao, Y.; Peng, X. Comparative study of inhibition mechanisms of structurally different flavonoid compounds on α-glucosidase and synergistic effect with acarbose. *Food Chem.* **2021**, *347*, 129056. [CrossRef]
38. Mabate, B.; Daub, C.D.; Malgas, S.; Edkins, A.L.; Pletschke, B.I. A Combination Approach in Inhibiting Type 2 Diabetes-Related Enzymes Using Ecklonia radiata Fucoidan and Acarbose. *Pharmaceutics* **2021**, *13*, 1979. [CrossRef]
39. Martin, A.; Montgomery, P. Acarbose: An α-glucosidase inhibitor. *Am. J. Health-Syst. Pharm.* **1996**, *53*, 2277–2290. [CrossRef]
40. Oboh, M.; Govender, L.; Siwela, M.; Mkhwanazi, B.N. Anti-Diabetic Potential of Plant-Based Pentacyclic Triterpene Derivatives: Progress Made to Improve Efficacy and Bioavailability. *Molecules* **2021**, *26*, 7243. [CrossRef]
41. Kan, L.; Capuano, E.; Fogliano, V.; Verkerk, R.; Mes, J.J.; Tomassen, M.M.M.; Oliviero, T. Inhibition of α-glucosidases by tea polyphenols in rat intestinal extract and Caco-2 cells grown on Transwell. *Food Chem.* **2021**, *361*, 130047. [CrossRef] [PubMed]
42. Zhou, J.; Qi, Q.; Wang, C.; Qian, Y.; Liu, G.; Wang, Y.; Fu, L. Surface plasmon resonance (SPR) biosensors for food allergen detection in food matrices. *Biosens. Bioelectron.* **2019**, *142*, 111449. [CrossRef] [PubMed]
43. Wu, Z.; Xu, H.; Wang, M.; Zhan, R.; Chen, W.; Zhang, R.; Kuang, Z.; Zhang, F.; Wang, K.; Gu, J. Molecular docking and molecular dynamics studies on selective synthesis of α-Amyrin and β-Amyrin by oxidosqualene cyclases from *Ilex asprella*. *Int. J. Mol. Sci.* **2019**, *20*, 3469. [CrossRef]
44. Zhang, L.; Wang, P.; Yang, Z.; Du, F.; Li, Z.; Wu, C.; Fang, A.; Xu, X.; Zhou, G. Molecular dynamics simulation exploration of the interaction between curcumin and myosin combined with the results of spectroscopy techniques. *Food Hydrocoll.* **2020**, *101*, 105455. [CrossRef]
45. Ni, M.; Hu, X.; Gong, D.; Zhang, G. Inhibitory mechanism of vitexin on α-glucosidase and its synergy with acarbose. *Food Hydrocoll.* **2020**, *105*, 105824. [CrossRef]
46. Chen, S.D.; Yong, T.Q.; Xiao, C.; Su, J.Y.; Zhang, Y.F.; Jiao, C.W.; Xie, Y.Z. Pyrrole alkaloids and ergosterols from *Grifola frondosa* exert anti-α-glucosidase and anti-proliferative activities. *J. Funct. Foods* **2018**, *43*, 196–205. [CrossRef]
47. Chen, S.D.; Yong, T.Q.; Xiao, C.; Gao, X.; Xie, Y.Z.; Hu, H.P.; Li, X.M.; Chen, D.L.; Pan, H.H.; Wu, Q.P. Inhibitory effect of triterpenoids from the mushroom *Inonotus obliquus* against α-glucosidase and their interaction: Inhibition kinetics and molecular stimulations. *Bioorg. Chem.* **2021**, *115*, 105276. [CrossRef] [PubMed]

48. Cai, S.; Wang, O.; Wang, M.; He, J.; Wang, Y.; Zhang, D.; Zhou, F.; Ji, B. In vitro inhibitory effect on pancreatic lipase activity of subfractions from ethanol extracts of fermented Oats (*Avena sativa* L.) and synergistic effect of three phenolic acids. *J. Agric. Food Chem.* **2012**, *60*, 7245–7251. [CrossRef]
49. Zhu, B.; Li, M.Y.; Lin, Q.; Liang, Z.; Xin, Q.; Wang, M.; He, Z.; Wang, X.; Wu, X.; Chen, G.G.; et al. Lipid oversupply induces CD36 sarcolemmal translocation via dual modulation of PKC zeta and TB1CD1: An early event prior to insulin resistance. *Theranostics* **2020**, *10*, 1332–1354. [CrossRef]

Article

Resolution of a Configurationally Stable Hetero[4]helicene

Michela Lupi [1], Martina Onori [1], Stefano Menichetti [1], Sergio Abbate [2], Giovanna Longhi [2] and Caterina Viglianisi [1,*]

[1] Department of Chemistry "Ugo Schiff" (DICUS), University of Florence, Via della Lastruccia 13, Sesto Fiorentino (FI), 50019 Florence, Italy; michela.lupi@unifi.it (M.L.); martinaonori@gmail.com (M.O.); stefano.menichetti@unifi.it (S.M.)

[2] Department of Molecular and Translational Medicine (DMMT), University of Brescia, V.le Europa 11 Brescia (BS), 25121 Brescia, Italy; sergio.abbate@unibs.it (S.A.); giovanna.longhi@unibs.it (G.L.)

* Correspondence: caterina.viglianisi@unifi.it

Abstract: We have developed an efficient chemical resolution of racemic hydroxy substituted dithia-aza[4]helicenes (DTA[4]H) **1(OH)** using enantiopure acids as resolving agents. The better diastereomeric separation was achieved on esters prepared with (1S)-(−)-camphanic acid. Subsequent simple manipulations produced highly optically pure (≥ 99% enantiomeric excess) (P) and (M)-**1(OH)** in good yields. The role of the position where the chiral auxiliary is inserted (*cape*- vs. *bay-zone*) and the structure of the enantiopure acid used on successful resolution are discussed.

Keywords: heterohelicene; chirality; resolution; enantiomers; chiroptical; screw-shaped compounds

1. Introduction

Chirality is one of the most crucial assets of nature and is of paramount importance in several areas of science, technology and medicine. Molecular chirality has been recognized for a long time and has provided guidance in the design of drugs and functional materials. Furthermore, a smart combination of chiral phenomena and supramolecular chemistry resulted in an emerging interdisciplinary field called supramolecular chirality [1].

Helicenes are compounds with a screw-shaped skeleton formed by ortho-condensed (hetero)aromatic rings with a non-planar structure due to the steric superimposition of terminal rings or/and the substituents on these rings [2], which force the molecule to adopt a helical conformation (Figure 1). This important class of axially chiral compounds has a barrier of interconversion between *M* and *P* enantiomers increasing with the increase of the ring number forming the helicene backbone.

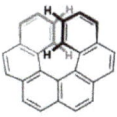

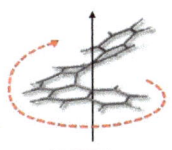 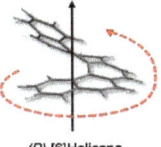

Figure 1. *M* and *P* enantiomers for a [6]carbohelicene.

Circular dichroism (CD) and circularly polarized luminescence (CPL) are just a few of the chiroptical proprieties that make helicenes valuable in potential applications such as advanced optical information storage, circularly polarized organic light-emitting diodes (CP-OLEDs), circularly polarized light detecting organic field effect transistors (CP-OFETs), chirality-induced spin selectivity (CISS) devices and stereoselective sensing chiroptical probes in biological processes [3–14].

Recently, increased attention has focused on the binding of small molecules to specific DNA structures to inhibit the biological functions in which these structures participate. Indeed, helicenes enantioselectivity offered a way to rationally design Z-DNA-depending inhibitors of biological functions [11].

Over the years, a variety of heterohelicenes and helical-shaped molecules, containing nitrogen, oxygen, sulfur, and other hetero-elements, have been synthesized and their unique properties studied. Great effort has been devoted to the setting of new simple and multi-gram synthetic procedures that allow for the isolation of helicenes in enantiopure form as required for various practical applications, including chirogenesis [7,15,16].

Dithia-aza[4]helicenes (DTA[4]H) **1** (Scheme 1) can be described as bis-phenothiazines with an aryl ring and a nitrogen atom in common forced into a helical shaped structure by the four long carbon–sulfur bonds. These [1,4]benzothiazino[2,3,4-*kl*]phenothiazines represent one of the attractive rare examples of geometrically stable [4]helicenes with racemization barriers higher than those measured for all carbon [5]helicenes.

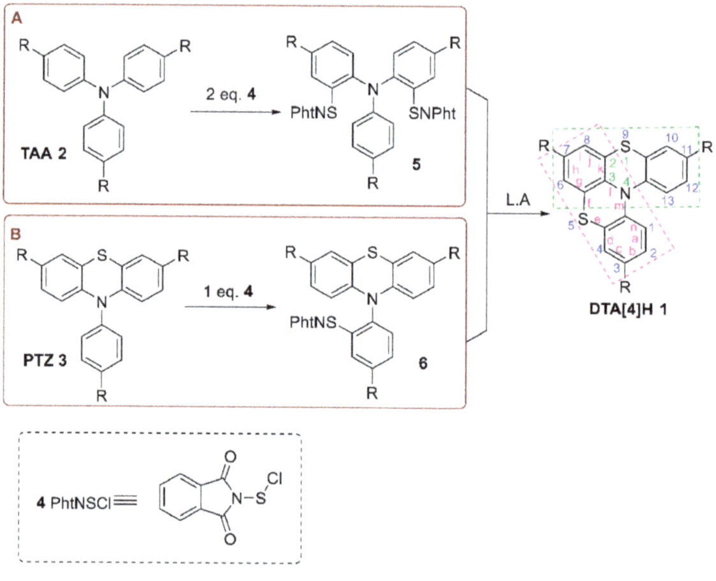

Scheme 1. Synthetic pathways (**A** or **B**) to dithia-aza[4]helicenes (DTA[4]H) **1**.

DTA[4]H are obtained [17–21], as racemic mixture, from properly substituted triarylamines (TAA) **2** or *N*-aryl phenothiazines (PTZ) **3** (Scheme 1, pathway A and B respectively), through regioselective sulfenylation(s) with two or one equivalents of phthalimidesulfenyl chloride (PhtNSCl (**4**), Pht = phthaloyl). Reacting the resulting sulfenylated derivatives **5** or **6** with a Lewis Acid (L.A.), typically BF_3OEt_2 or $AlCl_3$, causes two or one electrophilic intramolecular cyclization with formation of helicenes **1**.

Along with their peculiar helical-shaped structure, derivatives **1** show a very good one-electron donor ability, being easily, and reversibly, oxidized to the corresponding exceptionally stable crystalline chiral radical cations **1**$^{\bullet+}$ [18,21] (Scheme 2). We have also demonstrated that the oxidative process is extremely sensitive to the medium, and under acidic conditions, molecular oxygen becomes an efficient single electron transfer (SET) oxidant, giving rise to the formation of **1**$^{\bullet+}$. Furthermore, radical cations **1**$^{\bullet+}$ can be generated also via irradiation at 240–400 nm of helicenes in the presence of PhCl [21] (Scheme 2).

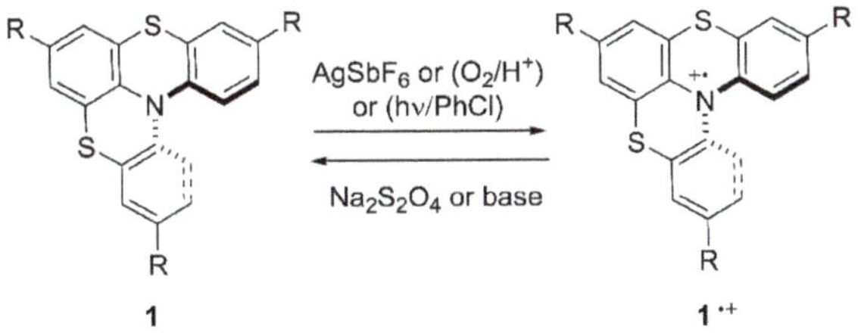

Scheme 2. Red-ox behavior of DTA[4]H **1**.

Indeed, we have also prepared, via ring-opening metathesis polymerization, dithia-aza[4]helicene functionalized polynorbornenes showing a pH depending reversible redox behavior as a new class of tunable material reversibly switchable by pH-triggered redox processes [22].

The availability of differently functionalized enantiomerically pure helicenes, avoiding the limitation associated with chiral HPLC resolution [17,23] is mandatory for the development of the appealing applications of these peculiar systems [21,22,24]. Therefore, in recent years we tried to set off regio-, stereo- and enantioselective synthetic approaches for the preparations of **1**.

Actually, the synthetic procedure depicted in Scheme 1, while failing to control the absolute stereochemistry of the process, allowed for the control of the regiochemistry during ring closure as well as the possibility of inserting different substituents in specific positions. Thus, we have studied the chemical resolutions of **1** using the classical temporary insertion of chiral auxiliaries.

The DTA[4]H **1** derivatives requested for the above described applications require the insertion, as an anchoring unit, of a hydroxyl group in different positions of the helical backbone. Thus, we decided to take advantage of these phenolic groups for the introduction of chiral auxiliaries through esterification with enantiopure acids **7** and in order to verify whether the diastereoisomeric mixture of esters **8** obtained can be separated. Herein we report how the helicene topology and chiral auxiliary structure could be matched to allow the resolution of phenolic DTA[4]H **1(OH)** (Scheme 3).

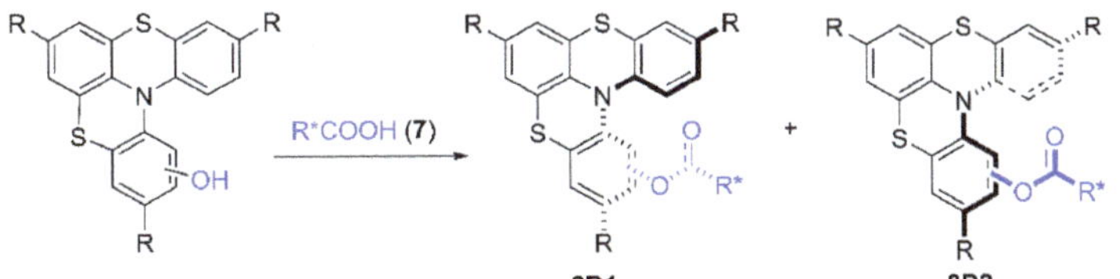

Scheme 3. Esterification with enantiopure acids **7**.

2. Results and Discussion

Selected DTA[4]H **1** can be resolved through HPLC in the chiral stationary phase as we previously reported [17,23]; however, this method is unsuitable to obtain enantiopure DTA[4]H in multigram scale. Instead, the diastereomeric process-based resolution has advantages in the the viewpoint of cost, generality and the amount of enantiopure products achieved. Thus, we decided to study the insertion of chiral auxiliaries, for example through esterification reactions, to verify whether the mixture of diastereomers obtained can be separated by flash chromatography allowing isolation of pure *M* and *P* DTA[4]H in relevant quantity.

We have demonstrated that *N*-arylphenothiazines PTZ **3** are suitable substrates for the synthesis of unsymmetrically hydroxy substituted helicenes **1(OH)** [20]. We selected helicene **1a(OH)** and **1b(OH)** (Scheme 4) to prepare diastereoisomeric esters using enantiopure acid **7** (Scheme 5).

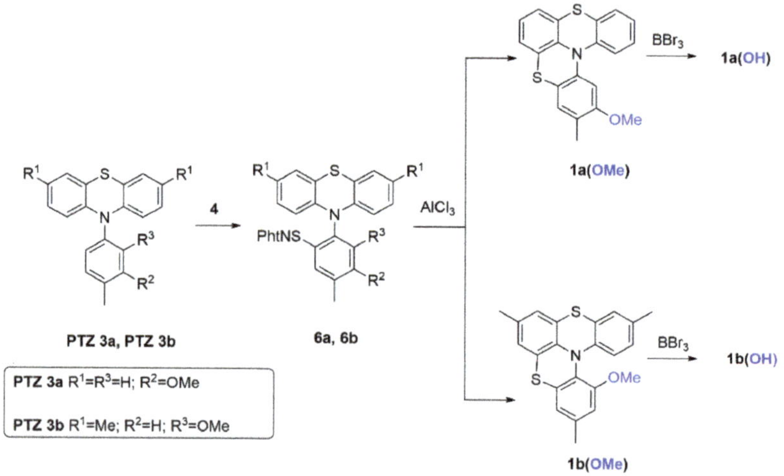

Scheme 4. Synthesis of hydroxy substituted helicenes **1a(OH)** and **1b(OH)**.

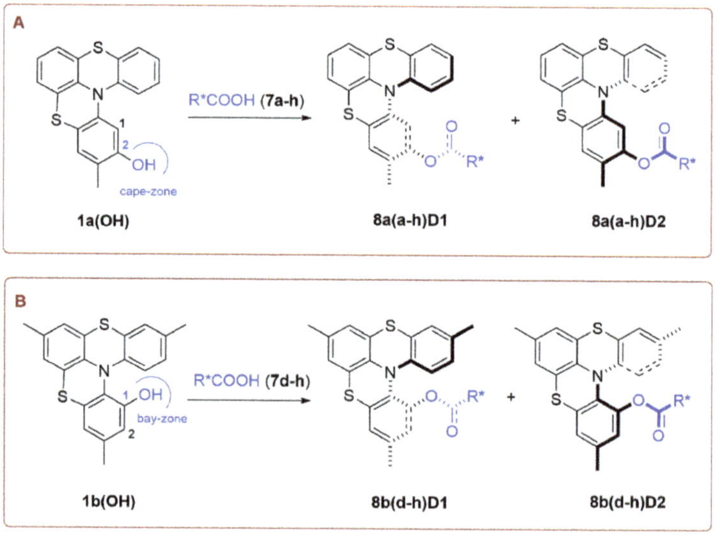

Scheme 5. Esterification of helicenes **1a(OH)** and **1b(OH)**, panels (**A**,**B**), respectively, with enantiopure acids **7**.

Firstly, we planned the introduction of chiral auxiliaries in 2-hydroxy-substituted ADT[4]H **1a(OH)** presenting a hydroxyl group in the 2-position (that we indicate as *cape-zone*) of the helicene, which was relatively easy to prepare [20].

Racemic **1a(OH)** was esterified with different enantiopure acids **7a–h** yielding a diastereomeric mixture (D1+D2) of esters **8a(a–h)**. Esterification reactions were carried out in presence of diisopropylcarbodiimide (DIC) and 4-dimethylaminopyridine (DMAP) as catalysts, in dry CH_2Cl_2 at room temperature (Scheme 5A).

Regardless, all chiral acids **7a–h** (Scheme 5 panel A and Table 1) allowed the formation of diastereomeric esters **8a(a–h)** (Scheme 5 panel A and Table 1) in good yields, in none of these cases it was possible to separate the diastereoisomeric mixtures by flash chromatography or crystallization.

Table 1. Diastereomeric esters **8** obtained reaction racemic phenols **1a(OH)** or **1b(OH)** with enantiopure acids **7a–h**.

Chiral Auxiliary	Product	Yield	Resolution
7a (S)-(−)-Perillic acid	8aaD1/8aaD2	60%	No resolution
7b Camphorsulfonyl chloride	8abD1/8abD2	56%	No resolution
7c (−)-o,o′-Dibenzoyl-L-tartaric acid mono(dimethyl amide)	8acD1/8acD2	56%	No resolution
7d (1S)-(+)-Ketopinic acid	8adD1/8adD2	79%	No resolution
	8bdD1/8bdD2	55%	No resolution
7e (S)-(+)-2-(6-Methoxy-2-naphthyl)propionic acid	8aeD1/8aeD2	94%	No resolution
	8beD1/8beD2	58%	No resolution
7f (-)-mono-(1R)-Menthylphthalate	8afD1/8afD2	54%	No resolution
	8bfD1/8bfD2	45%	No resolution
7g (S)-N-Boc pipecolic acid	8agD1/8agD2	69%	No resolution
	8bgD1/8bgD2	28%/15%	Chromatographic resolution is possible
7h (1S)-(−)-Camphanic acid	8ahD1/8ahD2	57%	No resolution
	8bhD1/8bhD2	37% / 33%	Chromatographic resolution is possible

We thought, as suggested by literature data [25–38], that functionalization of the *cape-zone* position keeps the chiral auxiliaries too far from the superimposition area of the terminal aryl rings vanishing separation. Thus, we moved to helicene **1b(OH)**, which allowed for the insertion of chiral auxiliaries on 1-position, *ortho* to the nitrogen atom, which we indicated as *bay-zone*, i.e., exactly in the area of terminal ring superimposition (Scheme 5B).

Using chiral acids **7d–h** (Scheme 5 panel B and Table 1), the corresponding diastereomeric esters **8b(d–h)** (Scheme 5 panel B and Table 1) were obtained in moderate yields generally lower than those of the corresponding esters prepared using phenol **1a(OH)**, indicating, as expected, a more difficult access to the OH group of **1b(OH)** laying the *bay-zone* (Table 1). For several esters **8b** it was possible to identify the presence of the two diastereomers (D1 and D2, chromatographic elution order) by ^{1}H and ^{13}C NMR, and, eventually, to reveal a slightly different chromatographic behavior on TLC (Figure 2).

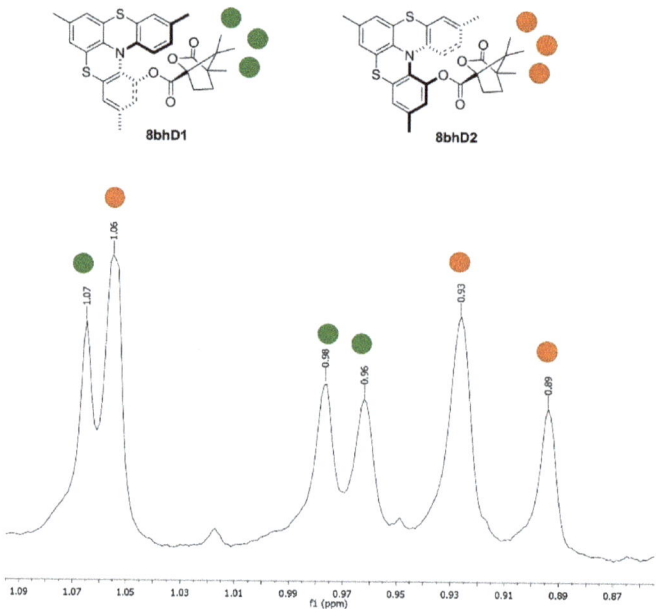

Figure 2. ^{1}H NMR spectrum of diastereomers (D1 and D2) **8bh**.

Despite the introduction of the chiral auxiliary in the *bay-zone* diastereoisomeric esters D1 and D2 of **8b(d–f)** were not separable by flash chromatography in spite of an accurate selection of eluent mixtures. However, esterification of **1b(OH)** with N-boc pipecolic acid **7g** allowed for the partial separation by flash chromatography on silica gel of diastereomers **8bgD1** and **8bgD2**, which were characterized by ^{1}H and ^{13}C NMR. Optical rotation of **8bgD1** was: $[\alpha]_D^{20}$ −157 (c 0.1, CH$_2$Cl$_2$), while for **8bgD2** was: $[\alpha]_D^{20}$ +49 (c 0.1, CH$_2$Cl$_2$). ^{1}H NMR spectra show that the product **8bgD1** was isolated as single diastereomer, while the product **8bgD2** was isolated as a roughly 3:1 mixture of the two diastereomers.

Esters **8bgD1** and **8bgD2** were hydrolysed with 3 eq. of NaOH in CH$_2$Cl$_2$/MeOH to give enantiomeric phenols (*M*)-**1b(OH)** and (*P*)-**1b(OH)**. Phenols were analysed by HPLC in the chiral stationary phase in order to calculate the enantiomeric ratio. Chromatograms showed that product (*M*)-**1b(OH)** $[\alpha]_D^{20}$ −161 (c 0.1, CH$_2$Cl$_2$) was obtained as single enantiomer (e.e. ≥ 99%), while helicene (*P*)-**1b(OH)** $[\alpha]_D^{20}$ +75 (c 0.1, CH$_2$Cl$_2$) exhibits the enantiomeric ratio 72:28 (e.e. = 44%).

Esterification of **1b(OH)** with (1*S*)-(−)-camphanic acid **7h** provided, with our great satisfaction, diastereomers **8bhD1** and **8bhD2** that were successfully separated by flash chromatography and characterized by ^{1}H and ^{13}C NMR spectroscopy (Figure 3 and Supplementary Materials).

Optical rotation was measured and gave $[\alpha]_D^{20}$ −129 (c 0.1, CH$_2$Cl$_2$) for **8bhD1** and $[\alpha]_D^{20}$ + 126, (c 0.1, CH$_2$Cl$_2$) for **8bhD2**. Hydrolysis of diastereomeric esters provided helicenes (*M*)-**1b(OH)** and (*P*)-**1b(OH)**, respectively. HPLC analysis with a chiral stationary phase showed that the first eluted product, (*P*)-**1b(OH)** $[\alpha]_D^{20}$ +166 (c 0.1, CH$_2$Cl$_2$), exhibits an enantiomeric ratio = 98:2 (e.e. = 96%), while the second eluted product, (*M*)-**1b(OH)** $[\alpha]_D^{20}$ −167 (c 0.1, CH$_2$Cl$_2$), was obtained as a single enantiomer (e.e. ≥ 99%), (Scheme 6).

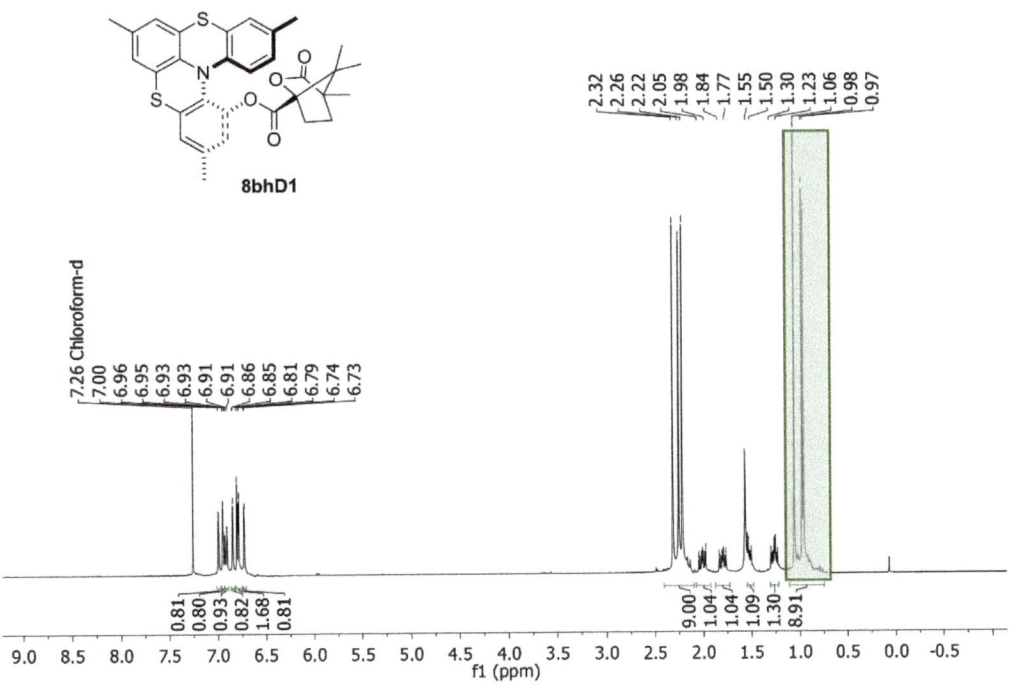

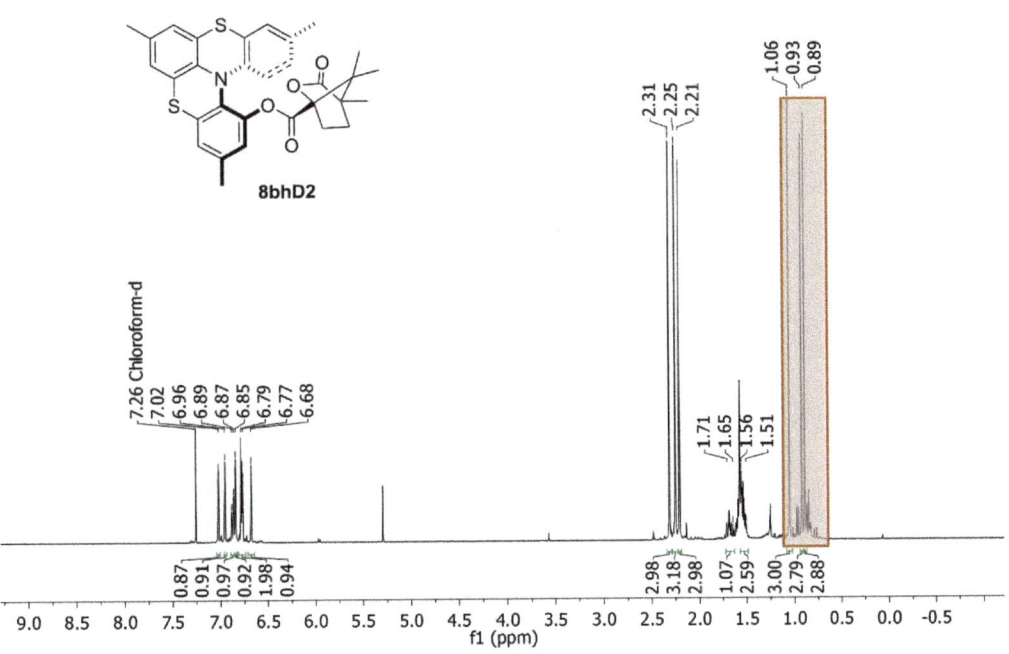

Figure 3. ^{1}H-NMR spectra of diastereomers **8bhD1** and **8bhD2**.

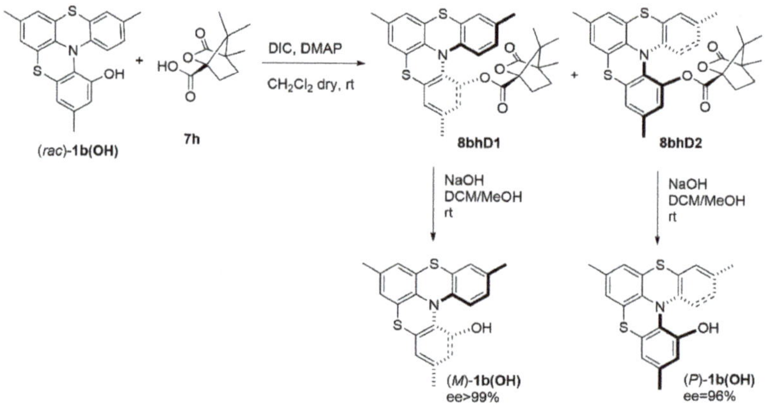

Scheme 6. Chemical resolution of helicene (*rac*)-**1b(OH)**.

The assignment of the absolute configuration of DTA[4]H **1** derivatives has been established as *P*-(+) and *M*-(−), which is typical for helicene systems; opposite assignment is quite unusual, as we have established in ref [17,19]. In this work the absolute configuration of **1b(OH)** was validated by comparison of the electronic circular dichroism (ECD) spectra of the two optical enantiomers-(+)-**1b(OH)** and (−)-**1b(OH)**, assigned to (*P*)-**1b(OH)** and (*M*)-**1b(OH)**, with the calculated spectrum of the *M* structure.

In order to assign the configuration, DFT and TD-DFT calculations have been conducted with the Gaussian16 package [39]. Two orientations are possible for the hydroxyl-group; the two optimized structures in the *M* configuration are reported in Figure 4 with their Boltzmann populations. Two functionals have been considered; the two differing in the amount of the exact exchange included M06 with 27% HF exchange and M06-2X with 54% HF exchange [40]. The solvent has been treated at the iefpcm level. Structural results are similar for the two functionals.

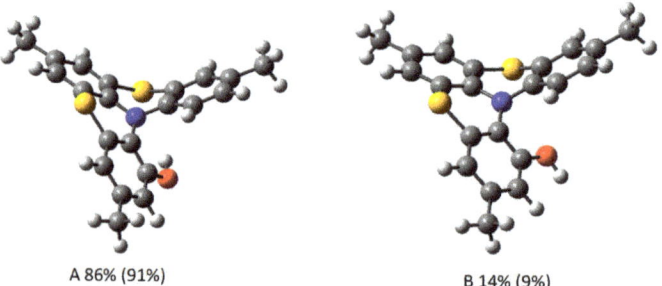

Figure 4. Optimized 3D-structures for the two possible conformers of (*M*)-**1b(OH)**. Percent population factors calculated at M06/cc-pvtz, iefpcm level; in parenthesis population factors calculated at M062X/cc-pvtz, iefpcm level.

CD and absorption spectra have been calculated at the same level of theory, a constant Gaussian 0.2 eV bandwidth was applied to each transition. The experimental CD and absorption spectra have been recorded for the two enantiomers in 4.2 mM dichloromethane solution in a 0.1 mm quartz cuvette using a JASCO-815SE instrument.

The comparison of experimental data with calculations are presented in Figure 5. In order to compare with data, +4 nm shift has been applied to the results obtained with M06, +26 nm with M06-2X; calculation of similarity index between experimental and calculated spectra suggested the best shift for the best correspondence, as reported in the Supplementary Materials paragraph. It is also shown that the two conformers give very similar spectra, while in Figure 5 the Boltzmann weighed average is presented. Both functionals permit confirmation of the configuration as *M*-(−) (correspondingly *P*-(+)). This conclusion agrees with what was obtained for the parent molecule triarylamine hetero[4]helicene examined in reference [19].

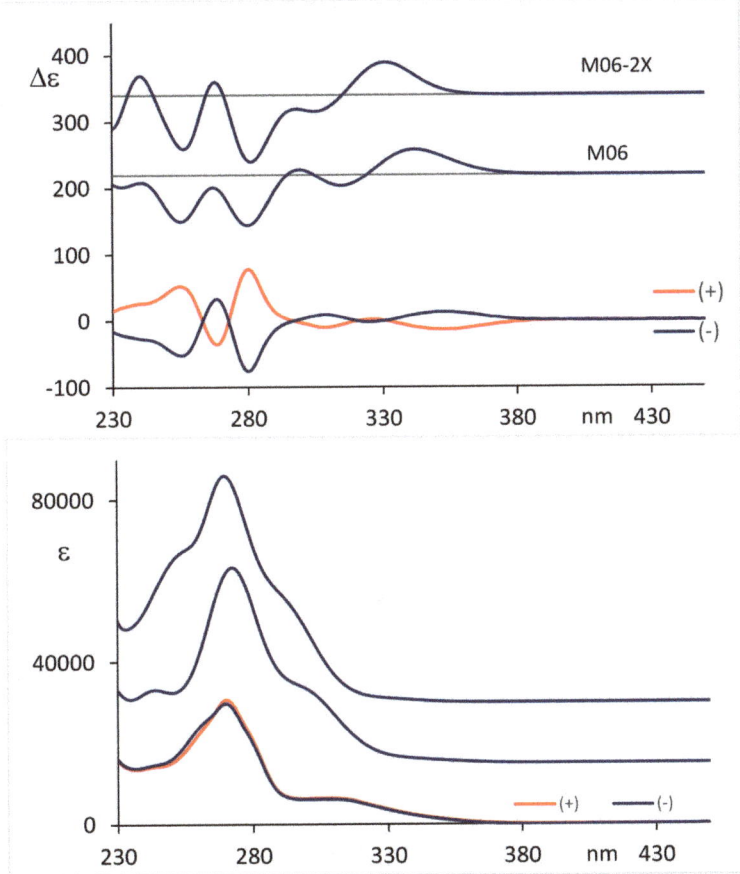

Figure 5. CD (top) and absorption (bottom) experimental and calculated spectra with two choices for the DFT functional (see text); calculation performed on *M*-**1b(OH)**. The calculated spectra are Boltzmann weighed averages of the two conformers A and B.

Overall, our results confirm that, on chemical resolution of helicenes, the position where the chiral auxiliaries are inserted is extremely important, being the *bay-zone* that allows for the higher effect on enantiomeric discrimination. At the same time we have confirmed previous studies reporting chromatographic resolutions of [7]carbo- and [7]heterohelicenes by means of tetra- and monocamphanate esters [25–31]. In each of these cases, the (1*S*)-camphanate of the (*P*)-helicenol moves more slowly upon chromatography on silica gel than the (1*S*)-camphanate of the (*M*)-helicenol [27].

3. Materials and Methods

^{1}H and ^{13}C NMR spectra were recorded with Varian Mercury Plus 400, Varian Inova 400, using CDCl$_3$ as solvent. Residual CHCl$_3$ at δ = 7.26 ppm and central line of CDCl$_3$ at δ = 77.16 ppm were used as the reference of ^{1}H-NMR spectra and ^{13}C NMR spectra, respectively. FT-IR spectra were recorded with a Spectrum Two FT-IR Spectrometer. ESI-MS spectra were recorded with a JEOL MStation JMS700. Melting points were measured with a Stuart SMP50 Automatic Melting Point Apparatus. Optical rotation measurements were performed on a JASCO DIP-370 polarimeter (JASCO, Easton, MD, USA) and the specific rotation of compounds was reported [41].

All the reactions were monitored by TLC on commercially available precoated plates (silica gel 60 F 254) and the products were visualized with acidic vanillin solution. Silica gel 60 (230–400 mesh) was used for column chromatography. Dry solvents were obtained by The PureSolv Micro Solvent Purification System. Chloroform was washed with water several times and stored over calcium chloride. Pyridine and TEA were freshly distilled from KOH. Phthalimide sulfenyl chloride was prepared from the corresponding disulfide (purchased from Chemper snc) as reported elsewhere. Helicenes **1a** and **1b** were described elsewhere [17].

General Procedure for the synthesis of diastereomeric esters from **1** by Steglich esterification: To a solution of **1** in dry CH_2Cl_2 (roughly 0.03–0.04 M), the enantiopure acid **7** (1.2 eq), DMAP (0.1 eq) and DIC (1–1.2 eq) were added at 0 °C. The solution was stirred at room temperature under a nitrogen atmosphere for 2–29 h, then was diluted with CH_2Cl_2 (60 mL), washed with a saturated solution of NH_4Cl (2 × 40 mL), with a saturated solution of $NaHCO_3$ (3 × 40) then with NH_4Cl (3 × 40 mL). The organic layer was dried over Na_2SO_4, filtered and evaporated under reduced pressure. The crude was purified by flash chromatography on silica gel.

Diastereoisomers **8aaD1** and **8aaD2**. (*M/P*)-3-methyl[1,4]benzothiazino[2,3,4-*kl*]pheno thiazine-2-yl (*S*)-perillate. Following the general *Steglich esterification* procedure from **1a(OH)** (60 mg, 0.18 mmol) and (*S*)-(−)-perillic acid **7a** (36 mg, 0.22 mmol), kept for 22 h at rt. The crude was purified by flash chromatography on silica gel (petroleum ether/ CH_2Cl_2 1:3, R_f 0.86) to afford the mixture of the two diastereomeric compounds **8aaD1** and **8aaD2** (52 mg, 60% yield) as a white solid (mp 105–115 °C). ^{1}H NMR (400 MHz, $CDCl_3$)* δ: 1.45-1.55 (m, 2H), 1.76 (s, 6H), 1.89–1.95 (m, 2H), 2.12 (s, 6H), 2.16–2.41 (m, 8H), 2.51–2.59 (m, 2H), 4.74 (bs, 2H), 4.78 (bs, 2H), 6.89 (bs, 2H), 6.92–7.05 (m, 8H), 7.06 (bs, 2H), 7.12–7.25 (m, 8H) ppm. ^{13}C NMR (100 MHz, $CDCl_3$)* δ: 15.8, 20.9, 24.76, 24.80, 27.1, 31.4, 40.1, 108.63, 108, 64, 110.0, 110.6, 114.7, 120.6, 124.1, 124.8, 125.0, 125.7, 125.75, 125.83, 126.0, 127.0, 127.3, 127.7, 128.0, 129.3, 129.5, 139.5, 141.2, 141.8, 142.5, 148.7, 149.2, 165.2, 165.3 ppm (34 signals for 58 different carbons). Elem. Anal. for $C_{29}H_{25}NO_2S_2$: Calcd. C 72.02, H 5.21, N 2.90; found C 71.80; H 5.21, N 2.89. *Et_3N was added to neutralize $CHCl_3$ acidity.

Diastereoisomers **8abD1** and **8abD2**. (*M/P*)-3-methyl[1,4]benzothiazino[2,3,4-*kl*]pheno thiazine-2-yl (1*S*)-10-camphorsulfonate. To a solution of **1a(OH)** (60 mg; 0.18 mmol) and TEA (22 mg, 0.22 mmol) in 4 mL of dry CH_2Cl_2, (1*S*)-(+)-10-camphorsulfonyl chloride **7b** (51 mg, 0.20 mmol) is added at 0 °C. After 10 min the solution was allowed to warm at room temperature and was stirred for 18 h under a nitrogen atmosphere. The mixture was diluted with AcOEt (25 mL) and washed with water (3 × 20 mL). The organic layer was dried over Na_2SO_4, filtered, and then evaporated under reduced pressure. The crude was purified by flash chromatography on silica gel (petroleum ether/AcOEt: 5/1, R_f 0.38) to afford the mixture of the two diastereomeric compounds **8abD1** and **8abD2** (55 mg, 56% yield) as a white solid (mp 190–195 °C). ^{1}H NMR (400 MHz, $CDCl_3$) δ: 0.87 (s, 3H), 0.88 (s, 3H), 1.11 (s, 6H), 1.40–1.47 (m, 2H), 1.64–1.72 (m, 2H), 1.92–1.98 (m, 2H), 2.00–2.13 (m, 4H), 2.31 (s, 3H), 2.32 (s, 3H), 2.36–2.57 (m, 4H), 3.13–3.18 (m, 2H), 3.73–3.78 (m, 2H), 6.93–7.25 (m, 18H) ppm. ^{13}C NMR (100 MHz, $CDCl_3$) δ: 11.6, 16.4, 16.5, 19.8, 20.02, 20.03, 25.3, 26.9, 27.0, 42.5, 43.0, 43.1, 46.3, 48.01, 48.03, 48.4, 48.6, 58.19, 58.24, 114.7, 114.8, 120.45, 120.50, 125.1, 125.2, 125.4, 125.5, 125.7, 125.8, 125.9, 126.0, 126.1, 126.2, 127.1, 127.8, 127.9, 128.11, 128.13, 128.2, 128.4, 130.0, 130.1, 139.2, 141.29, 141.32, 142.26, 142.29, 147.0, 147.1, 213.8, 213.9 ppm. Elem. Anal. for $C_{29}H_{27}NO_4S_3$: Calcd. C 63.36, H 4.95, N 2.55; found C 63.38, H 4.95, N 2. 54.

Diastereoisomers **8acD1** and **8acD2**. (*M/P*)-3-methyl[1,4]benzothiazino[2,3,4-*kl*]pheno thiazine-2-yl o,o′-dibenzoyl-L-tartrate. Following the general procedure from **1a(OH)**

(85 mg, 0.25 mmol) and (-)-o,o'-dibenzoyl-L-tartaric acid mono(dimethyl amide) **7c** (117 mg, 0.30 mmol), kept for 2 h at room temperature. The crude was purified by flash chromatography on silica gel (AcOEt/CH$_2$Cl$_2$ 1/20, Rf 0.65) to afford the mixture of the two diastereomeric compounds **8acD1** and **8acD2** (99 mg, 56% yield) as a white solid (mp 102–106 °C). ^{1}H NMR (400 MHz, CDCl$_3$) δ: 2.09 (s, 3H), 2.10 (s, 3H), 2.92 (s, 3H), 2.93 (s, 3H), 3.17 (s, 3H), 3.30 (s, 3H), 6.16 (d, 1H, J = 4.8 Hz), 6.20 (d, 1H, J = 5.2 Hz), 6.31 (d, 1H, J = 4.8 Hz), 6.33 (d, 1H, J = 5.2 Hz), 6.83–7.20 (m, 18H), 7.44–7.54 (m, 8H), 7.54–7.58 (m, 4H), 7.99–8.08 (m, 8H) ppm. ^{13}C NMR (100 MHz, CDCl$_3$) δ: 15.61, 15.63, 36.27, 36.31, 37.2, 69.5, 69.6, 70.9, 113.8, 114.0, 120.35, 120.41, 124.8, 124.9, 125.0, 125.1, 125.5, 125.6, 125.72, 125.74, 125.86, 125.92, 126.75, 126.84, 126.9, 127.6, 127.7, 127.9, 128.0, 128.35, 128.43, 128.52, 128.53, 128.58, 128.59, 128.60, 129.58, 129.59, 130.0, 130.09, 130.12, 130.15 133.82, 133.86, 139.12, 139.13, 141.1, 141.2, 142.1, 142.3, 148.09, 148.10, 165.0, 165.1, 165.3, 165.43, 165.44, 165.45, 165.50 ppm (59 signals for 78 carbons). IR (ATR solid) n: 3063, 2929, 1763, 1724, 1664, 1477, 1432, 1240 cm^{-1}. Elem. Anal. for C$_{39}$H$_{30}$N$_2$O$_7$S$_2$: Calcd. C 66.65, H 4.30, N 3.99; found C 66.61, H 4.28, N 4.00.

Diastereoisomers **8adD1** and **8adD2**. (M/P)-3-methyl[1,4]benzothiazino[2,3,4-kl]phenothiazine-2-yl (1S)-ketopinate. Following the general *Steglich esterification* procedure from **1a(OH)** (60 mg, 0.18 mmol) and (1S)-(+)-ketopinic acid **7d** (56 mg, 0.31 mmol), kept for 29 h at room temperature. The crude was purified by flash chromatography on silica gel (petroleum ether/CH$_2$Cl$_2$ 1/3, Rf 0.45) to afford the mixture of the two diastereomeric compounds **8adD1** and **8adD2** (71 mg, 79% yield) as a white solid (mp 130–133 °C). ^{1}H NMR (400 MHz, CDCl$_3$)* δ: 1.14 (s, 6H), 1.21 (s, 3H), 1.22 (s, 3H), 1.42–1.48 (m, 2H), 1.88–1.97 (m, 3H), 2.01–2.10 (m, 3H), 2.13–2.15 (m, 2H), 2.18 (s, 3H), 2.19 (s, 3H), 2.40–2.48 (m, 2H), 2.54–2.60 (m, 2H), 6.85 (s, 1H), 6.87 (s, 1H), 6.94–7.06 (m, 10H), 7.12–7.23 (m, 6H) ppm. ^{13}C NMR (100 MHz, CDCl$_3$)* δ: 16.2, 20.0, 21.50, 21.52, 26.49, 26.52, 26.7, 26.8, 44.02, 44.58, 44.61, 49.51, 49.56, 68.20, 68.24, 110.6, 114.70, 114.72, 120.41, 120.44, 124.6, 124.8, 124.9, 125.0, 125.70, 125.72, 125.8, 126.1, 126.8, 127.2, 127.3, 127.75, 127.82, 128.1, 129.59, 129.61, 139.4, 140.9, 141.0, 142.66, 142.71, 148.8, 168.1, 210.3 ppm (44 signals for 58 different carbons). Elem. Anal. for C$_{29}$H$_{25}$NO$_3$S$_2$: Calcd. C 69.71, H 5.04, N 2.80; found: C 69.73, H 5.01, N 2.77. *Et$_3$N was added to neutralize CHCl$_3$ acidity.

Diastereoisomers **8aeD1** and **8aeD2**. (M/P)-3-methyl[1,4]benzothiazino[2,3,4-kl]phenothiazine-2-yl (S)-2-(6-methoxy-2-naphthyl) propionate. Following the general *Steglich esterification* procedure from **1a(OH)** (60 mg, 0.18 mmol) and (S)-(+)-2-(6-methoxy-2-naphthyl) propionic acid **7e** (50 mg, 0.22 mmol), kept for 20 h at room temperature. The crude was purified by flash chromatography on silica gel (petroleum ether/CH$_2$Cl$_2$ 1/3, Rf 0.80) to afford the mixture of the two diastereomeric compounds **8aeD1** and **8aeD2** (94 mg, 94% yield) as an orange solid (mp 132–136 °C). ^{1}H NMR (400 MHz, CDCl$_3$)* δ: 1.39 (d, 6H, J = 6.5 Hz), 1.77 (s, 3H), 1.79 (s, 3H), 3.39 (s, 6H), 3.84 (q, 1H, J = 6.6 Hz), 4.12 (q, 1H, J = 6.7 Hz), 6.81 (bs, 2H), 6.91–7.05 (m, 10H), 7.10–7.20 (m, 10H), 7.44–7.47 (m, 2H), 7.67–7.73 (m, 6H) ppm. ^{13}C NMR (100 MHz, CDCl$_3$)* δ: 15.46, 15.48, 18.5, 18.6, 20.6, 21.2, 22.7, 23.6, 42.3, 45.56, 45.59, 55.4, 105.7, 108.6, 110.6, 114.4, 114.5, 119.3, 120.5, 124.3, 124.8, 125.0, 125.7, 125.8, 125.96, 126.01, 126.3, 126.4, 126.9, 127.1, 127.2, 127.4, 127.7, 127.8, 127.98, 128.01, 129.0, 129.4, 129.5, 133.9, 134.95, 134.99, 139.4, 141.0, 142.50, 142.54, 148.9, 157.9, 172.7 ppm (49 signals for 66 different carbons). Elem. Anal. for C$_{33}$H$_{25}$NO$_3$S$_2$: Calcd. C 72.37, H 4.60, N 2.56; found C 72.29; H 4.58, N 2.57. *Et$_3$N was added to neutralize CHCl$_3$ acidity.

Diastereoisomers **8afD1** and **8afD2**. (M/P)-3-methyl[1,4]benzothiazino[2,3,4-kl]phenothiazine-2-yl mono-(1R)-menthylphthalate. Following the general procedure from **1a(OH)** (70 mg, 0.21 mmol) and (-)mono(1R)-menthylphthalate **7f** (76 mg, 0.25 mmol), kept for 24 h at room temperature. The crude was purified by flash chromatography on silica gel (petroleum ether/ CH$_2$Cl$_2$ 2/1, Rf 0.78)) to afford the mixture of the two diastereomeric compounds **8afD1** and **8afD2** (70 mg, 54% yield) as a white solid (mp 140.8–146.8 °C). ^{1}HNMR (400 MHz, CDCl$_3$) δ: 0.73–0.69 (m, 6H), 0.83–0.89 (m, 14H), 0.97–1.16 (m, 4H), 1.36–1.51 (m, 4H), 1.64–1.71 (m, 4H), 1.85–1.98 (m, 2H), 2.05–2.11 (m, 2H), 2.21 (s, 3H), 2.22 (s, 3H), 4.82–4.89 (m, 2H), 6.93–7.00 (m, 6H), 7.03 (dd, 2H, J = 1.2 Hz, J = 7.5 Hz), 7.08 (bs,

1H), 7.09 (bs, 1H), 7.11–7.16 (m, 2H), 7.18 (d, 2H, J = 7.7 Hz), 7.24–7.29 (m, 2H), 7.54–7.60 (m, 4H), 7.68–7.73 (m, 2H), 7.85–7.90 (m, 2H) ppm. ^{13}C NMR (100 MHz, CDCl$_3$) δ: 15.19, 16.0, 16.3, 16.5, 20.89, 20.93, 22.15, 22.18, 23.5, 23.6, 26.3, 26.4, 31.55, 31.57, 34.37, 34.41, 40.6, 40.7, 47.23, 47.25, 75.95, 75.02, 114.47, 114.49, 120.60, 120.64, 124.5, 124.6, 124.9, 125.0, 125.71, 125.75, 125.77, 125.79, 126.07, 126.08, 126.95, 126.99, 127.33, 127.34, 127.71, 127.75, 127.98, 128.00, 128.8, 129.0, 129.51, 129.52, 129.56, 129.60, 130.6, 130.89, 130.94, 131.0, 131.8, 131.9, 133.5, 133.7, 139.5, 141.2, 142.50, 142.52, 148.97, 148.99, 165.3, 165.4, 166.8, 166.9 ppm (68 signals for 74 different carbons). IR (ATR solid) n: 2953, 2923, 2867, 1749, 1715, 1477, 1431, 1276, 1236 cm^{-1}. Elem. Anal. for C$_{37}$H$_{35}$NO$_4$S$_2$: Calcd. C 71.47, H 5.67, N 2.25; found C 71.36, H 5.65, N 2.26.

Diastereoisomers **8agD1** and **8agD2**. (*M*/*P*)-3-methyl[1,4]benzothiazino[2,3,4-*kl*]pheno thiazine-2-yl (S)-N-Boc pipecolinate. Following the general procedure from **1a(OH)** (61 mg, 0.18 mmol) and (S)-N-Boc pipecolic acid **7g** (50 mg, 0.22 mmol) kept for 22 h at room temperature. The crude was purified by flash chromatography (CH$_2$Cl$_2$, *Rf* 0.51) on silica gel to afford the mixture of the two diastereomeric compounds **8agD1** and **8agD2** (70 mg, 69%yield) as a white solid (mp 106–117 °C). ^{1}H NMR (400 MHz, CDCl$_3$)* δ: 1.37–1.44 (m, 22H), 1.65–1.80 (m, 8H), 2.13 (s, 6H), 2.29–2.33 (m, 2H), 2.93–3.09 (m, 2H), 3.89–3.95 (m, 1H), 4.04–4.09 (m, 2H), 4.92 (bs, 1H), 5.07 (bs, 1H), 7.21–6.77 (m, 18H) ppm. Elem. Anal. For C$_{30}$H$_{30}$N$_2$O$_4$S$_2$: Calcd. C 65.91, H 5.53, N 5.12; found: C 65.35, H 5.31, N 4.98. *Et$_3$N was added to neutralize CHCl$_3$ acidity.

Diastereoisomers **8ahD1** and **8ahD2**. (*M*/*P*)-3-methyl[1,4]benzothiazino[2,3,4-*kl*]pheno thiazine-2-yl (1S)-camphanate. Following a *Steglich esterification* procedure from **1a(OH)** (60 mg, 0.18 mmol) and (1S)-(−)-camphanic acid **7h** (43 mg, 0.22 mmol), kept for 17 h at room temperature. The crude was purified by flash chromatography on silica gel (petroleum ether/AcOEt/diethyl ether: 10:1:3, *Rf* 0.43)) to afford the mixture of the two diastereomeric compounds **8ahD1** and **8ahD2** (52 mg, 57% yield) as a white solid (mp 170 °C dec). ^{1}H NMR (400 MHz, CDCl$_3$)* δ: 1.04 (s, 3H), 1.06 (s, 3H), 1.11 (s, 6H), 1.13 (s, 3H), 1.14 (s, 3H), 1.69–1.77 (m, 2H), 1.92–1.99 (m, 2H), 2.11–2.19 (m, 8H), 2.45–2.54 (m, 2H), 6.84 (s, 1H), 6.86 (s, 1H), 6.93–7.00 (m, 6H), 7.02–7.08 (m, 4H), 7.12–7.21 (m, 6H) ppm. ^{13}C NMR (100 MHz, CDCl$_3$)* δ: 9.78, 9.81, 16.0, 16.2, 16.94, 16.96, 17.01, 17.04, 29.0, 31.2, 31.3. 53.6, 54.6, 54.97, 55.00, 90.89, 90.94, 114.16, 114.21, 120.4, 120.5, 125.0, 125.1, 125.24, 125.27, 125.6, 125.8, 125.9, 126.0, 126.1, 126.50, 126.52, 126.9, 127.0, 127.8, 128.1, 129.8, 139.3, 141.2, 141.3, 142.4, 142.5, 148.1, 148.2, 165.8, 165.9, 177.8 ppm (47 signals for 58 different carbons). Elem. Anal. for C$_{29}$H$_{25}$NO$_4$S$_2$: Calcd. C 67.55, H 4.89, N 2.72; found C 67.49, H 4.88, N 2.72. *Et$_3$N was added to neutralize CHCl$_3$ acidity.

Diastereoisomers **8bdD1, 8bdD2**. (*M*/*P*)-1,3,7-trimethyl[1,4]benzothiazino[2,3,4-*kl*] phenothiazine-2-yl (1S)-ketopinate. Following procedure from **1b(OH)** (40 mg, 0.11 mmol) and (1S)-(+)-ketopinic acid **7d** (20 mg, 0.11 mmol), kept for 18 h at room temperature. The crude was purified by flash chromatography on silica gel (petroleum eter/CH$_2$Cl$_2$ 1/2, *Rf* 0.38)) to afford the mixture of the two diastereomeric compounds **8bdD1** and **8bdD2** (32 mg, 55% yield) as a white solid (mp 190–199 °C). ^{1}H NMR (400 MHz, CDCl$_3$) δ: 0.76 (s,3H), 1.01 (s, 3H), 1.06 (s, 3H), 1.11 (s, 3H), 1.18–1.36 (m, 5H), 1.63–1.70 (m, 1H), 1.73–1.90 (m, 4H), 1.94–1.99 (m, 2H), 2.21–2.31 (m, 18H), 2.42–2.49 (m, 2H), 6.77–6.84 (m, 7H), 6.88–7.01 (m, 7H) ppm. ^{13}C NMR (100 MHz, CDCl3) δ: 19.61, 19.64, 20.51, 20.54, 20.61, 20.7, 20.78, 20.82, 20.9, 24.9, 25.3, 26.16, 26.24, 43.7, 43.8, 43.9, 44.0, 48.2, 48.5, 67.5, 67.6, 118.1, 118.6, 123.4, 123.8, 125.45, 125.55, 125.6, 125.78, 125.83, 125.9, 126.0, 126.3, 126.4, 127.0, 127.2, 127.5, 127.8, 128.3, 128.5, 129.4, 130.0, 130.7, 130.9, 133.6, 133.7, 134.8, 135.70, 135.74, 138.0, 138.1, 142.1, 142.4, 142.5, 167.2, 167.8, 210.54, 210.55 ppm (58 signals for 62 carbons). IR (ATR solid) n: 2961, 2920, 2888, 1762, 1737, 1480, 1448, 1313, 1270 cm^{-1}. Elem. Anal. for C$_{31}$H$_{29}$NO$_3$S$_2$: Calcd. C 70.56, H 5.54, N 2.65; found C 70.48, H 5.55, N 2.64.

Diastereoisomers **8beD1** and **8beD2**. (*M*/*P*)-1,3,7-trimethyl[1,4]benzothiazino[2,3,4-*kl*]phenothiazine-2-yl (S)-2-(6-methoxy-2-naphthyl) propionate. Following the general procedure from **1b(OH)** (48 mg, 0.13 mmol) and (S)-(+)-2-(6-methoxy-2-naphthyl) propionic acid **7e** (28 mg, 0.13 mmol), kept for 12 h at room temperature. The crude was purified by flash chromatography on silica gel (petroleum ether/CH$_2$Cl$_2$ 1/2, Rf 0.74)) to afford the mixture of the two diastereomeric compounds **8beD1** and **8beD2** (43 mg, 58% yield) as a white solid (mp 250 °C dec). ^{1}H NMR (400 MHz, CDCl$_3$) δ: 1.23 (d, 3H, J = 7.2 Hz), 1.39 (d, 3H, J = 7.2 Hz) 2.208 (s, 3H), 2.214 (s, 3H), 2.22 (s, 3H), 2.23 (s, 3H), 2.29 (s, 3H), 2.33 (s, 3H), 3.10 (q, 1H, J = 7.3 Hz), 3.20 (q, 1H, J = 7.2 Hz), 3.91 (s, 3H), 3.92 (s, 3H), 6.52 (bs, 1H), 6.65 (bs, 1H), 6.75–6.94 (m, 10H), 6.99 (bs, 1H), 7.04 (bs, 1H), 7.09–7.17 (m, 5H), 7.27–7.30 (m, 1H), 7.39 (bs, 1H), 7.53 (bs, 1H), 7.64–7.71 (m, 4H) ppm. Elem Anal. for C$_{35}$H$_{29}$NO$_3$S$_2$: Calcd. C 73.02, H 5.08, N 2.43; found C 73.03, H 5.09, N 2.43.

Diastereoisomers **8bfD1** and **8bfD2**. (*M/P*)-1,3,7-trimethyl[1,4]benzothiazino[2,3,4-*kl*]phenothiazine-2-yl mono(1*R*)-menthylphthalate. Following the general procedure from **1b(OH)** (40 mg, 0.11 mmol) and (-)mono(1*R*)-menthylphthalate **7f** (33 mg, 0.11 mmol), kept for 18 h at room temperature. The crude was purified by flash chromatography on silica gel (petroleum ether/CH$_2$Cl$_2$ 1/1, Rf 0.63)) to afford the mixture of the two diastereomeric compounds **8bfD1** and **8bfD2** (32 mg, 45% yield) as a white solid (mp 180–190 °C). ^{1}HNMR (400 MHz, CDCl$_3$) δ: 0.66 (d, 3H, J = 6.9 Hz), 0.78–0.90 (m, 17H), 0.98–1.11 (m, 4H), 1.40–1.53 (m, 4H), 1.65–1.71 (m, 4H), 1.90–1.99 (m, 2H), 2.05–2.26 (m, 14H), 2.53 (s, 6H), 4.85–4.93 (m, 2H), 6.70–6.85 (m, 10H), 6.95–7.00 (m, 4H), 7.07 (bs, 2H), 7.26–7.30 (m, 2H), 7.44–7.54 (m, 4H) ppm. ^{13}CNMR (100 MHz, CDCl$_3$) δ: 16.1, 16.4, 20.5, 20.6, 20.90, 20.92, 21.02, 22.1, 22.2, 23.4, 23.5, 26.2, 26.3, 31.5, 31.6, 34.4, 34.5, 40.5, 40.7, 47.3, 47.4, 75.7, 118.45, 118.50, 123.35, 123.40, 125.7, 125.76, 125.77, 125.79, 125.9, 126.26, 126.34, 126.4, 127.0, 127.46, 127.48, 128.05, 128.12. 128.16, 128.3, 129.50, 129.53, 129.6, 129.75, 129.77, 130.15, 130.18, 130.7, 130.8, 131.4, 131.5, 133.6, 133.9, 134.1, 134.7, 134.8, 135.36, 135.40, 138.1, 138.2, 141.45, 141.52, 141.98, 142.01, 163.82, 163.84, 167.28, 167.33 ppm (56 signals for 78 different carbons). Elem. Anal. for C$_{39}$H$_{39}$NO$_4$S$_2$: Calcd. C 72.08, H 6.05, N 2.16; found C 71.99, H 6.06, N 2.15.

Diastereoisomers **8bgD1** and **8bgD2**. (*M/P*)-1,3,7-trimethyl[1,4]benzothiazino[2,3,4-*kl*]phenothiazine-2-yl (S)-N-Boc pipecolate. Following the general procedure from **1b(OH)** (40 mg, 0.11 mmol) and (S)-N-Boc pipecolic acid **7g** (25 mg, 0.11 mmol), kept for 3 h at room temperature. The crude was purified by flash chromatography on silica gel (petroleum ether/CH$_2$Cl$_2$ 1/3, D1 Rf 0.27, D2 Rf 0.20) to afford the product **8bgD1** (17 mg, 28% yield) as a white solid (mp 79–82 °C) and the product **8bgD2** (9 mg, 15% yield) as a white solid (mp 121–125 °C). **8bgD1**: ^{1}HNMR (400Mz, CDCl$_3$) δ: 1.10–1.21 (m, 1H), 1.26–1.53 (m, 14H), 2.21 (s, 3H), 2.28 (s, 3H), 2.29 (s, 3H), 2.29 (s, 3H), 2.80–2.95 (m, 1H), 3.79–3.96 (m, 1H), 4.40–4.47 (m, 1H), 6.78–6.91 (m, 6H), 7.00 (bs, 1H) ppm. ^{13}C NMR (100 MHz, CDCl3) δ: 20.6, 20.7, 21.1, 21.4, 24.7, 25.0, 26.0, 28.5, 29.8, 41.2, 42.2, 54.3, 55.3, 77.2, 80.0, 80.1, 118.2, 122.6, 123.2, 125.5, 125.8, 126.3, 127.2, 127.7, 127.9, 128.1, 128.5, 129.5, 130.9, 131.3, 133.8, 134.2, 134.8, 134.9, 135.5, 138.1, 141.9, 142.4, 155.4, 155.7, 169.7, ppm. IR (ATR solid): n: 2973, 2924, 2860, 1764, 1689, 1480, 1448, 1364, 1252 cm^{-1}. $[α]_D^{20}$ −157, (c 0.1, CH$_2$Cl$_2$). **8bgD2**: ^{1}HNMR (400Mz, CDCl$_3$) δ: 0.79–0.99 (m, 1H), 1.26–1.45 (m, 14H), 2.14–2.29 (m, 9H), 2.46–2.94 (m, 1H), 3.65–3.92 (m, 1H), 4.39–4.63 (m, 1H), 6.78–6.91 (m, 6H), 7.00–7.02 (m, 1H) ppm. ^{13}C NMR (100 MHz, CDCl3) δ: 20.1, 20.2, 20.5, 20.6, 20.70, 20.74, 20.8, 21.0, 24.7, 24.9, 26.0, 26.1, 28.4, 28.6, 41.0, 42.0, 54.4, 55.3, 80.0, 80.3, 115.2, 118.1, 118.2, 122.7, 122.9, 125.5, 125.8, 126.4, 127.5, 128.1, 128.3, 128.9, 129.5, 131.5, 133.8, 134.9, 135.6, 138.1, 142.5, 155.7, 169.7, 169.9, ppm. $[α]_D^{20}$ +49 (c 0.1, CH$_2$Cl$_2$).

Diastereoisomers **8bhD1** and **8bhD2**. (*M/P*)-1,3,7-trimethyl[1,4]benzothiazino[2,3,4-*kl*]phenothiazine-2-yl (1*S*)-camphanate. Following the general procedure from **1b(OH)** (259 mg, 0.71 mmol) and (1*S*)-(−)-camphanic acid **7h** (170 mg, 0.86 mmol), kept for 22 h at room temperature. The crude was purified by flash chromatography on silica gel (petroleum ether/CH$_2$Cl$_2$ 1/3, D1 Rf 0.37, D2 Rf 0.26) to afford product **8bhD1** (143 mg, 37% yield) as a white solid (mp 70–72 °C) and product **8bhD2** (126 mg, 33% yield) as a white solid (mp 86–88 °C). **8bhD1**: ^{1}HNMR (400 MHz, CDCl$_3$) δ: 0.97 (s, 3H), 0.98 (s, 3H), 1.06 (s, 3H), 1.23–1.30 (m, 1H), 1.50–1.55 (m, 1H), 1.77–1.84 (m, 1H), 1.98–2.05 (m, 1H), 2.22 (s, 3H), 2.26 (s, 3H), 2.32 (s, 3H), 6.73 (bs, 1H), 6.79–6.81 (m, 2H), 6.85 (bs, 1H), 6.91–6.96 (m, 2H), 7.00 (bs, 1H) ppm. ^{13}C NMR (100 MHz, CDCl$_3$) δ: 9.8, 16.9, 17.0, 20.6, 20.7, 20.8, 29.0, 29.7, 54.5, 54.8, 90.8, 118.2, 122.7, 125.46, 125.49, 126.0, 126.2, 126.5, 127.1, 127.9, 128.7,

129.8, 131.6, 134.5, 135.0, 135.7, 137.9, 141.3, 141.6, 164.5, 177.9. IR (ATR solid) n: 2970, 2922, 2867, 1787, 1776, 1481, 1449, 1309, 1250 cm^{-1}. $[\alpha]_D^{20}$ −129 (c 0.1, CH$_2$Cl$_2$). **8bhD2** ^{1}HNMR (400 MHz, CDCl$_3$) δ: 0.89 (s, 3H), 0.93 (s, 3H), 1.06 (s, 3H), 1.51–1.56 (m, 3H), 1.65–1.71 (m, 1H), 2.21 (s, 3H), 2.25 (s, 3H), 2.31 (s, 3H), 6.68 (bs, 1H), 6.77–6.79 (m, 2H), 6.84–6.89 (m, 2H), 6.96 (bs, 1H), 7.02 (bs, 1H) ppm. ^{13}C NMR (100 MHz, CDCl$_3$) δ: 9.8, 16.9, 17.0, 20.6, 20.7, 20.8, 28.9 29.6, 54.3, 54.8, 90.8, 118.1, 122.7, 125.2, 125.9, 126.3, 126.6 (2C), 127.5, 128.0, 128.2, 129.9, 132.0, 133.9, 135.0, 135.7, 137.8, 141.6, 142.1, 164.5, 177.7 ppm. IR (ATR solid) n: 2967, 2922, 2865, 1776, 1482, 1449, 1309, 1250 cm^{-1}. $[\alpha]_D^{20}$ + 126 (c 0.1, CH$_2$Cl$_2$).

General Procedure for the hydrolysis: To a solution of ester **8** in CH$_2$Cl$_2$/MeOH: 10/1 (roughly 0.05 M) 3 eq. of NaOH was added, and the solution was stirred for 4–6 h at room temperature. The solution was diluted with water and HCl (1M) was added until the pH was neutral, then the mixture was extracted with CH$_2$Cl$_2$ (3 × 5 mL). The organic layer was dried over Na$_2$SO$_4$, filtered, and then evaporated under reduced pressure. The crude was purified by flash chromatography on silica gel (Petroleum Ether/CH$_2$Cl$_2$ 1/2) to afford the products (M)-**1b(OH)** or (P)-**1b(OH)** as a white solid in quantitative yield. (M)-**1b(OH)** $[\alpha]_D^{20}$ −161 (c 0.1, CH$_2$Cl$_2$) and (P)-**1b(OH)** $[\alpha]_D^{20}$ +166 (c 0.1, CH$_2$Cl$_2$).

Experimental HPLC Analytical (250 × 4.6 mm) column packed with Chiralpak IA chiral stationary phase was purchased from Chiral Technologies Europe. The HPLC resolution of products was performed on a HPLC Waters Alliance 2695 equipped with a 200 µL loop injector and a spectrophotometer UV Waters PDA 2996. The mobile phase, delivered at a flow rate of 1.2 mL/min, was hexane/CH$_2$Cl$_2$ 70/30 v/v + 1% MeOH.

4. Conclusions

In this paper we have reported that the fine matching of the structures of the chiral auxiliaries used and, above all, the topology of their insertion on the helical skeleton, *bay-zone* vs *cape-zone*, allow for the chemical resolution of DTA[4]H) **1**. Helicene **1b(OH)** allowed for the insertion of the chiral auxiliary on the 1-position, the area of terminal ring superimposition that we indicated as the *bay-zone*. Esterification of **1b(OH)** with (1S)-(−)-camphanic acid **7h** provided diastereomers **8bhD1** and **8bhD2** which were successfully separated by flash chromatography and hydrolyzed providing enantiopure helicenes (M)-**1b(OH)** and (P)-**1b(OH)**, respectively.

Supplementary Materials: The following supporting information is available online: HPLC Analysis, NMR spectra, DFT calculations of compound **1b(OH)**, Optimized structures' coordinates.

Author Contributions: Conceptualization, S.M. and C.V.; methodology, validation and investigation, M.O. and M.L.; data curation, M.L.; formal analysis, S.A. and G.L.; writing—original draft preparation, C.V.; writing—review and editing, C.V., S.M. and M.L.; supervision and project administration, C.V. and S.M.; funding acquisition, S.M. All authors have read and agreed to the published version of the manuscript.

Funding: This research received no external funding.

Institutional Review Board Statement: Not applicable.

Informed Consent Statement: Not applicable.

Data Availability Statement: Not applicable.

Acknowledgments: The authors thank MUR-Italy 'Progetto Dipartimenti di Eccellenza 2018–2022' allocated to the Department of Chemistry 'Ugo Schiff', University of Florence, Italy.

Conflicts of Interest: The authors declare no conflict of interest.

Sample Availability: Samples of the compounds are available from the authors.

References

1. Liu, M.; Zhang, L.; Wang, T. Supramolecular chirality in self-assembled systems. *Chem. Rev.* **2015**, *115*, 7304–7397. [CrossRef]
2. Shen, Y.; Chen, C.-F. Helicenes: Synthesis and applications. *Chem. Rev.* **2012**, *112*, 1463–1535. [CrossRef] [PubMed]

3. Mori, T. Chiroptical properties of symmetric double, triple, and multiple helicenes. *Chem. Rev.* **2021**, *121*, 2373–2412. [CrossRef] [PubMed]
4. Reiné, P.; Ortuño, A.M.; Resa, S.; de Cienfuegos, L.Á.; Ribagorda, M.; Mota, A.J.; Abbate, S.; Longhi, G.; Miguel, D.; Cuerva, J.M. Enantiopure double *ortho* oligophenylethynylene-based helical structures with circularly polarized luminescence activity. *ChemPhotoChem* **2021**, *6*, e202100160. [CrossRef]
5. Hong, J.; Xiao, X.; Liu, H.; Fu, L.; Wang, X.-C.; Zhou, L.; Wang, X.-Y.; Qiu, Z.; Cao, X.-Y.; Narita, A.; et al. X-shaped thiadiazole-containing double [7]heterohelicene with strong chiroptical response and π-stacked homochiral assembly. *Chem. Com.* **2021**, *57*, 5566–5569. [CrossRef] [PubMed]
6. Zhou, F.; Huang, Z.; Huang, J.; Cheng, R.; Yang, Y.; You, J. Triple Oxa[7]helicene with circularly polarized luminescence: Enhancing the dissymmetry factors via helicene subunit multiplication. *Org. Lett.* **2021**, *23*, 4559–4563. [CrossRef] [PubMed]
7. Hasan, M.; Borovkov, V. Helicene-based chiral auxiliaries and chirogenesis. *Symmetry* **2018**, *10*, 10. [CrossRef]
8. Neidle, S.; Waring, M. *Molecular Aspects of Anticancer Drug-DNA Interactions*; CRC Press: Boca Raton, FL, USA, 1993.
9. D'Incalci, M.; Sessa, C. DNA minor groove binding ligands: A new class of anticancer agents. *Expert Opin. Invest. Drugs* **1997**, *6*, 875–884. [CrossRef]
10. Honzawa, S.; Okubo, H.; Anzai, S.; Yamaguchi, M.; Tsumoto, K.; Kumagai, I. Chiral recognition in the binding of helicenediamine to double strand DNA: Interactions between low molecular weight helical compounds and a helical polymer. *Bioorg Med. Chem.* **2002**, *10*, 3213–3218. [CrossRef]
11. Xu, Y.; Zhang, Y.X.; Sugiyama, H.; Umano, T.; Osuga, H.; Tanaka, K. (P)-helicene displays chiral selection in binding to Z-DNA. *J. Am. Chem Soc.* **2004**, *126*, 6566–6567. [CrossRef]
12. Shinohara, K.; Sannohe, Y.; Kaieda, S.; Tanaka, K.; Osuga, H.; Tahara, H.; Xu, Y.; Kawase, T.; Bando, T.; Sugiyama, H. A chiral wedge molecule inhibits telomerase activity. *J. Am. Chem. Soc.* **2010**, *132*, 3778–3782. [CrossRef] [PubMed]
13. Leydecker, T.; Wang, Z.M.; Torricelli, F.; Orgiu, E. Organic-based inverters: Basic concepts, materials, novel architectures and applications *Chem. Soc. Rev.* **2020**, *49*, 7627–7670. [CrossRef] [PubMed]
14. Wang, J.; Wang, Y.; Xie, X.; Ren, Y.; Zhang, B.; He, L.; Zhang, J.; Wang, L.-D.; Wang, P. A helicene-based molecular semiconductor enables 85 °C stable perovskite solar cells. *ACS Energy Lett.* **2021**, *6*, 1764–1772. [CrossRef]
15. Gingras, M. One hundred years of helicene chemistry. Part 1: Non-stereoselective syntheses of carbohelicenes. *Chem. Soc. Rev.* **2013**, *42*, 968–1006. [CrossRef]
16. Gingras, M.; Félix, G.; Peresutti, R. One hundred years of helicene chemistry. Part 2: Stereoselective syntheses and chiral separations of carbohelicenes. *Chem. Soc. Rev.* **2013**, *42*, 1007–1050. [CrossRef]
17. Lamanna, G.; Faggi, C.; Gasparrini, F.; Ciogli, A.; Villani, C.; Stephens, P.J.; Devlin, F.J.; Menichetti, S. Efficient thia-bridged triarylamine heterohelicenes: Synthesis, resolution, and absolute configuration determination. *Chem. Eur. J.* **2008**, *14*, 5747–5750. [CrossRef]
18. Menichetti, S.; Cecchi, S.; Procacci, P.; Innocenti, M.; Becucci, L.; Franco, L.; Viglianisi, C. Thia-bridged triarylamine heterohelicene radical cations as redox-driven molecular switches. *Chem. Commun.* **2015**, *51*, 11452–11454. [CrossRef]
19. Longhi, G.; Castiglioni, E.; Villani, C.; Sabia, R.; Menichetti, S.; Viglianisi, C.; Devlin, F.; Abbate, S. Chiroptical properties of the ground and excited states of two thia-bridged triarylamine heterohelicenes. *J. Photochem. Photobiol. A Chem.* **2016**, *331*, 138–145. [CrossRef]
20. Menichetti, S.; Faggi, C.; Onori, M.; Piantini, S.; Ferreira, M.; Rocchi, S.; Lupi, M.; Marin, I.; Maggini, M.; Viglianisi, C. Thia-bridged triarylamine hetero[4]helicenes: Regioselective synthesis and functionalization C. *Eur. J. Org. Chem.* **2019**, *2019*, 168–175. [CrossRef]
21. Amorati, R.; Valgimigli, L.; Baschieri, A.; Guo, Y.; Mollica, F.; Menichetti, S.; Lupi, M.; Viglianisi, C. SET and HAT/PCET acid-mediated oxidation processes in helical shaped fused bis-phenothiazines. *ChemPhysChem* **2021**, *22*, 1446–1454. [CrossRef]
22. Lupi, M.; Menichetti, S.; Stagnaro, P.; Utzeri, R.; Viglianisi, C. Thia-bridged triarylamine[4]helicene-functionalized polynorbornenes as redox-active pH-sensitive polymers. *Synthesis* **2021**, *53*, 2602–2611. [CrossRef]
23. Gliemann, B.D.; Petrovic, A.G.; Zolnhofer, E.M.; Dral, P.O.; Hampel, F.; Breitenbruch, G.; Schulze, P.; Raghavan, V.; Meyer, K.; Polavarapu, P.L.; et al. Configurationally stable chiral dithia-bridged hetero[4]helicene radical cation: Electronic structure and absolute configuration. *Chem. Asian J.* **2017**, *12*, 31–35. [CrossRef] [PubMed]
24. Giaconi, N.; Sorrentino, A.L.; Poggini, L.; Lupi, M.; Polewczyk, V.; Vinai, G.; Torelli, P.; Magnani, A.; Sessoli, R.; Menichetti, S.; et al. Stabilization of an enantiopure sub-monolayer of helicene radical cations on a Au(111) surface through noncovalent interactions. *Angew. Chem.* **2021**, *133*, 15404–15408. [CrossRef]
25. Nuckolls, C.; Katz, T.J.; Katz, G.; Collings, P.J.; Castellanos, L. Synthesis and aggregation of a conjugated helical molecule. *J. Am. Chem. Soc.* **1999**, *121*, 79–88. [CrossRef]
26. Nuckolls, C.; Katz, T.J.; Castellanos, L. Aggregation of conjugated helical molecules. *J. Am. Chem. Soc.* **1996**, *118*, 3767–3768. [CrossRef]
27. Thongpanchang, T.; Paruch, K.; Katz, T.J.; Rheingold, A.L.; Lam, K.-C.; Liable-Sands, L. Why (1S)-camphanates are excellent resolving agents for helicen-1-ols and why they can be used to analyze absolute configurations. *J. Org. Chem.* **2000**, *65*, 1850–1856. [CrossRef] [PubMed]
28. Fox, J.M.; Goldberg, N.M.; Katz, T.J. Efficient synthesis of functionalized [7] helicenes. *J. Org. Chem.* **1998**, *63*, 7456–7462. [CrossRef]

29. Dreher, S.D.; Weix, D.J.; Katz, T.J. Easy synthesis of functionalized hetero [7] helicenes. *J. Org. Chem.* **1999**, *64*, 3671–3678. [CrossRef]
30. Dreher, S.D.; Paruch, K.; Katz, T.J. Application of the Russig-Laatsch reaction to synthesize a bis [5] helicene chiral pocket for asymmetric catalysis. *J. Org. Chem.* **2000**, *65*, 815–822. [CrossRef]
31. Dreher, S.D.; Katz, T.J.; Lam, K.-C.; Rheingold, A.L. First Friedel-Crafts diacylation of a phenanthrene as the basis for an efficient synthesis of nonracemic [7]helicenes. *J. Org. Chem.* **2000**, *65*, 7602–7608. [CrossRef]
32. Li, H.-Y.; Nehira, T.; Hagiwara, M.; Harada, N. Total synthesis and absolute stereochemistry of the natural atropisomer of the biflavone 4′,4‴,7,7″-tetra-O-methylcupressuflavone. *J. Org. Chem.* **1997**, *62*, 7222–7227. [CrossRef] [PubMed]
33. Fuji, K.; Sakurai, M.; Kinoshita, T.; Kawabata, T. Palladium-catalyzed asymmetric reduction of allylic esters with a new chiral monodentate ligand, 8-diphenylphosphino-8′-methoxy-1, 1′-binaphthyl. *Tetrahedron Lett.* **1998**, *39*, 6323–6326. [CrossRef]
34. Ohmori, K.; Kitamura, M.; Suzuki, K. From axial chirality to central chiralities: Pinacol cyclization of 2,2′-biaryldicarbaldehyde to trans-9,10-dihydrophenanthrene-9,10-diol. *Angew. Chem. Int. Ed. Engl.* **1999**, *38*, 1226–1229. [CrossRef]
35. Schaefer, T.; Penner, G.H. The conformational properties of some phenyl esters. Molecular orbital and nuclear magnetic resonance studies. *Can. J. Chem.* **1987**, *65*, 2175–2178. [CrossRef]
36. Schaefer, T.; Sebastian, R.; Penner, G.H. Long-range formyl proton coupling constants of 4-X-phenyl formats (X=H, F, CH$_3$, NO$_2$) and 2,6-dichlorophenyl formate. Conformations in solution. *Can. J. Chem.* **1988**, *66*, 1787–1793. [CrossRef]
37. Mohamadi, F.; Richards, N.G.J.; Guida, W.C.; Liskamp, R.; Lipton, M.; Caufield, C.; Chang, G.; Hendrickson, T.; Still, W.C. Macromodel—An integrated software system for modeling organic and bioorganic molecules using molecular mechanics. *J. Comput. Chem.* **1990**, *11*, 440–467. [CrossRef]
38. Allinger, N.L.; Yuh, Y.H.; Lii, J.-H. Molecular mechanics. The MM3 force field for hydrocarbons. 1. *J. Am. Chem. Soc.* **1989**, *111*, 8551–8566. [CrossRef]
39. Frisch, M.J.; Trucks, G.W.; Schlegel, H.B.; Scuseria, G.E.; Robb, M.A.; Cheeseman, J.R.; Scalmani, G.; Barone, V.; Petersson, G.A.; Nakatsuji, H.; et al. Gaussian 16, Revision C.01. Gaussian, Inc.: Wallingford, CT, USA, 2016. Available online: https://gaussian.com/citation/ (accessed on 28 December 2021).
40. Zhao, Y.; Truhlar, D.G. Truhlar The M06 suite of density functionals for main group thermochemistry, thermochemical kinetics, noncovalent interactions, excited states, and transition elements: Two new functionals and systematic testing of four M06-class functionals and 12 other functionals. *Theor. Chem. Acc.* **2008**, *120*, 215.
41. Coghill, A.M.; Garson, L.R. *The ACS Style Guide*, 3rd ed.; American Chemical Society: Washington, DC, USA, 2006; p. 274. [CrossRef]

Article

Terpenoid Hydrazones as Biomembrane Penetration Enhancers: FT-IR Spectroscopy and Fluorescence Probe Studies

Mariia Nesterkina [1,*], Serhii Smola [2], Nataliya Rusakova [2] and Iryna Kravchenko [1]

[1] Department of Organic and Pharmaceutical Technologies, Odessa National Polytechnic University, 65044 Odessa, Ukraine; kravchenko.pharm@gmail.com

[2] A.V. Bogatsky Physico-Chemical Institute, National Academy of Sciences of Ukraine, 65080 Odessa, Ukraine; sssmola@gmail.com (S.S.); natavrusakova@gmail.com (N.R.)

* Correspondence: mashaneutron@gmail.com; Tel.: +38-093-713-38-53

Abstract: Hydrazones based on mono- and bicyclic terpenoids (verbenone, menthone and carvone) have been investigated in vitro as potential biomembrane penetration enhancers. In this regard, liposomes composed of lecithin or cardiolipin as phospholipid phase components with incorporated fluorescence probes have been prepared using the thin-film ultrasonic dispersion method. The mean particle size of the obtained liposomes, established using laser diffraction, was found to be 583 ± 0.95 nm, allowing us to categorize them as multilamellar vesicles (MLVs) according to their morphology. Pursuant to fluorescence analysis, we may assume a reduction in microviscosity and, consequently, a decrease in the packing density of lecithin and cardiolipin lipids to be the major mechanism of action for terpenoid hydrazones **1–15**. In order to determine the molecular organization of the lipid matrix, lipids were isolated from rat strata cornea (SCs) and their interaction with tested compounds was studied by means of Fourier transform infrared spectroscopy. FT-IR examination suggested that these hydrazones fluidized the SC lipids via the disruption of the hydrogen-bonded network formed by polar groups of SC constituents. The relationship between the structure of terpenoid hydrazones and their ability to enhance biomembrane penetration is discussed.

Keywords: terpenes; hydrazones; penetration enhancers; liposomes; lipids; stratum corneum; laser diffraction; fluorescence probe; pyrene; FT-IR spectroscopy

Citation: Nesterkina, M.; Smola, S.; Rusakova, N.; Kravchenko, I. Terpenoid Hydrazones as Biomembrane Penetration Enhancers: FT-IR Spectroscopy and Fluorescence Probe Studies. *Molecules* 2022, 27, 206. https://doi.org/10.3390/molecules27010206

Academic Editors: Francesca Cardona, Camilla Parmeggiani and Camilla Matassini

Received: 30 November 2021
Accepted: 29 December 2021
Published: 29 December 2021

Publisher's Note: MDPI stays neutral with regard to jurisdictional claims in published maps and institutional affiliations.

Copyright: © 2021 by the authors. Licensee MDPI, Basel, Switzerland. This article is an open access article distributed under the terms and conditions of the Creative Commons Attribution (CC BY) license (https://creativecommons.org/licenses/by/4.0/).

1. Introduction

The development of novel drug molecules involves not only the interaction of the compounds with their potential pharmacological targets but also a mechanism of drug delivery that can overcome biological barriers such as the skin, blood–brain barrier and cell and nuclear membranes. Many pharmacologically active compounds exhibit low activity when administered orally due to their reduced bioavailability, largely caused by poor membrane permeability [1]. Transdermal delivery or topical applications are limited by the stratum corneum barrier, which is the skin's outermost layer, comprising cells compressed into a matrix of intercellular lipids [1,2]. In order to improve the membrane permeability of hydrophilic, low-molecular-weight compounds, the chemical modification of their structure can be applied. The prodrug strategy serves as an example of this approach, consisting of drug molecule derivatization, leading to the facilitation of the permeability of compounds and their subsequent enzymatic cleavage [3]. Such a method of increasing lipophilicity and prolonging action has been successfully employed for morphine, naltrexone, ketorolac, bupropion, theophylline and haloperidol [4–9]. More impactful, however, is a multidrug idea aimed at the transporting of hydrophilic drugs through membrane barriers through their conjugation with bioactive compounds possessing penetration enhancer (PE) properties. Among various PEs, terpenes and their oxygen-containing derivatives have received much attention due to their high efficiency, safety, low skin irritation and their ability to improve the permeability of both lipophilic and hydrophilic molecules [10]. Notably, terpenes demonstrate their own biological and pharmacological activity: antidepressant, antidiabetic, anticancer, anticonvulsant, antiviral, analgesic, anti-inflammatory and antioxidant effects [11–14]. Terpenoid scaffolds have been used to improve the membrane permeability of hydrophilic neurotransmitter amino acids (GABA and glycine); as a consequence, esters with multi-target activities and the ability to disrupt lipid packing have been synthesized [15,16].

The concept of multidrugs (or polypharmacology) was successfully implemented by our research group when searching for novel compounds capable of simultaneously affecting the central and peripheral nervous systems. In this respect, oxygen-containing terpenes, as agonists of transient receptor potential (TRP) channels and allosteric modulators of $GABA_A$ receptors [17,18], were conjugated with *para*-substituted phenoxyacetic acid hydrazides that also manifest anticonvulsant and nociceptive potentialities [19,20]. Therefore, hydrazones based on carvone, menthone and verbenone were obtained and tested in vivo as potential antiseizure and analgesic agents with prolonged action due to the enzymatically degraded azomethine –NH–N=C– group in their structure [21–23]. Bearing in mind the high pharmacological effect of the aforementioned terpenoid derivatives with oral administration and topical applications, studies of their impact on lipid molecular organization are feasible. Some main mechanisms of permeation enhancement involve the disruption of lipid bilayer packing, transient opening of tight junctions, complexation/carrier/ion pairing and disruption of the cellular protein structure [1]. In order to elucidate the mechanism of interaction between enhancers and artificial membranes or extracted lipids, the fluorescence probe method and Fourier transform infrared spectroscopy (FT-IR) have been extensively adopted [24,25].

Given the above, current paper is devoted to the investigation of the use of terpenoid hydrazones as biomembrane enhancers on models of artificial phospholipid membranes and lipids isolated from the stratum corneum (SC), exploiting fluorescence studies and Fourier transform infrared spectroscopy.

2. Results and Discussion

2.1. Pyrene Fluorescence Studies

According to the concept of polypharmacology, drug molecules may simultaneously interact with multiple targets, thereby interfering with multiple disease pathways [26]. This, in turn, contributes to avoiding the polypharmacy phenomenon that is defined as the use of multiple medications by patients with multimorbidity [27]. Developing the idea of polypharmacology (or multidrugs), we previously observed compounds contemporaneously affecting both the central and peripheral nervous systems. To that end, hydrazones based on verbenone (**1–5**), menthone (**6–10**) and carvone (**11–15**) containing the residues of para-substituted phenoxyacetic acid were synthesized (Figure 1).

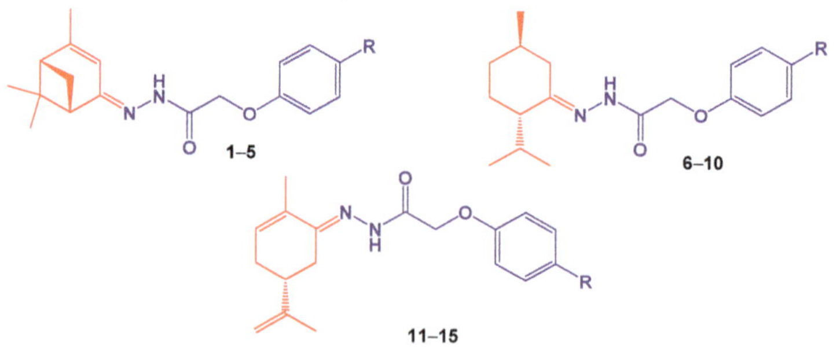

Figure 1. Structures of hydrazones based on (-)–verbenone (**1–5**), (-)–menthone (**6–10**) and (-)–carvone (**11–15**). R = H (**1, 6, 11**); R = Br (**2, 7, 12**); R = Cl (**3, 8, 13**); R = C(CH$_3$)$_3$ (**4, 9, 14**); R = O–C$_6$H$_5$ (**5, 10, 15**).

The aforementioned terpenoid derivatives were found to exhibit anticonvulsant activity on pentylenetetrazole and maximal electroshock seizure test results, along with analgesic potency on chemically and thermally-induced pain models [21–23]. Expanding the concept of polypharmacology, we attempted to design molecules capable of modulating different types of receptors and enhancing self-permeability across biological barriers after their oral administration or transdermal/topical delivery. Mono- and bicyclic oxygen-containing terpenes were selected as scaffolds for the synthesis of novel derivatives due to their high enhancement effect, safety and low skin irritation. In order to investigate the membrane permeability in vitro, liposomes are widely utilized as simple cell models because of their similarity to cellular membranes [28]. In the current study, liposomes composed of lecithin or cardiolipin as phospholipid phase components with incorporated fluorescence probes and terpenoid hydrazones **1–15** were prepared using the thin-film ultrasonic dispersion method.

In the present study, egg yolk lecithin, representing a multi-component system, was applied. Concerning its composition, the following constituents should be indicated: phospholipids, triglycerides, fatty acids, sterols and glycolipids [29]. In turn, phospholipids are divided into two main classes—sphingophospholipids and glycerophospholipids—depending on the structure of the backbone alcohol (sphingosine or glycerol, respectively). The phospholipid composition of the commercial egg yolk lecithin used in the liposome preparation is illustrated in Figure 2. As seen, the major glycerophospholipids comprise phosphatidylcholine (78.4%); phosphatidylethanolamine (17.6%); phosphatidylinositol (1.2%); phosphatidic acid (1.0%) and phosphatidylserine (1.8%).

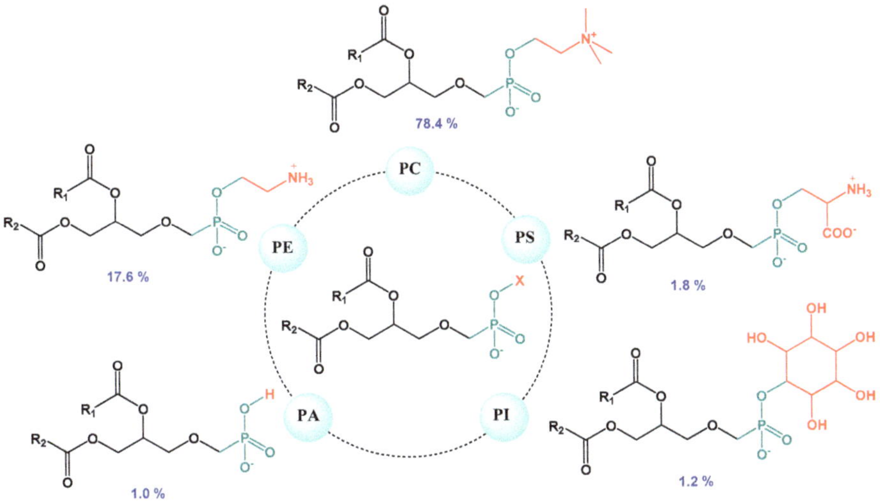

Figure 2. Phospholipid composition of egg yolk lecithin (%): PC—phosphatidylcholine, PE—phosphatidylethanolamine, PI—phosphatidylinositol, PA—phosphatidic acid, PS—phosphatidylserine.

The size of resulting liposomes was established using laser diffraction (LD), which is one of the ensemble methods for rapidly measuring the size and distribution of colloidal systems [30]. Laser diffraction particle size analysis showed that the mean particle size of lecithin and cardiolipin liposomes was 583 ± 0.95 nm (D_{10} 420 nm, D_{50} 583 nm and D_{90} 822 nm). Using conventional classifications, the obtained phospholipid liposomes can be attributed to multilamellar vesicles (MLVs) according to their morphology (>500 nm) [31].

In order to elucidate the ability of terpenoid hydrazones **1–15** to incorporate into biological membranes, pyrene was selected as a classical fluorescence probe due to its long decay time, its ability to form excimers and irs sensitivity towards the microenvironment [32]. Furthermore, pyrene molecules are predominantly localized at the level of hydrocarbon moieties of the lipid bilayer, which is the incorporation area of free terpenes [33]. The vibrational structure of the pyrene emission spectrum was characterized using five peaks, designated as I_1–I_5 at ~373, 379, 384, 394 and 410 nm, accordingly, which was due to the π → π* transitions. Given that the intensity of peak I_1 is increased in polar solvents, whereas the intensity of band I_3 is dramatically enhanced in hydrophobic environments, the ratio I_1/I_3 (I_{373}/I_{384}) was applied to detect the polarity of the pyrene vicinity. Additionally, for monomeric peaks, a broad band appears at longer wavelengths (from 425 to 550 nm, centered at 475 nm) when pyrene molecules in the excited state collide with ground-state pyrene rings. This interaction occurs at a distance ~10 Å between two pyrene forms, leading to the formation of excited-state dimers or "excimers" [34]. Pyrene excimerization clearly depends on the ability of fluorescent probes to laterally diffuse through the membrane. The extent of excimer formation is described mathematically by the excimer/monomer fluorescence intensity ratio (I_E/I_M, I_{475}/I_{394}), comparing the fluorescence intensity of the monomer peak at 394 nm (I_M) and the excimer band at 475 nm (I_E). The increase in the number of excimers formed and, consequently, in the I_{475}/I_{394} value indicates the low viscosity of the system, owing to the lateral diffusivity of the pyrene [35].

With a view to estimating the influence of terpenoid hydrazones **1–15** on lipid molecular organization, both the excimer-to-monomer (I_E/I_M) and the first-to-third (I_1/I_3) intensity ratios were calculated. The effect of hydrazones on membrane permeability was compared to those of initial terpenoids (verbenone, menthone and carvone) examined under the same experimental conditions.

The representative emission spectra of pyrene incorporated into the model lecithin liposomes in the presence of verbenone hydrazones (**1–5**) are illustrated in Figure 3. As shown, excimer formation was not observed at 475 nm in a sample comprising only lecithin phospholipids and membrane probes. However, the appearance of an excimer peak was detected when verbenone derivatives, along with menthone and carvone hydrazones **6–15**, were inserted into liposomal lipids.

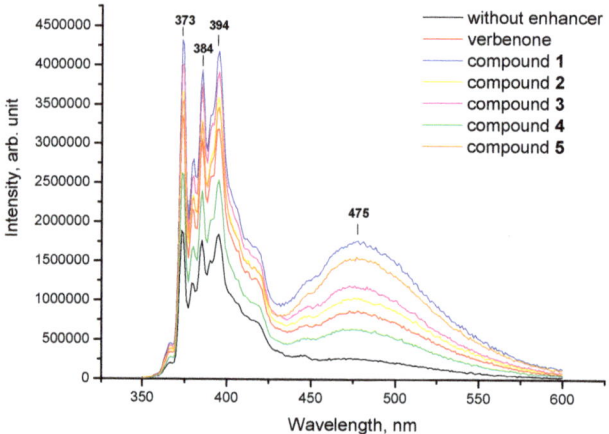

Figure 3. Representative fluorescent emission spectra of pyrene inserted into lecithin liposomes (control, black line) and in the presence of verbenone and its hydrazones **1–5**. All measurements were carried out at 25 °C; the excitation wavelength was 338 nm.

The excimer/monomer fluorescence intensity ratio of pyrene (I_{475}/I_{394}) has been extensively employed to evaluate the microviscosity of the hydrocarbon interiors of phospholipid membranes [36]. This parameter depends upon the rate of monomer lateral diffusion in lipid bilayers and was found to correlate with membrane fluidity [35,37]. The results of the I_E/I_M ratio calculation in lecithin liposomes containing verbenone, menthone and carvone, along with their hydrazones **1–15**, are summarized in Table 1.

Table 1. Pyrene excimer-to-monomer fluorescence intensity ratio I_E/I_M in lecithin liposomes with hydrazones **1–15** incorporated into the lipid membrane ($n = 3$).

Compound	I_E/I_M	Compound	I_E/I_M
Verbenone	0.269 ± 0.002	8	0.311 ± 0.010
Menthone	0.353 ± 0.008	9	0.264 ± 0.003
Carvone	0.291 ± 0.003	10	0.272 ± 0.005
1	0.426 ± 0.004	11	0.440 ± 0.009
2	0.293 ± 0.008	12	0.220 ± 0.011
3	0.413 ± 0.004	13	0.278 ± 0.004
4	0.248 ± 0.005	14	0.294 ± 0.008
5	0.305 ± 0.003	15	0.179 ± 0.004
6	0.352 ± 0.002	Control	0.142 ± 0.004
7	0.415 ± 0.009		

For all groups $p < 0.01$ compared to control.

The data reported in Table 1 indicate an increase in the excimer-to-monomer intensity ratio (from 0.179 to 0.440) for the liposome samples involving terpenoids and their derivatives **1–15**. In contrast, the I_E/I_M value for liposomes containing only fluorescent probes and lecithin phospholipids (control) was found to be 0.142 ± 0.004. As seen, I_E/I_M parameters for the initial terpenoids exceeded the control values two-fold and averaged 0.304. The highest I_E/I_M ratios were recorded for hydrazones **1**, **3**, **7** and **11**, at 0.426, 0.413, 0.415 and 0.440, respectively. It is worth noting the structure of the above-mentioned compounds that contain the residues of different terpenoids, along with H, Cl or Br atoms in the *para*-position of the benzene ring. Based on fluorescence spectroscopic measurements we may state the influence of verbenone, menthone, carvone and their derivatives **1–15** on excimer formation when these compounds were added to lecithin phospholipids. Thus, the incorporation of terpenoid hydrazones into the lecithin liposomes contributes to a decrease in lipid microviscosity; disruption of their packing density, resulting in pyrene displacement; and, consequently, excimer formation.

Since pyrene is an environmentally-sensitive fluorophore, it might be effectively used in order to estimate the micropolarity of the pyrene's vicinity by calculating the ratio of the first (373 nm) to the third (384 nm) vibronic band (I_1/I_3). The results of the determination of the I_1/I_3 values in lecithin liposomes with inserted mono-/bicycle terpenoids and hydrazones **1–15** are presented in Table 2.

Table 2. The ratio of pyrene monomer fluorescence intensity I_1/I_3 in lecithin liposomes with hydrazones **1–15** incorporated into the lipid membrane ($n = 3$).

Compound	I_1/I_3	Compound	I_1/I_3
Verbenone	1.105 ± 0.008 *	8	1.072 ± 0.012
Menthone	1.118 ± 0.004 **	9	1.086 ± 0.010
Carvone	1.134 ± 0.002 **	10	1.112 ± 0.006 **
1	1.104 ± 0.007 *	11	1.082 ± 0.011
2	1.113 ± 0.003 **	12	0.956 ± 0.008
3	1.072 ± 0.002	13	0.991 ± 0.003
4	1.101 ± 0.002 *	14	1.016 ± 0.004
5	1.095 ± 0.009	15	1.086 ± 0.009
6	1.122 ± 0.005 **	Control	1.067 ± 0.008
7	1.084 ± 0.011		

* $p < 0.05$, ** $p < 0.01$ compared to control.

For lecithin liposome samples containing only incorporated fluorescence probes, the I_1/I_3 parameter was shown to be 1.067 ± 0.008. The addition of initial terpenoids and hydrazones **1–15** to phospholipids during liposome preparation caused a slight change in the I_1/I_3 value from 0.956 to 1.134. Thus, an increase in the polarity of the fluorophore microenvironment was observed when adding initial terpenoids and their derivatives **1, 2, 4, 6, 10** to lecithin lipids, indicating no significant structure–property relationships.

Additionally, the impact of terpenoids and their hydrazones on the molecular organization of the lipid matrix was investigated using model liposomes based on cardiolipin. Figures 4 and 5 illustrate the representative emission spectra of pyrene inserted into the cardiolipin liposomes comprising verbenone hydrazones (**1–5**), menthone (**6–10**) and carvone (**11–15**) derivatives. As seen, an unstructured band appears at longer wavelengths (ranging from 425 to 550 nm) with a peak at 475 nm when verbenone hydrazones, along with menthone and carvone derivatives **6–15**, were incorporated into cardiolipin liposomes.

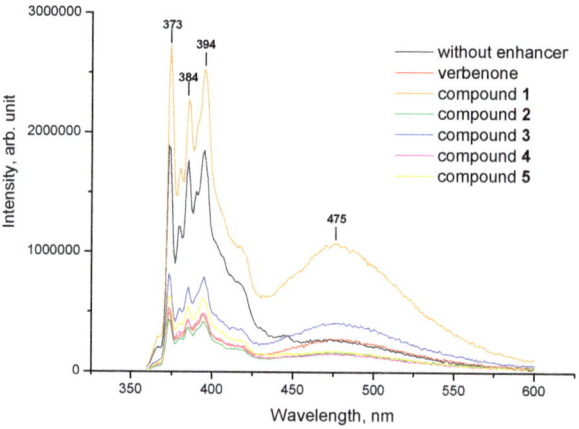

Figure 4. Representative fluorescent emission spectra of pyrene inserted into cardiolipin liposomes (control, black line) and in the presence of verbenone and its hydrazones **1–5**. All measurements were carried out at 25 °C; the excitation wavelength was 338 nm.

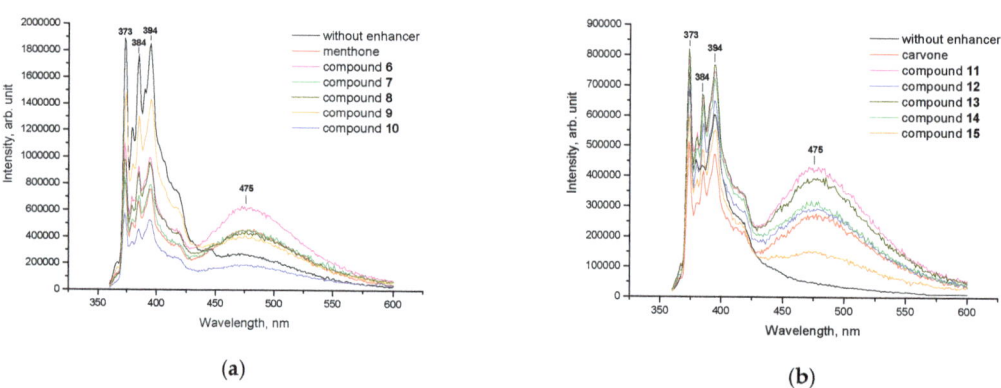

Figure 5. Representative fluorescent emission spectra of pyrene inserted into cardiolipin liposomes (control, black line) and in the presence of menthone **6–10**, (**a**) and carvone **11–15**, (**b**) hydrazones. All measurements were carried out at 25 °C; the excitation wavelength was 338 nm.

The inclusion of terpenoids and hydrazones **1–15** into cardiolipin liposomes was followed by an increase in pyrene fluorescence in the field of excimer formation (450–550 nm) compared to controls (Table 3). As highlighted, consistently high values of the I_E/I_M parameter are typical for initial terpenoids, at 0.574, 0.599 and 0.570 for verbenone, menthone and carvone, accordingly. Among hydrazones, the greatest impact on pyrene lateral diffusion was revealed for compounds **3, 6, 7, 11** and **13** with I_E/I_M ratios 0.529, 0.621, 0.571, 0.570 and 0.511, respectively. By analyzing the structure–property relationship, the following substituents in hydrazone structures should be emphasized: H, Cl or Br atoms in the *para*-position of the benzene ring, which is in accordance with the experiment involving lecithin liposomes. Hence, the incorporation of terpenoid hydrazones into cardiolipin liposomes leads to the enhancement of pyrene lateral mobility, indicating the excimerization of the fluorescence probe.

Table 3. Pyrene excimer-to-monomer fluorescence intensity ratio I_E/I_M in cardiolipin liposomes with hydrazones **1–15** incorporated into the lipid membrane ($n = 3$).

Compound	I_E/I_M	Compound	I_E/I_M
Verbenone	0.574 ± 0.004	8	0.448 ± 0.010
Menthone	0.599 ± 0.003	9	0.289 ± 0.002
Carvone	0.570 ± 0.006	10	0.374 ± 0.004
1	0.426 ± 0.002	11	0.570 ± 0.005
2	0.384 ± 0.004	12	0.444 ± 0.009
3	0.529 ± 0.009	13	0.511 ± 0.011
4	0.315 ± 0.011	14	0.437 ± 0.008
5	0.280 ± 0.009	15	0.287 ± 0.006
6	0.621 ± 0.006	Control	0.146 ± 0.004
7	0.571 ± 0.004		

For all groups $p < 0.01$ compared to control.

The I_1/I_3 ratio, which indicates an increase in the polarity of the pyrene microenvironment has been also calculated for cardiolipin liposomes. As shown in Table 4, a significant increase in the I_1/I_3 parameter was predominantly observed after the incorporation of hydrazones with bulky -C(CH$_3$)$_3$ and -O-C$_6$H$_5$ groups into lipid membranes. This growth of the I_1/I_3 value demonstrates an increase in the polarity of the fluorophore microenvironment, which might be caused by the appearance of hydrophilic clusters in cardiolipin lipid layers.

Table 4. The ratio of pyrene monomer fluorescence intensity I_1/I_3 in cardiolipin liposomes with hydrazones **1–15** incorporated into the lipid membrane ($n = 3$).

Compound	I_1/I_3	Compound	I_1/I_3
Verbenone	1.096 ± 0.010	8	1.122 ± 0.011 *
Menthone	1.260 ± 0.009 **	9	1.143 ± 0.008 **
Carvone	1.236 ± 0.002 **	10	1.262 ± 0.004 **
1	1.026 ± 0.011	11	1.197 ± 0.005 **
2	1.217 ± 0.008 **	12	1.212 ± 0.010 **
3	1.157 ± 0.004 **	13	1.217 ± 0.009 **
4	1.248 ± 0.008 **	14	1.196 ± 0.007 **
5	1.155 ± 0.006 **	15	1.232 ± 0.006 **
6	1.181 ± 0.007 **	Control	1.081 ± 0.002
7	1.163 ± 0.010 **		

* $p < 0.05$, ** $p < 0.01$ compared to control.

Thus, based on fluorescence analysis, we may suggest a reduction in microviscosity and, consequently, a decrease in the packing density of lecithin and cardiolipin lipids, as the major mechanism of action for terpenoid hydrazones with H, Cl or Br atoms in the *para*-position of the benzene ring. Hydrazones containing bulky -C(CH$_3$)$_3$ and -O-C$_6$H$_5$ groups were also found to increase the membrane polarity via the appearance of hydrophilic clusters or via the penetration of water molecules into the lipid layers of cardiolipin liposomes.

2.2. FT-IR Spectroscopy Investigation

Along with fluorescence probe studies, Fourier-transform infrared spectroscopy (FT-IR) refers to a significant technique for examination the molecular organization of an artificial lipid membrane or lipid isolated from the stratum corneum [38,39]. In our research, the FT-IR spectroscopy method was applied to suggest the mechanism of interaction between lipids isolated from rat strata cornea and terpenoid hydrazones. This measurement was based on the estimation of hydrogen-bonding interactions between SC lipids and terpenoid derivatives **1–15**. In this regard, the changes in the absorbance intensity of functional groups of SC constituents were analyzed.

The general structure of the SC involves an array of keratinized cells embedded in a lipid matrix [40]. Thus, the lipids of the stratum corneum consist of two groups: freely extractable intercellular lipids and covalently bound lipids of the corneocyte membrane. SC permeability is identified precisely by intercellular lipids belonging to the following classes: ceramides (Cer), cholesterol (Chol) and long-chain free fatty acids (FFA) that are extracted by means of a chloroform:methanol system [41,42]. Recently, we determined the lipid composition extracted from the SC via a mixture of aforementioned solvents and defined cholesterol, cholesterol oleate, fatty acids, triglycerides and ceramides as major components of the SC intercellular space [15]. Bearing in mind the chemical structures of the designated compounds, the most significant ones for FT-IR analysis were the absorption bands of functional groups that form a network of hydrogen bonds in the lipid matrix, namely, OH and C=O groups (see Supplementary Materials).

Three main peaks were observed in the FT-IR spectrum of pure lipids isolated from SCs: a band at 3393 cm^{-1} corresponding to stretching vibrations of the associated OH groups; two absorption bands at 1737 cm^{-1} and 1656 cm^{-1} of the carbonyl C=O group associated with the monomeric state (1737 cm^{-1}) and carboxylic acid dimers (1656 cm^{-1}). The values of peak intensity at 3393, 1737 and 1656 cm^{-1} for samples of SC lipids comprising hydrazones **1–15** along with initial terpenoids (verbenone, menthone and carvone) are summarized in Table 5. The intensity of the abovementioned bands was expressed as the percentage of incident light transmission (T), which is inversely proportional to the peak intensity. Strong hydrogen-bonding interactions between the hydroxyl groups of SC lipids was observed as a broad band at 3393 cm^{-1} with an intensity of transmission of 78%. A decrease in the intensity of OH stretching vibrations was recorded after the addition of terpenoids and their derivatives **1–15**; this phenomenon is associated with the disruption of hydrogen bonding in the SC lipid matrix. The most pronounced influence was revealed by incorporating hydrazones with H, Cl or Br atoms in the *para*-position of the benzene ring into SC lipids (with an average percentage of transmission of 15%–29%). The same trend was typical for the C=O stretching frequency of carboxylic acid dimers (1656 cm^{-1})—there was a reduction in the intensity of this band when mixing lipids with terpenoid derivatives **1–3, 6–8, 11–13**, whereas hydrazones containing bulky -C(CH$_3$)$_3$ and -O-C$_6$H$_5$ groups exhibited no effect in lipid liquefaction. Notably, both pure terpenoids and their hydrazones displayed a slight impact on C=O group vibrations related to its monomeric form (1737 cm^{-1}). Hence, the disruption of the hydrogen-bonded network formed by polar groups of SC lipids might be suggested as a mechanism of action for terpenoid hydrazones **1–15**.

In this manner, the results of the investigation of hydrazones as biomembrane penetration enhancers using fluorescence probe studies and FT-IR spectroscopy indicated the increased permeability of these compounds across membranes. These data further substantiate the high pharmacological effect of terpenoid derivatives and fully correlate with previously published data [21–23] regarding the analgesic and anticonvulsant activity of the abovementioned compounds. According to our study, the higher antinociceptive and antiseizure potency after topical delivery and oral administration, respectively, was displayed by terpenoid derivatives with H, Cl or Br atoms in the *para*-position of the benzene ring. This work is an example of compounds' designs simultaneously influencing their different pharmacological targets and enhancing their self-permeability.

Table 5. The intensity of absorption bands in the FT-IR spectra of samples containing hydrazones **1–15** and lipids isolated from the SC.

Compound	Intensity of Band, % of Transmission (T)/Absorbance		
	3393 cm^{-1}	1737 cm^{-1}	1656 cm^{-1}
Verbenone	31.5/0.502	68.6/0.164	30.9/0.510
Menthone	27.7/0.558	62.3/0.206	18.9/0.724
Carvone	26.2/0.582	64.4/0.191	24.2/0.616
1	16.2/0.790	54.2/0.266	21.9/0.660
2	21.3/0.672	70.1/0.154	27.6/0.559
3	27.6/0.559	76.8/0.115	30.7/0.513
4	40.6/0.391	93.2/0.031	56.2/0.250
5	30.0/0.523	40.2/0.396	20.3/0.693
6	15.6/0.807	36.1/0.442	25.0/0.602
7	25.6/0.592	64.6/0.190	27.8/0.556
8	28.0/0.553	57.3/0.242	24.9/0.604
9	44.1/0.356	76.8/0.115	41.6/0.381
10	47.9/0.320	89.1/0.050	45.7/0.340
11	16.0/0.796	48.2/0.317	24.3/0.614
12	22.8/0.642	56.7/0.246	29.1/0.536
13	29.8/0.526	62.8/0.202	27.5/0.561
14	46.2/0.335	72.8/0.138	49.1/0.309
15	51.3/0.290	66.9/0.175	42.7/0.370
Control	78.8/0.103	44.1/0.356	42.2/0.375

3. Materials and Methods

3.1. General

Egg yolk lecithin and cardiolipin from the bovine heart were obtained from Biolek (Kharkov, Ukraine). Pyrene and trypsin (from the bovine pancreas) were purchased from Merck (Darmstadt, Germany). All organic solvents and other chemicals used were of analytical grade. Terpenoid hydrazones **1–15** were synthesized and fully characterized in our previous studies [21–23]. Pure terpenoids were used as obtained from their commercial supplier: (−)-verbenone, (−)-menthone and (−)-carvone (TCI, Philadelphia, PA, USA).

3.2. Liposome Preparation

Liposomes based on lecithin or cardiolipin were prepared using the thin-film evaporation method. For this purpose, solutions of phospholipids (lecithin or cardiolipin, 0.02 mol/L), pure terpenoids or their hydrazones **1–15** (0.002 mol/L) and pyrene (0.0002 mol/L) were prepared in chloroform. One milliliter of each above-mentioned solution was taken and placed in a round-bottom flask. Then, the solvent was removed via slow evaporation under a vacuum at 40 °C. The dried mixture was resuspended in 10 mL of deionized water and vigorously stirred for 10 min. The resulting emulsion was then sonicated for 10 min at a frequency of 22 kHz. All liposomes were freshly prepared on the day of the experiment.

3.3. Determination of Liposome Size Distribution

The particle size distributions of both lecithin and cardiolipin liposomes were determined via laser diffraction using a Mastersizer 3000 equipped with a Hydro SM dispersion unit (Malvern Instruments, Malvern, UK). The stirrer speed was set at 2000 rpm; distilled water was used as a dispersant. The obscuration value was in the range of 1–5% for each analysis. Results are expressed as volume median diameters D_{10}, D_{50} and D_{90}.

3.4. Fluorescence Measurements

The steady-state fluorescence emission spectra of liposomes solutions containing pyrene were recorded on a Horiba Jobin-Yvon Fluorog-FL 3-22 spectrophotometer, equipped with a 450 W Xe lamp with the use of 1 cm path-length quartz cuvettes. The excitation wavelength for all samples containing pyrene was 338 nm. The slit width of both excitation and emission was set at 2 nm. The fluorescence intensity ratios of the first to third vibronic bands (I_1/I_3) were determined at 373 nm and 384 nm, respectively. The excimer/monomer emission intensity (I_E/I_M) ratio was calculated by measuring the relative intensities of pyrene excimer and monomer forms at 394 nm and 475 nm, respectively.

3.5. Experimental Animals

Skin samples were collected from male Wistar rats (150–180 g). All animals were kept under a 12 h light regime in a standard animal facility with free access to water and food, in compliance with the European Convention for the Protection of Vertebrate Animals Used for Experimental and Other Specific Purposes (Strasbourg, 1986), ARRIVE guidelines and the principles of the National Ukrainian Bioethics Congress (Kyiv, 2003). All animals were purchased from Odessa National Medical University, Ukraine. The Animal Ethics Committee (agreement No. 6/2021) of Odessa National Polytechnic University (Ukraine) approved the study.

3.6. Isolation of Stratum Corneum

Skin sites were selected from large, uniform body areas of the male Wistar rats. The rats were sacrificed with the inhalation of an excess of chloroform, followed by shaving and surgical removal of the abdominal and back regions. After the removal of subcutaneous fat tissue using a scalpel, the SC was separated from the epidermis by incubating it in trypsin solution (0.15% in PBS buffer, pH 7.4) for 24 h at 4 °C and thereafter for 4 h at 37 °C. Then, the SC was mechanically separated and its trypsinazation was terminated via the addition of trypsin inhibitor solution, with subsequent deionized water washes and drying.

3.7. Extraction of SC Lipids

SC sheets were homogenized, dipped into chloroform:methanol (2:1) solution and kept in the dark for 72 h. Then the extract was separated, washed twice with distilled water and the lower organic lipid-containing layer was evaporated to dryness under a nitrogen atmosphere below 40 °C.

3.8. FT-IR Spectroscopy

FT-IR spectra were recorded on a Frontier FT-IR spectrometer (Perkin-Elmer, Hopkinton, MA, USA). The samples for FT-IR studies were prepared by dissolving the lipids isolated from SCs in carbon tetrachloride (CCl_4), followed by addition of terpenoid hydrazones **1–15** (10% relative to lipids' mass). FT-IR spectra were measured for films obtained using a technique of slow solvent evaporation directly from undercover in a nitrogen atmosphere. All FT-IR investigations were performed at room temperature (25 °C).

3.9. Statistical Analysis

All results are expressed as mean ± standard error mean (SEM). One-way analysis of variance (ANOVA) was used to determine the statistical significance of the results, followed by Tukey's post hoc comparison. $p < 0.05$ was considered significant.

4. Conclusions

In this paper, we confirmed the impact of terpenoid hydrazones containing residues of *para*-substituted phenoxyacetic acid on the molecular organization of the lipid matrix. Fluorescence probe analysis with the use of lecithin and cardiolipin liposomes suggested that terpenoid derivatives decrease phospholipid microviscosity and disrupt their packing density. Considering the potential application of compounds **1–15** as topical agents, their influence on lipids isolated from rat stratum corneum was investigated by FT-IR spectroscopy. The disruption of the hydrogen-bonded network formed by polar groups of stratum corneum lipids was proposed as a mechanism of action for terpenoid hydrazones **1–15**. Hydrazones containing the residues of verbenone, carvone and menthone, along with H, Cl or Br atoms in the *para*-position of the benzene ring, were found to achieve a higher effect as biomembrane penetration enhancers; these results further substantiate the high pharmacological effects observed for terpenoid derivatives and correlate with the previously obtained data on the anticonvulsant and analgesic activity of the above-mentioned compounds. Thus, the influence of terpenoid hydrazones on the molecular packing of lipids substantiates the feasibility of their use both after oral administration and in transdermal delivery in vivo.

Supplementary Materials: The following are available online. Figures S1–S19, FT-IR spectra of pure stratum corneum (SC) lipids and samples containing hydrazones **1–15**, along with lipids isolated from SCs.

Author Contributions: Conceptualization, I.K. and N.R.; methodology, M.N.; software, S.S.; investigation, M.N. and S.S.; data curation, I.K. and N.R.; writing—original draft preparation, M.N.; writing—review and editing, I.K., N.R. and S.S.; visualization, M.N.; supervision, I.K. All authors have read and agreed to the published version of the manuscript.

Funding: This research received no external funding.

Institutional Review Board Statement: The study was conducted according to the European Convention for the Protection of Vertebrate Animals Used for Experimental and Other Specific Purposes (Strasbourg, 1986), ARRIVE guidelines and the principles of the National Ukrainian Bioethics Congress (Kyiv, 2003). The Animal Ethics Committee (agreement No. 6/2021) of Odessa National Polytechnic University (Ukraine) approved the study.

Informed Consent Statement: Not applicable.

Data Availability Statement: The data presented in this study are contained within the article.

Conflicts of Interest: The authors declare no conflict of interest.

Sample Availability: Samples are available from the authors.

References

1. Aungst, B.J. Absorption enhancers: Applications and advances. *AAPS J.* **2012**, *14*, 10–18. [CrossRef]
2. Prausnitz, M.; Langer, R. Transdermal drug delivery. *Nat. Biotechnol.* **2008**, *26*, 1261–1268. [CrossRef] [PubMed]
3. N'Da, D.D. Prodrug strategies for enhancing the percutaneous absorption of drugs. *Molecules* **2014**, *19*, 20780–20807. [CrossRef] [PubMed]
4. Wang, J.J.; Sung, K.C.; Huang, J.F.; Yeh, C.H.; Fang, J.Y. Ester prodrugs of morphine improve transdermal drug delivery: A mechanistic study. *J. Pharm. Pharmacol.* **2007**, *59*, 917–925. [CrossRef] [PubMed]
5. Stinchcomb, A.I.; Swaan, P.W.; Ekabo, O.; Harris, K.K.; Browe, J.; Hammell, D.C.; Cooperman, T.A.; Pearsall, M. Straight-chain naltrexone ester prodrugs: Diffusion and concurrent esterase biotransformation in human skin. *J. Pharm. Sci.* **2002**, *91*, 2571–2578. [CrossRef]
6. Qandil, A.; Al-Nabulsi, S.; Al-Taani, B.; Tashtoush, B. Synthesis of piperazinylalkyl ester prodrugs of ketorolac and their *in vitro* evaluation for transdermal delivery. *Drug Dev. Ind. Pharm.* **2008**, *34*, 1054–1063. [CrossRef]
7. Kiptoo, P.K.; Paudel, K.S.; Hammell, D.C.; Pinninti, R.R.; Chen, J.; Crooks, P.A.; Stinchcomb, A.L. Transdermal delivery of bupropion and its active metabolite, hydroxybupropion: A prodrug strategy as an alternative approach. *J. Pharm. Sci.* **2009**, *98*, 583–594. [CrossRef]
8. Kerr, D.; Roberts, W.; Tebbett, I.; Sloan, K.B. 7-Alkylcarbonyloxymethyl prodrugs of theophylline: Topical delivery of theophylline. *Int. J. Pharm.* **1998**, *167*, 37–48. [CrossRef]
9. Morris, A.P.; Brain, K.R.; Heard, C.M. Skin permeation and ex vivo skin metabolism of O-acyl haloperidol ester prodrugs. *Int. J. Pharm.* **2009**, *367*, 44–50. [CrossRef]
10. Chen, J.; Jiang, Q.-D.; Chai, Y.-P.; Zhang, H.; Peng, P.; Yang, X.-X. Natural terpenes as penetration enhancers for transdermal drug delivery. *Molecules* **2016**, *21*, 1709. [CrossRef]
11. Cox-Georgian, D.; Ramadoss, N.; Dona, C.; Basu, C. Therapeutic and medicinal uses of terpenes. In *Medicinal Plants*; Joshee, N., Dhekney, S., Parajuli, P., Eds.; Springer: Cham, Switzerland, 2019; pp. 333–359. [CrossRef]

12. Nóbrega de Almeida, R.; Agra, M.d.F.; Negromonte Souto Maior, F.; De Sousa, D.P. Essential oils and their constituents: Anticonvulsant activity. *Molecules* **2011**, *16*, 2726–2742. [CrossRef] [PubMed]
13. Wang, C.-Y.; Chen, Y.-W.; Hou, C.-Y. Antioxidant and antibacterial activity of seven predominant terpenoids. *Int. J. Food Prop.* **2019**, *22*, 230–238. [CrossRef]
14. De Sousa, D.P. Analgesic-like activity of essential oils constituents. *Molecules* **2011**, *16*, 2233–2252. [CrossRef]
15. Nesterkina, M.; Smola, S.; Kravchenko, I. Effect of esters based on terpenoids and GABA on fluidity of phospholipid membranes. *J. Liposome Res.* **2019**, *29*, 239–246. [CrossRef]
16. Nesterkina, M.; Kravchenko, I. Synthesis and pharmacological properties of novel esters based on monocyclic terpenes and GABA. *Pharmaceuticals* **2016**, *9*, 32. [CrossRef]
17. Premkumar, L.S. Transient receptor potential channels as targets for phytochemicals. *ACS Chem. Neurosci.* **2014**, *5*, 1117–1130. [CrossRef]
18. Manayi, A.; Nabavi, S.M.; Daglia, M.; Jafari, S. Natural terpenoids as a promising source for modulation of GABAergic system and treatment of neurological diseases. *Pharmacol. Rep.* **2016**, *68*, 671–679. [CrossRef]
19. Pages, N.; Maurois, P.; Bac, P.; Eynde, J.J.V.; Tamariz, J.; Labarrios, F.; Chamorro, G.; Vamecq, J. The α-asarone/clofibrate hybrid compound, 2-methoxy-4-(2-propenyl)phenoxyacetic acid (MPPA), is endowed with neuroprotective and anticonvulsant potentialities. *Biomed. Aging Pathol.* **2011**, *1*, 210–215. [CrossRef]
20. Turan-Zitouni, G.; Yurttaş, L.; Kaplancıklı, Z.A.; Can, Ö.D.; Özkay, Ü.D. Synthesis and anti-nociceptive, anti-inflammatory activities of new aroyl propionic acid derivatives including N-acylhydrazone motif. *Med. Chem. Res.* **2015**, *24*, 2406–2416. [CrossRef]
21. Nesterkina, M.; Barbalat, D.; Zheltvay, I.; Rakipov, I.; Atakay, M.; Salih, B.; Kravchenko, I. (2S,5R)-2-Isopropyl-5-methylcyclohexanone hydrazones. *Molbank* **2019**, *2019*, 1062. [CrossRef]
22. Nesterkina, M.; Barbalat, D.; Kravchenko, I. Design, synthesis and pharmacological profile of (−)-verbenone hydrazones. *Open Chem.* **2020**, *18*, 943–950. [CrossRef]
23. Nesterkina, M.; Barbalat, D.; Konovalova, I.; Shishkina, S.; Atakay, M.; Salih, B.; Kravchenko, I. Novel (-)-carvone derivatives as potential anticonvulsant and analgesic agents. *Nat. Prod. Res.* **2021**, *35*, 4978–4987. [CrossRef]
24. Ahad, A.; Aqil, M.; Ali, A. The application of anethole, menthone, and eugenol in transdermal penetration of valsartan: Enhancement and mechanistic investigation. *Pharm. Biol.* **2016**, *54*, 1042–1051. [CrossRef]
25. Suhonen, M.; Li, S.K.; Higuchi, W.I.; Herron, J.N. A liposome permeability model for stratum corneum lipid bilayers based on commercial lipids. *J. Pharm. Sci.* **2008**, *97*, 4278–4293. [CrossRef] [PubMed]
26. Reddy, A.S.; Zhang, S. Polypharmacology: Drug discovery for the future. *Expert. Rev. Clin. Pharmacol.* **2013**, *6*, 41–47. [CrossRef] [PubMed]
27. Masnoon, N.; Shakib, S.; Kalisch-Ellett, L.; Caughey, G.E. What is polypharmacy? A systematic review of definitions. *BMC Geriatr.* **2017**, *17*, 230. [CrossRef] [PubMed]
28. Routledge, S.J.; Linney, J.A.; Goddard, A.D. Liposomes as models for membrane integrity. *Biochem. Soc. Trans.* **2019**, *47*, 919–932. [CrossRef]
29. Palacios, L.E.; Wang, T. Egg-yolk lipid fractionation and lecithin characterization. *J. Amer. Oil Chem. Soc.* **2005**, *82*, 571–578. [CrossRef]
30. Pei, Y.; Hinchliffe, B.A.; Minelli, C. Measurement of the size distribution of multimodal colloidal systems by laser diffraction. *ACS Omega* **2021**, *6*, 14049–14058. [CrossRef]
31. Maja, L.; Željko, K.; Mateja, P. Sustainable technologies for liposome preparation. *J. Supercrit. Fluids* **2020**, *165*, 104984. [CrossRef]
32. Bains, G.; Patel, A.B.; Narayanaswami, V. Pyrene: A probe to study protein conformation and conformational changes. *Molecules* **2011**, *16*, 7909–7935. [CrossRef]
33. Sapra, B.; Jain, S.; Tiwary, A.K. Percutaneous permeation enhancement by terpenes: Mechanistic view. *AAPS J.* **2008**, *10*, 120. [CrossRef]
34. Bains, G.K.; Kim, S.H.; Sorin, E.J.; Narayanaswami, V. The extent of pyrene excimer fluorescence emission is a reflector of distance and flexibility: Analysis of the segment linking the LDL receptor-binding and tetramerization domains of apolipoprotein E3. *Biochemistry* **2012**, *51*, 6207–6219. [CrossRef]
35. Ando, Y.; Asano, Y.; Le Grimellec, C. Pyrene fluorescence: A potential tool for estimation of short-range lateral mobility in membranes of living renal epithelial cells. *Renal Physiol. Biochem.* **1995**, *18*, 246–253. [CrossRef]
36. Melnick, R.L.; Haspel, H.C.; Goldenberg, M.; Greenbaum, L.M.; Weinstein, S. Use of fluorescent probes that form intramolecular excimers to monitor structural changes in model and biological membranes. *Biophys. J.* **1981**, *34*, 499–515. [CrossRef]
37. Chaudhuri, A.; Haldar, S.; Chattopadhyay, A. Organization and dynamics in micellar structural transition monitored by pyrene fluorescence. *Biochem. Biophys. Res. Commun.* **2009**, *390*, 728–732. [CrossRef]
38. Lewis, N.A.H.R.; McElhaney, R.N. Membrane lipid phase transitions and phase organization studied by Fourier transform infrared spectroscopy. *Biochim Biophys. Acta Biomembr.* **2013**, *1828*, 2347–2358. [CrossRef] [PubMed]
39. Blume, A. Properties of lipid vesicles: FT-IR spectroscopy and fluorescence probe studies. *Curr. Opin. Colloid Interface Sci.* **1996**, *1*, 64–77. [CrossRef]
40. Wertz, P.W. Lipids and the permeability and antimicrobial barriers of the skin. *J. Lipids* **2018**, *2018*, 5954034. [CrossRef]

41. Mueller, J.; Schroeter, A.; Steitz, R.; Trapp, M.; Neubert, R.H. Preparation of a new oligolamellar stratum corneum lipid model. *Langmuir* **2016**, *32*, 4673–4680. [CrossRef] [PubMed]
42. Gooris, G.S.; Bouwstra, J.A. Infrared spectroscopic study of stratum corneum model membranes prepared from human ceramides, cholesterol, and fatty acids. *Biophys. J.* **2007**, *92*, 2785–2795. [CrossRef] [PubMed]

Article

Synthesis and Biological Evaluation of S-, O- and Se-Containing Dispirooxindoles

Maksim Kukushkin [1,2], Vladimir Novotortsev [1], Vadim Filatov [1], Yan Ivanenkov [3], Dmitry Skvortsov [1], Mark Veselov [3], Radik Shafikov [1], Anna Moiseeva [1], Nikolay Zyk [1], Alexander Majouga [1,4] and Elena Beloglazkina [1,*]

1. Department of Chemistry, Lomonosov Moscow State University, Leninskie Gory, 1/3, GSP-1, 119991 Moscow, Russia; lemmingg@mail.ru (M.K.); vladnov9216@rambler.ru (V.N.); nanovf@mail.ru (V.F.); skvorratd@mail.ru (D.S.); iltarn@mail.ru (R.S.); moiseeva@org.chem.msu.ru (A.M.); zyk@org.chem.msu.ru (N.Z.); alexander.majouga@gmail.com (A.M.)
2. Research Laboratory of Biophysics, National University of Science and Technology MISiS, 119049 Moscow, Russia
3. Laboratory of Medicinal Chemistry and Bioinformatic, Moscow Institute of Physics and Technology (MIPT), Institutski Pereulok 9, 141701 Dolgoprudny, Russia; smart_people@inbox.ru (Y.I.); veselovmark@gmail.com (M.V.)
4. Dmitry Mendeleev University of Chemical Technology of Russia, Miusskaya Sq. 9, 125047 Moscow, Russia
* Correspondence: beloglazki@mail.ru

Abstract: A series of novel S-, O- and Se-containing dispirooxindole derivatives has been synthesized using 1,3-dipolar cycloaddition reaction of azomethine ylide generated from isatines and sarcosine at the double C=C bond of 5-indolidene-2-chalcogen-imidazolones (chalcogen was oxygen, sulfur or selenium). The cytotoxicity of these dispiro derivatives was evaluated in vitro using different tumor cell lines. Several molecules have demonstrated a considerable cytotoxicity against the panel and showed good selectivity towards colorectal carcinoma HCT116 p53$^{+/+}$ over HCT116 p53$^{-/-}$ cells. In particular, good results have been obtained for LNCaP prostate cell line. The performed in silico study has revealed MDM2/p53 interaction as one of the possible targets for the synthesized molecules. However, in contrast to selectivity revealed during the cell-based evaluation and the results obtained in computational study, no significant p53 activation using a reporter construction in p53wt A549 cell line was observed in a relevant concentration range.

Keywords: dispirooxindoles; anticancer activity; cytotoxicity; 3D molecular docking; p53/MDM2 interaction

1. Introduction

Design and development of novel potent anticancer therapeutics are the most important tasks of synthetic organic and medicinal chemistry. Among the compounds with antitumor action, an important place is occupied by the spiro and dispiro derivatives of indolinones, due to the conformational rigidity of spiro scaffold which allows the introduction into the molecules of functional groups necessary for interaction with biological targets in the required arrangement, and the indolinone fragment simulates the tryptophan moiety, in many cases involved in such interactions [1–5]. So, spiro-oxindole alkaloids, which were firstly derived from the families *Apocynaceae* and *Rubiaceae* [6] and latter were found in a wide range of complex natural products [7–10] have shown significant anticancer activity. These compounds contain the spiro ring fusion at position 3 of the indolinone core, with different substitutions around the pyrrolidine and indolinone moieties.

A promising direction in the treatment of cancer is the development of compounds that affect the interaction of p53–MDM2 proteins. The p53 protein, which is a tumor suppressor, is one of the potential targets of antitumor therapy. Tumor suppressor p53, not being complexed with its MDM2 inhibitor, can trigger cell apoptosis [11,12]. In more than

50% of tumor cell cultures, the p53 protein is mutated [12], and its activation or restoration of its function may be effective in anticancer therapy due to apoptosis initiating or arresting cell growth [13]. Note that some small molecules inhibit MDM2/p53 interaction and now are undergoing preclinical or clinical trials against different types of cancer [14–18]. Among these molecules, compound nutlin-3a is one of the most known inhibitors of the p53–MDM2 protein–protein interaction; this compound is able to bind to the p53–MDM2 pocket and inhibit this protein interaction in nanomolar concentrations [19]. Nutlin-3a induced nongenotoxic stabilization of p53 protein and subsequent activation of a p53 pathway [20]. The molecule of nutlin-3a, which definitely binds to the site 1 of MDM2 protein, may be a template for the design of new p53–MDM2 inhibitor molecules [21]. Since the indole ring of Trp23 residue of p53 is located deep inside a hydrophobic pocket of MDM2 and its NH group forms a hydrogen bond with the backbone carbonyl in MDM2, Trp23 appears to be most crucial for binding of p53 to MDM2. Previously [22], Wang's group searched for chemical moieties that can mimic the Trp23 interaction with MDM2. In addition to the indole ring itself, they have found that oxindole can perfectly mimic the side chain of Trp23 for interaction with MDM2. These modeling studies also showed that compounds with a spiro-linked structure are capable of better binding to MDM2 by limiting the conformational mobility of the molecule (предыдущая) and the spiro-(oxindole-3,3′-pyrrolidine) core structure may be used as the starting point for the design of a new class of MDM2 inhibitors. The oxindole can closely mimic the Trp23 side chain in p53 in both hydrogen bond formation and hydrophobic interactions with MDM2, and the spiro-pyrrolidine ring provides a rigid scaffold.

We have recently described a series of novel spiro-oxindoles containing thiohydantoin [23–25], selenohydantoin [26] or hydantoin [25,27] moieties, presumably having an anticancer effect by inhibiting the p53/MDM2 protein interaction; one such derivative has recently successfully completed preclinical trials as a drug for the treatment of colorectal cancer [28]. The most active compounds of this type have shown cytotoxicity in the 4–11 μm range on cancer cell lines HepG2, MCF-7, SiHa and HCT116, [23], and some p53 activation by Western blotting (see Supplementary Information, Figure S1).

In this paper, we present a series of novel compounds of the dispiro-indolinone series with a modified spiro-oxindole core (Figure 1) with promising anticancer activity. Most of the previously investigated thiohydantoin-based spiro-oxindoles [23,25,28] (Figure 1) had in their structures the nitrogen atom of the central pyrrolidine ring directly attached to the carbon atom at spiro-conjugation; in the series of compounds described in this article, nitrogen atom of pyrrolidine ring is in the central position of spiro-conjugated cycle, similar to the MI series compounds, which demonstrated a significant cytotoxic effect on the prostate cancer cell line LNCap, with IC$_{50}$ = 86 nM, and on the colorectal cancer cell line HCTwt, with IC$_{50}$ = 22 μM, and were recognized as a selective inhibitor of p53–MDM2 interaction due to theirability to induce cell apoptosis in tumor cells without affecting healthy ones [29].

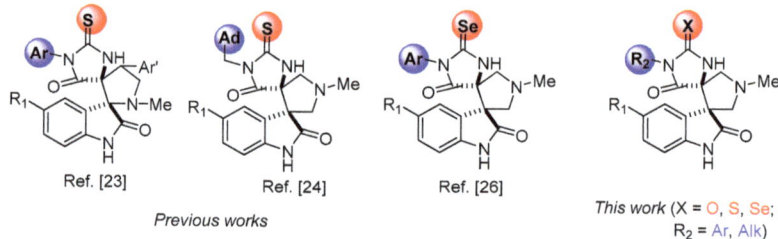

Figure 1. Compounds synthesized earlier and in this work.

In contrast to the compounds of such structural type, described in [24,26], this article presents dispiro derivatives with aryl and non-carcass alkyl substituents at N(3) position and with different exocyclic chalcogen atoms (oxygen, sulfur or selenium) in imidazolone fragment. Some results of cytotoxic action mechanisms studying for synthesized compounds are also presented, as well as the molecular docking data to evaluate their possible binding affinity toward MDM2.

2. Results and Discussion

2.1. Chemistry

To obtain the target dispyro derivatives **4, 6, 9**, the series of disubstituted thioureas **1a–j** or thiohydantoins **2a, b** (Scheme 1) was initially synthesized starting from aryl- or alkyl-amine and ethyl isothiocyanatoacetate. The reaction proceeded smoothly in ether at room temperature and furnished the desired intermediates with 61–98% yield.

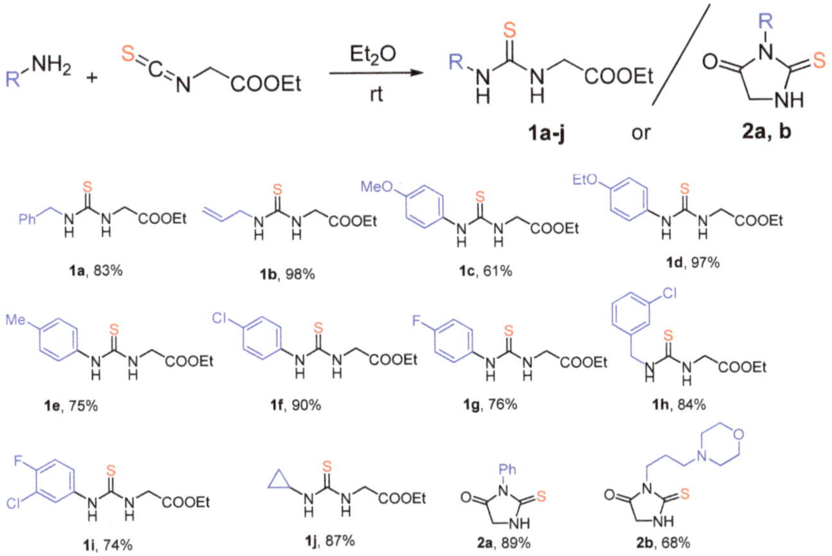

Scheme 1. Synthesis of the compounds **1a–k, 2**.

Compounds **1a–j** and **2a, b** were then treated with equimolar amount of isatin or 5-chloroisatin to obtain indolidene-thiohydantoins **3a–x** (Scheme 2), analogously to previously described reactions of substituted thioureas with aromatic aldehydes [30]. Finally, compounds **3a–x** were reacted with sarcosine and paraformaldehyde in toluene under reflux to obtain the desired substituted dispiroindolinones **4a–x** in a moderate-to-high yield. The reaction, apparently, proceeds according to the mechanism of 1,3-dipolar cycloaddition of azomethine ylide generated from isatin and sarcosine at the C=C double bond of indolidenehydantoins **3** [23]. According to the NMR spectroscopy data, the reactions in all cases proceed with the formation of single diastereomeric products **4a–x** with the relative S^*, R^*-configuration, which was confirmed by the data of X-ray crystallographic analysis for the compound **4s** (Figure 2).

For comparison, we have synthesized some O- and Se-containing analogs of the spirohydantoins **4**, namely for compounds **4g, 4h, 4q, 4r**. Some hydantoin derivatives containing spiro-linked indolinone fragments showed significant in vitro cytotoxic activity [25]. The ability of organoselenium compounds to exhibit antioxidant properties [31,32] mimicking the action of the glutathione peroxidase enzyme [33] makes it possible to use them in anticancer therapy as auxiliary antioxidants to neutralize the oxidizing agents produced by certain anticancer drugs.

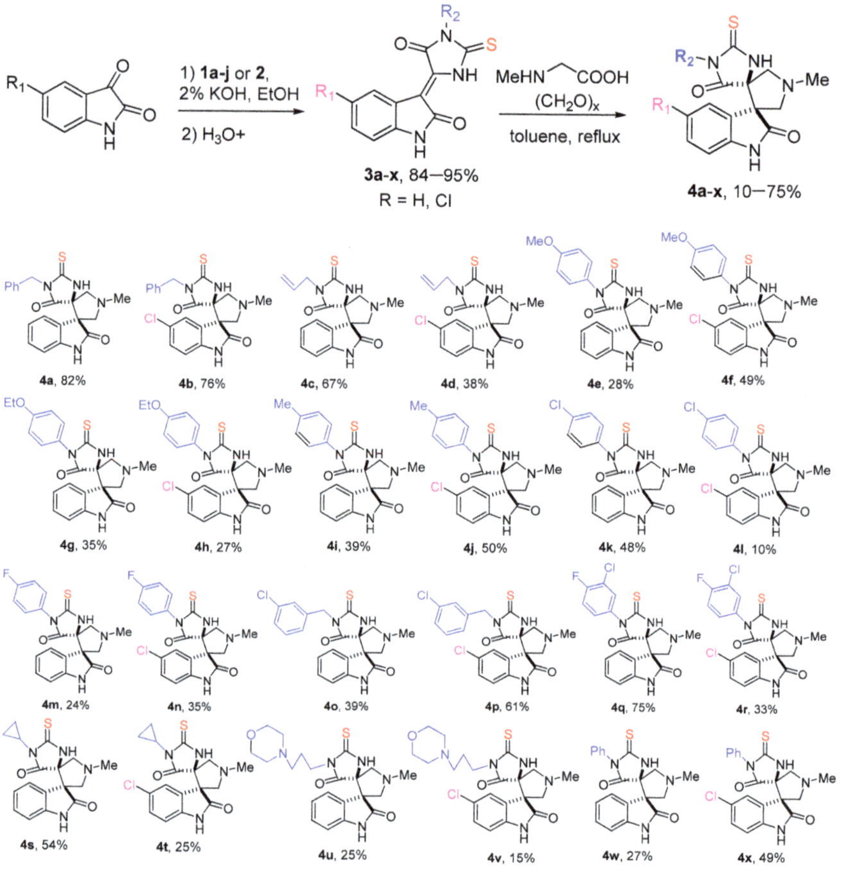

Scheme 2. Synthesis of 3-(5-oxo-2-thioxoimidazolidin-4-ylidene)indolin-2-ones **3a–x** and dispiroindolinines **4a–x**.

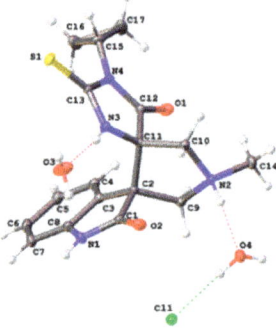

Figure 2. Molecular structure of compound **4s** (as a salt with HCl).

Selenohydantoin derivatives **6g**, **6h**, **6q**, **6r** may be readily obtained from sarcosine, paraformaldehyde and the corresponding indolidene-selenohydantoins **5g**, **5h**, **5q**, **5r** according to the modified method described in [26,34] (Scheme 3).

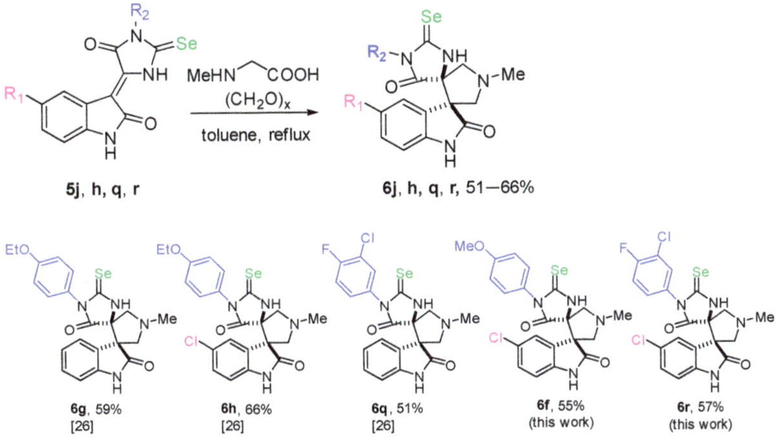

Scheme 3. Synthesis of Se-containing analogue of the compounds **4**.

O-containing derivatives **9** were synthesized from S-alkylated derivatives **7** of corresponding thiohydantoins **3** (Scheme 4). Thus, the starting compounds **3f**, **3h**, **3r** were vigorously stirred with MeI in KOH/EtOH at room temperature for 30 min and then were treated with HCl/EtOH under reflux conditions to provide the corresponding hydantoins **8** in good yield (about 70%). The desired products **9f**, **9h**, **9r** were obtained at the compounds **8** interaction with sarcosine and paraformaldehyde in 75% yields.

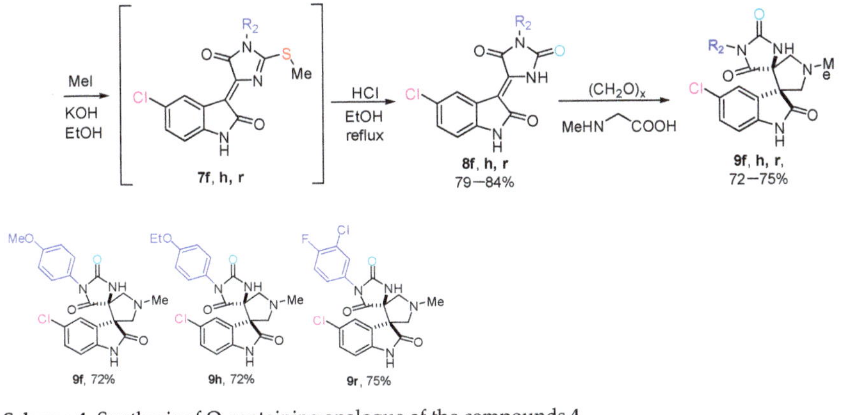

Scheme 4. Synthesis of O-containing analogue of the compounds **4**.

2.2. Biological Evaluation

2.2.1. Cytotoxicity

All the synthesized dispiro-oxindoles **4** and some of their selenium and oxygen analogues **6**, **9** have subsequently been tested on their in vitro anticancer efficiency against a panel of different tumor cell lines, on the assumption that they, like the previously described compounds of dispiro-thiohydantoins type [23], may be able to inhibit p53/MDM protein interaction. The used models included human prostate cancer cell lines LNCaP and PC3, breast cancer cell line (MCF-7), human colon cancer cells (HCT116$^{+/+}$, p53 positive, p53$^{+/+}$ and HCT116$^{-/-}$, p53 negative, p53$^{-/-}$), human lung adenocarcinoma epithelial cell line (A549), SV40-transformed normal human lung fibroblast cells (VA13), as well as human embryonic kidney 293 cells (Hek293) stably expressing SV40 large T antigen (Hek293T). The cytotoxicity of the evaluated molecules was properly assessed using a MTT assay based on the modified approach reported by Ferrari and colleagues [35]. Nutlin-3 [19], known as

P53/MDM2 interaction inhibitor, were also tested on some cell lines for comparison. The results of the cytotoxicity study are summarized in Table 1.

As shown in Table 1, the most potent compounds from the 4a–x series exhibited a CC_{50} value in the range of 1.1–12.6 µM against the used cells' panel. However, no relevant selectivity was observed among the cell types, although compound 4f, for example, exhibits a cytotoxic effect that exceeds the effect of the nutlinreference sample. Therefore, this compound demonstrated rather overt cytotoxic effect. Seven compounds (4a–f, 4u, 4f, 6r, 9r) were found to be selective on LNCaP cells over PC3 cells. Among them, the upper selectivity towards the remaining cells was observed for compounds 4e, 4f, 4u, 4w. Compounds 4e, 4f, 4u, 4w showed the best selectivity index (S value for cell line pairs is defined as ratio of its CC_{50} values) for LNCaP/PC3 cells (>30, 2.8, 15.7, 5.3, respectively). Compounds 4a, 4b, 4d, 4f showed moderate selectivity and efficiency. The negative response of PC3 cells to the treatment with the compounds having high S values possible may be associated with MDM2/p53 mode of action due to p53 tumor suppressor pathway in this cell type which is, in most cases, disrupted by human papilloma virus (HPV) [36]. The activated p53 induces the transcription of MDM2, which can directly interact with transactivation domain of p53 thereby inhibiting its transcription activity by targeting it for polyubiquitination and further proteasome-mediated degradation [37]. In many cancer cells, including HepG2, Hek and MCF-7, the overexpression of MDM2 gene is actually observed resulting in significant apoptosis attenuation. However, the obtained results do not allow to draw an unambiguous conclusion whether the studied compounds are actually involved in the direct activation of p53.

For instance, compound 4f inhibited the proliferation of LNCaP and PC3 cells with CC_{50} values of 4.5 ± 0.32 µM and 12.6 ± 2.3 µM, respectively, in contrast to its close structural analogue 4h with no activity at all (the difference is in p-position of the phenyl ring: compound 4f contains methoxy substituent while compound 4h ethoxy group). It can be primarily attributed to steric clashes; however, this hypothesis is under debate because of compounds 4i and 4j with methyl group in p-position weakly inhibited cell growth across the panel as compared to compound 4f. This may be partly explained by an additional hydrogen bond that can be provided by OMe group. To further elucidate the dominant mode of action, we used $HCT116^{+/+}$ and $HCT116^{-/-}$ cells by analogy with the paper published by Shangary and colleagues [38]. This isogenic cell line is commonly applied to investigate the p53/MDM2-dependent mode of action. Compounds, 4p, 4u and 4w showed a significant selectivity against $HCT116^{+/+}$ over $HCT116^{-/-}$, with absolute S values which are >2.9, >1.8 and >4.8, respectively. For the control sample, etoposide (a topoisomerase poison [39]), the selectivity index was 1.97, therefore compounds with S > 2 should be rather classified as having poor selectivity. Under the same conditions, nutlin-3, the knownp53/MDM2-interaction inhibitor [19] was found to be also active and selective against $HCT116^{+/+}$ cells over $HCT116^{-/-}$ cells showing CC_{50} values of 3.3 ± 0.13 µM and 35.12 ± 2.65 µM, respectively, and S = 10.6. Summarizing, based on the assay performed, only three thiohydantoine derivatives, 4p, 4u and 4w, can be reasonably regarded as the most promising candidates for further evaluation and optimization.

Table 1. Cytotoxicity of dispiro-oxindoles 4, 6, 9 against different cell lines (MTT test).

Compound	Cell Lines							
	A549	MCF7	VA13	Hek293T	HCT+/+	HCT−/−	LNCaP	PC3
	CC_{50}, µM							
4a	20.3 ± 3.7	31.3 ± 6.7	16.1 ± 2.5	9.8 ± 1.6	66.2 ± 11.74	40.11 ± 19.11	12.5 ± 2.1	30.1 ± 9.1
4b	9.1 ± 2.2	24.3 ± 2.3	21.4 ± 1.9	9.5 ± 1.7	60.73 ± 13.75	30.01 ± 15.21	10.3 ± 1.1	49.0 ± 12.7
4c	45.9 ± 13.4	na	na	41.8 ± 12.4	na	14.33 ± 6.65	9.8 ± 3.3	53.1 ± 15.8
4d	32.1 ± 6.7	68.4 ± 8.8	30.8 ± 4.5	11.5 ± 3	57.73 ± 18.41	40.11 ± 5.23	20.0 ± 3.6	50.0 ± 9.1
4e	18.9 ± 3.1	22.8 ± 5.6	13.2 ± 2.3	6.5 ± 0.9	27.9 ± 6.94	36.0 ± 12.55	3.45 ± 0.45	na
4f	2.8 ± 0.4	2.3 ± 0.5	2.3 ± 0.4	1.1 ± 0.1	1.95 ± 0.43	2.35 ± 0.95	4.5 ± 0.32	12.6 ± 2.3
6f	15.8 ± 1.5	20.4 ± 3.6	12.5 ± 1.5	14 ± 1.5	-	-	-	-
4g	38.2 ± 11.5	75.2 ± 29.3	73.6 ± 23.4	30.2 ± 8.2	na	na	na	na

Table 1. Cont.

Compound	A549	MCF7	VA13	Hek293T	HCT[+/+]	HCT[−/−]	LNCaP	PC3
				Cell Lines				
				CC_{50}, μM				
4h	6.2 ± 2.1	8.5 ± 1.2	6.3 ± 1.1	6.7 ± 0.9	na	na	na	na
4i	na	88.2 ± 32.2	na	na	na	na	na	na
4j	62 ± 8.3	13.7 ± 2.1	27.2 ± 5.3	17.4 ± 2	na	na	na	na
4k	67.5 ± 6.9	27.9 ± 6.9	59.1 ± 7.6	69 ± 3.7	52.7 ± 11.4	40.9 ± 20.5	5.2 ± 0.91	7.8 ± 1.4
4l	37 ± 6.5	19.1 ± 4.1	30.4 ± 3	34 ± 3.4	42.5 ± 9.1	57.11 ± 5.76	4.4 ± 0.67	2.2 ± 0.7
4m	na	105.9 ± 43.1	171.0 ± 15.8	na	23.3 ± 3.3	42.4 ± 4.7	3.4 ± 0.7	1.93 ± 0.91
4n	na	–	–	53.0 ± 9.5	na	na	na	na
4o	53.4 ± 11.2	23.2 ± 3.6	36.3 ± 3.2	48 ± 5.3	na	na	na	na
4p	11.3 ± 2.4	6.6 ± 1	12.3 ± 1.6	18.4 ± 1.7	34.2 ± 5.7	na	4.1 ± 0.1	2.2 ± 1.4
4q	na	62.2 ± 27.7	57.9 ± 18.3	69.1 ± 21	na	na	na	na
4r	14.1 ± 1.8	14.1 ± 1.2	13.9 ± 1.5	11.1 ± 1.3	na	na	na	na
6r	26.5 ± 5.4	–	–	11.9 ± 0.9	–	–	11.8 ± 1.5	>50
9r	35.3 ± 17.7	–	–	15.1 ± 1.2	–	–	19.6 ± 1.8	>50
4s	na	na	na	na	na	na	na	na
4t	na	na	na	na	na	na	na	na
4u	na	na	na	na	20.8 ± 4.8	na	3.4 ± 0.6	53.6 ± 17.3
4v	na	–	–	103.2 ± 27.7	na	na	na	na
4w	na	57.9 ± 17.1	94.1 ± 19.7	75.1 ± 8.6	54.7 ± 9.5	na	9.8 ± 1.8	52.2 ± 11.6
4x	na	–	–	na	na	na	na	na
Etoposide	0.3 ± 0.1	2.6 ± 0.9	1.1 ± 0.2	0.3 ± 0.1	0.43 ± 0.12	0.85 ± 0.22	–	–
Nutlin-3	–	–	–	–	3.3 ± 0.13	35.12 ± 2.65	–	–

na—not active (CC_{50} >100 μM); "–"—not tested. Etoposide [39] as a topoisomerase poison, and nutlin-3 [19] as P53/MDM2 interaction inhibitor, were also tested on some cell lines as positive control.

Interestingly, although compound **4r** was absolutely inactive against LNCaP and PC3 cells, both their analogues, as **6r** (selenohydantoin derivative) and as **9r** (hydantoin derivative) showed good activity vs. LNCaP line over PC3 cells.

2.2.2. Molecular Docking Study

To further investigate the possibility of the obtained spiro derivatives to inhibit the interaction of P53 and MDM2 proteins, we studied them using the molecular docking. Initially, we have collected a database of some analogues of the compounds **4** scaffold (**I–VI**, Figure 3) [3,4,14,40,41] and speculated that small-molecule MDM2 inhibitors are the most similar in structure to our series. All compounds selected for comparison contained an oxindole fragment spiro-conjugated with the pyrrolidine ring. For instance, compound **II** has shown an IC_{50} value of 22 μM (mitogenesis inhibition, dye assay, WST-8) against PC3 human prostate adenocarcinoma cells (p53-null) [41], while under the same conditions towards LNCaP cells (androgen-dependent), it has demonstrated IC_{50} = 18 ± 13 nM [41]. Cytotoxicity of the compound **II** has been evaluated also against HCT116[+/+] and HCT116[−/−] human colon carcinoma cells [41]. It has shown 80-fold selectivity towards p53-positive cells with IC_{50} values of 0.1 μM (vs. HCT116[+/+]) and 8 μM (vs. HCT116[−/−]), respectively. Compound **IV** demonstrated a CC_{50} value of 0.44 μM against HEK293 cells (24 h cytotoxicity, MTT assay) [14]. Compound **V** was reported to inhibit the growth and progression of HCT116 cells with an IC_{50} value of 90 nM (MTT assay) [4]. Molecule **VI** was evaluated against MCF7 cells (hormone-dependent) and PC3 cell line [42]. As a result, it showed high cytotoxicity and provided IC_{50} values of 40 nM and 0.41 μM, respectively. These reference activities are comparable with that observed for the most active compounds disclosed in the current work. The protein–protein interaction between MDM2 and p53 is observed via the first ~120 N-terminal amino acid (AA) residues of MDM2 and the first 30 N-terminal AAs of p53 [43]. Twenty years ago, the first high-resolution co-crystal structure of MDM2 with a p53 peptide (residues 15–29, PDB code 1YCR) was reported by Kussie and co-workers [44]. Since, more than 50 crystallographic complexes have been published for a variety of small-molecule MDM2/p53 inhibitors belonging to different classes, including spiro-oxindoles.

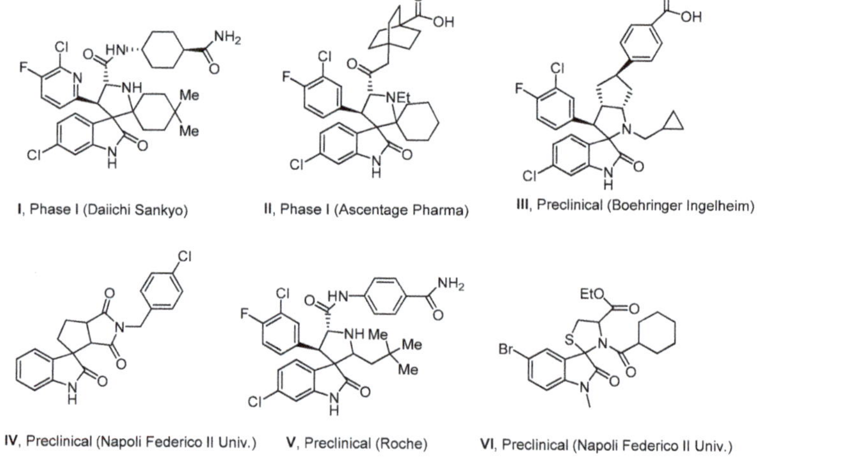

Figure 3. Examples of small-molecule MDM2 inhibitors with spiro-oxindole fragments in the molecules [40–44].

The analysis of MDM2/p53 binding interface revealed that MDM2-bound p53 peptide adopts an α-helical conformation and interacts with MDM2 primarily through the hydrophobic triad of Phe19, Trp23 and Leu26. These *"trident"* of i, i + 4, i + 7 binds tightly into a medium-sized pocket in the structure of MDM2. This compact and well-defined binding site has been used to design many small-molecule high-affinity MDM2 inhibitors which effectively block the MDM2/p53 interaction thereby blocking tumor growth and progression. To elucidate the possible binding affinity of the evaluated compounds toward MDM2, a static 3D molecular docking study was performed in ICM-Pro software [45] based on several available X-ray data, including 4JVR, 4MDQ, 4JWR as well as 5C5A, 5HMH and 4ZYF. The binding site for 3D-molecular docking study was constructed following the standard procedure of binding site preparation in ICM-Pro with default settings. The procedure included the following steps: converting PDB-file to ICM-Pro object, optimizing hydrogens, excluding water molecules, moving template ligand out from pocket, and constructing receptor maps also with default settings in ICM-Pro.

The binding site for 3D-molecular docking study was constructed following the standard procedure of binding site preparation in ICM-Pro with default settings. The procedure included the following steps: converting PDB-file to ICM-Pro object, optimizing hydrogens, excluding water molecules, moving template ligand out from pocket, and constructing receptor maps also with default settings in ICM-Pro. The binding site was then compared with the binding mode revealed recently for recombinant p53 binding domain (residues 17–125) [21].

The validation of the constructed docking model was performed using the reference compounds as they were found in PDB-files. For the 4JVR-based model particularly, we have provided the results of molecular docking study as compared to original X-ray data (Figure 4A, compound V docked into the binding site in the same conformation with RMSD = 0.23). The reference compounds were then docked into the constructed model starting from 2D or 3D structures with or without stereo assignment. The obtained results (Figure 4A) were well correlated with the published RSA data. The most promising dispiro compound **4u**, described in this work, was then docked into the static pocket using an extensive range of key force-fields, particularly describing hydrophobic interactions. As shown in Figure 4B, the selected molecule has a very similar to the reference compound binding mode. However, the predicted active conformation is distinct from that published for other MDM2 inhibitors. Thus, methyl group of compound **4u** is located in a deep cavity by analogy to most of the reported 5-halogen substituted oxindoles.

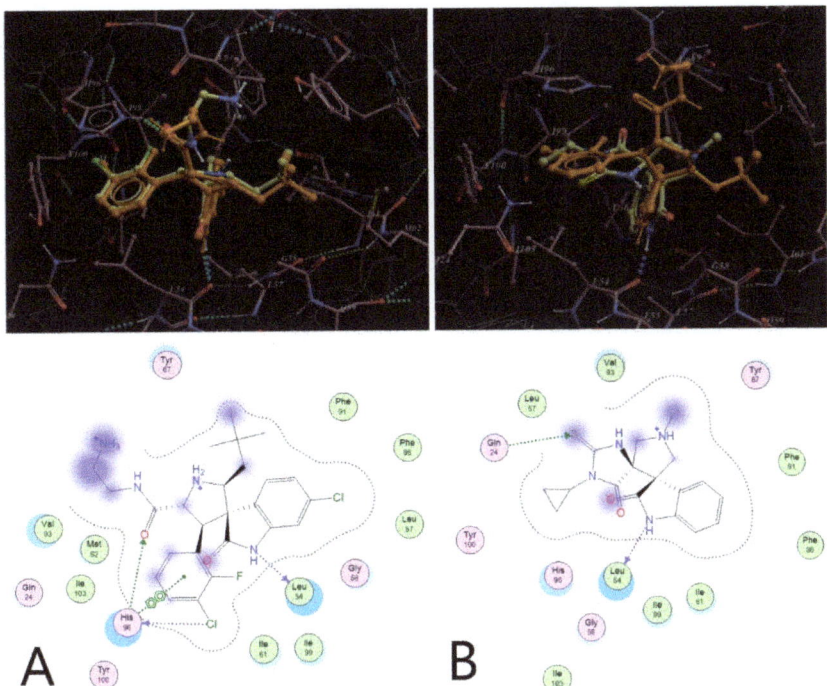

Figure 4. The results of moleculardocking study: (**A**) MDM2 inhibitor **V** (see Figure 3) bound in the target pocket (4JVR)—RSA data (orange), the predicted active conformation (yellow), $E^b = -88$ kcal/mol, RMSD = 0.23; (**B**) compound **4u** (orange, $E^b = -53$ kcal/mol) and the reference compound **V** (yellow) docked into the same binding site (the best conformations are shown).

Thus, although the mode of binding for the compound **4u** areambiguous, its scaffold has the 3D-pharmacophore elements critical for binding to the pre-defined MDM2 pocket as compared to the reported MDM2/p53 inhibitors, including other spiro-oxindoles.

2.2.3. P53 Activation

In an effort to further elucidate the underlying mechanism of action of the compounds **4–6** we have performed cell-based assay with the p53 reporter construction [46] particularly sensitive to MDM2 inhibitors. In general, the obtained results demonstrated that the activation of p53 was observed upon the treatment with high concentrations of all synthesized compounds (>100 μM). This concentration is close to highly cytotoxic range (only 7–10% of cells stayed alive). In the same concentrations, 2.1-fold p53 activation was observed for compound **9r**; under the same conditions, nutlin-3a showed from 3.6- and up to 5.1-fold increase in the p53 activation [47]. The effect of compounds **4** was slightly higher than the threshold value and could presumably be attributed to p53 activation primarily due to cell death and not vice versa.

Thus, although nutlin, chosen as a reference molecule, is comparable to the compound **4f** in terms of cytotoxicity, unlike nutlin, the MDM2 protein is apparently not the main target of the compounds described in this work

3. Materials and Methods

3.1. General Information

All common reagents were purchased from commercial suppliers and used as received. The melting points are uncorrected. ^{1}H-NMR spectra were recorded on Bruker

Avance 400 and Agilent MR-400 spectrometers at 400 MHz in CDCl$_3$ or DMSO-d$_6$. Chemical shifts were measured relative to solvent residual signals and referenced in part per million to TMS. Chemical shifts are reported in parts per million relative to TMS. High resolution mass spectra (HRMS) were recorded on an OrbitrapElite (Thermo Scientific) mass spectrometer with electrospray ionization (ESI) and orbital trap. To inject solutions with a concentration of 0.1 to 9 mg/mL (in 1% formic acid in acetonitrile), direct injection into the ion source using a syringe pump (5 mL/min) was used. The spray voltage was ±3.5 kV, the temperature of the capillary was 275 °C.

Compound **2a** was synthesized as described in [48,49].

X-ray study was performed on diffractometer Bruker APEX DUO (MoKα-radiation, graphite monochromator, φ-scan). The X-ray structure was solved by direct methods and refined using full-matrix anisotropic approximation for F^2_{hkl}. The location of the hydrogen atoms was predicted geometrically, their positions were well-adjusted using the "rider" model Uiso(H) = 1.5Ueq(C) formethyl groups and 1.2 Ueq(X) for the remaining H-atoms. All calculations were performed in SHELX software version 2015 [50] and OLEX-2 [51].

CCDC 2120852 contains the supplementary crystallographic data for this paper. These data can be obtained free of charge via www.ccdc.cam.ac.uk/data_request/cif (Embargoed Date 8 September 2022), or by emailing data_request@ccdc.cam.ac.uk, or by contacting The Cambridge Crystallographic Data Centre, 12 Union Road, Cambridge CB2 1EZ, UK; fax: +44 1223 336033.

3.2. Synthesis

3.2.1. General Procedure for the Synthesis of Thioureas (**1**) and Thiohydantoins (**2**)

Amine (1 equiv) was added to a solution of ethyl isothiocyanatoacetate (1 equiv) in ether. The resulting mixture was stirred for 1 hour at room temperature. After the reaction was completed (TLC control), the solvent was evaporated in vacuo and the formed precipitate was filtered off, washed with cold diethyl ether and dried in air.

Ethyl 2-(3-benzylthioureido)acetate (1a)

From 0.54 g (5.0 mmol) of benzylamine and 0.73 g (5.0 mmol) of ethyl isothiocyanatoacetate, compound **1a** (1.0 g, 83%) was obtained as a white solid. ^{1}H-NMR (400 MHz, CDCl$_3$) δ: 8.25 (bs, 1H, NH), 7.40 (d, J = 8.7 Hz, 2H), 7.26 (d, J = 8.7 Hz, 2H), 6.70 (bs, 1H, NH), 4.42 (s, 2H), 4.22 (q, J = 9.2 Hz, 2H), 1.29 (t, J = 9.2 Hz, 3H). HRMS (ESI+) m/z calcd. for (C$_{12}$H$_{16}$N$_2$O$_2$S, M + H): 253.1005, found: (M + H): 253.1015.

Ethyl 2-(3-allylthioureido)acetate (1b)

From 0.29 g (5.0 mmol) of allylamine and 0.73 g (5.0 mmol) of ethyl isothiocyanatoacetate, compound **1b** (0.93 g, 98%) was obtained as a yellow oil. ^{1}H-NMR (400 MHz, CDCl$_3$) δ: 6.82 (bs, 1H, NH), 6.74 (s, 1H, NH), 5.85 (m, 1H), 5.28 (d, J = 17.1 Hz, 1H), 5.20 (d, J = 10.2 Hz, 1H), 4.38 (d, J = 4.9 Hz, 1H), 4.21 (q, J = 7.2 Hz, 2H), 4.05 (s, 2H), 1.28 (t, J = 7.2 Hz, 3H). HRMS (ESI+) m/z calcd. for (C$_8$H$_{15}$N$_2$O$_2$S, M + H): 203.0849, found: (M + H): 203.0857.

Ethyl 2-(3-(4-methoxyphenyl)thioureido)acetate (1c)

From 0.62 g (5.0 mmol) of 4-methoxyaniline and 0.73 g (5.0 mmol) of ethyl isothiocyanatoacetate, compound **1c** (0.77 g, 61%) was obtained as a light yellow solid. M.p. 94–96 °C. ^{1}H-NMR (400 MHz, CDCl$_3$) δ: 8.12 (bs, 1H, NH), 7.23 (d, J = 8.1 Hz, 2H), 7.16 (d, J = 8.3 Hz, 2H), 6.60 (bs, 1H, NH), 4.41 (s, 2H), 4.20 (q, J = 7.2 Hz, 2H), 2.36 (s, 3H), 1.27 (t, J = 7.2 Hz, 3H). HRMS (ESI+) m/z calcd. for (C$_{12}$H$_{16}$N$_2$O$_3$S, M + H): 269.0954, found: (M + H): 269.0958.

Ethyl 2-(3-(4-ethoxyphenyl)thioureido)acetate (1d)

From 0.69 g (5.0 mmol) of 4-ethoxyaniline and 0.73 g (5.0 mmol) of ethyl isothiocyanatoacetate, compound **1d** (1.30 g, 97%) was obtained as a purple solid. M.p. 124–126 °C. ^{1}H-NMR (400 MHz, CDCl$_3$) δ: 7.90 (bs, 1H, NH), 7.19 (d, J = 9.1 Hz, 2H), 6.93 (d, J = 8.9 Hz,

2H), 6.44 (bs, 1H, NH), 4.41 (s, 2H), 4.20 (q, J = 7.2 Hz, 2H), 4.04 (q, J = 7.0 Hz, 2H), 1.42 (t, J = 7.0 Hz, 3H), 1.27 (t, J = 7.2 Hz, 3H). HRMS (ESI+) m/z calcd. for ($C_{13}H_{18}N_2O_3S$, M + H): 283.1111, found: (M + H): 283.1106.

Ethyl 2-(3-(p-tolyl)thioureido)acetate (1e)

From 0.54 g (5.0 mmol) of 4-methylaniline and 0.73 g (5.0 mmol) of ethyl isothiocyanatoacetate, compound **1e** (0.89 g, 75%) was obtained as a plum solid. M.p. 132–134 °C. ^{1}H-NMR (400 MHz, CDCl$_3$) δ: 7.91 (bs, 1H, NH), 7.21 (d, J = 8.9 Hz, 2H), 6.95 (d, J = 8.9 Hz, 2H), 6.45 (bs, 1H, NH), 4.41 (s, 2H), 4.20 (q, J = 7.2 Hz, 2H), 3.82 (s, 3H), 1.28 (t, J = 7.2 Hz, 3H).

Ethyl 2-(3-(4-chlorophenyl)thioureido)acetate (1f)

From 0.64 g (5.0 mmol) of 4-chloroaniline and 0.73 g (5.0 mmol) of ethyl isothiocyanatoacetate, compound **1f** (1.16 g, 90%) was obtained as a white solid. M.p. 154–156 °C. ^{1}H-NMR (400 MHz, CDCl$_3$) δ: 8.25 (bs, 1H, NH), 7.40 (d, J = 8.7 Hz, 2H), 7.25 (d, J = 8.7 Hz, 2H), 6.70 (bs, 1H, NH), 4.42 (s, 2H), 4.22 (q, J = 7.2 Hz, 2H), 1.29 (t, J = 7.2 Hz, 3H). HRMS (ESI+) m/z calcd. for ($C_{11}H_{13}ClN_2O_2S$, M + H): 273.0459, found: (M + H): 273.0468.

Ethyl 2-(3-(4-fluorophenyl)thioureido)acetate (1g)

From 0.56 g (5.0 mmol) of 4-fluoroaniline and 0.73 g (5.0 mmol) of ethyl isothiocyanatoacetate, compound **1g** (0.97 g, 76%) was obtained as a white solid. M.p. 119–120 °C. ^{1}H-NMR (400 MHz, CDCl$_3$) δ: 8.15 (bs, 1H, NH), 7.33–7.25 (m, 2H), 7.14 (t, J = 8.5 Hz, 2H), 6.56 (bs, 1H, NH), 4.42 (s, 2H), 4.21 (q, J = 7.1 Hz, 2H), 1.28 (t, J = 7.1 Hz, 3H). HRMS (ESI+) m/z calcd. for ($C_{11}H_{13}FN_2O_2S$, M + H): 257.0755, found: (M + H): 257.0766.

Ethyl 2-(3-(3-chlorobenzyl)thioureido)acetate (1h)

From 0.74 g (5.0 mmol) of e-chlorobenzylamine and 0.73 g (5.0 mmol) of ethyl isothiocyanatoacetate. compound **1h** (1.12 g, 84%) was obtained as a white solid. M.p. 122–123 °C. 1H-NMR(400 MHz, CDCl$_3$) δ: 7.39–7.19 (m, 5H), 4.67 (bs, 2H), 4.38 (s, 2H), 4.18 (q, J = 7.1 Hz, 2H), 1.27 (t, J = 7.1 Hz, 3H). HRMS (ESI+) m/z calcd. for ($C_{12}H_{15}ClN_2O_2S$, M + H): 287.0616, found: (M + H): 287.0628.

Ethyl 2-(3-(3-chloro-4-fluorophenyl)thioureido)acetate (1i)

From 0.73 g (5.0 mmol) of 4-fluoro,3-chloroaniline and 0.73 g (5.0 mmol) of ethyl isothiocyanatoacetate, compound **1i** (1.08 g, 74%) was obtained as a white solid. M.p. 111–112 °C. ^{1}H-NMR (400 MHz, CDCl$_3$) δ: 8.26 (bs, 1H, NH), 7.42 (dd, J_1 = 2.4 Hz, J_2 = 6.4 Hz, 1H), 7.26–7.17 (m, 2H), 6.71 (bs, 1H, NH), 4.42 (s, 2H), 4.22 (q, J = 7.1 Hz, 2H), 1.30 (t, J = 7.2 Hz, 3H). HRMS (ESI+) m/z calcd. for ($C_{11}H_{12}ClFN_2O_2S$, M + H): 291.0365, found: (M + H): 291.0377.

Ethyl 2-(3-cyclopropylthioureido)acetate (1j)

From 0.29 g (5.0 mmol) of cyclopropylamine and 0.73 g (5.0 mmol) of ethyl isothiocyanatoacetate, compound **1j** (0.8 g, 87%) was obtained as a white solid. M.p. 129–130 °C. ^{1}H-NMR (400 MHz, CDCl$_3$) δ: 6.86 (bs, 1H, NH), 6.56 (bs, 1H, NH), 4.45 (d, J = 4.5 Hz, 2H), 4.27 (q, J = 7.0 Hz, 2H), 2.54 (bs, 1H), 1.32 (t, J = 7.2 Hz, 1H), 0.92–0.85 (m, 2H), 0.76–0.69 (m, 2H).

3-(3-Morpholinopropyl)-2-thioxoimidazolidin-4-one (2b)

From 0.72 g (5.0 mmol) of 3-(N-morpholino)propylamine and 0.73 g (5.0 mmol) of ethyl isothiocyanatoacetate, compound **1k** (0.82 g, 68%) was obtained as a pink solid. M.p. 149–151 °C. ^{1}H-NMR (400 MHz, CDCl$_3$) δ: 8.33 (bs, 1H, NH), 4.08 (s, 2H), 3.89 (t, J = 6.9 Hz, 2H), 3.82–3.74 (m, 4H), 2.72–2.55 (m, 6H), 2.04–1.93 (m, 2H).

3.2.2. General Procedure for the Synthesis of 5-Substituted-2-thiohydantoins 3a–x

Thioureidoacetate **1** or 2-thioxoimidazolidine **2** (1 equiv) was dissolved in 2% KOH/EtOH; then isatin or 5-chloroisatin (1 equiv) was added. The resulting mixture was stirred for 30 min. After the reaction was completed (TLC control), the mixture was poured into water and neutralized with HCl. The formed precipitate was filtered off, washed with cold water, then washed with cold diethyl ether and dried in air.

(Z)-3-(1-Benzyl-5-oxo-2-thioxoimidazolidin-4-ylidene)indolin-2-one (3a)

From **1a** (0.36 g, 1.5 mmol) and isatin (0.22 g, 1.5 mmol), compound 3a (0.48 g, 94%) was obtained as a red solid. M.p. > 300 °C. ^{1}H-NMR (400 MHz, DMSO-d$_6$) δ: 11.56 (s, 1H, NH), 11.10 (s, 1H, NH), 8.53 (d, J = 7.8 Hz, 1H), 7.43–7.26 (m, 6H), 7.03 (td, J_1 = 1.0 Hz, J_2 = 7.7 Hz, 1H), 6.93 (d, J = 7.8 Hz, 1H), 5.06 (s, 2H). HRMS (ESI+) m/z calcd. for (C$_{18}$H$_{13}$N$_3$O$_2$S, M + H): 336.0801, found: (M + H): 336.0797.

(Z)-3-(1-Benzyl-5-oxo-2-thioxoimidazolidin-4-ylidene)-5-chloroindolin-2-one (3b)

From **1a** (0.36 g, 1.5 mmol) and 5-chloroisatin (0.27 g, 1.5 mmol), compound **3b** (0.51 g, 92%) was obtained as a dark red solid. M.p. > 300 °C. ^{1}H-NMR (400 MHz, DMSO-d$_6$) δ: 11.63 (s, 1H, NH), 11.18 (s, 1H, NH), 8.55 (m, 1H), 7.43–7.25 (m, 6H), 6.92 (d, J = 8.3 Hz, 1H), 5.05 (s, 2H). HRMS (ESI+) m/z calcd. for (C$_{18}$H$_{12}$ClN$_3$O$_2$S, M + H): 370.0411, found: (M + H): 370.0411.

(Z)-3-(1-Allyl-5-oxo-2-thioxoimidazolidin-4-ylidene)indolin-2-one (3c)

From **1b** (0.28 g, 1.5 mmol) and isatin (0.22 g, 1.5 mmol,) compound **3c** (0.36 g, 83%) was obtained as a red solid. M.p. 257–259 °C. ^{1}H-NMR (400 MHz, DMSO-d$_6$) δ: 11.50 (bs, 1H, NH), 11.07 (s, 1H, NH), 8.52 (d, J = 7.8 Hz, 1H), 7.32 (td, J_1 = 1.0 Hz, J_2 = 7.7 Hz, 1H), 7.04 (td, J_1 = 0.7 Hz, J_2 = 7.7 Hz, 1H), 6.92 (d, J = 7.8 Hz, 1H), 5.86 (m, 1H), 5.20 (dd, J_1 = 1.0 Hz, J_2 = 9.7 Hz, 1H), 5.17 (m, 1H), 4.45 (d, J = 5.0 Hz, 2H). HRMS (ESI+) m/z calcd. for (C$_{14}$H$_{11}$N$_3$O$_2$S, M + H): 286.0644, found: (M + H): 286.0643.

(Z)-3-(1-Allyl-5-oxo-2-thioxoimidazolidin-4-ylidene)-5-chloroindolin-2-one (3d)

From **1b** (0.28 g, 1.5 mmol) and 5-chloroisatin (0.27 g, 1.5 mmol), compound **3d** (0.40 g, 83%) was obtained as a dark red solid. M.p. 258–260 °C. ^{1}H-NMR (400 MHz, DMSO-d$_6$) δ: 11.56 (bs, 1H, NH), 11.19 (s, 1H, NH), 8.55 (d, J = 2.0 Hz, 1H), 7.36 (dd, J_1 = 2.2 Hz, J_2 = 8.3 Hz, 1H), 6.93 (d, J = 8.31 Hz, 1H), 5.86 (m, 1H), 5.25–5.16 (m, 2H), 4.45 (d, J = 5.1 Hz, 2H). HRMS (ESI+) m/z calcd. for (C$_{14}$H$_{10}$ClN$_3$O$_2$S, M + H): 320.0255, found: (M + H): 320.0252.

(Z)-3-(1-(4-Methoxyphenyl)-5-oxo-2-thioxoimidazolidin-4-ylidene)indolin-2-one (3e)

From **1c** (0.38 g, 1.5 mmol) and isatin (0.22 g, 1.5 mmol), compound **3e** (0.49 g, 93%) was obtained as a red solid. M.p. > 300 °C. ^{1}H-NMR (400 MHz, DMSO-d$_6$) δ: 11.60 (bs, 1H, NH), 10.95 (bs, 1H, NH), 8.56 (d, J = 7.70 Hz, 1H), 7.32 (d, J = 8.1 Hz, 2H), 7.29–7.22 (m, 3H), 6.97 (t, J = 7.6 Hz, 1H), 6.90 (d, J = 7.8 Hz, 1H), 2.38 (s, 3H). HRMS (ESI+) m/z calcd. for (C$_{18}$H$_{13}$N$_3$O$_3$S, M + H): 352.0750, found: (M + H): 352.0771.

(Z)-5-Chloro-3-(1-(4-methoxyphenyl)-5-oxo-2-thioxoimidazolidin-4-ylidene)indolin-2-one (3f)

From **1c** (0.38 g, 1.5 mmol) and 5-chloroisatin (0.27 g, 1.5 mmol), compound **3f** (0.54 g, 94%) was obtained as a dark red solid. M.p. > 300 °C. ^{1}H-NMR (400 MHz, DMSO-d$_6$) δ: 11.69 (bs, 1H, NH), 11.24 (bs, 1H, NH), 8.54 (d, J = 2.2 Hz, 1H), 7.36–7.29 (m, 5H), 6.96 (dd, J_1 = 2.2 Hz, J_2 = 8.3 Hz, 1H), 2.39 (s, 3H). HRMS (ESI+) m/z calcd. for (C$_{18}$H$_{12}$ClN$_3$O$_3$S, M + H): 386.0360, found: (M + H): 386.0387.

(Z)-3-(1-(4-Ethoxyphenyl)-5-oxo-2-thioxoimidazolidin-4-ylidene)indolin-2-one (3g)

From **1d** (0.40 g, 1.5 mmol) and isatin (0.22 g, 1.5 mmol), compound **3g** (0.49 g, 89%) was obtained as a red solid. M.p. > 300 °C. ^{1}H-NMR (400 MHz, DMSO-d$_6$) δ: 11.62 (bs, 1H, NH), 11.14 (s, 1H, NH), 8.50 (d, J = 7.6 Hz, 1H), 7.37–7.29 (m, 3H), 7.06 (d, J = 8.9 Hz, 2H), 7.02 (t, J = 7.7 Hz, 1H), 6.95 (d, J = 8.0 Hz, 1H), 4.09 (q, J = 6.9 Hz, 2H), 1.36 (t, J = 6.9 Hz, 3H). HRMS (ESI+) m/z calcd. for (C$_{19}$H$_{15}$N$_3$O$_3$S, M + H): 366.0906, found: (M + H): 366.0895.

(Z)-5-Chloro-3-(1-(4-ethoxyphenyl)-5-oxo-2-thioxoimidazolidin-4-ylidene)indolin-2-one (3h)

From **1d** (0.40 g, 1.5 mmol) and 5-chloroisatin (0.27 g, 1.5 mmol), compound **3h** (0.52 g, 88%) was obtained as a dark red solid. M.p. > 300 °C. ^{1}H-NMR (400 MHz, DMSO-d$_6$) δ: 11.66 (bs, 1H, NH), 11.23 (s, 1H, NH), 8.55 (s, 1H), 7.37 (dd, J_1 = 2.2 Hz, J_2 = 8.4 Hz, 1H), 7.33 (d, J = 8.7 Hz, 2H), 7.07 (d, J = 8.6 Hz, 2H), 6.96 (d, J = 8.3 Hz, 1H), 4.09 (q, J = 6.8 Hz, 2H), 1.36 (t, J = 6.8 Hz, 3H). HRMS (ESI+) *m/z* calcd. for (C$_{19}$H$_{15}$ClN$_3$O$_3$S, M + H): 400.0517, found: (M + H): 400.0496.

(Z)-3-(5-oxo-2-thioxo-1-(p-tolyl)imidazolidin-4-ylidene)indolin-2-one (3i)

From **1e** (0.36 g, 1.5 mmol) and isatin (0.22 g, 1.5 mmol), compound **3i** (0.48 g, 95%) was obtained as a red solid. M.p. > 300 °C. ^{1}H-NMR (400 MHz, DMSO-d$_6$) δ: 11.62 (s, 1H, NH), 11.14 (s, 1H, NH), 8.50 (d, J = 7.8 Hz, 1H), 7.38–7.30 (m, 3H), 7.09 (d, J = 8.8 Hz, 2H), 7.03 (t, J = 7.6 Hz, 1H), 6.95 (d, J = 7.7 Hz, 1H), 3.83 (s, 3H). HRMS (ESI+) *m/z* calcd. for (C$_{18}$H$_{13}$N$_3$O$_2$S, M + H): 336.0801, found: (M + H): 336.0796.

(Z)-5-Chloro-3-(5-oxo-2-thioxo-1-(p-tolyl)imidazolidin-4-ylidene)indolin-2-one (3j)

From **1e** (0.36 g, 1.5 mmol) and 5-chloroisatin (0.27 g, 1.5 mmol), compound **3j** (0.51 g, 92%) was obtained as a dark red solid. M.p. > 300 °C. ^{1}H-NMR (400 MHz, DMSO-d$_6$) δ: 11.67 (bs, 1H, NH), 11.25 (s, 1H, NH), 8.55 (d, J = 1.4 Hz, 1H), 7.40–7.32 (m, 3H), 7.09 (d, J = 8.8 Hz, 2H), 6.96 (d, J = 8.3 Hz, 1H), 3.83 (s, 3H). HRMS (ESI+) *m/z* calcd. for (C$_{18}$H$_{12}$ClN$_3$O$_2$S, M + H): 370.0411, found: (M + H): 370.0414.

(Z)-3-(1-(4-Chlorophenyl)-5-oxo-2-thioxoimidazolidin-4-ylidene)indolin-2-one (3k)

From **1f** (0.39 g, 1.5 mmol) and isatin (0.22 g, 1.5 mmol), compound **3k** (0.45 g, 84%) was obtained as a red solid. M.p. > 300 °C. ^{1}H-NMR (400 MHz, DMSO-d$_6$) δ: 11.69 (bs, 1H, NH), 11.14 (bs, 1H, NH), 8.48 (d, J = 7.7 Hz, 1H), 7.63 (d, J = 8.4 Hz, 2H), 7.48 (d, J = 8.4 Hz, 2H), 7.32 (t, J = 7.7 Hz, 1H), 7.02 (t, J = 7.7 Hz, 1H), 6.94 (d, J = 7.7 Hz, 1H). HRMS (ESI+) *m/z* calcd. for (C$_{17}$H$_{10}$ClN$_3$O$_2$S, M + H): 356.0255, found: (M + H): 356.0257.

(Z)-5-Chloro-3-(1-(4-chlorophenyl)-5-oxo-2-thioxoimidazolidin-4-ylidene)indolin-2-one (3l)

From **1f** (0.39 g, 1.5 mmol) and 5-chloroisatin (0.27 g, 1.5 mmol), compound **3l** (0.51 g, 86%) was obtained as a dark red solid. M.p. > 300 °C. ^{1}H-NMR (400 MHz, DMSO-d$_6$) δ: 11.76 (bs, 1H, NH), 11.24 (s, 1H, NH), 8.53 (d, J = 2.0 Hz, 1H), 7.65 (d, J = 8.6 Hz, 2H), 7.49 (d, J = 8.6 Hz, 2H), 7.38 (dd, J_1 = 2.1 Hz, J_2 = 8.3 Hz, 1H), 6.96 (d, J = 8.3 Hz, 1H). HRMS (ESI+) *m/z* calcd. for (C$_{17}$H$_9$Cl$_2$N$_3$O$_2$S, M + H): 389.9865, found: (M + H): 389.9846.

(Z)-3-(1-(4-Fluorophenyl)-5-oxo-2-thioxoimidazolidin-4-ylidene)indolin-2-one (3m)

From **1g** (0.38 g, 1.5 mmol) and isatin (0.22 g, 1.5 mmol), compound **3m** (0.45 g, 89%) was obtained as a red solid. M.p. > 300 °C. ^{1}H-NMR (400 MHz, DMSO-d$_6$) δ: 11.68 (s, 1H, NH), 11.13 (s, 1H, NH), 8.50 (d, J = 7.0 Hz, 1H), 7.58–7.46 (m, 2H), 7.40 (t, J = 8.0 Hz, 2H), 7.33 (t, 7.5 Hz, 1H), 7.03 (t, J = 7.0 Hz, 1H), 6.95 (d, J = 7.0 Hz, 1H). HRMS (ESI-) *m/z* calcd. for (C$_{17}$H$_{10}$FN$_3$O$_2$S, M-H): 338.0394, found: (M-H): 338.0405.

(Z)-5-Chloro-3-(1-(4-fluorophenyl)-5-oxo-2-thioxoimidazolidin-4-ylidene)indolin-2-one (3n)

From **1g** (0.38 g, 1.5 mmol) and 5-chloroisatin (0.27 g, 1.5 mmol), compound **3n** (0.51 g, 91%) was obtained as a red solid. M.p. > 300 °C. ^{1}H-NMR (400 MHz, DMSO-d$_6$) δ: 11.75 (s, 1H, NH), 11.24 (s, 1H, NH), 8.56–8.52 (m, 1H), 7.55–7.48 (m, 2H), 7.46–7.35 (m, 3H), 6.96 (dd, J_1 = 3.6 Hz, J_2 = 8.3 Hz, 1H). HRMS (ESI-) *m/z* calcd. for (C$_{17}$H$_9$ClFN$_3$O$_2$S, M-H): 372.0004, found: (M-H): 372.0017.

(Z)-3-(1-(3-Chlorobenzyl)-5-oxo-2-thioxoimidazolidin-4-ylidene)indolin-2-one (3o)

From **1h** (0.43 g, 1.5 mmol) and isatin (0.22 g, 1.5 mmol), compound **3o** (0.48 g, 87%) was obtained as a red solid. M.p. 296–298 °C. ^{1}H-NMR (400 MHz, DMSO-d_6) δ: 11.59 (bs, 1H, NH), 11.09 (s, 1H, NH), 8.52 (d, J = 7.7 Hz, 1H), 7.46 (s, 1H), 7.40–7.29 (m, 4H), 7.03 (t, J = 7.6 Hz, 1H), 6.93 (d, J = 7.8 Hz, 1H), 5.05 (s, 2H). HRMS (ESI-) m/z calcd. for ($C_{18}H_{12}ClN_3O_2S$, M-H): 368.0266, found: (M-H): 368.0255.

(Z)-5-Chloro-3-(1-(3-chlorobenzyl)-5-oxo-2-thioxoimidazolidin-4-ylidene)indolin-2-one (3p)

From **1h** (0.43 g, 1.5 mmol) and 5-chloroisatin (0.27 g, 1.5 mmol), compound **3p** (0.53 g, 87%) was obtained as a red solid. M.p. > 300 °C. ^{1}H-NMR (400 MHz, DMSO-d_6) δ: 11.66 (bs, 1H, NH), 11.18 (s, 1H, NH), 8.56 (d, J = 1.7 Hz, 1H), 7.46 (s, 1H), 7.41–7.32 (m, 4H), 6.94 (d, J = 8.3 Hz, 1H), 5.05 (s, 2H). HRMS (ESI-) m/z calcd. for ($C_{18}H_{11}Cl_2N_3O_2S$, M-H): 401.9865, found: (M-H): 401.9879.

(Z)-3-(1-(3-Chloro-4-fluorophenyl)-5-oxo-2-thioxoimidazolidin-4-ylidene)indolin-2-one (3q)

From **1i** (0.44 g, 1.5 mmol) and isatin (0.22 g, 1.5 mmol), compound **3q** (0.47 g, 84%) was obtained as a red solid. M.p. > 300 °C. ^{1}H-NMR (400 MHz, DMSO-d_6) δ: 11.75 (bs, 1H, NH), 10.92 (s, 1H, NH), 8.56 (d, J = 7.8 Hz, 1H), 7.73 (dd, J_1 = 1.88 Hz, J = 6.5 Hz, 1H), 7.61 (t, J = 9.0 Hz, 1H), 7.48 (m, 1H), 7.27 (t, J = 7.3 Hz, 1H), 6.98 (t, J = 7.6 Hz, 1H), 6.90 (d, J = 7.6 Hz, 1H). HRMS (ESI-) m/z calcd. for ($C_{17}H_9ClFN_3O_2S$, M-H): 372.0004, found: (M-H): 372.0017.

(Z)-5-Chloro-3-(1-(3-chloro-4-fluorophenyl)-5-oxo-2-thioxoimidazolidin-4-ylidene)indolin-2-one (3r)

From **1i** (0.44 g, 1.5 mmol) and 5-chloroisatin (0.27 g, 1.5 mmol), compound **3r** (0.53 g, 86%) was obtained as a red solid. M.p. > 300 °C. ^{1}H-NMR (400 MHz, DMSO-d_6) δ: 11.81 (bs, 1H, NH), 11.24 (s, 1H, NH), 8.51 (d, J = 1.8 Hz, 1H), 7.77 (dd, J_1 = 2.3 Hz, J_2 = 6.7 Hz, 1H), 7.65 (t, J = 8.99 Hz, 1H), 7.52 (m, 1H), 7.37 (dd, J_1 = 2.02 Hz, J_2 = 8.38 Hz, 1H), 6.95 (d, J = 8.3 Hz, 1H). HRMS (ESI-) m/z calcd. for ($C_{17}H_8Cl_2FN_3O_2S$, M-H): 405.9615, found: (M-H): 405.9628.

(Z)-3-(1-Cyclopropyl-5-oxo-2-thioxoimidazolidin-4-ylidene)indolin-2-one (3s)

From **1j** (0.30 g, 1.5 mmol) and isatin (0.22 g, 1.5 mmol), compound **3s** (0.26 g, 92%) was obtained as a red solid. M.p. 277–279 °C (decomp.). ^{1}H-NMR (400 MHz, DMSO-d_6) δ: 11.35 (bs, 1H, NH), 11.08 (s, 1H, NH), 8.52 (d, J = 7.8 Hz, 1H), 7.32 (t, J = 7.6 Hz, 1H), 7.04 (t, J = 7.6 Hz, 1H), 6.92 (d, J = 7.8 Hz, 1H), 2.80 (m, 1H), 1.04–0.98 (m, 4H). HRMS (ESI-) m/z calcd. for ($C_{14}H_{11}N_3O_2S$, M-H): 284.0488, found: (M-H): 284.0499.

(Z)-5-Chloro-3-(1-cyclopropyl-5-oxo-2-thioxoimidazolidin-4-ylidene)indolin-2-one (3t)

From **1j** (0.30 g, 1.5 mmol) and 5-chloroisatin (0.27 g, 1.5 mmol), compound **3t** (0.29 g, 92%) was obtained as a red solid. M.p. 286–288 °C (decomp.). ^{1}H-NMR (400 MHz, DMSO-d_6) δ: 11.40 (s, 1H, NH), 11.19 (s, 1H, NH), 8.56 (m, 1H), 7.36 (d, J = 8.3 Hz, 1H), 6.94 (dd, J_1 = 2.1 Hz, J = 8.3 Hz, 1H), 2.80 (m, 1H), 1.04–0.99 (m, 4H). HRMS (ESI-) m/z calcd. for ($C_{14}H_{10}ClN_3O_2S$, M-H): 318.0099, found: (M-H): 318.0112.

(Z)-3-(1-(3-Morpholinopropyl)-5-oxo-2-thioxoimidazolidin-4-ylidene)indolin-2-one hydrochloride (3u)

From **2a** (0.37 g, 1.5 mmol) and isatin (0.22 g, 1.5 mmol), compound **3u** (0.57 g, 93%) was obtained as a red solid. M.p. 270–272 °C. ^{1}H-NMR (400 MHz, DMSO-d_6) δ: 11.54 (s, 1H), 11.18 (s, 1H), 11.07 (bs, 1H), 8.54 (d, J = 8.0 Hz, 1H), 7.33 (t, J = 7.7 Hz, 1H), 7.05 (t, J = 7.7 Hz, 1H), 6.95 (d, J = 7.7 Hz, 1H), 3.96–3.88 (m, 4H), 3.78 (t, J = 11.8 Hz, 2H), 3.36 (d,

J = 11.8 Hz, 2H), 3.22–3.13 (m, 2H), 3.07–2.95 (m, 2H), 2.19–2.09 (m, 2H). HRMS (ESI+) m/z calcd. for ($C_{18}H_{20}N_4O_3S$, M + H): 373.1328, found: (M + H): 373.1322.

(Z)-5-Chloro-3-(1-(3-morpholinopropyl)-5-oxo-2-thioxoimidazolidin-4-ylidene)indolin-2-one hydrochloride (3v)

From **2a** (0.37 g, 1.5 mmol) and 5-chloroisatin (0.27 g, 1.5 mmol), compound **3v** (0.63 g, 95%) was obtained as a dark red solid. M.p. 249–251 °C. ^{1}H-NMR (400 MHz, DMSO-d$_6$) δ: 11.61 (s, 1H), 11.31 (s, 1H), 11.09 (bs, 1H), 8.58 (d, J = 2.0 Hz, 1H), 7.37 (dd, J_1 = 2.2 Hz, J_2 = 8.4 Hz, 1H), 6.97 (d, J = 8.3 Hz, 1H), 3.96–3.88 (m, 4H), 3.85–3.75 (m, 4H), 3.35 (d, J = 11.6 Hz, 2H), 3.22–3.13 (m, 2H), 3.07–2.95 (m, 2H), 2.20–2.10 (m, 2H). HRMS (ESI+) m/z calcd. for ($C_{18}H_{19}Cl_1N_4O_3S$, M + H): 407.0939, found: (M + H): 407.0917.

(Z)-3-(5-Oxo-1-phenyl-2-thioxoimidazolidin-4-ylidene)indolin-2-one (3w)

From **2b** (0.29 g, 1.5 mmol) and isatin (0.22 g, 1.5 mmol), compound **3w** (0.45 g, 93%) was obtained as a red solid. M.p. > 300 °C. ^{1}H-NMR (400 MHz, DMSO-d$_6$) δ: 11.57 (s, 1H, NH), 10.87 (s, 1H, NH), 8.46 (d, J = 7.8 Hz, 1H), 7.52–7.41 (m, 3H), 7.37–7.33 (m, 2H), 7.21 (td, J_1 = 1.1 Hz, J_2 = 7.7 Hz, 1H), 6.91 (td, J_1 = 1.0 Hz, J_2 = 7.7 Hz, 1H), 6.86 (d, J = 7.7 Hz, 1H). HRMS (ESI-) m/z calcd. for ($C_{17}H_{11}N_3O_2S$, M-H): 320.0488, found: (M-H): 320.0506.

(Z)-5-Chloro-3-(5-oxo-1-phenyl-2-thioxoimidazolidin-4-ylidene)indolin-2-one (3x)

From **2b** (0.29 g, 1.5 mmol) and 5-chloroisatin (0.27 g, 1.5 mmol), compound three times (0.50 g, 94%) was obtained as a red solid. M.p. > 300 °C. ^{1}H-NMR (400 MHz, DMSO-d$_6$) δ: 11.74 (bs, 1H, NH), 11.25 (s, 1H, NH), 8.55 (m, 1H), 7.59–7.49 (m, 3H), 7.47–7.42 (m, 2H), 7.37 (m, 1H), 6.96 (d, J = 8.4 Hz, 1H). HRMS (ESI-) m/z calcd. for ($C_{17}H_{10}ClN_3O_2S$, M-H): 354.0099, found: (M-H): 354.0113.

3.2.3. General Procedure for the Synthesis of Dispiroindolinones 4a–x

Corresponding 5-indolidene-2-thioxoimidazolidin **3** (1 equiv) and sarcosine (4 equiv) were dissolved in toluene and the mixture heated to a boiling point. After that, paraformaldehyde (4 equiv) was added. The resulting mixture was refluxed for 5–8 hours (TLC control). After the reaction was completed, the solvent was evaporated in vacuo. The product was then purified using column chromatography (silica gel 60, 0.04–0.063 mm/230–400 mesh, CHCl$_3$:MeOH/50:1) to afford products as a yellow or pink solid. This solid was washed with acetone to yield corresponding dispirooxindole as white crystalline solid.

1'-Methyl-1-benzyl-2-thioxodispiro[imidazolidine-4,3'-pyrrolidine-4',3''-indoline]-2'',5-dione (4a)

From **3a** (0.22 g, 0.65 mmol), sarcosine (0.23 g, 2.6 mmol) and paraformaldehyde (0.08 g, 2.6 mmol), compound **4a** (0.21 g, 82%) was obtained as a white solid. M.p. 189–190 °C. ^{1}H-NMR (400 MHz, DMSO-d$_6$) δ: 10.42 (bs, 1H, NH), 7.26 (t, J = 7.6 Hz, 1H), 7.23–7.10 (m, 4H), 7.08 (d, J = 7.7 Hz, 1H), 6.86–6.74 (m, 3H), 4.74 (d, J = 15.3 Hz, 1H), 4.66 (d, J = 15.3 Hz, 1H), 4.38 (d, J = 12.6 Hz, 1H), 4.23 (d, J = 12.6 Hz, 1H), 3.40–3.30 (m, 3H), 3.25 (dd, J_1 = 6.7 Hz, J_2 = 9.4 Hz, 2H), 3.06 (d, J = 9.9 Hz, 1H), 2.44 (s, 3H), 2.17 (s, 6H). HRMS (ESI+) m/z calcd. for ($C_{21}H_{20}N_4O_2S$, M + H): 393.1379, found: (M + H): 393.1364.

5''-Chloro-1'-methyl-1-benzyl-2-thioxodispiro[imidazolidine-4,3'-pyrrolidine-4',3''-indoline]-2'',5-dione (4b)

From **3b** (0.24 g, 0.65 mmol), sarcosine (0.23 g, 2.6 mmol) and paraformaldehyde, (0.08 g, 2.6 mmol) compound **4b** (0.21 g, 76%) was obtained as a white solid. M.p. 152–153 °C. ^{1}H-NMR (400 MHz, DMSO-d$_6$) δ: 10.54 (bs, 1H, NH), 7.34 (dd, J_1 = 2.0 Hz, J_2 = 8.3 Hz, 1H), 7.29 (d, J = 2.0 Hz, 1H), 7.25 (t, J = 7.5 Hz, 1H), 7.20–7.12 (m, 4H), 7.09 (d, J = 8.6 Hz, 1H), 6.75 (d, J = 6.5 Hz, 2H), 4.74 (d, J = 15.5 Hz, 1H), 4.67 (d, J = 15.5 Hz, 1H), 4.37 (d, J = 12.7 Hz, 1H), 4.20 (d, J = 12.7 Hz, 1H), 3.40–3.30 (m, 3H), 3.19 (t, J = 10.8 Hz, 2H), 3.07 (d, J = 9.7 Hz, 1H), 2.43 (s, 3H), 2.12 (s, 6H). HRMS (ESI+) m/z calcd. for ($C_{21}H_{19}ClN_4O_2S$, M + H): 427.0990, found: (M + H): 427.0976.

1′-Methyl-1-allyl-2-thioxodispiro[imidazolidine-4,3′-pyrrolidine-4′,3″-indoline]-2″,5-dione (4c)

From **3c** (0.19 g, 0.65 mmol), sarcosine (0.23 g, 2.6 mmol) and paraformaldehyde (0.08 g, 2.6 mmol), compound **4c** (0.15 g, 67%) was obtained as a white solid. M.p. 281–283 °C(decomp.). ^{1}H-NMR (400 MHz, DMSO-d$_6$) δ: 10.57 (s, 1H, NH), 10.44 (s, 1H, NH), 7.19 (d, J = 7.7 Hz, 1H), 7.15 (d, J = 8.1 Hz, 1H), 6.86 (t, J = 7.6 Hz, 1H), 6.79 (d, J = 7.7 Hz, 1H), 5.47 (m, 1H), 4.89 (d, J = 10.4 Hz, 1H), 4.59 (d, J = 17.4 Hz, 1H), 4.18–4.03 (m, 2H), 3.40 (d, J = 9.8 Hz, 1H), 3.32 (d, J = 9.8 Hz, 1H), 3.16 (d, J = 9.9 Hz, 1H), 3.05 (d, J = 10.0 Hz, 1H), 2.45 (c, 3H). HRMS (ESI+) m/z calcd. for (C$_{17}$H$_{18}$N$_4$O$_2$S, M + H): 343.1223, found: (M + H): 343.1207.

5″-Chloro-1′-methyl-1-allyl-2-thioxodispiro[imidazolidine-4,3′-pyrrolidine-4′,3″-indoline]-2″,5-dione (4d)

From **3d** (0.21 g, 0.65 mmol), sarcosine (0.23 g, 2.6 mmol) and paraformaldehyde (0.08 g, 2.6 mmol), compound **4d** (0.09 g, 38%) was obtained as a white solid. **M.p.** 141–142 °C. ^{1}H-NMR (400 MHz, DMSO-d$_6$) δ: 10.49 (bs, 1H, NH), 7.33 (dd, J_1 = 1.9 Hz, J_2 = 8.4 Hz, 1H), 7.26 (d, J = 1.7 Hz, 1H), 7.09 (d, J = 8.4 Hz, 1H), 5.47 (m, 1H), 4.88 (d, J = 10.2 Hz, 1H), 4.55 (d, J = 17.2 Hz, 1H), 4.38 (d, J = 12.6 Hz, 1H), 4.22 (d, J = 12.6 Hz, 1H), 4.20–4.02 (m, 2H), 3.30–3.21 (m, 3H), 3.04 (d, J = 10.0 Hz, 1H), 2.44 (s, 3H), 2.19 (s, 6H). HRMS (ESI+) m/z calcd. for (C$_{17}$H$_{17}$ClN$_4$O$_2$S, M + H): 377.0833, found: (M + H): 377.0822.

1′-Methyl-1-(4-methoxyphenyl)-2-thioxodispiro[imidazolidine-4,3′-pyrrolidine-4′,3″-indoline]-2″,5-dione (4e)

From **3e** (0.23 g, 0.65 mmol), sarcosine (0.23 g, 2.6 mmol) and paraformaldehyde (0.08 g, 2.6 mmol), compound **4e** (0.08 g, 28%) was obtained as a white solid. M.p. 155–156 °C. ^{1}H-NMR (400 MHz, DMSO-d$_6$) δ: 10.67–10.57 (m, 2H, NH), 7.26 (t, J = 7.5 Hz, 1H), 7.20 (d, J = 8.1 Hz, 2H), 7.12 (d, J = 7.5 Hz, 1H), 6.93 (t, J = 7.5 Hz, 1H), 6.86 (d, J = 7.7 Hz, 1H), 6.71 (d, J = 8.1 Hz, 2H), 3.49 (d, J = 9.9 Hz, 1H), 3.38 (m, 1H), 3.32 (m, 1H), 3.07 (d, J = 10.2 Hz, 1H), 2.47 (s, 3H), 2.31 (s, 3H). HRMS (ESI+) m/z calcd. for (C$_{21}$H$_{20}$N$_4$O$_3$S, M + H): 409.1328, found: (M + H): 409.1323.

5″-Cchloro-1′-methyl-1-(4-methoxyphenyl)-2-thioxodispiro[imidazolidine-4,3′-pyrrolidine-4′,3″-indoline]-2″,5-dione (4f)

From **3f** (0.25 g, 0.65 mmol), sarcosine (0.23 g, 2.6 mmol) and paraformaldehyde (0.08 g, 2.6 mmol), compound **4f** (0.14 g, 49%) was obtained as a white solid. M.p. 289–290 °C. ^{1}H-NMR (400 MHz, DMSO-d$_6$) δ: 10.79 (s, 1H, NH), 10.74 (s, 1H, NH), 7.34 (dd, J_1 = 2.0 Hz, J_2 = 8.3 Hz, 1H), 7.23 (d, J = 8.1 Hz, 2H), 7.12 (d, J = 1.8 Hz, 1H), 6.88 (d, J = 8.3 Hz, 1H), 6.74 (d, J = 8.1 Hz, 2H), 3.45–3.30 (m, 3H), 3.09 (d, J = 10.2 Hz, 1H), 2.47 (s, 3H), 2.32 (s, 3H). HRMS (ESI+) m/z calcd. for (C$_{21}$H$_{19}$ClN$_4$O$_3$S, M + H): 443.0939, found: (M + H): 443.0940.

1′-Methyl-1-(4-ethoxyphenyl)-2-thioxodispiro[imidazolidine-4,3′-pyrrolidine-4′,3″-indoline]-2″,5-dione (4g)

From **3g** (0.24 g, 0.65 mmol), sarcosine (0.23 g, 2.6 mmol) and paraformaldehyde (0.08 g, 2.6 mmol), compound **4g** (0.10 g, 35%) was obtained as a white solid. M.p. 240–241 °C. ^{1}H-NMR (400 MHz, DMSO-d$_6$) δ: 10.63 (s, 1H, NH), 10.60 (s, 1H, NH), 7.26 (t, J = 7.7 Hz, 1H), 7.13 (d, J = 7.5 Hz, 1H), 6.98–6.89 (m, 3H), 6.68 (d, J = 7.7 Hz, 1H), 6.73 (d, J = 8.6 Hz, 2H), 4.03 (q, J = 7.0 Hz, 2H), 3.49 (d, J = 10.0 Hz, 1H), 3.40–3.28 (m, 2H), 3.07 (d, J = 10.0 Hz, 1H), 2.47 (s, 3H), 1.32 (t, J = 7.0 Hz, 3H). HRMS (ESI+) m/z calcd. for (C$_{22}$H$_{22}$N$_4$O$_3$S, M + H): 423.1485, found: (M + H): 423.1480.

5″-Chloro-1′-methyl-1-(4-ethoxyphenyl)-2-thioxodispiro[imidazolidine-4,3′-pyrrolidine-4′,3″-indoline]-2″,5-dione (4h)

From **3h** (0.26 g, 0.65 mmol), sarcosine (0.23 g, 2.6 mmol) and paraformaldehyde (0.08 g, 2.6 mmol), compound **4h** (0.08 g, 27%) was obtained as a white solid. M.p. 268–271 °C. ^{1}H-

NMR (400 MHz, DMSO-d_6) δ: 10.80–10.61 (bs, 2H, NH), 7.33 (dd, J_1 = 1.6 Hz, J_2 = 8.2 Hz, 1H), 7.13 (d, J = 1.6 Hz, 1H), 6.95 (d, J = 8.8 Hz, 2H), 6.87 (d, J = 8.3 Hz, 1H), 6.76 (d, J = 8.4 Hz, 2H), 4.04 (q, J = 7.0 Hz, 2H), 3.45–3.34 (m, 3H), 3.40–3.28 (m, 2H), 3.09 (d, J = 10.3 Hz, 1H), 2.47 (s, 3H), 1.32 (t, J = 7.0 Hz, 3H). HRMS (ESI+) m/z calcd. for ($C_{22}H_{21}ClN_4O_3S$, M + H): 457.1095, found: (M + H): 457.1090.

1'-Methyl-1-(p-tolyl)-2-thioxodispiro[imidazolidine-4,3'-pyrrolidine-4',3''-indoline]-2'',5-dione (4i)

From **3i** (0.22 g, 0.65 mmol), sarcosine (0.23 g, 2.6 mmol) and paraformaldehyde (0.08 g, 2.6 mmol), compound **4i** (0.10 g, 39%) was obtained as a white solid. M.p. 155–157 °C. ^{1}H-NMR (400 MHz, DMSO-d_6) δ: 10.61 (bs, 1H, NH), 10.57 (bs, 1H, NH), 7.26 (dd, J_1 = 1.0 Hz, J_2 = 7.7 Hz, 1H), 7.14 (d, J = 7.2 Hz, 1H), 6.97–6.91 (m, 3H), 6.86 (d, J = 7.7 Hz, 1H), 6.75 (d, J = 8.8 Hz, 2H), 3.76 (s, 3H), 3.49 (d, J = 10.1 Hz, 1H), 3.37 (d, J = 10.1 Hz, 1H), 3.32 (m, 1H), 3.07 (d, J = 10.1 Hz, 1H), 2.47 (s, 3H). HRMS (ESI+) m/z calcd. for ($C_{21}H_{20}N_4O_2S$, M + H): 393.1379, found: (M + H): 393.1384.

5''-Chloro-1'-methyl-1-(p-tolyl)-2-thioxodispiro[imidazolidine-4,3'-pyrrolidine-4',3''-indoline]-2'',5-dione (4j)

From **3j** (0.24 g, 0.65 mmol), sarcosine (0.23 g, 2.6 mmol) and paraformaldehyde (0.08 g, 2.6 mmol), compound **4j** (0.14 g, 50%) was obtained as a white solid. M.p. 159–160 °C. ^{1}H-NMR (400 MHz, DMSO-d_6) δ: 10.76 (s, 1H, NH), 10.69 (bs, 1H, NH), 7.33 (dd, J_1 = 2.2 Hz, J_2 = 8.3 Hz, 1H), 7.14 (d, J = 2.2 Hz, 1H), 6.99–6.94 (m, 2H), 6.87 (d, J = 8.3 Hz, 1H), 6.80–6.75 (m, 2H), 3.78 (s, 3H), 3.44–3.32 (m, 3H), 3.09 (d, J = 10.2 Hz, 1H), 2.47 (s, 3H). HRMS (ESI+) m/z calcd. for ($C_{21}H_{19}ClN_4O_2S$, M + H): 427.0990, found: (M + H): 427.0987.

5''-Chloro-1'-methyl-1-(4-chlorophenyl)-2-thioxodispiro[imidazolidine-4,3'-pyrrolidine-4',3''-indoline]-2'',5-dione (4k)

From **3k** (0.23 g, 0.65 mmol), sarcosine (0.23 g, 2.6 mmol) and paraformaldehyde (0.08 g, 2.6 mmol), compound **4k** (0.13 g, 48%) was obtained as a white solid. M.p. 163–164 °C. ^{1}H-NMR (400 MHz, DMSO-d_6) δ: 10.72 (bs, 1H, NH), 10.62 (s, 1H, NH), 7.53–7.48 (m, 2H), 7.26 (td, J_1 = 1.2 Hz, J_2 = 7.7 Hz, 1H), 7.12 (d, J = 7.4 Hz, 1H), 6.93 (td, J_1 = 0.9 Hz, J_2 = 7.7 Hz, 1H), 6.91–6.87 (m, 2H), 6.86 (d, J = 7.9 Hz, 1H), 3.49 (d, J = 10.1 Hz, 1H), 3.39–3.33 (m, 2H), 3.08 (d, J = 10.0 Hz, 1H), 2.48 (s, 3H). HRMS (ESI+) m/z calcd. for ($C_{20}H_{17}ClN_4O_2S$, M + H): 413.0833, found: (M + H): 413.0829.

5''-Chloro-1'-methyl-1-(4-chlorophenyl)-2-thioxodispiro[imidazolidine-4,3'-pyrrolidine-4',3''-indoline]-2'',5-dione (4l)

From **3l** (0.25 g, 0.65 mmol), sarcosine (0.23 g, 2.6 mmol) and paraformaldehyde (0.08 g, 2.6 mmol), compound **4l** (0.03 g, 10%) was obtained as a white solid. M.p. 239–240 °C. ^{1}H-NMR (400 MHz, DMSO-d_6) δ: 10.84 (bs, 1H, NH), 10.77 (s, 1H, NH), 7.55–7.50 (m, 2H), 7.26 (dd, J_1 = 2.2 Hz, J_2 = 8.3 Hz, 1H), 7.12 (d, J = 2.1 Hz, 1H), 6.94–6.89 (m, 2H), 6.87 (d, J = 8.3 Hz, 1H), 3.49 (d, J = 10.1 Hz, 1H), 3.44–3.38 (m, 2H), 3.33 (m, 1H), 3.10 (d, J = 10.4 Hz, 1H), 2.48 (s, 3H). HRMS (ESI+) m/z calcd. for ($C_{20}H_{16}Cl_2N_4O_2S$, M + H): 447.0443, found: (M + H): 447.0433.

1'-Methyl-1-(4-fluorophenyl)-2-thioxodispiro[imidazolidine-4,3'-pyrrolidine-4',3''-indoline]-2'',5-dione (4m)

From **3m** (0.22 g, 0.65 mmol), sarcosine (0.23 g, 2.6 mmol) and paraformaldehyde (0.08 g, 2.6 mmol), compound **4m** (0.06 g, 24%) was obtained as a white solid. M.p. 273–274 °C. ^{1}H-NMR (400 MHz, DMSO-d_6) δ: 10.91 (bs, 1H, NH), 10.79 (s, 1H, NH), 7.55 (t, J = 9.0 Hz, 1H), 7.43 (dd, J_1 = 2.1 Hz, J_2 = 8.3 Hz, 1H), 7.13–7.07 (m, 2H), 6.94 (m, 1H), 6.89 (d, J = 8.3 Hz, 1H), 3.45–3.37 (m, 2H), 3.32 (d, J = 10.2 Hz, 1H), 3.10 (d, J = 10.2 Hz, 1H), 2.48 (s, 3H). HRMS (ESI+) m/z calcd. for ($C_{20}H_{17}FN_4O_2S$, M + H): 397.1129, found: (M + H): 397.1115

5″-Chloro-1′-methyl-1-(4-fluorophenyl)-2-thioxodispiro[imidazolidine-4,3′-pyrrolidine-4′,3″-indoline]-2″,5-dione (4n)

From **3n** (0.24 g, 0.65 mmol), sarcosine (0.23 g, 2.6 mmol) and paraformaldehyde (0.08 g, 2.6 mmol), compound **4n** (0.10 g, 35%) was obtained as a white solid. M.p. 273–274 °C. ^{1}H-NMR (400 MHz, DMSO-d_6) δ: 10.82 (bs, 1H, NH), 10.78 (bs, 1H, NH), 7.37–7.26 (m, 3H), 7.12 (s, 1H), 6.97–6.85 (m, 3H), 3.43–3.38 (m, 2H), 3.32 (d, J = 10.2 Hz, 1H), 3.10 (d, J = 10.2 Hz, 1H), 2.48 (s, 3H). HRMS (ESI+) m/z calcd. for ($C_{20}H_{16}FClN_4O_2S$, M + H): 431.0739, found: (M + H): 431.0760.

1′-Methyl-1-(3-chlorobenzyl)-2-thioxodispiro[imidazolidine-4,3′-pyrrolidine-4′,3″-indoline]-2″,5-dione (4o)

From **3o** (0.24 g, 0.65 mmol), sarcosine (0.23 g, 2.6 mmol) and paraformaldehyde (0.08 g, 2.6 mmol), compound **4o** (0.24 g, 87%) was obtained as a white solid. M.p. 273–274 °C. ^{1}H-NMR (400 MHz, DMSO-d_6) δ: 10.59 (bs, 1H, NH), 10.57 (bs, 1H, NH), 7.28 (d, J = 7.6 Hz, 1H), 7.23–7.11 (m, 2H), 7.05 (s, 1H), 7.02 (d, J = 7.6 Hz, 1H), 6.78 (d, J = 7.7 Hz, 1H), 6.74 (d, J = 7.6 Hz, 1H), 6.67 (t, J = 7.6 Hz, 1H), 4.76 (d, J = 15.5 Hz, 1H), 4.66 (d, J = 15.5 Hz, 1H), 3.39 (d, J = 10.1 Hz, 1H), 3.20 (d, J = 10.0 Hz, 1H), 3.05 (d, J = 10.0 Hz, 1H), 2.44 (s, 3H). HRMS (ESI+) m/z calcd. for ($C_{21}H_{19}ClN_4O_2S$, M + H): 427.0995, found: (M + H): 427.0981.

5″-Chloro-1′-methyl-1-(3-chlorobenzyl)-2-thioxodispiro[imidazolidine-4,3′-pyrrolidine-4′,3″-indoline]-2″,5-dione (4p)

From **3p** (0.26 g, 0.65 mmol), sarcosine (0.23 g, 2.6 mmol) and paraformaldehyde (0.08 g, 2.6 mmol), compound **4p** (0.18 g, 61%) was obtained as a white solid. M.p. 261–262 °C. ^{1}H-NMR (400 MHz, DMSO-d_6) δ: 10.71 (s, 1H, NH), 10.68 (bs, 1H, NH), 7.29–7.20 (m, 2H), 7.18 (t, J = 7.9 Hz, 1H), 7.12 (s, 1H), 6.98 (s, 1H), 6.80 (d, J = 8.3 Hz, 1H), 6.67 (d, J = 7.3 Hz, 1H), 4.77 (d, J = 15.7 Hz, 1H), 4.68 (d, J = 15.7 Hz, 1H), 3.30 (d, J = 11.0 Hz, 1H), 3.24 (d, J = 8.8 Hz, 2H), 3.07 (d, J = 9.9 Hz, 1H), 2.44 (s, 3H). HRMS (ESI+) m/z calcd. for ($C_{21}H_{18}Cl_2N_4O_2S$, M + H): 461.0600, found: (M + H): 461.0610.

1′-Methyl-1-(3-chloro-4-fluorophenyl)-2-thioxodispiro[imidazolidine-4,3′-pyrrolidine-4′,3″-indoline]-2″,5-dione (4q)

From **3q** (0.24 g, 0.65 mmol), sarcosine (0.23 g, 2.6 mmol) and paraformaldehyde (0.08 g, 2.6 mmol), compound **4q** (0.21 g, 75%) was obtained as a white solid. M.p. 252–254 °C. ^{1}H-NMR (400 MHz, DMSO-d_6) δ: 10.81 (s, 1H, NH), 10.64 (s, 1H, NH), 7.51 (t, J = 8.9 Hz, 1H), 7.27 (t, J = 7.6 Hz, 1H), 7.12 (d, J = 7.3 Hz, 2H), 6.93 (t, J = 7.5 Hz, 1H), 6.89–6.81 (m, 3H), 3.47 (d, J = 10.2 Hz, 1H), 3.37 (d, J = 10.8 Hz, 2H), 3.08 (d, J = 10.2 Hz, 1H), 2.48 (s, 3H). HRMS (ESI+) m/z calcd. for ($C_{20}H_{16}ClFN_4O_2S$, M + H): 431.0739, found: (M + H): 431.0721.

5″-Chloro-1′-methyl-1-(3-chloro-4-fluorophenyl)-2-thioxodispiro[imidazolidine-4,3′-pyrrolidine-4′,3″-indoline]-2″,5-dione (4r)

From **3r** (0.27 g, 0.65 mmol), sarcosine (0.23 g, 2.6 mmol) and paraformaldehyde (0.08 g, 2.6 mmol), compound **4r** (0.10 g, 33%) was obtained as a white solid. M.p. 177–180 °C. ^{1}H-NMR (400 MHz, DMSO-d_6) δ: 10.91 (bs, 1H, NH), 10.79 (s, 1H, NH), 7.55 (t, J = 9.0 Hz, 1H), 7.34 (dd, J_1 = 2.1 Hz, J_2 = 8.3 Hz, 1H), 7.13–7.07 (m, 2H), 6.94 (m, 1H), 6.89 (d, J = 8.3 Hz, 1H), 3.45–3.37 (m, 2H), 3.32 (d, J = 10.2 Hz, 1H), 3.10 (d, J = 10.2 Hz, 1H), 2.48 (s, 3H). HRMS (ESI+) m/z calcd. for ($C_{20}H_{15}Cl_2FN_4O_2S$, M + H): 465.0350, found: (M + H): 465.0341.

1′-Methyl-1-cyclopropyl-2-thioxodispiro[imidazolidine-4,3′-pyrrolidine-4′,3″-indoline]-2″,5-dione (4s)

From **3s** (0.19 g, 0.65 mmol), sarcosine (0.23 g, 2.6 mmol) and paraformaldehyde (0.08 g, 2.6 mmol), compound **4s** (0.12 g, 54%) was obtained as a white solid. M.p. 272–273 °C (decomp.). ^{1}H-NMR (400 MHz, DMSO-d_6) δ: 10.50 (s, 1H, NH), 10.28 (s, 1H, NH), 7.19 (t, J = 7.7 Hz, 1H), 7.11 (d, J = 7.5 Hz, 1H), 6.89 (t, J = 7.6 Hz, 1H), 6.77 (d, J = 7.7 Hz, 1H), 3.30 (d, J = 10.0 Hz, 1H), 3.25 (d, J = 10.0 Hz, 1H), 3.14 (d, J = 10.0 Hz, 1H), 3.01 (d, J = 10.0 Hz,

1H), 2.45 (m, 1H), 2.43 (s, 3H), 0.82–0.70 (m, 2H), 0.62 (m, 1H), 0.11 (m, 1H). HRMS (ESI+) m/z calcd. for ($C_{17}H_{18}N_4O_2S$, M + H): 343.1223, found: (M + H): 343.1241.

5″-Chloro-1′-methyl-1-cyclopropyl-2-thioxodispiro[imidazolidine-4,3′-pyrrolidine-4′,3″-indoline]-2″,5-dione (4t)

From **3t** (0.21 g, 0.65 mmol), sarcosine (0.23 g, 2.6 mmol) and paraformaldehyde (0.08 g, 2.6 mmol), compound **4t** (0.06 g, 25%) was obtained as a white solid. M.p. 275–276 °C. ^{1}H-NMR (400 MHz, DMSO-d_6) δ: 10.65 (s, 1H, NH), 10.41 (s, 1H, NH), 7.26 (m, 1H), 7.10 (m, 1H), 6.79 (dd, J_1 = 2.1 Hz, J_2 = 8.3 Hz, 1H), 3.26–3.16 (m, 3H), 3.02 (d, J = 10.0 Hz, 1H), 2.43 (s, 3H), 0.89–0.74 (m, 2H), 0.59 (m, 1H), 0.11 (m, 1H). HRMS (ESI+) m/z calcd. for ($C_{17}H_{17}ClN_4O_2S$, M + H): 377.0834, found: (M + H): 377.0851.

1′-Methyl-1-(3-morpholinopropyl)-2-thioxodispiro[imidazolidine-4,3′-pyrrolidine-4′,3″-indoline]-2″,5-dione (4u)

From **3u** (0.24 g, 0.65 mmol), sarcosine (0.23 g, 2.6 mmol) and paraformaldehyde (0.08 g, 2.6 mmol), compound **4u** (0.08 g, 25%) was obtained as a white solid. M.p. 216–217 °C. ^{1}H-NMR (400 MHz, DMSO-d_6) δ: 10.55 (s, 1H, NH), 10.39 (s, 1H, NH), 7.22–7.14 (m, 2H), 6.88 (t, J = 7.6 Hz, 1H), 6.78 (d, J = 7.8 Hz, 1H), 3.61–3.43 (m, 6H), 3.38 (d, J = 7.1 Hz, 1H), 3.31 (d, J = 10.6 Hz, 1H), 3.15 (d, J = 10.0 Hz, 1H), 3.04 (d, J = 10.0 Hz, 1H), 2.44 (s, 3H), 2.28–2.18 (m, 4H), 2.05 (t, J = 6.7 Hz, 2H), 1.42–1.33 (m, 2H). HRMS (ESI+) m/z calcd. for ($C_{21}H_{27}N_5O_3S$, M + H): 430.1907, found: (M + H): 430.1904.

5″-Chloro-1′-methyl-1-(3-morpholinopropyl)-2-thioxodispiro[imidazolidine-4,3′-pyrrolidine-4′,3″-indoline]-2″,5-dione (4v)

From **3v** (0.26 g, 0.65 mmol), sarcosine (0.23 g, 2.6 mmol) and paraformaldehyde (0.08 g, 2.6 mmol), compound **4v** (0.05 g, 15%) was obtained as a white solid. M.p. 224–225 °C. ^{1}H-NMR (400 MHz, DMSO-d_6) δ: 10.69 (s, 1H, NH), 10.49 (s, 1H, NH), 7.25 (dd, J_1 = 1.8 Hz, J_2 = 8.5 Hz, 1H), 7.19 (s, 1H), 6.80 (d, J = 8.1 Hz, 1H), 3.64–3.46 (m, 6H), 3.30–3.18 (m, 3H), 3.06 (d, J = 9.9 Hz, 1H), 2.43 (s, 3H), 2.28–2.19 (m, 4H), 2.10–2.03 (m, 2H), 1.44–1.28 (m, 2H). HRMS (ESI+) m/z calcd. for ($C_{21}H_{26}ClN_5O_3S$, M + H): 464.1517, found: (M + H): 464.1519.

1′-Methyl-1-phenyl-2-thioxodispiro[imidazolidine-4,3′-pyrrolidine-4′,3″-indoline]-2″,5-dione (4w)

From **3w** (0.21 g, 0.65 mmol), sarcosine (0.23 g, 2.6 mmol) and paraformaldehyde (0.08 g, 2.6 mmol), compound **4w** (0.07 g, 27%) was obtained as a white solid. M.p. 273–274 °C. ^{1}H-NMR (400 MHz, DMSO-d_6) δ: 10.65 (bs, 1H, NH), 10.63 (s, 1H, NH), 7.45–7.36 (m, 3H), 7.27 (t, J = 7.7 Hz, 1H), 7.15 (d, J = 7.5 Hz, 1H), 6.94 (t, J = 7.6 Hz, 1H), 6.89–6.82 (m, 3H), 3.51 (d, J = 10.0 Hz, 1H), 3.37 (d, J = 10.0 Hz, 2H), 3.08 (d, J = 10.2 Hz, 1H), 2.48 (s, 3H). HRMS (ESI+) m/z calcd. for ($C_{20}H_{18}N_4O_2S$, M + H): 379.1223, found: (M + H): 379.1236.

5″-Chloro-1′-methyl-1-phenyl-2-thioxodispiro[imidazolidine-4,3′-pyrrolidine-4′,3″-indoline]-2″,5-dione (4x)

From **three times** (0.23 g, 0.65 mmol), sarcosine (0.23 g, 2.6 mmol) and paraformaldehyde (0.08 g, 2.6 mmol), compound four times (0.14 g, 49%) was obtained as a white solid. M.p. 268–269 °C. ^{1}H-NMR (400 MHz, DMSO-d_6) δ: 10.79 (bs, 1H, NH), 10.78 (bs, 1H, NH), 7.47–7.39 (m, 3H), 7.34 (dd, J_1 = 1.6 Hz, J_2 = 8.3 Hz, 1H), 7.14 (s, 1H), 6.91–6.85 (m, 3H), 3.43 (d, J = 10.2 Hz, 1H), 3.39 (d, J = 10.2 Hz, 1H), 3.34 (d, J = 10.2 Hz, 1H), 3.10 (d, J = 10.2 Hz, 1H), 2.48 (s, 3H). HRMS (ESI+) m/z calcd. for ($C_{20}H_{17}ClN_4O_2S$, M + H): 413.0834, found: (M + H): 413.0850.

3.2.4. General Procedure for the Synthesis of Dispiroindolinones **6**

Corresponding 5-indolidene-2-selenoxoimidazolidin **5** (1 equiv) and sarcosine (4 equiv) were dissolved in toluene and the mixture heated to a boiling point. After that paraformaldehyde (4 equiv) was added. The resulting mixture was refluxed for 5–8 hours (TLC control).

After the reaction was completed, the solvent was evaporated in vacuo. The product was then purified using column chromatography (silica gel 60, 0.04–0.063 mm/230–400 mesh, CHCl$_3$:MeOH/50:1) to afford products as a white solid. This solid was washed with cold methanol to yield corresponding dispirooxindole as light brown crystalline solid.

5″-Chloro-1-(3-chloro-4-fluorophenyl)-1′-methyl-2-selenoxodispiro[imidazolidine-4,3′-pyrrolidine-4′,3″-indoline]-2″,5-dione (6r)

From 5r (0.30 g, 0.65 mmol), sarcosine (0.23 g, 2.6 mmol) and paraformaldehyde (0.08 g, 2.6 mmol), compound 6r (0.19 g, 57%) was obtained as light brown solid. M.p. 193–194 °C (decomp.). ^{1}H-NMR (400 MHz, DMSO-d$_6$) δ: 11.62 (s, 1H, NH), 10.81 (s, 1H, NH), 7.57 (t, J = 9.0 Hz, 1H), 7.36 (dd, J_1 = 1.8 Hz, J_2 = 8.3 Hz, 1H), 7.12 (m, 1H), 7.10 (m, 1H), 6.96 (m, 1H), 6.90 (d, J = 8.3 Hz, 1H), 3.43 (s, 2H), 3.32 (m, 1H), 3.12 (d, J = 10.2 Hz, 1H), 2.49 (s, 3H). HRMS (ESI+) m/z calcd. for (C$_{20}$H$_{17}$ClN$_4$O$_2$S, M + H): 512.9800, found: (M + H): 513.0050.

5″-Chloro-1-(4-methoxyphenyl)-1′-methyl-2-selenoxodispiro[imidazolidine-4,3′-pyrrolidine-4′,3″-indoline]-2″,5-dione (6f)

From 5f (0.28 g, 0.65 mmol), sarcosine (0.23 g, 2.6 mmol) and paraformaldehyde (0.08 g, 2.6 mmol), compound 6f (0.19 g, 71%) was obtained as a white solid. M.p. 284–285 °C (decomp.). ^{1}H-NMR (400 MHz, DMSO-d$_6$) δ: 11.41 (bs, 1H, NH), 10.79 (s, 1H, NH), 7.33 (dd, J_1 = 1.9 Hz, J_2 = 8.3 Hz, 1H), 7.13 (d, J = 1.9 Hz, 1H), 6.96 (d, J = 8.7 Hz, 2H), 6.87 (d, J = 8.3 Hz, 1H), 6.78 (d, J = 8.6 Hz, 2H), 3.77 (s, 3H), 3.45–3.37 (m, 2H), 3.31 (m, 1H), 3.10 (d, J = 10.2 Hz, 1H), 2.48 (s, 3H). HRMS (ESI+) m/z calcd. for (C$_{21}$H20ClN$_4$O$_3$Se, M + H): 491.0384, found: (M + H): 491.0385.

3.2.5. General Procedure for the Synthesis of 5-Substituted Hydantoins 8

Corresponding 5-substituted-2-thiohydantoin 4 (1 equiv) was added to solution of the potassium hydroxide (1.05 equiv) in EtOH at room temperature (~4 mL EtOH for 100 mg of 3). After that, MeI (1.5 eq) was added and the reaction mixture was stirred at room temperature overnight. Then EtOH:HCl conc.(1:1) was added to the reaction (~4 mL EtOH for 100 mg of 3) and refluxed for 2 hours. Further, the reaction cooled to room temperature and formed precipitate was filtered off, washed with ethanol and dried in air. All compounds were obtained as red crystalline powders.

(Z)-5-Chloro-3-(1-(4-methoxyphenyl)-5-oxo-2-thioxoimidazolidin-4-ylidene)indolin-2-one (8f)

From 3f (0.116 g, 0.30 mmol), KOH (0.018 g, 0.32 mmol) and MeI (0.064 g, 0.45 mmol), compound 8f (0.089 g, 80%) was obtained as a red solid. M.p. > 300 °C. ^{1}H-NMR (400 MHz, DMSO-d$_6$) δ: 11.08 (s, 1H, NH), 11.06 (s, 1H, NH), 8.59 (d, J = 2.0 Hz, 1H), 7.95 (s, 1H, Ar), 7.39 (d, J = 8.8 Hz, 2H), 7.33 (dd, J_1= 2.2 Hz, J_2 = 8.4 Hz, 1H), 7.08 (d, J = 9.0 Hz, 2H), 6.94 (d, J = 8.4 Hz, 1H), 3.82 (s, 3H). HRMS (ESI+) m/z calcd. for (C$_{18}$H$_{12}$ClN$_3$O$_4$ M + H): 370.0595, found: (M + H): 370.0588.

(Z)-5-Chloro-3-(1-(4-ethoxyphenyl)-5-oxo-2-thioxoimidazolidin-4-ylidene)indolin-2-one (8h)

From 3h (0.120 g, 0.30 mmol), KOH (0.018 g, 0.32 mmol) and MeI (0.064 g, 0.45 mmol), compound 8h (0.097 g, 84%) was obtained as a white solid. M.p. > 300 °C. ^{1}H-NMR (400 MHz, DMSO-d$_6$) δ: 11.66 (s, 1H, NH), 11.24 (s, 1H, NH), 8.54 (m, 1H), 7.39–7.70 (m, 3H), 7.09–7.04 (m, 2H), 6.94 (m, 1H) 4.08 (q, J = 6.9 Hz, 2H), 1.36 (t, J = 6.9 Hz, 3H). HRMS (ESI+) m/z calcd. for (C$_{19}$H$_{14}$ClN$_3$O$_4$, M + H): 084.0751, found: (M + H): 413.0781.

(Z)-5-Chloro-3-(1-(3-chloro-4-fluorophenyl)-5-oxo-2-thioxoimidazolidin-4-ylidene)indolin-2-one (8r)

From 3r (0.122 g, 0.30 mmol), KOH (0.018 g, 0.32 mmol) and MeI (0.064 g, 0.45 mmol), compound 8r (0.093 g, 79%) was obtained as a white solid. M.p. > 300 °C. ^{1}H-NMR (400 MHz, DMSO-d$_6$) δ: 11.22 (s, 1H, NH), 11.10 (s, 1H, NH), 8.54 (m, 1H), 7.77 (dd,

J_1 = 2.2 Hz, J_2 = 6.7 Hz, 1H), 7.64 (m, 1H), 7.54 (m, 1H), 7.36 (m, 1H), 6.95 (t, J = 9.0 Hz, 1H). HRMS (ESI+) m/z calcd. for ($C_{17}H_9Cl_2FN_3O_3$, M + H): 392.0005, found: (M + H): 491.0998.

3.2.6. General Procedure for the Synthesis of Dispiroindolinones 9

Corresponding imidazolidin 8 (1 equiv) and sarcosine (4 equiv) were dissolved in toluene and the mixture heated to a boiling point. After that, paraformaldehyde (4 equiv) was added. The resulting mixture was refluxed for 5–8 hours (TLC control). After the reaction was completed, the solvent was evaporated in vacuo. The product was then purified using column chromatography (silica gel 60, 0.04–0.063 mm/230–400 mesh, $CHCl_3$:MeOH/50:1) to afford products as a white solid. This solid was washed with acetone to yield corresponding dispirooxindole as a white crystalline solid.

5″-Chloro-1-(4-methoxyphenyl)-1′-methyldispiro[imidazolidine-4,3′-pyrrolidine-4′,3″-indoline]-2,2″,5-trione (9f)

From **8f** (0.074 g, 0.20 mmol), sarcosine (0.071 g, 0.80 mmol) and paraformaldehyde (0.026 g, 0.80 mmol), compound **9f** (0.061 g, 72%) was obtained as a white solid. M.p. 295–296 °C. ^{1}H-NMR (400 MHz, DMSO-d_6) δ: 10.74 (s, 1H NH), 8.80 (s, 1H, NH), 7.33 (dd, J_1 = 2.2 Hz, J_2 = 8.3 Hz, 1H), 7.17 (d, J = 2.0 Hz, 1H), 6.98 (d, J = 9.0 Hz, 2H), 6.90 (d, J = 8.9 Hz, 2H), 6.87 (m, 1H), 3.44–3.36 (m, 2H), 3.31 (d, J = 10.2 Hz, 1H), 3.08 (d, J = 10.3 Hz, 1H), 2.47 (s, 3H). HRMS (ESI+) m/z calcd. for ($C_{21}H_{19}ClN_4O_4$, M + H): 427.1173, found: (M + H): 427.1177.

5″-Chloro-1-(4-ethoxyphenyl)-1′-methyldispiro[imidazolidine-4,3′-pyrrolidine-4′,3″-indoline]-2,2″,5-trione (9f)

From **8h** (0.076 g, 0.20 mmol), sarcosine (0.071 g, 0.80 mmol) and paraformaldehyde (0.026 g, 0.80 mmol), compound **9h** (0.055 g, 72%) was obtained as a white solid. M.p. 273–274 °C. ^{1}H-NMR (400 MHz, DMSO-d_6) δ: 10.74 (s, 1H, NH), 8.80 (s, 1H, NH), 7.32 (m, 1H), 7.17 (s, 1H), 7.00–6.92 (m, 2H), 6.92–6.82 (m, 3H), 4.08–3.98 (m, 2H), 3.45–3.36 (m, 2H), 3.29 (m, 1H), 3.07 (m, 1H), 2.46 (s, 3H), 1.37–1.27 (m, 3H). HRMS (ESI+) m/z calcd. for ($C_{22}H_{21}ClN_4O_4$ M + H): 441.1330, found: (M + H): 441.1295.

5″-Chloro-1-(3-chloro-4-fluorophenyl)-1′-methyldispiro[imidazolidine-4,3′-pyrrolidine-4′,3″-indoline]-2,2″,5-trione (9r)

From **8r** (0.078 g, 0.20 mmol), sarcosine (0.071 g, 0.80 mmol) and paraformaldehyde (0.026 g, 0.80 mmol), compound **9r** (0.067 g, 75%) was obtained as a white solid. M.p. 197–198 °C. ^{1}H-NMR (400 MHz, DMSO-d_6) δ: 10.75 (s, 1H, NH), 8.99 (s, 1H, NH), 7.55 (t, J = 9.0 Hz, 1H), 7.33 (dd, J_1 = 2.2 Hz, J_2 = 8.3 Hz, 1H), 7.24 (dd, J_1 = 2.5 Hz, J_2 = 6.7 Hz, 1H), 7.15 (d, J = 2.0 Hz, 1H), 7.06 (m, 1H), 6.87 (d, J = 8.3 Hz, 1H), 3.41–3.34 (m, 3H), 3.08 (d, J = 10.2 Hz, 1H), 2.46 (s, 3H). HRMS (ESI+) m/z calcd. for ($C_{20}H_{16}Cl_2FN_4O_3$, M + H): 449.0578, found: (M + H): 449.0589.

3.3. Biological Evaluation

MTT test. The MTT assay was carried out according to [35] with few modifications; 3000 Cells (for HEK293T, A549 and MCF7cell lines) or 4000 cells (for VA13 cell line) were seeded in each well of a 96-well plate. After 20 h incubation, the tested compounds diluted in culture medium were added to the cells and incubated 72 h at 37 °C under CO_2 (5%) atmosphere. Assays were performed in triplicates. The MTT (3-[4,5-dimethylthiazol-2-yl]-2,5 diphenyl-tetrazolium bromide) reagent was then added to the cells up to final concentration of 0.5 g/L (10X stock solution in PBS was used) and incubated for 2 h at 37 °C (5% CO_2). The MTT solution was then discarded and 140 µL of DMSO was added. The plates were swayed on a shaker (60 rpm) to solubilize the formazan. The absorbance was measured using a microplate reader at a wavelength of 565 nm. The analysis of cytotoxicity and the estimation of IC50 values were carried out with the built-in functions in the GraphPad Prism program (GraphPad Software, Inc., San Diego, CA) P53 activation.

The β-galactosidase reporter construction equipped with the p53 promotor frame [46] was used to assess the p53 expression level in p53wt A549 cell line. The compounds were tested in the concentration range of 0.5–120 µM with triple dilution steps. The incubation time was 24 h. To take into account the toxic effect of the molecules, the output signal was normalized considering the number of the cells estimated by MTT test with the same incubation tome (24 h). The output was statistically significant if the background signal was exceeded two or more times.

4. Conclusions

In the present study, a series of novel dispiro-oxindole derivatives of 2-chaicogen-imidazol-4-ones has been described. The synthesized molecules have key 3D-pharmacophore features essential for binding into the major MDM2 pocket as it has been predicted during the molecular docking study. However, these compounds have an alternative binding mode in contrast to other MDM2 inhibitors, therefore, they should be cautiously regarded as having this mechanism of action. MTT test with different cell lines, including p53 positive and negative, has not provided unambiguous results on the mechanism of action for this series although some signs of p53 activation have been observed. Nevertheless, the most active compounds from this series show fairly good cytotoxicity values (2.2–9.8 µM) on various cell lines in the MTT test, which makes them promising for further optimization and research.

Supplementary Materials: The following are available online. Figure S1: NMR spectra of synthesized compounds.

Author Contributions: Conceptualization, M.K. and Y.I.; Data curation, D.S.; Funding acquisition, D.S., Y.I. and E.B.; Investigation, V.N., V.F. and R.S.; Methodology, M.K. and V.N.; Project administration, N.Z.; Resources, D.S., Y.I. and E.B.; Supervision, E.B.; Visualization, A.M. (Anna Moiseeva) and M.V.; Writing—original draft, Y.I. and M.V.; Writing—review and editing, A.M. (Alexander Majouga) and E.B. All authors have read and agreed to the published version of the manuscript.

Funding: This research was funded by Russian Science Foundation, grant number 21-13-00023 (synthesis), Russian Foundation for Basic Research, grant numbers 19-03-00201 and 18-29-08060, and Applied Genetic Resource Faculty of MITP (Support Grant 075-15-2021-684) (biological evaluation).

Institutional Review Board Statement: Not applicable.

Informed Consent Statement: Not applicable.

Data Availability Statement: The data presented in this study are openly available in file mail.ru repository at https://cloud.mail.ru/public/svf7/4u5R7wxqq, accessed on 20 November 2021.

Acknowledgments: The-NMR and X-ray studies of this work were supported by the M.V. Lomonosov Moscow State University Program of Development.

Conflicts of Interest: The authors declare no conflict of interest.

Sample Availability: Samples of the compounds are not available from the authors.

References

1. Tsukano, C.; Takemoto, Y. Synthetic Approaches to Spiro-oxindoles and Iminoindolines Based on Formation of C2AC3 Bond. *Heterocycles* **2014**, *89*, 2271–2302. [CrossRef]
2. Shu, L.; Li, Z.; Gu, C.; Fishlock, D. Synthesis of a SpiroindolinonePyrrolidinecarboxamide MDM2 Antagonist. *Org. Process. Res. Dev.* **2013**, *17*, 247–256. [CrossRef]
3. Gollner, A.; Rudolph, D.; Arnhof, H.; Bauer, M.; Blake, S.M.; Boehmelt, G.; Cockroft, X.-L.; Dahmann, G.; Ettmayer, P.; Gerstberger, T.; et al. Discovery of Novel Spiro[3 H-Indole-3,2-pyrrolidin]-2(1H)-one Compounds as Chemically Stable and Orally Active Inhibitors of the MDM2–p53 Interaction. *J. Med. Chem.* **2016**, *59*, 10147–10162. [CrossRef]
4. Zhang, Z.; Ding, Q.; Liu, J.-J.; Zhang, J.; Jiang, N.; Chu, X.-J.; Bartkovitz, D.; Luk, K.-C.; Janson, C.; Tovar, C.; et al. Discovery of Potent and Selective Spiroindolinone MDM2 Inhibitor, RO8994, for Cancer Therapy. *Bioorg. Med. Chem.* **2014**, *22*, 4001–4009. [CrossRef] [PubMed]

5. Ding, K.; Lu, Y.; Nikolovska-Coleska, Z.; Wang, G.; Qiu, S.; Shangary, S.; Gao, W.; Qin, D.; Stuckey, J.; Krajewski, K.; et al. Structure-Based Design of Spiro-Oxindoles as Potent, SpecificSmall-Molecule Inhibitors of the MDM2p53 Interaction. *J. Med. Chem.* **2006**, *49*, 3432–3435. [CrossRef]
6. Galliford, C.V.; Scheidt, K.A. Pyrrolidinyl-Spirooxindole Natural Products as Inspirations for the Development of Potential Therapeutic Agents. *Angew. Chem. Int. Ed.* **2007**, *46*, 8748–8758. [CrossRef] [PubMed]
7. Poulos, Z. *Spirooxindole Alkaloids*; Lambert Academic Publishing: Weinheim, Germany, 2014; pp. 1–280.
8. Pavlovska, T.L.; Redkin, R.; Lipson, V.V.; Atamanuk, D.V. Molecular Diversity of Spirooxindoles. Synthesis and Biological Activity. *Mol. Divers.* **2016**, *20*, 299–344. [CrossRef] [PubMed]
9. Yu, B.; Yu, D.-Q.; Liu, H.-M. Spirooxindoles: Promising Scaffolds for Anticancer Agents. *Eur. J. Med. Chem.* **2015**, *97*, 673–698. [CrossRef]
10. Santos, M.M.M. Recent Advances in the Synthesis of Biologically Active Spirooxindoles. *Tetrahedronn* **2014**, *70*, 9735–9757. [CrossRef]
11. Wu, X.; Bayle, J.H.; Olson, D.; Levine, A.J. Thep53-mdm-2 autoregulatory feedback loop. *Genes Dev.* **1993**, *7*, 1126–1132. [CrossRef]
12. Vogelstein, B.; Lane, D.P.; Levine, A.J. Surfing the p53 network. *Nature* **2000**, *408*, 307–310. [CrossRef]
13. Levine, A.J. p53, the Cellular Gatekeeper for Growth and Division. *Cell* **1997**, *88*, 323–331. [CrossRef]
14. Gomez-Monterrey, I.; Bertamino, A.; Portaб, A.; Carotenuto, A.; Musella, S.; Aquino, C.; Granata, I.; Sala, M.; Brancaccio, D.; Picone, D.; et al. Identification of the Spiro(Oxindole-3,30-thiazo-lidine)-Based Derivatives as Potential p53 Activity Modulators. *J. Med. Chem.* **2010**, *53*, 8319–8329. [CrossRef]
15. Nakamaru, K.; Seki, T.; Tazaki, K.; Tse, A. Preclinical Characterization of a Novel Orally-Available MDM2 Inhibitor DS-3032b: Anti-Tumor Profile and Predictive Biomarkers for Sensitivity. *Mol. Cancer Ther.* **2015**, *14*, B5. [CrossRef]
16. Wang, S.; Sun, W.; Zhao, Y.; McEachern, D.; Meaux, I.; Barrierem, C.; Stuckey, J.A.; Meagher, J.L.; Bai, L.; Liu, L.; et al. SAR405838: An Optimized Inhibitor of MDM2-p53 Interaction That Induces Complete and Durable Tumor Regression. *Cancer Res.* **2014**, *74*, 5855–5865. [CrossRef] [PubMed]
17. Huang, W.; Cai, L.; Chen, C.; Xie, X.; Zhao, Q.; Zhao, X.; Zhou, H.; Han, B.; Peng, C. Computational Analysis of Spiro-Oxindole Inhibitors of the MDM2-p53 Interaction: Insights and Selection of Novel Inhibitors. *J. Biomol. Struct. Dyn.* **2016**, *34*, 341–351. [CrossRef]
18. Ding, Q.; Zhang, Z.; Liu, J.-J.; Jiang, N.; Zhang, J.; Ross, T.M.; Chu, X.-J.; Bartkovitz, D.; Podlaski, F.; Janson, C.; et al. Discovery of RG7388, a Potent and Selective p53–MDM2 Inhibitor in Clinical Development. *J. Med. Chem.* **2013**, *56*, 5979–5983. [CrossRef]
19. Shangary, S.; Qin, D.; McEachern, D.; Liu, M.; Miller, R.S.; Qiu, S.; Nikolovska-Coleska, Z.; Ding, K.; Wang, G.; Chen, J.; et al. Temporal activation of p53 by a specific MDM2 inhibitor is selectively toxic to tumors and leads to complete tumor growth inhibition. *Proc. Natl. Acad. Sci. USA* **2008**, *105*, 3933–3938. [CrossRef] [PubMed]
20. Nguyen, D.; Liao, W.; Zeng, S.X.; Lu, H. Reviving the guardian of the genome: Small molecule activators of p53. *Pharmacol. Ther.* **2017**, *178*, 92–108. [CrossRef] [PubMed]
21. Anil, B.; Riedinger, C.; Endicott, J.A.; Noble, M.E. The structure of an MDM2-Nutlin-3a complex solved by the use of a validated MDM2 surface-entropy reduction mutant. *Acta Crystallogr. D Biol. Crystallogr.* **2013**, *69*, 1358–1366. [CrossRef]
22. Ding, K.; Lu, Y.; Nikolovska-Coleska, Z.; Qiu, S.; Ding, Y.; Gao, W.; Stuckey, J.; Krajewski, K.; Roller, P.P.; Tomita, Y.; et al. Structure-based design of potent non-peptide MDM2 inhibitors. *J. Am. Chem. Soc.* **2005**, *127*, 10130–10131. [CrossRef]
23. Ivanenkov, Y.A.; Vasilevski, S.V.; Beloglazkina, E.K.; Kukushkin, M.E.; Machulkin, A.E.; Veselov, M.S.; Chufarova, N.V.; Vanzcool, A.; Zyk, N.V.; Skvortsov, D.A.; et al. Design, Synthesis and Biological Evaluation of Novel Potent MDM2/p53 Small-Molecule Inhibitors. *Bioorg. Med. Chem. Lett.* **2015**, *25*, 404–409. [CrossRef]
24. Kukushkin, M.E.; Skvortsov, D.A.; Kalinina, M.A.; Tafeenko, V.A.; Burmistrov, V.V.; Butov, G.M.; Zyk, N.V.; Majouga, A.G.; Beloglazkina, E.K. Synthesis and cytotoxicity of oxindoles dispiro derivatives with thiohydantoin and adamantane fragments. *Phosphorus Sulfur Silicon Relat. Elem.* **2020**, *195*, 544–555. [CrossRef]
25. Beloglazkina, A.A.; Karpov, N.A.; Mefedova, S.R.; Polyakov, V.S.; Skvortsov, D.A.; Kalinina, M.A.; Tafeenko, V.A.; Majouga, A.G.; Zyk, N.V.; Beloglazkina, E.K. Synthesis of dispirooxindoles containing n-unsubstituted heterocyclic moieties and study of their anticancer activity. *Russ. Chem. Bull.* **2019**, *68*, 1006–1013. [CrossRef]
26. Novotortsev, V.K.; Kukushkin, M.E.; Tafeenko, V.A.; Skvortsov, D.A.; Kalinina, M.A.; Timoshenko, R.V.; Chmelyuk, N.S.; Vasilyeva, L.A.; Tarasevich, B.N.; Gorelkin, P.V.; et al. Dispirooxindoles based on 2-selenoxo-imidazolidin-4-ones: Synthesis, cytotoxicity and ROS generation ability. *IJMS* **2021**, *22*, 2613. [CrossRef] [PubMed]
27. Beloglazkina, A.; Barashkin, A.; Polyakov, V.; Kotovsky, G.; Karpov, N.; Mefedova, S.; Zagribelny, B.; Ivanenkov, Y.; Kalinina, M.; Skvortsov, D.; et al. Synthesis and biological evaluation of novel dispiro compounds based on 5-arylidenehydantoins and isatins as inhibitors of p53–mdm2 protein–protein interaction. *Chem. Heterocycl. Compd.* **2020**, *56*, 5613. [CrossRef]
28. Majouga, A.G.; Beloglazkina, E.K.; Beloglazkina, A.A.; Kukushkin, M.E.; Ivanenkov, Y.A.; Veselov, M.S. New Dispiro-Indolinones, Inhibitors of MDM2/p53 Interaction, Method of Synthesis and Application. Patent #RU2629750C2, 9 April 2015.
29. Shangary, S.; Ding, K.; Qiu, S.; Nikolovska-Coleska, Z.; Bauer, J.A.; Liu, M.; Wang, G.; Lu, Y.; McEachern, N.; Bernard, D.; et al. Reactivation of p53 by a specific MDM2 antagonist (MI-43) leads to p21-mediated cell cycle arrest and selective cell death in colon cancer. *Mol. Cancer Ther.* **2008**, *7*, 1533–1542. [CrossRef]

30. Kuznetsova, O.Y.; Antipin, R.L.; Udina, A.U.; Krasnovskaya, O.O.; Beloglazkina, E.K.; Terenin, V.I.; Oteliansky, V.E.; Zyk, N.V.; Majouga, A.G. An improved protocol for synthesis of 3-substituted 5-arylidene-2-thiohydantoins: Two-step procedure alternative to classical methods. *J. Heterocyclic Chem.* **2016**, *53*, 1570–1577. [CrossRef]
31. Wirth, T. Small Organoselenium Compounds: More than just Glutathione Peroxidase Mimics. *Angew. Chem. Int. Ed.* **2015**, *54*, 10074–10076. [CrossRef]
32. Müller, A.; Cadenas, E.; Graf, P.; Sies, H. A novel biologically active seleno-organic compound-1. Glutathione peroxidase-like activity in vitro and antioxidant capacity of PZ 51 (Ebselen). *Biochem. Pharmacol.* **1984**, *33*, 3235–3239. [CrossRef]
33. Schewe, T. Molecular actions of ebselen-an antiinflammatory antioxidant. *Gen. Pharmacol.* **1995**, *26*, 1153–1169. [CrossRef]
34. Novotortsev, V.K.; Kukushkin, M.E.; Tafeenko, V.A.; Zyk, N.V.; Beloglazkina, E.K. New spiro-linked indolinonepyrrolidineselenoxoimidazolones. *Mendeleev Commun.* **2020**, *30*, 320–321. [CrossRef]
35. Ferrari, M.; Fornasiero, M.C.; Isetta, A.M. MTT colorimetric assay for testing macrophage cytotoxic activity in vitro. *J. Immunol. Methods* **1990**, *131*, 165–172. [CrossRef]
36. Hietanen, S.; Lain, S.; Krausz, E.; Blattner, C.; Lane, D.P. Activation of p53 in cervical carcinoma cells by small molecules. *Proc. Natl. Acad. Sci. USA* **2000**, *97*, 8501–8506. [CrossRef]
37. Haupt, Y.; Maya, R.; Kazaz, A.; Oren, M. Mdm2 promotes the rapid degradation of p53. *Nature* **1997**, *387*, 296–299. [CrossRef] [PubMed]
38. Arriola, E.L.; Lopez, A.R.; Chresta, C.M. Differential regulation of p21waf-1/cip-1 and Mdm2 by etoposide: Etoposide inhibits the p53-Mdm2 autoregulatory feedback loop. *Oncogene* **1999**, *18*, 1081–1091. [CrossRef]
39. Vassilev, L.T.; Vu, B.T.; Graves, B.; Carvajal, D.; Podlaski, F.; Filipovic, Z.; Kong, N.; Kammlott, U.; Lukacs, C.; Klein, C.; et al. In vivo activation of the p53 pathway by small-molecule antagonists of MDM2. *Science* **2004**, *303*, 844–848. [CrossRef]
40. DiNardo, C.D.; Rosenthal, J.; Andreeff, M.; Zernovak, O.; Kumar, P.; Gajee, R.; Chen, S.; Rosen, M.; Song, S.; Kochan, J.; et al. Phase 1 Dose Escalation Study of MDM2 Inhibitor DS-3032b in Patients with Hematological Malignancies-Preliminary Results. *Blood* **2016**, *128*, 593. [CrossRef]
41. Aguilar, A.; Lu, J.; Liu, L.; Du, D.; Bernard, D.; McEachern, D.; Przybranowski, S.; Li, X.; Luo, R.; Wen, B.; et al. Discovery of 4-((3′R,4′S,5′R)-6″-Chloro-4′-(3-chloro-2-fluorophenyl)-1′-ethyl-2″-oxodispiro[cyclohexane-1,2′-pyrrolidine-3′,3″-indoline]-5′-carboxamido)bicyclo[2.2.2]octane-1-carboxylic Acid (AA-115/APG-115): A Potent and Orally Active Murine Double Minute 2 (MDM2) Inhibitor in Clinical Development. *J. Med. Chem.* **2017**, *60*, 2819–2839. [CrossRef] [PubMed]
42. Bertamino, A.; Soprano, M.; Musella, S.; Rusciano, M.R.; Sala, M.; Vernieri, E.; Di Sarno, V.; Limatola, A.; Carotenuto, A.; Cosconati, S.; et al. Synthesis, in Vitro, and in Cell Studies of a New Series of [Indoline-3,2′-thiazolidine]-Based p53 Modulators. *J. Med. Chem.* **2013**, *56*, 5407–5421. [CrossRef] [PubMed]
43. Capoulade, C.; Bressac-de Pailleret, B.; Lefrere, I.; Ronsi, M.; Feunteun, J.; Tursz, T.; Wiels, J. Overexpression of MDM2, due to enhanced translation, results in inactivation of wild-type p53 in Burkitt's lymphoma cells. *Oncogene* **1998**, *16*, 1603–1610. [CrossRef]
44. Kussie, P.H.; Gorina, S.; Marechal, V.; Elenbaas, B.; Moreau, J.; Levine, A.J.; Pavletich, N.P. Structure of the MDM2 Oncoprotein Bound to the p53 Tumor Suppressor Transactivation Domain. *Science* **1996**, *274*, 948–953. [CrossRef] [PubMed]
45. Molsoft L.L.C. Available online: https://www.molsoft.com (accessed on 31 January 2018).
46. Kravchenko, J.E.; Ilyinskaya, G.V.; Komarov, P.G.; Agapova, L.S.; Kochetkov, D.V.; Strom, E.; Frolova, E.I.; Kovriga, I.; Gudkov, A.V.; Feinstein, E.; et al. Small-molecule RETRA suppresses mutant p53-bearing cancer cells through a p73-dependent salvage pathway. *Proc. Natl. Acad. Sci. USA* **2008**, *105*, 6302–6307. [CrossRef] [PubMed]
47. Graves, B.; Thompson, T.; Xia, M.; Janson, C.; Lukacs, C.; Deo, D.; Di Lello, P.; Fry, D.; Garvie, C.; Huang, K.S.; et al. Activation of the p53 pathway by small-molecule-induced MDM2 and MDMX dimerization. *Proc. Natl. Acad. Sci. USA* **2012**, *109*, 11788–11793. [CrossRef] [PubMed]
48. Majouga, A.G.; Beloglazkina, E.K.; Vatsadze, S.Z.; Moiseeva, A.A.; Moiseev, F.S.; Butin, K.P.; Zyk, N.V. The first example of a reversibly reducible Co-II complex with an anionic 2-thiohydantoin-type ligand. *Mend. Commun.* **2005**, *15*, 48–50. [CrossRef]
49. Beloglazkina, E.K.; Majouga, A.G.; Moiseeva, A.A.; Zyk, N.V.; Zefirov, N.S. Oxidation of triphenylphosphine and norbornene by nitrous oxide in the presence of $Co^{II}LCl_2$ [L = 3-phenyl-5-(2-pyridylmethylidene)-2-thiohydantoin]: The first example of CoII-catalyzed alkene oxidation by N2O. *Mend. Commun.* **2009**, *19*, 69–71. [CrossRef]
50. Sheldrick, G.M. SHELXT—Integrated space-group and crystal-structure determination. *Acta Cryst. C* **2015**, *71*, 3–8. [CrossRef]
51. Dolomanov, O.V.; Bourhis, L.J.; Gildea, R.J.; Howard, J.A.K.; Puschmann, H. OLEX2: A complete structure solution, refinement and analysis program. *J. Appl. Cryst.* **2009**, *42*, 339–341. [CrossRef]

Article

A Galactosidase-Activatable Fluorescent Probe for Detection of Bacteria Based on BODIPY

Xi Chen [1,2], Yu-Cong Liu [2], Jing-Jing Cui [2], Fang-Ying Wu [1,*] and Qiang Xiao [2,*]

1. College of Chemistry, Nanchang University, Nanchang 330031, China; chenxi2016@email.ncu.edu.cn
2. Key Laboratory of Organic Chemistry in Jiangxi Province, Institute of Organic Chemistry, Jiangxi Science & Technology Normal University, Nanchang 330013, China; liuyucong0522@163.com (Y.-C.L.); jingvsling@126.com (J.-J.C.)
* Correspondence: fywu@ncu.edu.cn (F.-Y.W.); xiaoqiang@tsinghua.org.cn (Q.X.); Tel./Fax: +86-791-83969882 (F.-Y.W.); +86-791-86422903 (Q.X.)

Abstract: Pathogenic *E. coli* infection is one of the most widespread foodborne diseases, so the development of sensitive, reliable and easy operating detection tests is a key issue for food safety. Identifying bacteria with a fluorescent medium is more sensitive and faster than using chromogenic media. This study designed and synthesized a β-galactosidase-activatable fluorescent probe BOD-Gal for the sensitive detection of *E. coli*. It employed a biocompatible and photostable 4,4-difluoro-3a,4a-diaza-s-indancene (BODIPY) as the fluorophore to form a β-*O*-glycosidic bond with galactose, allowing the BOD-Gal to show significant on-off fluorescent signals for in vitro and in vivo bacterial detection. This work shows the potential for the use of a BODIPY based enzyme substrate for pathogen detection.

Keywords: BODIPY; β-galactosidase activity; PET; fluorescent

1. Introduction

Foodborne disease is a widespread and ever-increasing public health problem, especially in developing countries and *Escherichia coli (E. coli)*, *Listeria monocytogenes* and *Salmonella* are the most widespread pathogenic microorganisms that contaminate raw foods and drinking water [1]. Individuals infected by pathogenic bacteria show diarrhea, fever and nausea, with neurological disorders, multiple organ failure and death in severe cases [2], so monitoring and control of food pathogens are vitally important.

Although most *E. coli* strains are harmless, some types of pathogenic *E. coli*, such as Shiga toxin-producing *E. coli*, *E. coli* O157 and O104 can be life threatening [3]. As the key to control and prevent foodborne disease, various pathogen detection methods were developed [4] of which one of the most important uses is fluorogenic culture media to identify the pathogens. Fluorogenic media based on specific enzyme substrates have clear advantages over other detection methods [5], as it eliminates the need for subculture and further biochemical testing to establish the identity of certain microorganisms [6], simplifying the identification procedure and shortening the detection time. It is also more sensitive than chromogenic media since the change of light emission is more visually perceptible [7]. The principle of fluorogenic media detection is to design fluorogenic substrates, which can be specifically metabolized by pathogens of interest, resulting in the generation of a new fluorogenic entity through a reversible/irreversible process, causing associated changes in a fluorescence spectrum [8]. As an example, β-galactoside and β-glucuronide are the two target enzyme substrates which can be metabolized by *E. coli* [9]. The classical molecule 4-methylumbelliferyl-β-D-glucuronide (MUG) and its derivatives were designed to detect *E. coli* [8,10–12], but it is pH sensitive and its emitting wavelength region is near the autofluorescence of microorganisms [13]. A fluorescein substrate based on the hydrolysis of esterase has also been developed to detect *E. coli.* in drinking water [14], but esterase was metabolized by most of the organisms, thus, its selectivity was a potential problem. There is therefore a need to develop a highly sensitive and specific probe that is non-susceptible to pH change and can be used in fluorogenic media for the pathogens detection.

In this study, a β-D-galactosidase (β-Gal) activatable fluorogenic probe (BOD-Gal) was developed based on a 4,4-difluoro-3a,4a-diaza-s-indancene (BODIPY) fluorophore (Scheme 1). A BODIPY fluorophore possessing low toxicity and high biocompatibility can be modified in multiple sites and a slightly structural modification can tune its emitting light [15,16]. It has the merits of high extinction coefficient, high quantum yield, excellent photophysical stability and pH resistance [17], so BODIPY fluorophore and β-D-galactose can be linked by a β-O-glycosidic bond, which could be hydrolyzed by E. coli producing β-Gal and generate BODIPY fluorophore 1 in situ. This phenoxy residue could trigger a strong photoinduced electron transfer (PET) processing in BODIPY dyes [18] and BOD-Gal would show a highly sensitive response to β-Gal by an on-off fluorescent response in the PBS buffer. In addition, the PET mechanism was rationalized by the density functional theory (DFT) calculations. When applied to living E. coli samples, it can also successfully indicate the presence of pathogens on a media plate.

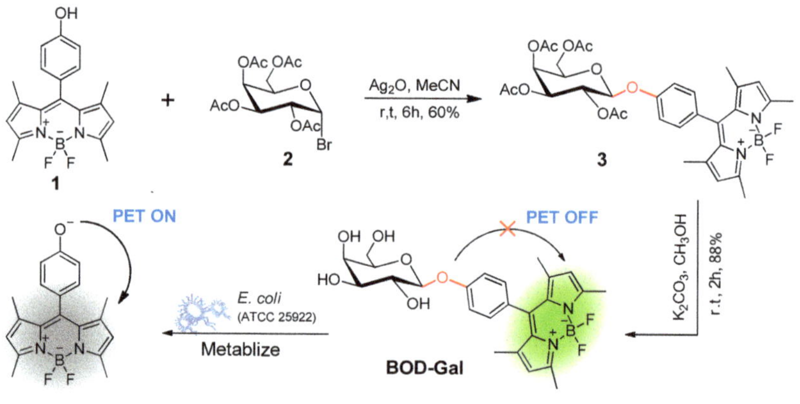

Scheme 1. Synthesis Bod-Gal and the proposed sensing mechanism for E. coli (ATCC 25922).

2. Results

2.1. In Vitro Spectrum Study

The sensing ability of BOD-Gal was first evaluated in vitro using β-Gal. The fluorescence and absorption responses of BOD-Gal (20 μM) in the absence and presence of β-Gal was recorded in aqueous solutions of phosphate-buffered saline (PBS) with dimethyl sulfoxide (DMSO) at a ratio of 49:1 (*v:v*) at a pH of 7.4, shown in Figure 1. The BOD-Gal displayed a strong green emission with a maximum at 516 nm, upon excitation at 470 nm. After BOD-Gal was incubated at 37 °C with β-Gal for 35 min, the fluorescence was reduced significantly as seen in Figure 1a. The UV spectra of BOD-Gal were investigated in the absence or presence of β-Gal. As shown in Figure 1b, the maximum absorption peaks were both at 498 nm, demonstrating that BOD-Gal enabled the assay of β-Gal based on a turn-off fluorescence mode.

Considering that a BOD-Gal probe displayed distinct fluorescence response changes to β-Gal in aqueous solutions, the influence of incubation time (Figure 1c) and the enzyme concentration (Figure 1d) on BOD-Gal were studied. After 8 U β-Gal was added to 10 μM BOD-Gal in PBS solution, the fluorescence intensity decreased with time and became stable after 35 min, indicating that the hydrolysis reaction was complete, so in the subsequent assay, the detection limit was set to 35 min. Fluorescence changes of BOD-Gal to different concentrations of β-Gal from 0 to 12 U were also investigated and the emission intensity decreased sharply with the increase of the concentration of β-Gal. In addition, fluorescence intensity is linearly correlated to the enzyme concentration in the range of 0 U–10 U (the insert of Figure 1d, R^2 = 0.9904). The limit of detection (LOD = 3σ/slope) for BOD-Gal toward β-Gal was calculated to be 0.038 U/mL.

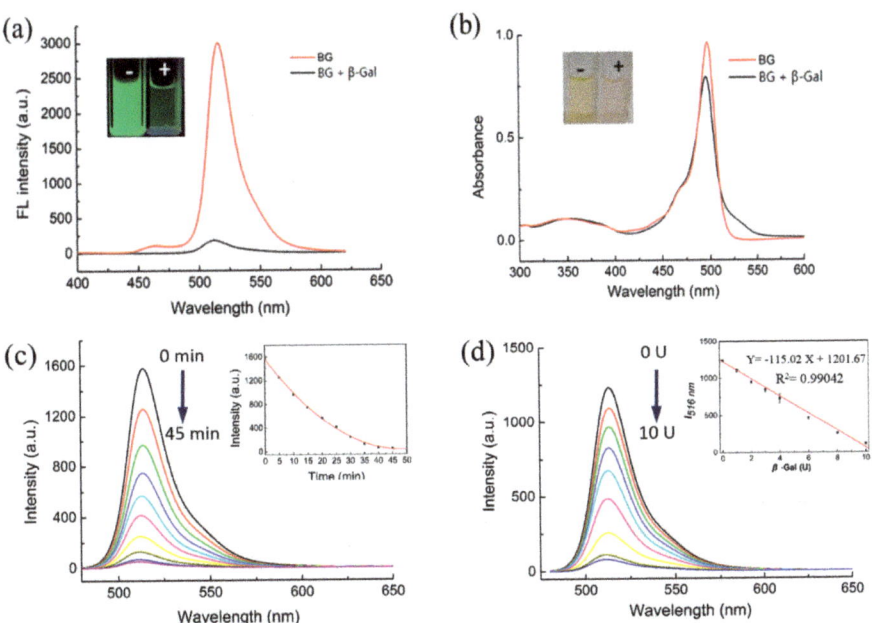

Figure 1. Fluorescence and absorption changes of BOD-Gal to β-Gal (8 U) in DMSO/PBS solution (PBS / DMSO = 49:1 v:v, pH = 7.4). "-" indicated the absence of β-Gal, "+" indicated the presence of β-Gal. (**a**) Fluorescence changes, λ_{ex} = 470 nm. (**b**) Absorption changes. (**c**) Time dependence of fluorescence spectra (0–45 min, λ_{ex} = 470 nm). Inset: Curve of fluorescence intensity versus time. (**d**) Fluorescence changes of BOD-Gal to different concentration of β-Gal (0 U–12 U), λ_{ex} = 470 nm. Inset: The relationship between I_{516} nm and the β-Gal concentration.

The potential interference of biological analytes toward BOD-Gal was investigated next. As shown in Figure 2, various enzyme species, amino acids and biomolecules, such as β-Gal, cellulase, lysozyme, trypsin, Cys, Hcy, GSH, DTT, NADPH, Vc, NaHS, $Na_2S_2O_3$, H_2O_2 and NaClO, were reacted with BOD-Gal. Only β-Gal produced an obvious reduction in fluorescent intensity compared with only subtle changes for up to 100 equiv. of the other competitive analytes, showing that BOD-Gal demonstrated high selectivity for the detection of β-Gal.

2.2. Kinetics Studies of Enzymatic Reaction

To evaluate the affinity of BOD-Gal toward β-Gal, its Km value was calculated by the Hanes–Woolf method. The enzymatic hydrolysis rate was measured by the formation of hydrolysate **1** with high-performance liquid chromatography (HPLC), and the standard curve of compound **1** is shown in Figure S1a. A series concentration of BOD-Gal (5, 10, 15, 20, 25 and 30 µM) was hydrolysed by β-Gal in a PBS buffer for 10 min. After inactivation, the samples were submitted for analysis by HPLC, where the reactant BOD-Gal was detected at 2.21 min. A new peak appearing at 10.34 min (Figure S2) was verified as compound **1** through standard sampling (Figure S2). The HPLC results further proved that the sensing mechanism was the hydrolysis of BOD-Gal upon enzyme triggered glycosylic bond cleavage. From the Hanes–Woolf plot (Figure S1b), the Km value was calculated to be 9.5×10^{-6} mol/L, which was significantly lower than that of 5-Bromo-4-chloro-3-indolyl-β-D-galactoside (X-Gal) at 2.6×10^{-4} mol/L [19], showing that BOD-Gal had a much higher affinity for β-Gal than commercially available X-Gal.

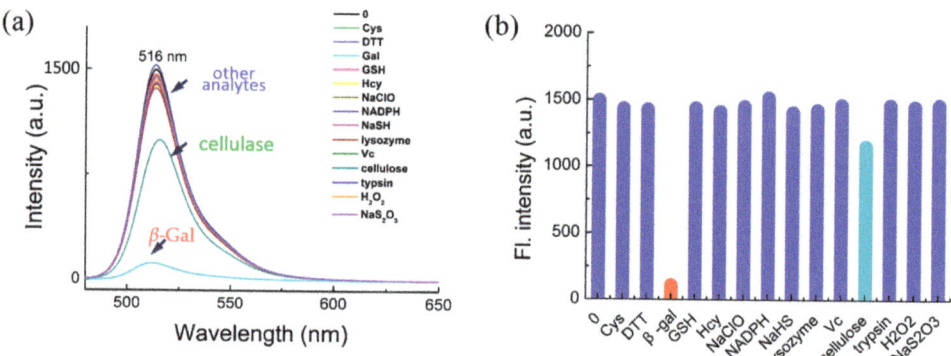

Figure 2. The selectivity of BOD-Gal for β-Gal. (**a**) Fluorescence spectra of BOD-Gal to various analytes in an aqueous system (PBS/DMSO = 49:1 v:v; pH = 7.4, 37 °C) with β-Gal (8U), Cys (1 mM), DTT (1 mM), GSH (1 mM), Hcy (1 mM), NaClO (1 mM), NADPH (1 mM), NaSH (1 mM), lysozyme (1 kU), Vc (1 mM), cellulose (1 kU), trypsin (1 kU), H_2O_2 (1 mM) and $Na_2S_2O_3$ (1 mM), λex = 470 nm. (**b**) Corresponding fluorescence intensity.

2.3. Theoretical Calculations

The DFT theoretical calculations were carried out at B3LYP/6-31 G*(d, p) level using Guassian 16 (considered as gas phase). The molecular structures were divided into "Acceptor" and "Donor" to discuss which was widely used as a computational model for the PET mechanism [20,21]. The results indicated that the highest occupied molecular orbital (HOMO, −5.38 eV) and lowest unoccupied molecular orbital (LUMO, −2.42 eV) energy levels of the acceptor (BODIPY unit) were ranged in between the HOMO (−2.04 eV) and LUMO (−6.40 eV) energy levels of donor **1** (Figure 3a), which implied that intramolecular charge transfer was forbidden, that is, PET progress was off. However, after the glyosidic bond of BOD-Gal was hydrolyzed by β-Gal to transform into a phenol anion subunit, the acceptor PET (a-PET) progress could be activated, where the HOMO energy level of donor **2** rose to −3.73 eV between the HOMO and LUMO energy of the acceptor part (Figure 3b) [22]. The optimized geometry and atom list were shown in Figures S3 and S4, Tables S1 and S2. Thus, the theoretical calculations gave a reasonable explanation of fluorescence on-off based on the PET mechanism.

2.4. Biological Activity

Based on the above results, BOD-Gal was applied to a standard formulation of growth media, where petri dishes were poured and inoculated with *E. coli*.

2.4.1. Biological Toxicity to Bacterial Propagation

To assess its biological toxicity and biocompatibility, the influence of various concentrations of BOD-Gal (0, 50, 100, 150 and 200 μM) on the growth of bacteria was investigated. The colony numbers are shown in Figure 4 and the corresponding data are shown in Table S3. When cultured at 50 μM, the growth of *E. coli* was not affected, but at higher concentrations (100, 150 and 200 μM), the growth rates were reduced to about 70%. This data indicated that the concentration of BOD-Gal had better not exceed 50 μM in bacterial culturing.

2.4.2. Fluorescence on Agar

The sensing effects of the enzyme (β-Gal) and pathogenic *E. coli* were compared on LB agar containing BOD-Gal. X-Gal was selected as the control, which is a widely used substrate for blue/white selection of β-Gal in laboratory and bioengineering. The results are shown in Figure 5.

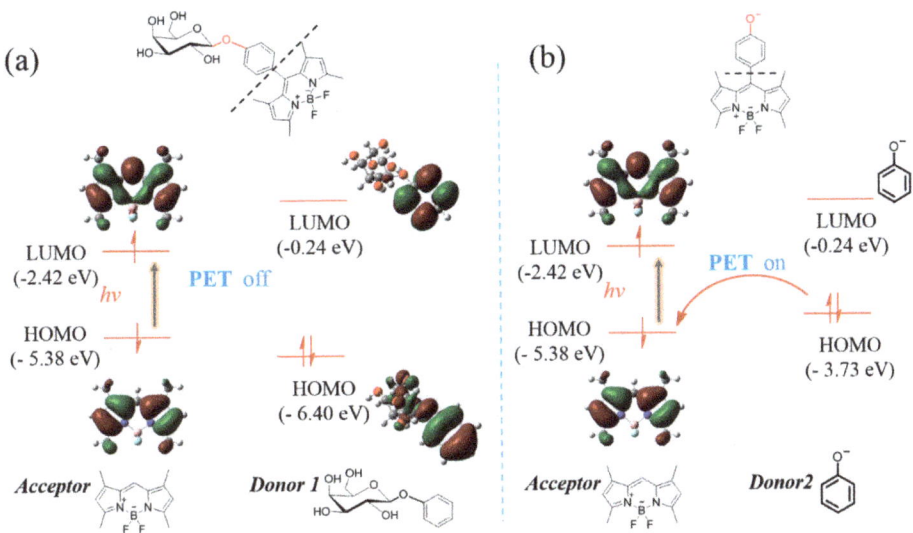

Figure 3. The calculated energy levels of acceptor and donor (**a**) before and (**b**) after hydrolysis reaction according to the theory of the PET mechanism-based on DFT at the B3LYP/6-31 G*(d, p) level.

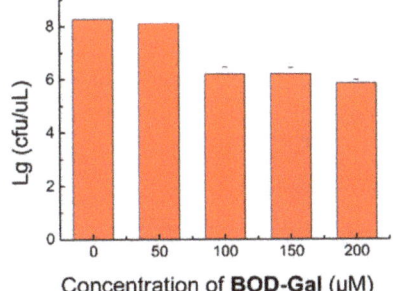

Figure 4. Standard plate count method (SPC) assay of *E. coli* was treated in the presence of BOD-Gal (0–200 μM) incubated for 8 h.

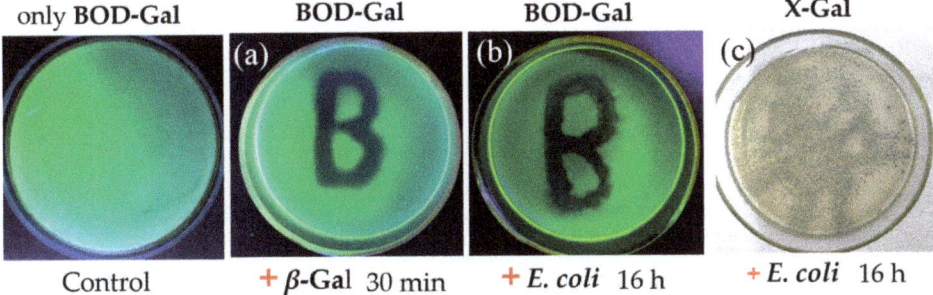

Figure 5. Comparison of different substrates on plate. (**a**) Painting β-Gal and (**b**) Inoculating *E. coli* (ATCC 25922) on the plate containing BOD-Gal. (**c**) Inoculating *E. coli* on the plate containing X-Gal (267 μg/mL). UV irradiation was achieved using a hand-held UV lamp, X-Gal staining appearance was in the ambient light condition.

The β-Gal was initially painted on the LB agar containing BOD-Gal. After culture at 37 °C for 30 min, the green fluorescence of the β-Gal-exposed region had significantly faded under UV illumination at 365 nm (Figure 5a), which indicated that BOD-Gal could be rapidly hydrolyzed by β-Gal on the agar medium. Next, E. coli (ATCC 25922) was inoculated on the agar that contained the BOD-Gal and X-Gal, respectively, and incubated at 37 °C for 16 h. From Figure 5b,c, it showed that the plate with BOD-Gal was more clearly distinguished under visual inspection, showing that BOD-Gal exhibited more sensitive detection than traditionally used X-Gal.

3. Materials and Methods

3.1. General Information

Unless otherwise stated, all the reagents were obtained from commercial sources and used as received without further purification. The stock solution of BOD-Gal was prepared in DMSO at the concentration of 1 mM. β-Gal and the other analytes were dissolved in deionized water and diluted to required concentrations. The NMR spectra were acquired on an AM-400 spectrometer (Bruker Co., Ltd., Karlsruhe, Germany) at room temperature with $CDCl_3$ or DMSO-d_6 as the solvent and TMS used as an internal standard. High-resolution mass spectrometry data were performed with an AB SCIEX TOF 4600 (AB Sciex Pte. Ltd., Framingham, MA, USA). Fluorescence spectra measurements were recorded on an F-7100 fluorescence spectrophotometer (Hitachi, Ltd., Tokyo, Japan) and UV/Vis spectra measurements were recorded on a UV-2501 spectrometer (Shimadzu Co., Ltd., Yamanashi, Japan). High-performance liquid chromatography (HPLC) analysis was performed on an Agilent 1220 Infinity (Agilent Technologies Inc., Santa Clara, CA, USA).

3.2. Enzyme Assay In Vitro

The BOD-Gal was used at a final concentration of 20 μM unless noted. Absorption and fluorescence spectra of BOD-Gal with β-Gal sourced from E. coli (Sangon Biotech Co., Ltd., Shanghai, China) were performed at 37 °C in a 2 mL total volume of a PBS buffer of 0.2 M at pH of 7.4 with a 1 cm cuvette.

3.3. HPLC Analysis

A series of different concentrations (5, 10, 15, 20, 25, 30, 35, 40 μM) of BOD-Gal was hydrolyzed by 0.8 U β-Gal in a PBS buffer at 0.2 M and pH = 7.4 for 10 min at 37 °C, then inactivated in a 100 °C water bath for 1 min. The samples were prepared by adding equal volumes of DMSO into the reaction solution, to completely dissolve the components. The samples were analysed by HPLC at ambient temperature, using water and acetonitrile as the mobile phase at a ratio of 48:52 (v:v) and detected at 496 nm. The peak corresponding to BOD-Gal and hydrolysate **1** was integrated (Figure S1).

3.4. Biological Experiment

3.4.1. The Culture of Bacteria

The E. coli (ATCC 25922) were inoculated from frozen stock into the sterile LB broth, by adding 2.5 g LB broth powder into 100 mL deionized water, autoclaving at 120 °C for 20 min and left at room temperature (r.t.) in a shaken flask. Depending on the experiments, after 8 to 16 h of vigorous shaking at 150 rpm at 37 °C in the dark, 200 μL isopropyl-beta-D-thiogalactopyranoside (IPTG) at a concentration of 0.1 g/L in sterilized deionized water was added and remixed by shaking at 37 °C to induce E. coli expressing β-Gal.

3.4.2. Preparation of Culture Media

The plate was prepared by adding 2 g agar (Solarbio and Technology Co., Ltd., Beijing, China) and 2.5 g LB broth (Hope Bio-Technology Co., Ltd., Qingdao, China) into 100 mL deionized water, autoclaving at 120 °C for 20 min and leaving to cool at 50 °C. Depending on the experiment, various volumes of stock solution of BOD-Gal that had been filtered through 0.22 μM Millipore were added to the warm agar to obtain the different concentrations of BOD-Gal in agar and then poured into sterile petri plates and cooled at room temperature over night to coagulate.

3.4.3. Antibacterial Evaluation

The antibacterial activity of BOD-Gal was studied by the standard plate count method (SPC). The original suspension bacterial solution was initially diluted to 10^{-2} dilution and plated on the culture media at different concentrations of BOD-Gal (0, 50, 100, 150 and 200 µM), respectively, with each concentration performed three times in parallel. This process was repeated for the 10^{-4} and 10^{-6} dilutions. The plates were inverted, incubated at 37 °C for 8 h and the petri plates containing between 30 to 300 colonies counted via a colony counter (Icount 20, Shineso, Hangzhou, China).

3.5. Synthesis

The synthesis of BOD-Gal is shown in Scheme 1. The starting material, compound **1**, was synthesized by a routine procedure used in the construction of the BODIPY core [23]. The tetra-O-acetyl-galactose bromide **2** was obtained according to the previous method [24]. Compound **3** was synthesized with a 60% yield by Koenigs–Knorr glycosylation of compound **1** and **2**. Zemplén deprotection in all the acetates with K_2CO_3/CH_3OH gave BOD-Gal in 88% yield. All intermediates and BDBH were well characterized by ^1H NMR spectroscopy, ^{13}C NMR spectroscopy and high-resolution electrospray ionization mass spectrometry (HR-ESI-MS).

3.5.1. Synthesis of Compound **1**

4-Hydroxybenzaldehyde (0.49 g, 4 mmol) and 2,4-dimethylpyrrole (0.76 g, 8 mmol) were dissolved in anhydrous CH_2Cl_2 (600 mL). Two drops of trifluoroacetic acid (TFA) were added and the resulting mixture was stirred in the dark for 12 h under N_2 at room temperature. After TLC showed the complete consumption of aldehyde, 2, 3-dichloro-5, 6-dicyano-1, 4-benzoquinone (DDQ) (1.09 g, 4.8 mmol) was added. After the mixture was stirred for 1 h, diisopropylethylamine (DIPEA, 5 mL) and $BF_3 \cdot OEt_2$ (5 mL) were added. The resulting mixture was further stirred for another 1 h, then concentrated and filtered. After the filtrate was washed twice with water and brine, the organic layer was collected, dried over anhydrous $MgSO_4$ and concentrated under reduced pressure. The obtained crude product was purified by column chromatography (R_f = 0.2, PE/EA = 3:1, eluent: PE/EA = 30/1–4/1, v/v) to give compound **1** (0.38 g, 28% yield) as a yellow-red powder. ^1H NMR (400 MHz, $CDCl_3$) δ (ppm): 7.12 (d, J = 8.4 Hz, 2H), 6.94 (d, J = 8.4 Hz, 2H), 5.98 (s, 2H), 5.30–5.26 (m, 1H), 2.55 (s, 6H), 1.44 (S, 6H); ^{13}C NMR (100 MHz, $CDCl_3$) δ (ppm): 156.3, 155.3, 143.2, 141.8, 131.8, 129.4, 127.2, 121.2, 116.1, 14.6; ^{19}F NMR (376 MHz, $CDCl_3$) δ (ppm): −146.06 (m, 2F). HRMS-ESI (m/z): [M]⁻ Calc. for ($C_{19}H_{18}BF_2N_2O$), 339.1480, found: 339.1489.

3.5.2. Synthesis of Compound **3**

Compound **1** (98 mg, 0.3 mmol), tetra-O-acetyl-α-D-galactose bromide **2** (148 mg, 0.36 mmol), and Ag_2O (104 mg, 0.45 mmol) were suspended in dry acetonitrile (5 mL). After the mixture was stirred for 6 h at r.t under argon and filtered, the solvent was removed in vacuum. The residue was purified by silica gel column (R_f = 0.5, PE/EA = 2:1, eluent: PE/EA =10/1–3/1, v/v) to give compound **3** (121 mg, 60% yield) as an orange solid. ^1H NMR (400 MHz, $CDCl_3$) δ (ppm): δ 7.19 (d, J = 8.4 Hz, 2H), 7.12 (d, J = 8.4 Hz, 2H), 5.97 (s, 2H), 5.54 to 5.52 (m, 1H), 5.48 (d, J = 3.2 Hz,1H), 5.16 to 5.12 (m, 2H), 4.28 to 4.23 (m, 1H), 4.18 to 4.09 (m, 2H), 2.53 (s, 6H), 2.18 (s, 3H), 2.10 (s, 3H), 2.04 (s, 3H), 2.02 (s, 3H), 1.39 (s, 6H); ^{13}C NMR (100 MHz, $CDCl_3$) δ (ppm): 170.36, 170.26, 170.17, 169.40, 157.46, 155.68, 143.08, 141.06, 131.70, 129.94, 129.52, 121.38, 117.61, 99.59, 71.36, 70.85, 68.73, 66.96, 61.45, 20.83, 20.71, 20.65, 14.68; ^{19}F NMR (376 MHz, $CDCl_3$) δ (ppm): −146.17 (m,2F). HRMS-ESI (m/z): [M+H]⁺ Calc. for ($C_{33}H_{38}BF_2N_2O_{10}$), 671.2588, found: 671.2587.

3.5.3. Synthesis of BOD-Gal

Compound 3 (78 mg, 0.1 mmol) was dissolved in anhydrous methanol (3 mL), then K_2CO_3 (2 mg, 0.02 mmol) was added. The resulting mixture was stirred at r.t for 2 h and the pH was adjusted to 6~7 with 1 M aqueous HCl. After the reaction mixture was filtered, the filtrate was concentrated in vacuum. The residue was purified by silica gel column (R_f = 0.3, DCM/MeOH = 5:1, eluent: DCM/MeOH = 15:1, v/v) to afford BOD-Gal (52 mg, 88% yield, m.p. 143–145 °C) as an orange powder. $[\alpha]_D^{25}$ = +33.33. ^{1}H NMR (400 MHz, DMSO-d_6) δ (ppm): δ 7.26 (d, J = 8.4 Hz, 2H), 7.20 (d, J = 8.4 Hz, 2H), 6.17 (s, 2H), 5.23 (s, 1H), 4.90 (d, J = 8.0 Hz, 2H), 4.67 (s, 1H), 4.54 (s, 1H), 3.72 (s, 1H), 3.63–3.43 (m, 5H), 2.44 (s, 6H), 1.40 (s, 6H); ^{13}C NMR (100 MHz, DMSO-d_6) δ (ppm): 156.25, 154.66, 142.70, 142.01, 131.03, 128.98, 127.06, 121.26, 116.84, 101.11, 75.52, 73.23, 70.35, 68.06, 60.29, 14.19; ^{19}F NMR (376 MHz, DMSO-d_6) δ (ppm): −143.57 (m,2F). HRMS-ESI (m/z): [M+Na]$^+$ Calc. for ($C_{25}H_{30}BF_2N_2O_6Na$), 525.1984, found:525.1959.

4. Conclusions

This study designed, synthesized and characterized a novel fluorescent substrate, BOD-Gal, for the detection of β-galactosidase activity. It showed a distinct fluorescence reduction after hydrolysis by β-Gal, when other biologically competitive enzyme species, amino acids, and small molecules caused only faint fluorescent change. The results of HPLC chromatography confirmed that BOD-Gal was exclusively dissociated by β-Gal to a phenoxy residue, which induced a PET mechanism. In addition, BOD-Gal displayed a higher affinity and faster response to β-Gal than commercial X-Gal as shown on a Hanes–Woolf plot. For bacteria detection, BOD-Gal also had a significant response to pathogenic *E. coli* on an agar growth medium with low toxicity. In view of its convenience, sensitivity and speed, this substrate has the potential to promote the development of fluorescent media in pathogen detection.

Supplementary Materials: The following are available online. Scheme S1, Figure S1–S11, Table S1–S3.

Author Contributions: Q.X. and F.-Y.W. designed the study; X.C. and Y.-C.L. performed the experiments and analyzed the data; methodological support J.-J.C.; X.C. writing original draft; Q.X. and F.-Y.W. review and editing. All authors have read and agreed to the published version of the manuscript.

Funding: This work is financially supported by Natural Science Foundation of China (No. 21765014, 21505067 and 21365014) and Science and Technology project of Jiangxi Provincial Education Office (GJJ160793).

Institutional Review Board Statement: Not applicable.

Informed Consent Statement: Not applicable.

Data Availability Statement: Not applicable.

Acknowledgments: We wish to thank the timely help given by Jiang Bai of Jiangxi Science & Technology Normal University in the analyzing samples.

Conflicts of Interest: The authors declare no conflict of interest.

Sample Availability: Samples of all the compounds are available from the authors.

References

1. Todd, E.C.D. Epidemiology of food borne diseases: A worldwide review. *World Health Stat. Q.* **1997**, *50*, 30–50. Available online: https://www.who.int/iris/handle/10665/54782 (accessed on 20 August 2021).
2. Altekruse, S.F.; Swerdlow, D.L. The Changing Epidemiology of Foodborne Diseases. *Am. J. Med. Sci.* **1996**, *311*, 23–29. [CrossRef]
3. Yang, S.C.; Lin, C.H.; Aljuffali, I.A.; Fang, J.Y. Current pathogenic Escherichia coli foodborne outbreak cases and therapy development. *Arch. Microbiol.* **2017**, *199*, 811–825. [CrossRef]
4. Velusamy, V.; Arshak, K.; Korostynska, O.; Oliwa, K.; Adley, C. An overview of foodborne pathogen detection: In the perspective of biosensors. *Biotechnol. Adv.* **2010**, *28*, 232–254. [CrossRef]

5. Perry, J.D. A Decade of Development of Chromogenic Culture Media for Clinical Microbiology in an Era of Molecular Diagnostics. *Clin. Microbiol. Rev.* **2017**, *30*, 449–479. [CrossRef]
6. Manafi, M. New developments in chromogenic and fluorogenic culture media. *Int. J. Food Microbiol.* **2000**, *60*, 205–218. [CrossRef]
7. Varadi, L.; Gray, M.; Groundwater, P.W.; Hall, A.J.; James, A.L.; Orenga, S.; Perry, J.D.; Anderson, R.J. Synthesis and evaluation of fluorogenic 2-amino-1,8-naphthyridine derivatives for the detection of bacteria. *Org. Biomol. Chem.* **2012**, *10*, 2578–2589. [CrossRef]
8. Perry, J.D.; James, A.L.; Morris, K.A.; Oliver, M.; Chilvers, K.F.; Reed, R.H.; Gould, F.K. Evaluation of novel fluorogenic substrates for the detection of glycosidases in Escherichia coli and enterococci. *J. Appl. Microbiol.* **2006**, *101*, 977–985. [CrossRef] [PubMed]
9. Geissler, K.; Amoro's, M.M.; Alonso, J.L. Quantitative determination of total coliforms and Escherichia coli in marine waters with chromogenic and fluorogenic media. *J. Appl. Microbiol.* **2000**, *88*, 280–285. [CrossRef]
10. Perry, J.D.; Morris, K.A.; James, A.L.; Oliver, M.; Gould, F.K. Evaluation of novel chromogenic substrates for the detection of bacterial beta-glucosidase. *J. Appl. Microbiol.* **2007**, *102*, 410–415. [CrossRef] [PubMed]
11. Feng, P.C.S.; Hartman, P.A. Fluorogenic Assays for Immediate Confirmation of Escherichia coli. *Appl. Environ. Microbiol.* **1982**, *43*, 1320–1329. [CrossRef]
12. Martinez, C.R.; Rodriguez, T.L.; Zhurbenko, R.; Valdes, I.A.; Gontijo, S.M.; Gomes, A.D.; Suarez, D.F.; Sinisterra, R.D.; Cortes, M.E. Development of a calcium phosphate nanocomposite for fast fluorogenic detection of bacteria. *Molecules* **2014**, *19*, 13948–13964. [CrossRef]
13. Mihalcescu, I.; Van-Melle Gateau, M.; Chelli, B.; Pinel, C.; Ravanat, J.L. Green autofluorescence, a double edged monitoring tool for bacterial growth and activity in micro-plates. *Phys. Biol.* **2015**, *12*, 066016. [CrossRef]
14. Guilini, C.; Baehr, C.; Schaeffer, E.; Gizzi, P.; Rufi, F.; Haiech, J.; Weiss, E.; Bonnet, D.; Galzi, J.L. New fluorescein precursors for live bacteria detection. *Anal. Chem.* **2015**, *87*, 8858–8866. [CrossRef]
15. Wang, T.; Douglass, E.F., Jr.; Fitzgerald, K.J.; Spiegel, D.A. A "turn-on" fluorescent sensor for methylglyoxal. *J. Am. Chem. Soc.* **2013**, *135*, 12429–12433. [CrossRef]
16. Ucuncu, M.; Emrullahoglu, M. A BODIPY-based reactive probe for the detection of Au(III) species and its application to cell imaging. *Chem. Commun.* **2014**, *50*, 5884–5886. [CrossRef] [PubMed]
17. Ulrich, G.; Ziessel, R.; Harriman, A. The Chemistry of Fluorescent Bodipy Dyes: Versatility Unsurpassed. *Angew. Chem. Int. Ed.* **2008**, *47*, 1184–1201. [CrossRef] [PubMed]
18. Bozdemir, O.A.; Sozmen, F.; Buyukcakir, O.; Guliyev, R.; Cakmak, Y.; Akkaya, E.U. Reaction-Based Sensing of Fluoride Ions Using Built-In Triggers for Intramolecular Charge Transfer and Photoinduced Electron Transfer. *Org. Lett.* **2010**, *12*, 1400–1403. [CrossRef] [PubMed]
19. Gu, K.; Xu, Y.; Li, H.; Guo, Z.; Zhu, S.; Shi, P.; James, T.D.; Tian, H.; Zhu, W.H. Real-Time Tracking and In Vivo Visualization of β-Galactosidase Activity in Colorectal Tumor with a Ratiometric NIR Fluorescent Probe. *J. Am. Chem. Soc.* **2016**, *138*, 5334–5340. [CrossRef]
20. Bai, T.; Chu, T. Exploring the Simultaneous Biothiols-differentiating Detecting Feature of a BODIPY Chemosensor with DFT/TDDFT. *J. Mol. Liq.* **2020**, *309*, 113145. [CrossRef]
21. Ueno, T.; Urano, Y.; Kojima, H.; Nagano, T. Mechanism-based molecular design of highly selective fluorescence probes for nitrative stress. *J. Am. Chem. Soc.* **2006**, *128*, 10640–10641. [CrossRef]
22. Wang, X.; Bai, T.; Chu, T. A molecular design for a turn-off NIR fluoride chemosensor. *J. Mol. Model.* **2021**, *27*, 104. [CrossRef]
23. Jiao, L.; Yu, C.; Li, J.; Wang, Z.; Wu, M.; Hao, E. Beta-formyl-BODIPYs from the Vilsmeier-Haack reaction. *J. Org. Chem.* **2009**, *74*, 7525–7528. [CrossRef] [PubMed]
24. Tian, J.; Ouyang, W.; He, Y.; Ning, Q.; Bai, J.; Ding, H.; Xiao, Q. Practical Synthesis of the Fluorogenic Enzyme Substrate 4-Methylumbelliferyl α-l-Idopyranosiduronic Acid. *Synlett* **2020**, *31*, 1083–1086. [CrossRef]

Article

New 4-Aminoproline-Based Small Molecule Cyclopeptidomimetics as Potential Modulators of $\alpha_4\beta_1$ Integrin

Andrea Sartori [1], Kelly Bugatti [1], Elisabetta Portioli [1], Monica Baiula [2], Irene Casamassima [2], Agostino Bruno [1], Francesca Bianchini [3], Claudio Curti [1], Franca Zanardi [1] and Lucia Battistini [1,*]

[1] Department of Food and Drug, University of Parma, Parco Area delle Scienze 27/A, 43124 Parma, Italy; andrea.sartori@unipr.it (A.S.); kelly.bugatti@unipr.it (K.B.); elisabetta.portioli@studenti.unipr.it (E.P.); agostino.bruno@unipr.it (A.B.); claudio.curti@unipr.it (C.C.); franca.zanardi@unipr.it (F.Z.)

[2] Department of Pharmacy and Biotechnology, University of Bologna, Via Irnerio 48, 40126 Bologna, Italy; monica.baiula@unibo.it (M.B.); irene.casamassima@unibo.it (I.C.)

[3] Department of Experimental and Clinical Biomedical Sciences, University of Florence, Viale G.B. Morgagni 50, 50134 Firenze, Italy; francesca.bianchini@unifi.it

* Correspondence: lucia.battistini@unipr.it; Tel.: +39-0521-906040

Citation: Sartori, A.; Bugatti, K.; Portioli, E.; Baiula, M.; Casamassima, I.; Bruno, A.; Bianchini, F.; Curti, C.; Zanardi, F.; Battistini, L. New 4-Aminoproline-Based Small Molecule Cyclopeptidomimetics as Potential Modulators of $\alpha_4\beta_1$ Integrin. *Molecules* **2021**, *26*, 6066. https://doi.org/10.3390/molecules26196066

Academic Editors: Camilla Matassini, Francesca Cardona and Camilla Parmeggiani

Received: 30 July 2021
Accepted: 4 October 2021
Published: 7 October 2021

Publisher's Note: MDPI stays neutral with regard to jurisdictional claims in published maps and institutional affiliations.

Copyright: © 2021 by the authors. Licensee MDPI, Basel, Switzerland. This article is an open access article distributed under the terms and conditions of the Creative Commons Attribution (CC BY) license (https://creativecommons.org/licenses/by/4.0/).

Abstract: Integrin $\alpha_4\beta_1$ belongs to the leukocyte integrin family and represents a therapeutic target of relevant interest given its primary role in mediating inflammation, autoimmune pathologies and cancer-related diseases. The focus of the present work is the design, synthesis and characterization of new peptidomimetic compounds that are potentially able to recognize $\alpha_4\beta_1$ integrin and interfere with its function. To this aim, a collection of seven new cyclic peptidomimetics possessing both a 4-aminoproline (Amp) core scaffold grafted onto key $\alpha_4\beta_1$-recognizing sequences and the (2-methylphenyl)ureido-phenylacetyl (MPUPA) appendage, was designed, with the support of molecular modeling studies. The new compounds were synthesized through SPPS procedures followed by in-solution cyclization maneuvers. The biological evaluation of the new cyclic ligands in cell adhesion assays on Jurkat cells revealed promising submicromolar agonist activity in one compound, namely, the *c*[Amp(MPUPA)Val-Asp-Leu] cyclopeptide. Further investigations will be necessary to complete the characterization of this class of compounds.

Keywords: aminoproline scaffold; integrin targeting; ligand design; peptidomimetic synthesis; leukocyte integrins

1. Introduction

Integrins constitute a major class of cell adhesion receptors in mammals and play a vital role in cell–cell and cell–extracellular environment communication by regulating crucial aspects of cellular functions, including migration, adhesion, differentiation, growth, and survival. They are expressed in almost all cell types with varied distribution pattern [1,2]. Given their fundamental contribution in human physiology, specific integrin dysregulation phenomena are linked to the pathogenesis of many disease states (including cancer, thrombosis, vascular diseases, autoimmune pathologies, osteoarthritis, osteoporosis), and this renders them attractive targets for biomedical research [3–5].

The integrin family comprises 24 different heterodimeric subtypes, classified according to the specific, non-covalent combination between α and β subunits. Among these, the $\alpha_4\beta_1$ and $\alpha_4\beta_7$ subtypes, as well as the β_2 integrin subclass, belong to the leukocyte-specific integrin family and are involved in the modulation of immune functions. In particular, the $\alpha_4\beta_1$ integrin, also known as very late antigen-4 (VLA-4), raised much attention due to its being constitutively expressed on the surface of lymphocytes and most leukocytes, and being involved in coordinating leukocyte homing in various tissues [6].

As a fruitful consequence of intense investigation on integrins, several integrin antagonists have been validated as drugs. For example, diverse small molecules and antibodies, including eptifibatide, tirofiban, and abciximab, which target the platelet-specific integrin $\alpha_{IIb}\beta_3$, are effectively used as therapeutic agents in the treatment of acute coronary syndromes and prevention of myocardial infarct following coronary intervention [7]. On the other hand, the known roles of leukocyte-specific integrins in events such as inflammation and host defense has prompted parallel anti-integrin strategies, yielding effective therapeutic anti-inflammatory agents [8,9]. Indeed, targeting α_4 integrins has proven to be effective for the treatment of inflammatory diseases, including multiple sclerosis and Crohn's disease, with some monoclonal antibodies being approved for clinical practice [10,11].

Inflammatory responses are crucial for host defense and are subjected to a complex system of control, aiming to prevent tissue damage and dangerous consequences. Since many inflammatory diseases are characterized by an influx of lymphocytes and leukocytes in the inflamed tissue, there is a keen interest in finding and testing compounds that have the potential to modulate these processes [12]. In this context, integrin activation during the different steps of the leukocyte adhesion cascade is the result of a fine-tuned orchestra of activation pathways and local regulatory networks at the site of inflammation, whose malfunctioning may cause severe disease patterns. The diseases associated with $\alpha_4\beta_1$ (and $\alpha_4\beta_7$) integrins are mainly of inflammatory and autoimmune nature, implying a pathological accumulation of activated leukocytes in the affected tissues such as, for example, inflammatory bowel disease, Crohn's disease, rheumatoid arthritis, asthma, multiple sclerosis, dry eye disease and allergic conjunctivitis [10,13]. Moreover, the strict correlation between inflammation and cancer is well established at present, and immunomodulation is recognized as a useful tool not only in the treatment of inflammatory and autoimmune pathologies, but also as an adjuvant in tumor therapy. It is known that chronic inflammatory states and tumor development are closely related and mutually supportive [14]. Indeed, during chronic inflammation, the release of chemokines and growth factors supports tumor development, while, on the other hand, the tumor state can induce the upregulation of immunosuppressive molecules and the dysregulated T-cell-mediated host responses. In addition, the $\alpha_4\beta_1$ integrin was demonstrated to play a pivotal role in tumor angiogenesis associated with chronic inflammation, a condition that may promote the angiogenetic switch in tumors [15]. Integrin $\alpha_4\beta_1$ is also involved in the recruitment of progenitor cells (multipotent cells derived from bone marrow stem cells), in the transendothelial tumor cell migration and, due to its overexpression in melanoma cells, $\alpha_4\beta_1$ is also considered a marker of metastatic risk [16,17].

In this complex scenario, the possibility to interfere with integrin activity is of great interest and α_4 integrins have become a target for fine modulation by interaction with small-molecule ligands, based on the emerging idea that *antagonist* ligands may interfere in leukocyte primary functions while, on the other hand, *agonist* ligands can serve to promote some useful integrin functions. Enhancement of cell adhesion, for example, impairing cell detachment, may prevent tumor cell migration and metastasis processes, or may induce progenitor cell retention for stem cell therapy [18].

The natural ligands of α_4 integrins comprise the vascular adhesion molecule-1 (VCAM-1) and the alternatively spliced connecting segment 1 (CS1) region of fibronectin (FN). In particular, FN is recognized through the Leu–Asp–Val (LDV) binding epitope [19], while VCAM-1 interacts with its receptor via the homologous and essentially isosteric binding sequence Gln–Ile–Asp–Ser–Pro–Leu (QIDSPL) [20]. The discovery that these short amino acidic sequences are minimal recognition motifs has prompted the research of small-molecule peptidomimetics resembling the natural binding epitope and fitting into the groove at the α and β subunit interface [10,21]. Figure 1 collects some notable results in the discovery of linear peptidomimetic ligands, reminiscent of the LDV sequence and targeting the $\alpha_4\beta_1$ receptor.

In 1999, the Adams' research group reported the synthesis of BIO1211 (compound **1**, Figure 1) [22], a potent and selective $\alpha_4\beta_1$ antagonist, which was shown to inhibit the

$\alpha_4\beta_1$/VCAM-1 interaction with an IC$_{50}$ of 4 nM (Jurkat cell adhesion assay) and to possess a marked selectivity for $\alpha_4\beta_1$ as compared to $\alpha_4\beta_7$ integrin (IC$_{50}$ $\alpha_4\beta_7$ = 2 µM).

BIO1211 is based on the peptide sequence Leu–Asp–Val–Pro (LDVP) substituted at the amino terminus with the 4-[(N-2-MethylPhenyl)Ureido]PhenylAcetyl group (MPUPA). The introduction of this last moiety was demonstrated to produce a substantial increase in both potency and enzymatic stability as compared to the LDV peptide precursor [23]; for this reason, BIO1211 is commonly used as a reference compound in many studies aiming to developing new $\alpha_4\beta_1$ ligands. The in vitro efficacy and potency of this compound were also confirmed in vivo: when administered as an aerosol, it showed prophylactic efficacy in a sheep model of allergic bronchoconstriction, electing this nonsteroidal compound as the first small-molecule $\alpha_4\beta_1$ antagonist to enter clinical trials. However, the residual peptide nature of BIO1211 caused a certain enzymatic instability. To overcome this behavior, a number of bioactive peptidomimetics have been prepared (Figure 1), which share common structural features, including an aromatic cap at the N-terminus, a suitable spacer, and a carboxylic group mimetic of the Asp residue, with BIO1211 [10]. Compound LLP2A (2, Figure 1), proposed by Peng et al. in 2006 [24], was identified in a competitive cell-based screening under a high concentration of soluble BIO1211. It showed an exceptionally high affinity toward $\alpha_4\beta_1$ receptor (IC$_{50}$ = 2 pM, Jurkat cell assay) without any effect on the cell proliferation and survival of $\alpha_4\beta_1$–positive cells. For this reason, it was differently functionalized with NIR-fluorescence probes, or labelled with radionuclides (^{111}In, ^{64}Cu, ^{99m}Tc and ^{18}F) to image several different tumors, including melanoma [25,26]. Recently, due to its high binding affinity to integrin $\alpha_4\beta_1$, which is highly expressed on mesenchymal stem cells (MSCs) and regulates MSC homing, adhesion, migration and differentiation, LLP2A has also been exploited for tissue engineering and regenerative medicine applications [27].

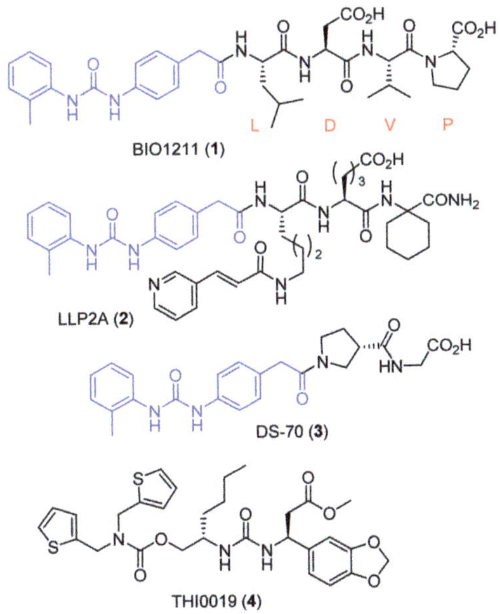

Figure 1. Examples of linear peptidomimetics targeting the $\alpha_4\beta_1$ integrin receptor, some of which have been advanced in preclinical studies [13,18,22,24]. The MPUPA moiety is depicted in blue.

According to a common trend in bioactive peptide research, the introduction of cyclic scaffolds, including proline derivatives and other five-membered heterocycles, has been exploited by many researchers as amide bond isosteres and conformational restraints in the design and synthesis of peptidomimetic integrin ligands [28]. Along this line, the insertion of a D-configured β2-proline scaffold into a peptidomimetic structure led to the development of compound DS-70 (**3**, Figure 1), which demonstrated a high binding affinity for human $\alpha_4\beta_1$ integrin and potent antagonist activity of α_4–mediated cell adhesion. Additionally, it was successfully tested in a guinea pig preclinical model of allergic conjunctivitis [13]. Lastly, compound THI0019 (**4**, Figure 1) [18] was the first $\alpha_4\beta_1$ agonist designed and synthesized starting from a potent $\alpha_4\beta_1$ antagonist as a template [29]. THI0019 was generated by introducing two structural modifications into a previously identified $\alpha_4\beta_1$ antagonist. As a result, THI0019 enhanced the rolling, spreading, adhesion, and migration of endothelial progenitor cells in vitro in a $\alpha_4\beta_1$-dependent fashion; the authors suggested that compound **4** could temporarily occupy the ligand binding pocket, inducing a small conformational change in the receptor that favors agonist displacement and binding of natural ligand, thus opening opportunities for stem cell therapy [18].

Despite the relevant results obtained in the preclinical evaluation of these molecules as targeting motifs in the construction of imaging probes, potential treatments in ocular diseases, or innovative materials for regenerative medicine, there is still ample room for the development of new and structurally varied binders, which may enrich the pool of existing $\alpha_4\beta_1$ ligands.

In recent years, the exploitation of the *cis*-4-amino-L-proline residue (Amp) as a conformation-inducing scaffold led to the development of novel classes of RGD-based cyclopeptide ligands of type **5** and **6** (Figure 2) targeting $\alpha_V\beta_3$, $\alpha_V\beta_5$, and/or $\alpha_V\beta_6$ integrin receptors with a good-to-high-affinity and selectivity [30–32]. These integrins are known to be directly involved in the evolution and diffusion of metastatic tumor cells and angiogenesis, as well as in the development of organ fibrosis.

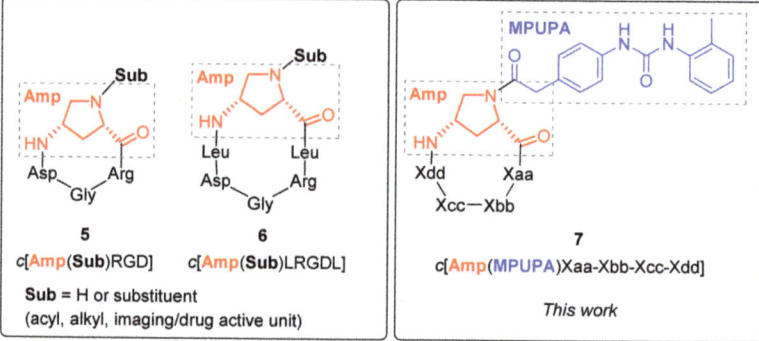

Figure 2. Amp-based cyclopeptides directed to RGD-recognizing $\alpha_V\beta_3$, $\alpha_V\beta_5$, and $\alpha_V\beta_6$ integrins (compounds **5** and **6**) [30–40], and general structure of cyclopeptidomimetics **7** designed and synthesized to target $\alpha_4\beta_1$ integrin in the present study.

The Amp scaffold is a new-to-nature, yet nature-reminiscent small-molecular entity, which can be grafted onto the peptide sequence of interest and impart proper ligand conformation [30], while conferring stability toward enzymatic degradation. Moreover, the Amp nucleus possesses a N^α-proline site free for covalent bonding to useful functional units; indeed, the Amp-based cyclopeptide cores were covalently conjugated to either fluorescent tags, chelating units, or established therapeutic drugs to obtain hybrid dual-active structures and nanoparticles [33–40].

In the present study, the Amp scaffold was selected as the core unit for building up a new class of cyclic small-molecule peptidomimetics by linking it to proper pharmacophoric groups, aiming to target the $\alpha_4\beta_1$ integrin receptor. To explore this possibility, we designed and synthesized a small collection of cyclic aminoproline-based peptidomimetics of general formula c[Amp(MPUPA)Xaa-Xbb-Xcc-Xdd] **7** (Figure 2), in which the Amp scaffold was grafted onto suitable peptide sequences (LDV motif and analogues) and functionalized at the N^α-proline site with the well-known α_4-integrin targeting MPUPA moiety.

In this work, we report the molecular modelling-driven design, the synthesis, and the chemical characterization of a collection of seven tetra- and pentacyclopeptidomimetics of type **7**, as well as the evaluation of their binding competence towards the $\alpha_4\beta_1$ integrin receptor by cell adhesion assays using Jurkat cells in the presence of VCAM-1, with the aim to preliminarily assess their ability to bind $\alpha_4\beta_1$ integrin and possibly serve as modulators of integrin function.

2. Results

2.1. Design of Novel $\alpha_4\beta_1$ Ligands

The study of the interactions between ligands and their biological targets greatly benefits from the availability of ligand-receptor crystallographic insights; since no X-ray analyses exist to date on the crystal structure of the $\alpha_4\beta_1$ receptor, or of the same receptor in complex with its small-molecule ligands, the design of a new class of cyclic Amp-based peptidomimetics required the generation and validation of a $\alpha_4\beta_1$ receptor model by molecular modelling studies [41].

The work started from the atomic coordinates of the single α_4 and β_1 domains, which were available from the $\alpha_4\beta_7$ integrin complex (PDB code: 3V4V) [42] and the $\alpha_5\beta_1$ integrin complex (PDB code: 3VI4) [43], respectively. In fact, these integrins possess a high degree of structural conservation, a large, solvent-exposed ligand-binding site at the α/β interface, and a divalent cation (Mg^{2+}) at the metal ion-dependent adhesion site (MIDAS), which may be involved in a coordinated bond with a carboxylate group of the ligand. Using the $\alpha_5\beta_1$ integrin complex as a template, the α_4 subunit was aligned with α_5 bound to β_1; then, the α_5 subunit was removed, giving a preliminary $\alpha_4\beta_1$ complex. The integrin complex thus obtained was refined and optimized by a minimization protocol (Figure 3) and then subjected to a validation procedure through docking studies.

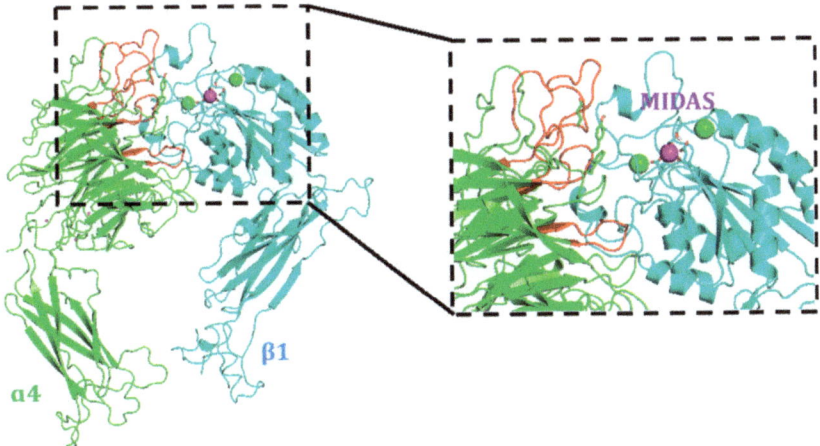

Figure 3. The developed $\alpha_4\beta_1$ complex shows the classical structural organization of the integrin family; α_4 subunit (green cartoon), β_1 subunit (cyan cartoon); the Ca^{2+} and Mg^{2+} ions of MIDAS are depicted as green and purple spheres, respectively. The portion of the β-propeller domain (α_4 subunit) involved in ligand binding region is evidenced in red.

To this aim, eight known $\alpha_4\beta_1$ integrin antagonists were selected, namely compound BIO1211 (**1**, Figure 1), compounds **8a**, **8b**, **9**, **10a**, **10b**, **11a**, and **11b** (Figure 4), along with one novel Amp-based cyclic candidate (compound c[Amp(MPUPA)-Leu-Asp-Val-Gly] **12**, Figure 4). This small collection of known peptidic and peptidomimetic structures showed a certain level of molecular diversity and inhibitory potencies towards $\alpha_4\beta_1$ integrin, ranging from micromolar to low nanomolar values [44–46].

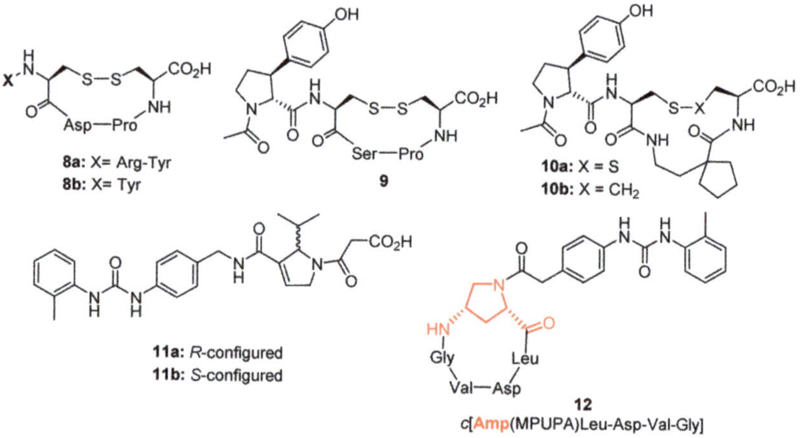

Figure 4. Known cyclic peptides (**8a**, **8b**, and **9**), their spirocyclic analogues (**10a**, **10b**), the dehydro-β-proline peptidomimetics **11a** and **11b**, and the novel Amp-based cyclopentapeptidomimetic **12** used for docking studies in this work.

In particular, the low-nanomolar cyclic peptides **8a**, **8b**, and **9**, containing the Cys-Asp-Pro-Cys or Cys-Ser-Pro-Cys core structures, and their spirocyclic analogues **10a** and **10b**, constrained by a disulfide (or a thioether) bridge, were conceived to mimic the essential $\alpha_4\beta_1$ IDS or LDV binding sequences [44,45]. Compound **11a** represents one of the linear analogues of BIO1211 obtained by a retro-sequence strategy and containing a dehydro-β-proline ring which, similarly to compounds **8–10**, showed a potent inhibitory activity of $\alpha_4\beta_1$/VCAM interaction with IC_{50} in the nanomolar range [46], accompanied by a superior enzymatic stability respect to cyclic peptides **8–10**.

Figure 5 shows the binding poses of compounds BIO1211 and **11a** within the $\alpha_4\beta_1$ binding site, as well as their overlapping structures. The analysis of these binding poses revealed that both compounds can interact with Mg^{2+} cation in the β subunit and share a similar disposition within the binding pocket of the receptor, with the common functional groups interacting with the same amino acid residues in the α subunit. Notably: (i) the ureido group within MPUPA of both compounds establishes a bidentate interaction with Glu124 (E124) residue, (ii) the terminal aromatic ring of the ureido group establishes a cation-π interaction with Lys156 (K156) residue, (iii) the isopropyl group of **11a** adopts a spatial orientation similar to the leucine side chain of BIO1211. This last observation would explain the experimental evidence showing that compound **11b** (the enantiomer of **11a**) is considerably less active on $\alpha_4\beta_1$ [46]. Furthermore, BIO1211 establishes a H-bond with Tyr187 (Y187), a crucial interaction, as highlighted by reported mutagenesis studies [47]. The rationalization of the binding poses of the selected compounds proved to be in agreement with the SAR studies reported in the literature [44,46], supporting the reliability of the developed receptor model.

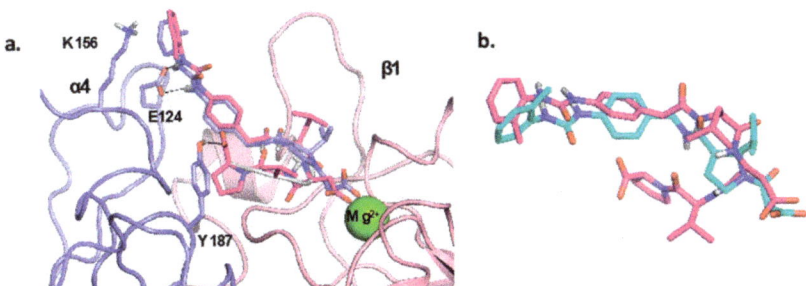

Figure 5. (a) Binding poses of BIO1211 (purple sticks) and **11a** (blue sticks) in the binding site of $\alpha_4\beta_1$ integrin (α_4 subunit blue ribbon; β_1 subunit pink ribbon). The Mg^{2+} cation is reported as a green sphere. (b) Overlapping of compounds BIO1211 (purple sticks) and **11a** (cyan sticks) obtained by docking studies.

The validated model was used in the subsequent docking studies, where the same experimental protocol was applied, to identify the binding modes of Amp-bearing $\alpha_4\beta_1$-ligands, and to predict possible structural modifications improving affinity toward the $\alpha_4\beta_1$ integrin. To this end, the docking procedure was used to evaluate c[Amp(MPUPA)Leu-Asp-Val-Gly] (**12**) as a new potential $\alpha_4\beta_1$ ligand. In Figure 6, the binding pose of compound **12** is shown and compared to that of BIO1211.

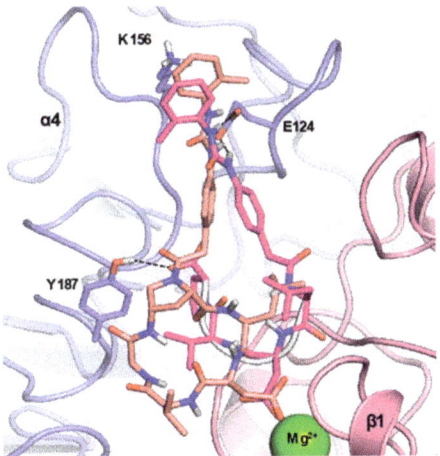

Figure 6. Binding poses of Amp-cyclopeptide **12** (pink sticks) and BIO1211 (purple sticks) in the binding site of $\alpha_4\beta_1$ integrin (α_4 subunit blue ribbon; β_1 subunit pink ribbon).

From the analysis of the docking poses, we noticed that cyclopeptide **12** (pink sticks) would be able to establish some comparable interactions to BIO1211 (purple sticks) in the binding pocket of the $\alpha_4\beta_1$ receptor model. Compound **12** seems to be particularly able to (i) chelate the divalent cation (Mg^{2+}) through the carboxylate group of the Asp residue, (ii) interact with the amino acidic residues Tyr187, Lys156 and Glu124 in a similar way as BIO1211; (iii) its Val residue seems to assume the favorable spatial orientation that was observed for BIO1211, and (iv) the MPUPA moiety of both compounds occupies the same region.

Starting from compound **12**, six additional cyclic Amp-based cyclopeptide derivatives were designed, namely, compounds **13–18** (Figure 7), to be launched in the synthesis program. The design was rooted in the following considerations: (i) substitution of the

Glu residue for Asp could further favor the interaction of the carboxylate group of the side chain with the divalent cation of MIDAS (e.g., compound **12** vs. **13**, **14** vs. **15**, **16** vs. **17**), (ii) restriction of the cyclopeptide ring via Gly depletion could provide insights about the influence of ring size and constrain on binding affinity (e.g., pentapeptide compounds **12–13** vs. tetrapeptide analogues **14–18**); (iii) exploitation of retro-sequences could expand exploration of the pharmacophoric space (e.g., VDL-based compound **16** vs. LDV counterpart **14**, and VEL-based compound **17** vs. LEV counterpart **15**), (iv) substitution of the RGD sequence for LDV would generate derivative **18**, which could likely be used as a negative control in $\alpha_4\beta_1$-directed biological assays.

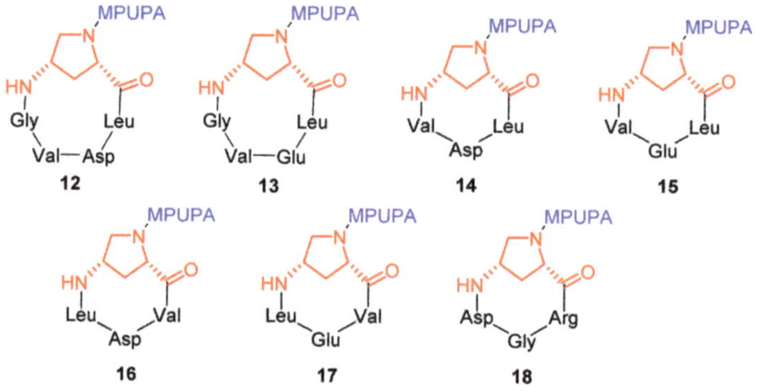

Figure 7. The collection of Amp-MPUPA-bearing cyclopeptidomimetics **12–18** designed and synthesized in this study.

2.2. Synthesis of Novel $\alpha_4\beta_1$ Ligands

The synthesis of the designed compounds **12–18** began with the chemoselective in-solution N^α deprotection of commercially available Fmoc-4-amino-1-Boc-pyrrolidine-2-carboxylic acid (**19**) to **20** (Scheme 1) and subsequent functionalization with the 4-[(N-2-methylphenyl)ureido]phenylacetyl (MPUPA) moiety **21**, to provide the N-Fmoc-Amp(MPUPA)-OH scaffold **22** (79% yield) to be used in the following SPPS procedures. The MPUPA moiety **21** was instead synthesized, with good yield, starting from the commercially available precursors, o-tolyl isocyanate and 4-aminophenylacetic acid, following literature procedure [22]. The synthesis of compound **22** entailed the preliminary activation of the carboxylic function within MPUPA unit **21** by means of HATU/HOAt/collidine coupling system in dry DMF, followed by the addition of **20**.

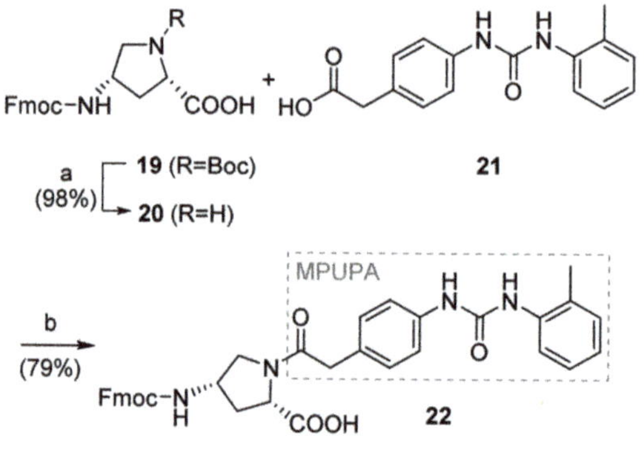

Scheme 1. Synthesis of N-Fmoc-Amp(MPUPA)-OH scaffold **22**. Reagents and conditions: (**a**) TFA, dry DCM, rt; (**b**) HATU, HOAt, collidine, dry DMF.

For the synthesis of the linear precursors of targeted cyclopeptides **12–18**, the Fmoc-based SPPS strategy was adopted, followed by in-solution cyclization and deprotection protocols (Scheme 2). All the linear peptide sequences were prepared, starting from the proper acid-labile chlorotrityl chloride resin preloaded with one of the three different starting amino acid residues, Asp(*t*Bu), Glu(*t*Bu) or Gly. Within all the designed peptide sequences, the aminoproline scaffold played a critical role; in fact, in all instances, this unit was in a central position within the linear peptides, creating a local constraint that would likely pre-organize the terminal chains toward the final macrocyclization step. The synthesis of the designed sequences required the stepwise addition of Fmoc-protected amino acids to the growing peptides, with alternating coupling steps (in the presence of HATU/HOAt/collidine) and Fmoc-removal procedures (by using 20% piperidine/DMF solution); then, the linear peptide sequences were readily cleaved from the resin using the conventional AcOH/TFE/DCM mixture. The crude linear peptides **23–29** were obtained in yields ranging from 78% to 99% for the entire solid phase sequences.

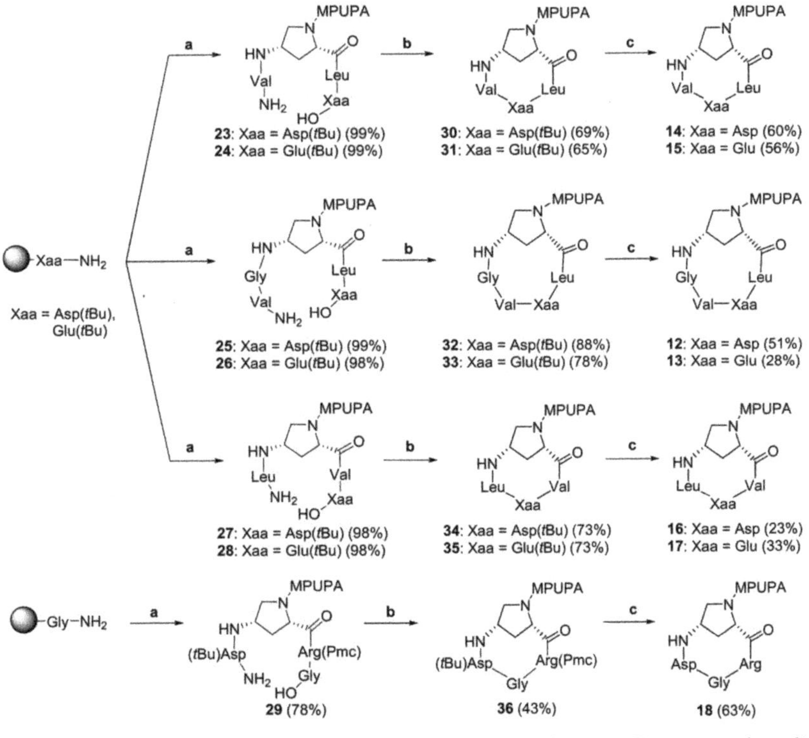

Scheme 2. Synthesis of Amp(MPUPA)-based target compounds **12–18**. Reagents and conditions: (**a**) Fmoc-SPPS strategy. *Coupling*: Fmoc-amino acid, HATU, HOAt, collidine, DMF, rt. *Fmoc cleavage*: 20% piperidine in DMF, rt. *Sequence of addition to obtain compounds*: (**23, 24**) Fmoc-Leu-OH, Fmoc-Amp(MPUPA)-OH (**22**), Fmoc-Val-OH; (**25, 26**) Fmoc-Leu-OH, Fmoc-Amp(MPUPA)-OH (**22**), Fmoc-Gly-OH, Fmoc-Val-OH; (**27, 28**) Fmoc-Val-OH, Fmoc-Amp(MPUPA)-OH (**22**), Fmoc-Leu-OH; (**29**) Fmoc-Arg(Pmc)-OH, Fmoc-Amp(MPUPA)-OH (**22**), Fmoc-Asp(*t*Bu)-OH. *Resin cleavage*: AcOH/TFE/DCM (1:1:3), rt. (**b**) Cyclization: HATU, HOAt, collidine, DCM/DMF (15:1), 1–3 mM, rt. (**c**) Deprotection: TFA/TIS/H₂O (95:2.5:2.5), rt.

The linear peptides **23–29** were then subjected to delicate, in-solution, head-to-tail cyclization. The cyclization reactions were carried out under diluted conditions (1–3 mM) in a solution of dry DCM/DMF solvent mixture, in a 15:1 ratio. The crude cyclized peptides were purified by automated flash chromatography, furnishing the protected cyclized peptides **30–36** with yields ranging from 43% to 88%. Finally, side-chain deprotection of the cyclic peptides was carried out under acidic conditions (TFA/TIS/H$_2$O 95:2.5:2.5). Compounds **12–18** were recovered as TFA salts after RP-HPLC purification, in yields ranging from 23% to 63%, with overall yields ranging from 21% to 44%. Target compounds **12–18** were fully characterized by high-resolution ESI mass spectrometry as well as various NMR techniques.

2.3. Biological Evaluation

To investigate the ability of the newly synthesized cyclopeptidomimetics **12–18** to recognize and bind $\alpha_4\beta_1$ integrin, cell adhesion assays were performed on VCAM-1. The compounds were evaluated for their ability to interfere with $\alpha_4\beta_1$ integrin-mediated cell adhesion by using Jurkat cells, which are known to constitutively express this integrin [46,48–51]. Compound BIO1211 (**1**) was included as a reference antagonist ligand, which is able to significantly reduce Jurkat cell adhesion to VCAM-1. The results of cell adhesion assays are summarized in Table 1.

Table 1. Effects of the new Amp-based cyclopeptides on $\alpha_4\beta_1$ integrin-mediated cell adhesion.

Compound	Structure [a]	EC$_{50}$/IC$_{50}$ (μM) [b]
12	c[Amp(MPUPA)Leu-Asp-Val-Gly]	>100
13	c[Amp(MPUPA)Leu-Glu-Val-Gly]	>100
14	c[Amp(MPUPA)Leu-Asp-Val]	>100
15	c[Amp(MPUPA)Leu-Glu-Val]	>100
16	c[Amp(MPUPA)Val-Asp-Leu]	0.37 ± 0.09 *agonist*
17	c[Amp(MPUPA)Val-Glu-Leu]	>100
18	c[Amp(MPUPA)Arg-Gly-Asp]	>100
1	MPUPA-Leu-Asp-Val-Pro	0.0046 ± 0.0030 [c] *antagonist*

[a] MPUPA = 4-[(N-2-Methylphenyl)ureido]phenylacetyl. [b] Data are presented as EC$_{50}$ for compounds enhancing cell adhesion (*agonists*) and as IC$_{50}$ for compounds reducing cell adhesion (*antagonists*) (μM). Cell adhesion mediated by $\alpha_4\beta_1$ integrin was measured by assaying Jurkat cell adhesion to VCAM-1 (2 μg/mL). Values are the means ± SD, n = 3. [c] Value determined in this assay. For a previously reported value, see ref. [13].

Under the adopted experimental conditions, most of the synthesized compounds were unable to compete with VCAM-1 for the binding to the $\alpha_4\beta_1$ receptor expressed on Jurkat cells and no effect was detected on the impairing or promoting of cell adhesion (anti-adhesive or pro-adhesive effect) at the tested concentrations (ranging from 0.1 nM to 100 μM). The new cyclopeptidomimetic **16**, instead, was able to modulate $\alpha_4\beta_1$ integrin-mediated cell adhesion, with an interesting potency in the submicromolar range; in particular, it behaved as an agonist, as it was able to increase Jurkat cell adhesion to VCAM-1 as compared to the control. More specifically, this compound, which features a constrained cyclotetrapeptide ring containing the retro-sequence Val-Asp-Leu, showed a dose-dependent enhancement in cell adhesion with an EC$_{50}$ of 0.37 μM, and, for this reason, it was referred to as an agonist.

In an attempt to rationalize our results, two additional experiments were envisaged to evaluate the possible competition for VCAM-1 binding site between Amp-based cyclotetrapeptide **16** and BIO1211, which is described as a potent noncovalent antagonist of $\alpha_4\beta_1$/VCAM-1 interaction in both receptor-binding studies and cell adhesion assays [22]. In the first set of experiments, Jurkat cells were pre-incubated with BIO1211 (1 μM) for 30 min and then incubated with compound **16** (100 μM), before being plated in VCAM-1 coated wells. As expected, BIO1211, acting as an antagonist, significantly decreased Jurkat cell adhesion to VCAM-1. Moreover, compound **16** was not able to modify the reduced cell adhesion induced by BIO1211 (Figure 8a). Similarly, when Jurkat cells were pre-incubated with compound **16**, BIO1211 did not modify the increased cell adhesion.

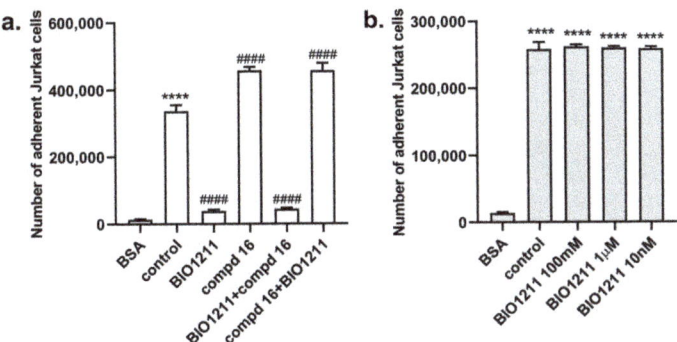

Figure 8. Jurkat cell adhesion to wells coated with VCAM-1 [2 µg/mL; panel (**a**)] or compound **16** [10 µg/mL; panel (**b**)] to evaluate any possible ligand binding competition. (**a**). The antagonist BIO1211 (1 µM) significantly reduced Jurkat cell adhesion to VCAM-1; on the contrary compound **16** (100 µM) behaved as an $\alpha_4\beta_1$ integrin agonist, increasing the adhesion of Jurkat cells to VCAM-1. When Jurkat cells were pre-incubated with BIO1211 and then with compound **16**, the latter was not able to modify the reduction in adhesion induced by BIO1211. Similarly, BIO1211 was not able to revert the effect induced by pre-incubation with compound **16**. (**b**). Even in absence of VCAM-1, compound **16** was able to induce Jurkat cell adhesion; BIO1211 (10 nM–100 mM) was not able to reduce the increment of cell adhesion induced by compound **16**. Control cells were not pre-incubated with any compound. Jurkat cells plated in wells coated with 10 µg/mL bovine serum albumin (BSA) were considered as negative control. Each value is the mean ± SD from four separate experiments carried out in quadruplicate. **** $p < 0.0001$ vs. BSA-coated wells; #### $p < 0.0001$ vs. control (Newman–Keuls test after ANOVA).

In a second set of experiments, the ability of compound **16** to increase cell adhesion was tested in the absence of VCAM-1. Wells were coated by passive adsorption with compound **16** or BSA (both at 10 µg/mL) as a negative control and Jurkat cell adhesion was measured (Figure 8b). Compound **16** produced a significant adhesion of Jurkat cells, even in the presence of different concentrations of BIO1211.

With the present data, a more detailed analysis of the structure–activity relationship within our ligand collection would be unwise.

3. Discussion

Two main aspects emerge from the experimental results given above, which may deserve comment: first, a discrepancy was observed between the computational-driven design and the experimental results, and second, an agonist behavior emerged in one candidate instead of the "expected" antagonist activity of the modeled structures.

Regarding the first point, it has to be underlined that, lacking sound structural details of the $\alpha_4\beta_1$ integrin, a reliable model for the design of potential $\alpha_4\beta_1$ ligands remains elusive, although several molecular modeling studies, computational screenings and 3D models have been reported to date [52–55]. Additionally, the high degree of conformational flexibility featuring the targeted receptor was not considered in this study, and this could have played a decisive role in decreasing the predictability potential of the molecular modeling studies.

The in vitro biological evaluation showed that, among the seven candidates, compound c[Amp(MPUPA)Val-Asp-Leu] (**16**) exhibited a low-micromolar (0.37 µM) agonist activity in Jurkat cell adhesion assay. The fact that the rational design based on antagonist ligands consigned an agonist product must not come as a surprise, as this was testified by notable precedents even in the field of $\alpha_4\beta_1$ ligands [18,24]. It has been demonstrated that even small structural variations in the integrin ligand core can cause the shift from antagonist to agonist behavior [18,49].

Competition experiments involving compound **16** and BIO1211 revealed that the respective agonist and antagonist behavior were not reciprocally modified. These results seem to exclude the competition between compound **16** and BIO1211 for the same binding site, and the agonist activity of **16** could be ascribed to an interaction of this compound in a different region of the receptor. This behavior has already been observed for other ligands of this integrin; for example, for known compound LLP2A (**2**, Figure 1), whose binding site on $\alpha_4\beta_1$ integrin receptor was claimed to be different, and close to (or only partially overlapping) with the binding site of VCAM-1 [24].

Finally, in contrast to RGD-dependent integrins, the binding regions of α_4 integrins (in particular the $\alpha_4\beta_7$ binding site) have been described as long and wide crevices, open at both ends and capable of the lengthwise accommodation of differently shaped binders [42]. This fact could explain that compounds **16** and BIO1211 do not seem to share the same binding region and could provide a reason for the difficulty encountered in the rational design.

4. Materials and Methods

4.1. Docking Studies

4.1.1. Protein Setup

$\alpha_5\beta_1$ (PDB code: 3VI4) [43] and $\alpha_4\beta_7$ (PDB code: 3V4V) [42] crystal structures were used for generation of the $\alpha_4\beta_1$ complex: the α_4 subunit was obtained by the $\alpha_4\beta_7$ complex, while the β_1 subunit was derived from $\alpha_5\beta_1$ receptor. Using the $\alpha_5\beta_1$ integrin complex as a template, the α_4 subunit was aligned with α_5 bound to β_1 ($\alpha_5\beta_1$), then the α_5 subunit was removed, giving a preliminary $\alpha_4\beta_1$ complex. The complex was prepared by using the Protein Preparation Wizard tool of Maestro 9.1 (https://www.schrodinger.com/; accessed on 9 November 2018) minimized by a multi-step protocol in which the harmonic restraints were gradually scaled. The complex obtained was used for the following docking studies.

4.1.2. Ligand Docking Calculations

All docking studies were carried out using the same experimental protocol. The structures of the different antagonists were prepared from the fragment-building tool available in Maestro 9.1 and the geometries were optimized using the force field OPLS-2005 [56]. The docking grid was centered on the Mg^{2+} atom and a grid size of $12 \times 12 \times 12$ Å was used. Docking studies were performed using Glide as software, the SP method and the enhanced sampling method for conformational exploration of the different ligands. The remaining docking parameters were used as default.

4.2. Chemistry

General Information

H-Gly-2-ClTrt resin (loading 0.63 mmol/g), H-Asp(*t*Bu)-2-ClTrt resin (loading 0.74 mmol/g), H-Glu(*t*Bu)-2-ClTrt resin (loading 0.85 mmol/g) were purchased from Novabiochem, (2*S*,4*S*)-Fmoc-4-amino-1-Boc-pyrrolidine-2-carboxylic acid from PolyPeptide and all other reagents from Alfa Aesar, TCI and Sigma-Aldrich. Automated flash column chromatography was carried out with the Biotage Isolera One system using Biotage KP-C18-HS (reverse phase). ESI-mass spectra were recorded on UHPLC/ESI-MS system (ACQUITY Ultra Performance LC; ESI, positive ions, Single Quadrupole analyzer) and are reported in the form of (m/z). HPLC purifications were performed on a Prostar 210 apparatus (Varian, UV detection) equipped with C_{18}-10 μm column (Discovery BIO Wide Pore 10×250 mm or 21.2×250 mm). Routine NMR spectra were recorded on Avance 300 or 400 (Bruker) NMR spectrometers. Chemical shifts (δ) are reported in parts per million (ppm) with CD_2HOD resonance peak set at 3.31 ppm. Multiplicities are indicated as s (singlet), d (doublet), t (triplet), q (quartet), m (multiplet), and b (broad). Coupling constants, *J*, are reported in Hertz. ^{1}H NMR assignments are corroborated by 1D and 2D experiments (gCOSY sequences). Optical rotations were measured using a Perkin–Elmer model 341 polarimeter at ambient temperature using a 100 mm cell with a 1 mL capacity and are given in units of 10^{-1} deg cm^2 g^{-1}. High resolution mass analysis

(ESI) was performed on LTQ ORBITRAP XL Thermo apparatus and are reported in the form of (*m*/*z*). (2*S*,4*S*)-4-*N*-(9-Fluorenylmethoxycarbonyl)aminoproline (**20**) and 4-[[[(2-methylphenyl)amino] carbonyl]amino]phenylacetic acid (MPUPA-OH) (**21**) were prepared according to the literature procedures [22,32].

4.3. Experimental Synthetic Procedures and Characterization Data

4.3.1. (2*S*,4*S*)-1-(MPUPA)-4-(Fmoc)aminoproline [Fmoc-(MPUPA)Amp-OH] (**22**)

To a stirred solution of MPUPA-OH **21** (107 mg, 0.38 mmol, 1.1 equiv), HATU (144 mg, 0.38 mmol, 1.1 equiv) and HOAt (51 mg, 0.38 mmol, 1.1 equiv) in dry DMF (2 mL), 2,4,6-collidine (95 µL, 0.72 mmol, 2.1 equiv) was added and the system left to stir for 30 min, under argon at room temperature. A solution of compound **20** and 2,4,6-collidine (46 µL, 0.34 mmol, 1 equiv) in dry DMF (8 mL) was then added dropwise to the reaction mixture, over 20 min. The reaction reached completion in 40 min and was then treated with an aqueous solution of HCl (0.1 N) to precipitate the product. The crude residue was filtered and purified by reverse phase flash chromatography [H_2O (0.1% TFA)/MeCN: linear gradient 80:20 to 20:80] furnishing compound **22** as withe glassy solid (168.2 mg, yield 79%). ^{1}H NMR (MeOD, 400 MHz): 7.76 (d, *J* = 7.5 Hz, 2H, ArH Fmoc), 7.66–7.56 (m, 3H, ArH Fmoc and MPUPA), 7.42–7.33 (m, 4H, ArH Fmoc and MPUPA), 7.32–7.24 (bm, 2H, ArH Fmoc), 7.23–7.11 (m, 4H, ArH MPUPA), 7.01 (ddd, *J* = 7.4, 7.4, 1.0 Hz, 1H, H6′ MPUPA), 4.40 (t, *J* = 6.8 Hz, 1H, H2 Amp), 4.31 (m, 2H, H1″ Fmoc), 4.24–4.14 (bm, 2H, H2″ Fmoc and H4 Amp), 3.94–3.86 (m, 1H, H5 Amp), 3.66 (d, *J* = 15.9 Hz, 1H, H1′ MPUPA), 3.65 (d, *J* = 15.8 Hz, 1H, H1′ MPUPA), 3.35 (m, 1H, H5 Amp), 2.60–2.49 (bm, 1H, H3 Amp), 2.25 (s, 3H, H8′ MPUPA), 1.95–1.84 (bm, 1H, H3 Amp). ^{13}C NMR (MeOD, 400 MHz): 174.0, 171.3, 156.6, 154.4, 143.8, 141.2, 138.1, 136.5, 130.1, 129.7, 129.2, 128.3, 127.4, 126.8, 126.1, 124.8, 124.0, 123.0, 119.6, 119.0, 66.4, 57.8, 52.1, 51.3, 50.2, 40.4, 34.2, 16.7.

4.3.2. General Procedure for Fmoc-Based SPPS

Linear peptides **23–29** were prepared according to the following general procedure, using the preloaded resins: (i) H-Asp(*t*Bu)-2-ClTrt resin (loading 0.74 mmol/g) (**23, 25, 27**); (ii) H-Glu(*t*Bu)-2-ClTrt resin (loading 0.85 mmol/g) (**24, 26, 28**); (iii) H-Gly-2-ClTrt resin (loading 0.63 mmol/g) (**29**). *Resin swelling*. The desired resin (1 equiv) was swollen in a solid phase reaction vessel with dry DMF (2 mL) under mechanical stirring; after 40 min the solvent was drained and the resin was washed with DCM (2×) and DMF (2×). *Peptide coupling*. A preformed solution of Fmoc-AA-OH (1.5 equiv) in dry DMF (2 mL) was treated with HATU (2 equiv), HOAt (2 equiv) and 2,4,6-collidine (2 equiv), and stirred for 10 min before adding to the resin. The mixture was shaken at room temperature for 5 h. Completion of the reaction was checked by the Kaiser test. The solution was drained and the resin was washed several times with DMF (2×), *i*PrOH, (2×), Et_2O (2×), DCM (2×). The resin was then treated with 20% piperidine in DMF (2 mL) and the mixture was stirred for 30 min (*Fmoc cleavage*). The solution was drained and the resin was washed with DMF (2×), *i*PrOH, (3×), Et_2O (2×), DCM (2×). The couplings of the further amino acids, in the proper sequence, were carried out under the same conditions. *Resin cleavage*. After coupling of the last Fmoc-AA-OH, the resin was treated with 2 mL of the cleavage mixture DCM/TFE/glacial AcOH (3:1:1) and kept under mechanical stirring for 20 min at room temperature. The solution was recovered and the resin was carefully washed with DCM (2×). This protocol was repeated twice. The combined solution was evaporated under reduced pressure affording the desired linear peptide, which was used in the following synthetic step without further purification.

4.3.3. General Procedure for Cyclization Reaction

Protected cyclic peptides **30–36** were prepared according to the following general procedure. A solution of linear peptide (1 equiv) and 2,4,6-collidine (3 equiv) in dry DCM/DMF solvent mixture (15:1 ratio) was prepared. The mixture was stirred under argon at room temperature and added dropwise to a solution of HATU (3 equiv) and

HOAt (3 equiv) in dry DCM/DMF solvent mixture (15:1 ratio). The reaction mixture was degassed by argon/vacuum cycles (3×) and left to stir under argon at room temperature for 5 h. After reaction completion, the solution was concentrated under vacuum. The crude product was purified by RP-flash chromatography [H_2O (0.1% TFA)/MeCN: linear gradient 80:20 to 20:80] furnishing the protected cyclic peptide as a glassy solid.

4.3.4. General Procedure for Deprotection Reaction

Final cyclic peptides **12–18** were prepared according to the following general procedure. The protected cyclic intermediate (1 equiv) was dissolved in TFA/TIS/H_2O (95:2.5:2.5) mixture and stirred at room temperature for 1 h. Then, the solvent was evaporated, and the crude residue was thoroughly washed with Et_2O (4×) and petroleum ether (2×). Preparative RP-HPLC purification was performed [C_{18}-10 μm, 21.2 × 250 mm column, solvent A: H_2O (0.1% TFA) and solvent B: MeCN, flow rate 8.0 mL/min; detection at 254 nm] using the following elution gradient: 0–1 min 10% B, 1–18 min 10–45% B, 18–25 min 45% B.

4.3.5. H-Val-1-(MPUPA)Amp-Leu-Asp(*t*Bu)-OH (**23**)

The synthesis of linear tetrapeptide **23** was performed following the SPPS general procedure, using the preloaded H-Asp(*t*Bu)-2-ClTrt resin (60.0 mg, 0.044 mmol, 1 equiv) and the following Fmoc-amino acids: Fmoc-Leu-OH (23.0 mg, 0.06 mmol, 1.5 equiv), Fmoc-(MPUPA)Amp-OH **22** (40.1 mg, 0.06 mmol), Fmoc-Val-OH (22.0 mg, 0.06 mmol). The linear tetrapeptide **23** (34.0 mg, yield 99%) was obtained as a white glassy solid, and used in the following synthetic step without further purification. MS (ESI$^+$) m/z 780.4 [M + H]$^+$.

4.3.6. H-Val-1-(MPUPA)Amp-Leu-Glu(*t*Bu)-OH (**24**)

The synthesis of linear tetrapeptide **24** was performed following the SPPS general procedure, using the preloaded H-Glu(*t*Bu)-2-ClTrt resin (51.1 mg, 0.043 mmol, 1 equiv) and the following Fmoc-amino acids: Fmoc-Leu-OH (23.0 mg, 0.06 mmol, 1.5 equiv), Fmoc-(MPUPA)Amp-OH **22** (40.2 mg, 0.06 mmol, 1.5 equiv), Fmoc-Val-OH (22.0 mg, 0.06 mmol, 1.5 equiv). The linear tetrapeptide **24** (34.1 mg, yield 99%) was obtained as a white glassy solid, and used in the following synthetic step without further purification. MS (ESI$^+$) m/z 794.4 [M + H]$^+$.

4.3.7. H-Val-Gly-1-(MPUPA)Amp-Leu-Asp(*t*Bu)-OH (**25**)

The synthesis of linear tetrapeptide **25** was performed following the SPPS general procedure, using the preloaded H-Asp(*t*Bu)-2-ClTrt resin (60.1 mg, 0.044 mmol, 1 equiv) and the following Fmoc-amino acids: Fmoc-Leu-OH (23.0 mg, 0.06 mmol, 1.5 equiv), Fmoc-(MPUPA)Amp-OH **22** (40.1 mg, 0.06 mmol, 1.5 equiv), Fmoc-Gly-OH (19.1 mg, 0.06 mmol, 1.5 equiv), Fmoc-Val-OH (22.2 mg, 0.06 mmol, 1.5 equiv). The linear tetrapeptide **25** (36.0 mg, yield 99%) was obtained as a white glassy solid, and used in the following synthetic step without further purification. MS (ESI$^+$) m/z 837.4 [M + H]$^+$.

4.3.8. H-Val-Gly-1-(MPUPA)Amp-Leu-Glu(*t*Bu)-OH (**26**)

The synthesis of linear tetrapeptide **26** was performed following the SPPS general procedure, using the preloaded H-Glu(*t*Bu)-2-ClTrt resin (51.2 mg, 0.043 mmol, 1 equiv) and the following Fmoc-amino acids: Fmoc-Leu-OH (23.1 mg, 0.06 mmol, 1.5 equiv), Fmoc-(MPUPA)Amp-OH **22** (40.0 mg, 0.06 mmol, 1.5 equiv), Fmoc-Gly-OH (19.1 mg, 0.06 mmol, 1.5 equiv), Fmoc-Val-OH (22.2 mg, 0.06 mmol, 1.5 equiv). The linear tetrapeptide **26** (36.0 mg, yield 98%) was obtained as a white glassy solid, and used in the following synthetic step without further purification. MS (ESI$^+$) m/z 794.4 [M + H]$^+$.

4.3.9. H-Leu-1-(MPUPA)Amp-Val-Asp(*t*Bu)-OH (**27**)

The synthesis of linear tetrapeptide **27** was performed following the SPPS general procedure, using the preloaded H-Asp(*t*Bu)-2-ClTrt resin (60.0 mg, 0.044 mmol, 1 equiv) and the following Fmoc-amino acids: Fmoc-Val-OH (22.1 mg, 0.06 mmol, 1.5 equiv), Fmoc-(MPUPA)Amp-OH **22** (40.2 mg, 0.06 mmol, 1.5 equiv), Fmoc-Leu-OH (23.2 mg, 0.06 mmol, 1.5 equiv). The linear tetrapeptide **27** (33.2 mg, yield 98%) was obtained as a white glassy solid, and used in the following synthetic step without further purification. MS (ESI$^+$) m/z 780.4 [M + H]$^+$.

4.3.10. H-Leu-1-(MPUPA)Amp-Val-Glu(*t*Bu)-OH (**28**)

The synthesis of linear tetrapeptide **28** was performed following the SPPS general procedure, using the preloaded H-Glu(*t*Bu)-2-ClTrt resin (51.0 mg, 0.043 mmol, 1 equiv) and the following Fmoc-amino acids: Fmoc-Val-OH (22.1 mg, 0.06 mmol, 1.5 equiv), Fmoc-(MPUPA)Amp-OH **22** (40.0 mg, 0.06 mmol, 1.5 equiv), Fmoc-Leu-OH (23.1 mg, 0.06 mmol, 1.5 equiv). The linear tetrapeptide **28** (34.2 mg, yield 98%) was obtained as a white glassy solid, and used in the following synthetic step without further purification. MS (ESI$^+$) m/z 794.4 [M + H]$^+$.

4.3.11. H-Asp(*t*Bu)-1-(MPUPA)Amp-Arg(Pmc)-Gly-OH (**29**)

The synthesis of linear tetrapeptide **29** was performed following the SPPS general procedure, using the preloaded H-Gly-2-ClTrt resin (70.1 mg, 0.044 mmol, 1 equiv) and the following Fmoc-amino acids: Fmoc-Arg(Pmc)-OH (43.2 mg, 0.06 mmol, 1.5 equiv), Fmoc-(MPUPA)Amp-OH **22** (40.0 mg, 0.06 mmol, 1.5 equiv), Fmoc-Asp(*t*Bu)-OH (27.0 mg, 0.06 mmol, 1.5 equiv). The linear tetrapeptide **29** (35 mg, yield 78%) was obtained as a white glassy solid, and used in the following step without further purification. MS (ESI$^+$) m/z 1047.5 [M + H]$^+$.

4.3.12. c[(MPUPA)Amp-Leu-Asp-Val-Gly] (**12**)

Compound **32** was prepared according to the cyclization general procedure. A solution of linear peptide **25** (16.9 mg, 0.020 mmol) and 2,4,6-collidine (8.0 µL, 0.061 mmol) in dry DCM/DMF solvent mixture (25.0 mL/2.0 mL) was added dropwise to a solution of HATU (23.1 mg, 0.061 mmol) and HOAt (8.3 mg, 0.061 mmol) in dry DCM/DMF solvent mixture (15.0 mL/1.0 mL). After RP-flash chromatography, the protected cyclic peptide **32** (14.6 mg, yield 88%) was obtained as a white solid. MS (ESI$^+$) m/z 818.4 [M + H]$^+$. Compound **12** was prepared according to the deprotection general procedure. The protected cyclic intermediate **32** (14.6 mg, 0.018 mmol) was treated with 0.89 mL of TFA/TIS/H$_2$O (95:2.5:2.5) mixture, then, after RP-flash chromatography, cyclic peptide **12** (7.0 mg, yield 51%) was obtained as a yellowish glassy solid. MS (ESI$^+$) m/z 763.4 [M + H]$^+$. ^{1}H-NMR (MeOD, 400 MHz): δ 7.64 (d, J = 8.0 Hz, 1H, H4′ MPUPA), 7.41 (d, J = 8.5 Hz, 2H, H3′ MPUPA), 7.21 (d, J = 8.6 Hz, 2H, H2′ MPUPA), 7.18 (dd, J = 7.9, 7.9 Hz, 1H, H5′ MPUPA), 7.13 (d, J = 8.5 Hz, 1H, H7′ MPUPA), 7.04 (dd, J = 7.3, 7.3 Hz, 1H, H6′ MPUPA), 4.70 (dd, J = 10.3, 2.6 Hz, 1H, H2 Amp), 4.60 (dd, J = 6.9, 4.8 Hz, 1H, Hα Asp), 4.60 (m, 1H, H4 Amp), 4.13 (m, 1H, Hα Gly), 4.10 (m, 1H, Hα Leu), 4.02 (m, 1H, Hα Val), 3.95 (dd, J = 11.2, 6.6 Hz, 1H, H5 Amp), 3.72 (d, J = 15.3 Hz, 1H, H1′ MPUPA), 3.69 (d, J = 15.6 Hz, 1H, H1′ MPUPA), 3.54 (m, 1H, Hα Gly), 3.48 (m, 1H, H5 Amp), 3.05 (dd, J = 17.1, 7.2 Hz, 1H, Hβ Asp), 2.95 (dd, J = 17.0, 4.9 Hz, 1H, Hβ Asp), 2.62 (m, 1H, H3 Amp), 2.31 (s, 3H, H8′ MPUPA), 2.24 (m, 1H, H3 Amp), 1.71 (m, 3H, Hβ Leu, Hγ Leu), 0.99 (m, 12H, CH$_3$ Leu, CH$_3$ Val). $[\alpha]_D^{25}$: -37 (c 1.0, MeOH).

4.3.13. c[(MPUPA)Amp-Leu-Glu-Val-Gly] (**13**)

Compound **33** was prepared according to the cyclization general procedure. A solution of linear peptide **26** (18.2 mg, 0.03 mmol) and 2,4,6-collidine (8.7 µL, 0.066 mmol) in dry DCM/DMF solvent mixture (26.0 mL/2.0 mL) was added dropwise to a solution of HATU (24.9 mg, 0.07 mmol) and HOAt (8.9 mg, 0.066 mmol) in dry DCM/DMF solvent mixture (16.0 mL/1.0 mL). After RP-flash chromatography, the protected cyclic peptide **33** (14.2 mg, yield 78%) was obtained as a white solid. MS (ESI$^+$) m/z 833.5 [M + H]$^+$. Compound **13** was prepared according to the deprotection general procedure. The protected cyclic intermediate **33** (14.2 mg, 0.017 mmol) was treated with 0.85 mL of TFA/TIS/H$_2$O (95:2.5:2.5) mixture and, after RP-flash chromatography, cyclic peptide **13** (3.7 mg, yield 28%) was obtained as a white-yellow glassy solid. MS (ESI$^+$) m/z 777.4 [M + H]$^+$. ^{1}H-NMR (MeOD, 400 MHz): δ 7.64 (d, J = 8.0 Hz, 1H, H4′ MPUPA), 7.41 (d, J = 8.5 Hz, 2H, H3′ MPUPA), 7.21 (d, J = 8.6 Hz, 2H, H2′ MPUPA), 7.18 (dd, J = 7.9, 7.9 Hz, 1H, H5′ MPUPA), 7.13 (d, J = 8.5 Hz, 1H, H7′ MPUPA), 7.04 (dd, J = 7.3, 7.3 Hz, 1H, H6′ MPUPA), 4.70 (dd, J = 10.6, 3.3 Hz, 1H, H2 Amp), 4.62 (m, 1H, H4 Amp), 4.38 (dd, J = 9.1, 3.8 Hz, 1H, Hα Glu), 4.21 (d, J = 17.0 Hz, 1H, Hα Gly), 4.16 (dd, J = 9.4, 5.9 Hz, 1H, Hα Leu), 4.03 (dd, J = 11.0, 7.3 Hz, 1H, H5 Amp), 3.76 (d, J = 7.0 Hz, 1H, Hα Val), 3.71 (d, J = 3.9 Hz, 2H, H1′ MPUPA), 3.47 (d, J = 17.1 Hz, 1H, Hα Gly), 3.45 (m, 1H, H5 Amp), 2.70 (ddd, J = 14.2, 9.4, 9.4 Hz, 1H, H3 Amp), 2.39 (m, 2H, Hγ Glu), 2.35 (m, 1H, H3 Amp), 2.31 (s, 3H, H8′ MPUPA), 2.26 (m, 1H, Hβ Glu), 2.14 (m, 1H, Hβ Val), 2.05 (m, 1H, Hβ Glu), 1.77 (m, 1H, Hγ Leu), 1.69 (m, 2H, Hβ Leu), 1.02 (m, 6H, 2CH$_3$ Val), 0.97 (m, 6H, 2CH$_3$ Leu). [α]$_D^{25}$: $-$38 (c 1.0, MeOH).

4.3.14. c[(MPUPA)Amp-Leu-Asp-Val] (14)

Compound **30** was prepared according to the cyclization general procedure. A solution of linear peptide **23** (14.4 mg, 0.012 mmol) and 2,4,6-collidine (7.3 µL, 0.055 mmol) in dry DCM/DMF solvent mixture (10.0 mL/1.0 mL) was added dropwise to a solution of HATU (21.1 mg, 0.055 mmol) and HOAt (7.5 mg, 0.055 mmol) in dry DCM/DMF solvent mixture (22.4 mL/1.6 mL). After RP-flash chromatography, the protected cyclic peptide **30** (9.7 mg, yield 69%) was obtained as a white solid. MS (ESI$^+$) m/z 762.4 [M + H]$^+$. Compound **14** was prepared according to the deprotection general procedure. The protected cyclic intermediate **30** (9.7 mg, 0.013 mmol, 1 equiv) was treated with 0.64 mL of TFA/TIS/H$_2$O (95:2.5:2.5) mixture and, after RP-flash chromatography, cyclic peptide **14** (5.3 mg, yield 60%) was obtained as a white glassy solid. MS (ESI$^+$) m/z 706.4 [M + H]$^+$. ^{1}H-NMR (MeOD, 400 MHz): δ 7.64 (dd, J = 8.0, 1.4 Hz, 1H, H4′ MPUPA), 7.40 (m, 2H, H3′ MPUPA), 7.22 (m, 2H, H2′ MPUPA), 7.18 (m, 1H, H5′ MPUPA), 7.13 (d, J = 8.5 Hz, 1H, H7′ MPUPA), 7.04 (ddd, J = 7.5, 7.5, 1.0 Hz, 1H, H6′ MPUPA), 4.67 (m, 1H, H4 Amp), 4.63 (d, J = 9.4 Hz, 1H, H2 Amp), 4.48 (t, J = 7.6 Hz, 1H, Hα Asp), 4.07 (t, J = 8.0 Hz, 1H, Hα Leu), 3.91 (dd, J = 11.9, 6.1 Hz, 1H, H5 Amp), 3.78 (d, J = 6.5 Hz, 1H, Hα Val), 3.70 (s, 2H, H1′ MPUPA), 3.68 (m, 1H, H5 Amp), 3.00 (dd, J = 16.6, 7.5 Hz, 1H, Hβ Asp), 2.90 (dd, J = 16.6, 7.5 Hz, 1H, Hβ Asp), 2.50 (m, 1H, Hβ Val), 2.46 (m, 1H, H3 Amp), 2.31 (s, 3H, H8′ MPUPA), 2.08 (d, J = 14 Hz, 1H, H3 Amp), 1.73 (m, 1H, Hγ Leu), 1.58 (m, 2H, Hβ Leu), 0.96 (m, 12H, 2CH$_3$ Val, 2CH$_3$ Leu). [α]$_D^{25}$: $-$38 (c 1.0, MeOH).

4.3.15. c[(MPUPA)Amp-Leu-Glu-Val] (15)

Compound **31** was prepared according to the cyclization general procedure. A solution of linear peptide **24** (15.4 mg, 0.019 mmol) and 2,4,6-collidine (7.7 µL, 0.058 mmol) in dry DCM/DMF solvent mixture (13.4 mL/0.6 mL) was added dropwise to a solution of HATU (22.1 mg, 0.058 mmol) and HOAt (7.9 mg, 0.058 mmol) in dry DCM/DMF solvent mixture (10.0 mL/1.0 mL). After RP-flash chromatography, the protected cyclic peptide **31** (9.8 mg, yield 65%) was obtained as a white solid. MS (ESI$^+$) m/z 776.4 [M + H]$^+$. Compound **15** was prepared according to the deprotection general procedure. The protected cyclic intermediate **31** (9.8 mg, 0.01 mmol) was treated with 0.55 mL of TFA/TIS/H$_2$O (95:2.5:2.5) mixture and then, after RP-flash chromatography, cyclic peptide **15** (5.1 mg, yield 56%) was obtained as a white glassy solid. MS (ESI$^+$) m/z 720.4 [M + H]$^+$. ^{1}H-NMR (MeOD, 400 MHz): δ 7.64 (d, J = 8.0 Hz, 1H, H4′ MPUPA), 7.41 (d, J = 8.5 Hz, 2H, H3′ MPUPA), 7.21 (d, J = 8.6 Hz, 2H, H2′ MPUPA), 7.18 (dd, J = 7.9, 7.9 Hz, 1H, H5′ MPUPA), 7.13 (d, J = 8.5 Hz, 1H, H7′ MPUPA), 7.04 (dd, J = 7.3, 7.3 Hz, 1H, H6′ MPUPA), 4.71 (m, 1H, H2 Amp), 4.65 (m, 1H, H4 Amp), 4.60 (m, 1H, Hα Glu), 4.03 (m, 2H, Hα Leu, Hα Val), 3.93 (dd, J = 11.4, 6.3 Hz, 1H, H5 Amp), 3.70 (s, 2H, H1′ MPUPA), 3.67 (m, 1H, H5 Amp), 2.47 (m, 1H, H3 Amp), 2.47 (m, 1H, Hβ Val), 2.36 (m, 2H, Hγ Glu), 2.31 (s, 3H, H8′ MPUPA), 2.18 (m, 2H, Hβ Glu), 2.07 (d, J = 14.2 Hz, 1H, H3 Amp), 1.69 (m, 3H, Hβ Leu, Hγ Leu), 0.96 (m, 12H, CH$_3$ Leu, CH$_3$ Val). [α]$_D^{25}$: $-$19 (c 1.0, MeOH).

4.3.16. c[(MPUPA)Amp-Val-Asp-Leu] (16)

Compound **34** was prepared according to the cyclization general procedure. A solution of linear peptide **27** (16.1 mg, 0.021 mmol) and 2,4,6-collidine (8.2 μL, 0.062 mmol) in dry DCM/DMF solvent mixture (10 mL/1.0 mL) was added dropwise to a solution of HATU (23.5 mg, 0.062 mmol) and HOAt (8.4 mg, 0.062 mmol) in dry DCM/DMF solvent mixture (15 mL/1 mL). After RP-flash chromatography, the protected cyclic peptide **34** (11.4 mg, yield 73%) was obtained as a white solid. MS (ESI$^+$) m/z 762.4 [M + H]$^+$. Compound **16** was prepared according to the deprotection general procedure. The protected cyclic intermediate **34** (11.4 mg, 0.015 mmol) was treated with 0.75 mL of TFA/TIS/H$_2$O (95:2.5:2.5) mixture and, after RP-flash chromatography (R_t = 24.3 min), cyclic peptide **16** (4.4 mg, yield 23%) was obtained as a white glassy solid. MS (ESI$^+$) m/z 706.4 [M + H]$^+$. ^{1}H-NMR (400 MHz, MeOD): δ 7.64 (dd, J = 8.0, 1.4 Hz, 1H, H4′ MPUPA), 7.40 (m, 2H, H3′ MPUPA), 7.22 (m, 2H, H2′ MPUPA), 7.18 (m, 1H, H5′ MPUPA), 7.13 (d, J = 8.5 Hz, 1H, H7′ MPUPA), 7.04 (ddd, J = 7.5, 7.5, 1.0 Hz, 1H, H6′ MPUPA), 4.69 (d, J = 9.6 Hz, 1H, H4 Amp), 4.68 (m, 1H, Hα Asp), 4.62 (dd, J = 8.9, 6.5 Hz, 1H, H2 Amp), 3.90 (dd, J = 11.9, 6.4 Hz, 1H, Hβ Asp), 3.82 (m, 1H, Hα Leu), 3.74 (dd, J = 8.5, 4.0 Hz, 1H, Hα Val), 3.69 (s, 2H, H1′ MPUPA), 3.67 (d, J = 11.9 Hz, 1H, Hβ Asp), 3.00 (dd, J = 17.0, 8.8 Hz, 1H, H5 Amp), 2.69 (dd, J = 17.0, 6.4 Hz, 1H, H5 Amp), 2.48 (ddd, J = 14.3, 10.3, 6,7 Hz, 1H, H3 Amp), 2.31 (s, 3H, H8′ MPUPA), 2.31 (m, 1H, Hβ Leu), 2.11 (d, J = 14.2 Hz, 1H, H3 Amp), 1.98 (m, 1H, Hβ Val), 1.87 (m, 1H, Hβ Leu), 1.70 (m, 1H, Hγ Leu), 1.10 (d, J = 6.8 Hz, 3H, CH$_3$ Val), 1.02 (d, J = 6.8 Hz, 3H, CH$_3$ Val), 0.92 (d, J = 6.8 Hz, 6H, 2CH$_3$ Leu). [α]$_D^{25}$: −24 (c 1.0, MeOH).

4.3.17. c[(MPUPA)Amp-Val-Glu-Leu] (17)

Compound **35** was prepared according to the cyclization general procedure. A solution of linear peptide **28** (17.0 mg, 0.02 mmol) and 2,4,6-collidine (8.7 μL, 0.064 mmol) in dry DCM/DMF solvent mixture (10.0 mL/2 mL) was added dropwise to a solution of HATU (24.4 mg, 0.064 mmol) and HOAt (8.7 mg, 0.064 mmol) in dry DCM/DMF solvent mixture (16.0 mL/1 mL). After RP-flash chromatography, the protected cyclic peptide **35** (12.0 mg, yield 73%) was obtained as a white solid. MS (ESI$^+$) m/z 776.4 [M + H]$^+$. Compound **17** was prepared according to the deprotection general procedure. The protected cyclic intermediate **35** (12.0 mg, 0.016 mmol) was treated with 0.77 mL of TFA/TIS/H$_2$O (95:2.5:2.5) mixture and, after RP-flash chromatography, cyclic peptide **17** (13.7 mg, yield 33%) was obtained as a white glassy solid. MS (ESI$^+$) m/z 720.3 [M + H]$^+$. ^{1}H-NMR (MeOD, 400 MHz): δ 7.64 (dd, J = 8.0, 1.4 Hz, 1H, H4′ MPUPA), 7.40 (m, 2H, H3′ MPUPA), 7.22 (m, 2H, H2′ MPUPA), 7.18 (m, 1H, H5′ MPUPA), 7.13 (d, J = 8.5 Hz, 1H, H7′ MPUPA), 7.04 (ddd, J = 7.3, 7.3, 1.1 Hz, 1H, H6′ MPUPA), 4.74 (m, 1H, H4 Amp), 4.68 (d, J = 9.8 Hz, 1H, H2 Amp), 4.31 (t, J = 8.4 Hz, 1H, Hα Glu), 3.93 (dd, J = 11.4, 6.3 Hz, 1H, H5 Amp), 3.85 (m, 1H, Hα Leu), 3.71 (m, 1H, Hα Val), 3.69 (s, 2H, H1′ MPUPA), 3.69 (m, 1H, H5 Amp), 2.48 (m, 1H, H3 Amp), 2.31 (s, 3H, H8′ MPUPA), 2.31 (m, 2H, Hγ Glu), 2.07 (m, 1H, H3 Amp), 2.02 (m, 2H, Hβ Glu), 2.01 (m, 1H, Hβ Val), 1.77 (m, 2H, Hβ Leu), 1.61 (m, 1H, Hγ Leu), 1.10 (m, 6H, 2CH$_3$ Val), 0.93 (m, 6H, 2CH$_3$ Leu). [α]$_D^{25}$: −38 (c 1.0, MeOH).

4.3.18. c[(MPUPA)Amp-Arg-Gly-Asp] (18)

Compound **36** was prepared according to the cyclization general procedure. A solution of linear peptide **29** (10.1 mg, 0.001 mmol) and 2,4,6-collidine (3.8 μL, 0.029 mmol) in dry DCM/DMF solvent mixture (11.2 mL/0.8 mL) was added dropwise to a solution of HATU (10.9 mg, 0.04 mmol) and HOAt (3.9 mg, 0.029 mmol) in dry DCM/DMF solvent mixture (6.2 mL/0.4 mL). After RP-flash chromatography, the protected cyclic peptide **36** (5.0 mg, yield 43%) was obtained as a white solid. MS (ESI$^+$) m/z 1028.5 [M + H]$^+$. Compound **18** was prepared according to the deprotection general procedure. The protected cyclic intermediate **36** (5.0 mg, 0.01 mmol) was treated with 0.22 mL of TFA/TIS/H$_2$O (95:2.5:2.5) mixture and, after RP-flash chromatography, cyclic peptide **18** (3.5 mg, yield 63%) was obtained as a yellowish glassy solid. MS (ESI$^+$) m/z 707.3 [M + H]$^+$. ^{1}H-NMR (MeOD, 400 MHz): δ 7.64 (d, J = 8.0 Hz, 1H, H4′ MPUPA), 7.41 (d, J = 8.5 Hz, 2H, H3′ MPUPA),

7.21 (d, J = 8.6 Hz, 2H, H2′ MPUPA), 7.18 (dd, J = 7.9, 7.9 Hz, 1H, H5′ MPUPA), 7.13 (d, J = 8.5 Hz, 1H, H7′ MPUPA), 7.04 (dd, J = 7.3, 7.3 Hz, 1H, H6′ MPUPA), 4.71 (t, J = 5.6 Hz, 1H, Hα Asp), 4.66 (d, J = 9.5 Hz, 1H, H2 Amp), 4.59 (t, J = 6.0 Hz, 1H, H4 Amp), 4.19 (d, J = 13.8 Hz, 1H, Hα Gly), 4.09 (t, J = 6.8 Hz, 1H, Hα Arg), 3.93 (dd, J = 11.8, 6.4 Hz, 1H, H5 Amp), 3.70 (m, 3H, H1′ MPUPA, H5 Amp), 3.42 (d, J = 13.8 Hz, 1H, Hα Gly), 3.20 (m, 2H, Hδ Arg), 2.81 (d, J = 5.7 Hz, 2H, Hβ Asp), 2.50 (m, 1H, H3 Amp), 2.31 (s, 3H, H8′ MPUPA), 2.18 (d, J = 14.4, 1H, H3 Amp), 1.76 (m, 2H, Hβ Arg), 1.65 (m, 2H, Hγ Arg). $[\alpha]_D^{25}$: −38 (c 1.0, MeOH).

4.4. Biology

4.4.1. Cell Culture

Jurkat E6.1 cells were purchased from American Type Culture Collection (ATCC, Rockville, MD, USA) and were routinely cultured in RPMI-1640 (LifeTechnologies, Milan, Italy) supplemented with 10% FBS (fetal bovine serum, Life Technologies) and 2 mM glutamine. Cells were kept at 37 °C under 5% CO_2 humidified atmosphere. Jurkat cells are a widely used cell model to study potential agonist or antagonist ligands able to modulate integrin-mediated cell adhesion [46,48–51]. Jurkat cells endogenously express $\alpha_4\beta_1$ integrin [13].

4.4.2. Cell Adhesion Assays

The assays were performed as previously described [51]. In brief, black 96 well plates were coated with VCAM-1 (2 µg/mL) overnight at 4 °C; then, non-specific hydrophobic binding sites were blocked with 1% BSA (bovine serum albumin, Sigma-Aldrich) in HBSS (Life Technologies) for 30 min at 37 °C. Jurkat cells were labelled with CellTracker green CMFDA (12.5 µM, Life Technolgies) and pre-incubated with various concentration (10^{-4}–10^{-10} M) of each new cyclopeptidomimetic or with the vehicle (DMSO) for 30 min at 37 °C. Afterwards, Jurkat cells were plated on VCAM-1-coated wells and incubated for 30 min at 37 °C. Then, wells were washed three times with 1% BSA in HBSS and Jurkat cells were lysed with 0.5% Triton X-100 in PBS for 30 min at 4 °C. Green fluorescence (Ex485 nm/Em 535 nm) was measured in an EnSpire Multimode Plate Reader (PerkinElmer, Waltham, MA, USA). Experiments were performed in quadruplicate and repeated at least three times. The number of adherent cells was determined by comparison with a standard curve made with a known concentration of labelled Jurkat cells. Data analysis and IC_{50}/EC_{50} values were calculated using GraphPad Prism 6.0 (GraphPad Software, San Diego, CA, USA).

In another set of experiments, Jurkat cells were plated (500,000 cells/well) in 96-wells plate previously coated by passive absorption with VCAM-1 (2 µg/mL) or with compound **16** (10 µg/mL), the most effective compound under examination, or with BSA (10 µg/mL, as negative control). To investigate any potential ligand binding competition, 30 min before plating cells on VCAM-1-coated wells, BIO1211 (1 µM) was added to the cells pre-incubated with compound **16** (100 µM) or compound **16** was used to treat to the cells pre-incubated with BIO1211. Moreover, Jurkat cells pre-incubated with different concentrations of BIO1211 (100 mM–10 nM) were plated in compound **16**-coated wells. The number of adherent cells was determined as described above.

5. Conclusions

The central role played by $\alpha_4\beta_1$ integrin in inflammatory and autoimmune pathologies and tumor-related diseases has been widely explored, so the search for potent and selective $\alpha_4\beta_1$ integrin binders has been and remains a topic of interest in current biomedical research.

The present study, addressing the synthesis of new potential $\alpha_4\beta_1$ integrin ligands, is rooted in this challenging field of research. Based on some initial computational suggestions, seven new cyclic peptidomimetics, all bearing a common aminoproline core scaffold and an MPUPA hexocyclic motif, were synthesized and structurally characterized. Preliminary in vitro biological evaluation revealed that one of these candidates, compound **16**, featuring the constrained c(Amp-VDL) cyclotetrapeptide structure, showed a moderate ability to enhance Jurkat cell adhesion to VCAM-1, and further biological evidence pointed to the exclusion of competition with the known antagonist BIO1211 for the same receptor binding site.

Further biological investigations will be necessary for a complete characterization of the agonist behavior of our compounds, to assess integrin selectivity and possibly define structural requirements for agonist vs. antagonist activity, with the ultimate intention of contributing to the expanding knowledge in the field of small-molecule integrin ligands.

Author Contributions: A.S. and K.B. equally contributed. Conceptualization and methodology, A.S., F.Z., L.B.; In silico calculations, A.B.; Synthesis and characterization of compounds, K.B., E.P., A.S., C.C.; Biological evaluation, M.B., I.C., F.B., Supervision and data curation, A.S., F.Z., L.B.; Writing, review, editing, K.B., F.Z., L.B. All authors have given approval to the final version of the manuscript.

Funding: This research was funded by University of Parma (BATTISTINI_L_FIL, ZNRFNC_RICERCA_IST).

Institutional Review Board Statement: Not applicable.

Informed Consent Statement: Not applicable.

Data Availability Statement: Authors will release data of docking complex results upon request. Please, contact the corresponding author.

Acknowledgments: Thanks are due to Centro Interdipartimentale Misure "G. Casnati" (University of Parma, Italy) for instrumental facilities. We gratefully acknowledge Professor Gabriele Costantino (Department of Food and Drug—University of Parma) for software facilities employed in the preliminary in silico studies.

Conflicts of Interest: The authors declare no conflict of interest.

Sample Availability: Samples of the final compounds **12–18** are available from the authors.

References

1. Hynes, R.O. Integrins: Bidirectional, allosteric signaling machines. *Cell* **2002**, *110*, 673–687. [CrossRef]
2. Arnaout, M.A.; Goodman, S.L.; Xiong, J.P. Structure and mechanics of integrin-based cell adhesion. *Curr. Opin. Cell Biol.* **2007**, *19*, 495–507. [CrossRef]
3. Cox, D.; Brennan, M.; Moran, N. Integrins as therapeutic targets: Lessons and opportunities. *Nat. Rev. Drug Discov.* **2010**, *9*, 804–820. [CrossRef]
4. Ley, K.; Rivera-Nieves, J.; Sandborn, W.J.; Shattil, S. Integrin-based therapeutics: Biological basis, clinical use and new drugs. *Nat. Rev. Drug Discov.* **2016**, *15*, 173–183. [CrossRef] [PubMed]
5. Hamidi, H.; Pietilä, M.; Ivaska, J. The complexity of integrins in cancer and new scopes for therapeutic targeting. *Br. J. Cancer* **2016**, *115*, 1017–1023. [CrossRef] [PubMed]
6. Abram, C.L.; Lowell, C.A. The ins and outs of leukocyte integrin signaling. *Annu. Rev. Immunol.* **2009**, *27*, 339–362. [CrossRef] [PubMed]
7. Huang, J.; Li, X.; Shi, X.; Zhu, M.; Wang, J.; Huang, S.; Huang, X.; Wang, H.; Li, L.; Deng, H.; et al. Platelet integrin αIIbβ3: Signal transduction, regulation, and its therapeutic targeting. *J. Hematol. Oncol.* **2019**, *12*, 26. [CrossRef]
8. Vanderslice, P.; Woodside, D.G. Integrin antagonists as therapeutics for inflammatory diseases. *Expert Opin. Investig. Drugs* **2006**, *15*, 1235–1255. [CrossRef]
9. Allen, S.J.; Moran, N. Cell adhesion molecules: Therapeutic targets for inhibition of inflammatory states. *Semin. Thromb. Hemost.* **2015**, *41*, 563–571. [CrossRef]
10. Baiula, M.; Spampinato, S.; Gentilucci, L.; Tolomelli, A. Novel ligands targeting α4β1 integrin: Therapeutic applications and perspectives. *Front. Chem.* **2019**, *7*, 489. [CrossRef]
11. Biswas, S.; Bryant, R.V.; Travis, S. Interfering with leukocyte trafficking in Crohn's disease. *Best Pract. Res. Clin. Gastroenterol.* **2019**, *38*, 101617. [CrossRef] [PubMed]
12. Herter, J.; Zarbock, A. Integrin regulation during leukocyte recruitment. *J. Immunol.* **2013**, *190*, 4451–4457. [CrossRef] [PubMed]
13. Dattoli, S.D.; Baiula, M.; de Marco, R.; Bedini, A.; Anselmi, M.; Gentilucci, L.; Spampinato, S. DS-70, a novel and potent α4 integrin antagonist, is an effective treatment for experimental allergic conjunctivitis in guinea pigs. *Br. J. Pharmacol.* **2018**, *175*, 3891–3910. [CrossRef] [PubMed]
14. Murata, M. Inflammation and cancer. *Environ. Health Prev. Med.* **2018**, *23*, 50. [CrossRef] [PubMed]
15. Avraamides, C.J.; Garmy-Susini, B.; Varner, J.A. Integrins in angiogenesis and lymphangiogenesis. *Nat. Rev. Cancer* **2008**, *8*, 604–617. [CrossRef]

16. Kuphal, S.; Bauer, R.; Bosserhoff, A.-K. Integrin signaling in malignant melanoma. *Cancer Metastasis Rev.* **2005**, *24*, 195–222. [CrossRef]
17. Klemke, M.; Weschenfelder, T.; Konstandin, M.H.; Samstag, Y. High affinity interaction of integrin α4β1 (VLA-4) and vascular cell adhesion molecule 1 (VCAM-1) enhances migration of human melanoma cells across activated endothelial cell layers. *J. Cell. Physiol.* **2007**, *212*, 368–374. [CrossRef]
18. Vanderslice, P.; Biediger, R.J.; Woodside, D.G.; Brown, W.S.; Khounlo, S.; Warier, N.D.; Gundlach, C.W.; Caivano, A.R.; Bornmann, W.G.; Maxwell, D.S.; et al. Small molecule agonist of very late antigen-4 (VLA-4) integrin induces progenitor cell adhesion. *J. Biol. Chem.* **2013**, *288*, 19414–19428. [CrossRef]
19. Komoriya, A.; Green, L.; Mervic, M.; Yamada, S.; Yamada, K.; Humphries, M. The minimal essential sequence for a major cell type-specific adhesion site (CS1) within the alternatively spliced type III connecting segment domain of fibronectin is leucine-aspartic acid-valine. *J. Biol. Chem.* **1991**, *266*, 15075–15079. [CrossRef]
20. Wang, J.H.; Pepinsky, R.B.; Stehle, T.; Liu, J.H.; Karpusas, M.; Browning, B.; Osborn, L. The crystal structure of an N-terminal two-domain fragment of vascular cell adhesion molecule 1 (VCAM-1): A cyclic peptide based on the domain 1 C-D loop can inhibit VCAM-1-alpha 4 integrin interaction. *Proc. Natl. Acad. Sci. USA* **1995**, *92*, 5714–5718. [CrossRef]
21. Jackson, D.Y. Alpha 4 Integrin Antagonists. *Curr. Pharm. Des.* **2002**, *8*, 1229–1253. [CrossRef] [PubMed]
22. Lin, K.-C.; Ateeq, H.S.; Hsiung, S.H.; Chong, L.T.; Zimmerman, C.N.; Castro, A.; Lee, W.-C.; Hammond, C.E.; Kalkunte, S.; Chen, L.-L.; et al. Selective, tight-binding inhibitors of integrin α4β1 that inhibit allergic airway responses. *J. Med. Chem.* **1999**, *42*, 920–934. [CrossRef] [PubMed]
23. Lin, K.-C.; Castro, A.C. Very late antigen 4 (VLA4) antagonists as anti-inflammatory agents. *Curr. Opin. Chem. Biol.* **1998**, *2*, 453–457. [CrossRef]
24. Peng, L.; Liu, R.; Marik, J.; Wang, X.; Takada, Y.; Lam, K.S. Combinatorial chemistry identifies high-affinity peptidomimetics against α4β1 integrin for in vivo tumor imaging. *Nat. Chem. Biol.* **2006**, *2*, 381–389. [CrossRef] [PubMed]
25. Walker, D.; Li, Y.; Roxin, Á.; Schaffer, P.; Adam, M.J.; Perrin, D.M. Facile synthesis and 18 F-radiolabeling of α 4 β 1 -specific LLP2A-aryltrifluoroborate peptidomimetic conjugates. *Bioorg. Med. Chem. Lett.* **2016**, *26*, 5126–5131. [CrossRef]
26. Beaino, W.; Anderson, C.J. PET Imaging of very late antigen-4 in melanoma: Comparison of 68Ga- and 64Cu-labeled NODAGA and CB-TE1A1P-LLP2A conjugates. *J. Nucl. Med.* **2014**, *55*, 1856–1863. [CrossRef]
27. Hao, D.; Ma, B.; He, C.; Liu, R.; Farmer, D.; Lam, K.S.; Wang, A. Surface modification of polymeric electrospun scaffolds via a potent and high-affinity integrin α4β1 ligand improved the adhesion, spreading and survival of human chorionic villus-derived mesenchymal stem cells: A new insight for fetal tissue engineering. *J. Mater. Chem. B* **2020**, *8*, 1649–1659. [CrossRef]
28. De Marco, R.; Tolomelli, A.; Juaristi, E.; Gentilucci, L. Integrin ligands with α/β-hybrid peptide structure: Design, bioactivity, and conformational aspects. *Med. Res. Rev.* **2016**, *36*, 389–424. [CrossRef] [PubMed]
29. Vanderslice, P.; Woodside, D.G.; Caivano, A.R.; Decker, E.R.; Munsch, C.L.; Sherwood, S.J.; LeJeune, W.S.; Miyamoto, Y.J.; McIntyre, B.W.; Tilton, R.G.; et al. Potent in vivo suppression of inflammation by selectively targeting the high affinity conformation of integrin α4β1. *Biochem. Biophys. Res. Commun.* **2010**, *400*, 619–624. [CrossRef]
30. Zanardi, F.; Burreddu, P.; Rassu, G.; Auzzas, L.; Battistini, L.; Curti, C.; Sartori, A.; Nicastro, G.; Menchi, G.; Cini, N.; et al. Discovery of subnanomolar arginine-glycine-aspartate-based αVβ3/αVβ5 integrin binders embedding 4-aminoproline residues. *J. Med. Chem.* **2008**, *51*, 1771–1782. [CrossRef] [PubMed]
31. Battistini, L.; Burreddu, P.; Carta, P.; Rassu, G.; Auzzas, L.; Curti, C.; Zanardi, F.; Manzoni, L.; Araldi, E.M.V.; Scolastico, C.; et al. 4-Aminoproline-based arginine-glycine-aspartate integrin binders with exposed ligation points: Practical in-solution synthesis, conjugation and binding affinity evaluation. *Org. Biomol. Chem.* **2009**, *7*, 4924. [CrossRef] [PubMed]
32. Bugatti, K.; Bruno, A.; Arosio, D.; Sartori, A.; Curti, C.; Augustijn, L.; Zanardi, F.; Battistini, L. Shifting towards α V β 6 integrin ligands using novel aminoproline-based cyclic peptidomimetics. *Chem. Eur. J.* **2020**, *26*, 13468–13475. [CrossRef] [PubMed]
33. Maggi, V.; Bianchini, F.; Portioli, E.; Peppicelli, S.; Lulli, M.; Bani, D.; Del Sole, R.; Zanardi, F.; Sartori, A.; Fiammengo, R. Gold nanoparticles functionalized with RGD-semipeptides: A simple yet highly effective targeting system for αVβ3 integrins. *Chem. Eur. J.* **2018**, *24*, 12093–12100. [CrossRef] [PubMed]
34. Pilkington-Miksa, M.; Arosio, D.; Battistini, L.; Belvisi, L.; De Matteo, M.; Vasile, F.; Burreddu, P.; Carta, P.; Rassu, G.; Perego, P.; et al. Design, synthesis, and biological evaluation of novel cRGD–paclitaxel conjugates for integrin-assisted drug delivery. *Bioconjug. Chem.* **2012**, *23*, 1610–1622. [CrossRef]
35. Battistini, L.; Burreddu, P.; Sartori, A.; Arosio, D.; Manzoni, L.; Paduano, L.; D'Errico, G.; Sala, R.; Reia, L.; Bonomini, S.; et al. Enhancement of the uptake and cytotoxic activity of doxorubicin in cancer cells by novel cRGD-semipeptide-anchoring liposomes. *Mol. Pharm.* **2014**, *11*, 2280–2293. [CrossRef] [PubMed]
36. Sartori, A.; Bianchini, F.; Migliari, S.; Burreddu, P.; Curti, C.; Vacondio, F.; Arosio, D.; Ruffini, L.; Rassu, G.; Calorini, L.; et al. Synthesis and preclinical evaluation of a novel, selective 111In-labelled aminoproline-RGD-peptide for non-invasive melanoma tumor imaging. *MedChemComm* **2015**, *6*, 2175–2183. [CrossRef]
37. Sartori, A.; Portioli, E.; Battistini, L.; Calorini, L.; Pupi, A.; Vacondio, F.; Arosio, D.; Bianchini, F.; Zanardi, F. Synthesis of novel c(AmpRGD)–sunitinib dual conjugates as molecular tools targeting the αvβ3 integrin/VEGFR2 couple and impairing tumor-associated angiogenesis. *J. Med. Chem.* **2017**, *60*, 248–262. [CrossRef]

38. Bianchini, F.; Portioli, E.; Ferlenghi, F.; Vacondio, F.; Andreucci, E.; Biagioni, A.; Ruzzolini, J.; Peppicelli, S.; Lulli, M.; Calorini, L.; et al. Cell-targeted c(AmpRGD)-sunitinib molecular conjugates impair tumor growth of melanoma. *Cancer Lett.* **2019**, *446*, 25–37. [CrossRef]
39. Sartori, A.; Corno, C.; De Cesare, M.; Scanziani, E.; Minoli, L.; Battistini, L.; Zanardi, F.; Perego, P. Efficacy of a selective binder of αVβ3 Integrin linked to the tyrosine kinase inhibitor sunitinib in ovarian carcinoma preclinical models. *Cancers* **2019**, *11*, 531. [CrossRef]
40. Bianchini, F.; De Santis, A.; Portioli, E.; Krauss, I.R.; Battistini, L.; Curti, C.; Peppicelli, S.; Calorini, L.; D'Errico, G.; Zanardi, F.; et al. Integrin-targeted AmpRGD sunitinib liposomes as integrated antiangiogenic tools. *Nanomed. Nanotechnol. Biol. Med.* **2019**, *18*, 135–145. [CrossRef]
41. Capuano, A.; Fogolari, F.; Bucciotti, F.; Spessotto, P.; Nicolosi, P.A.; Mucignat, M.T.; Cervi, M.; Esposito, G.; Colombatti, A.; Doliana, R. The α4β1/EMILIN1 interaction discloses a novel and unique integrin-ligand type of engagement. *Matrix Biol.* **2018**, *66*, 50–66. [CrossRef]
42. Yu, Y.; Zhu, J.; Mi, L.-Z.; Walz, T.; Sun, H.; Chen, J.; Springer, T.A. Structural specializations of α4β7, an integrin that mediates rolling adhesion. *J. Cell Biol.* **2012**, *196*, 131–146. [CrossRef]
43. Nagae, M.; Re, S.; Mihara, E.; Nogi, T.; Sugita, Y.; Takagi, J. Crystal structure of α5β1 integrin ectodomain: Atomic details of the fibronectin receptor. *J. Cell Biol.* **2012**, *197*, 131–140. [CrossRef] [PubMed]
44. Jackson, D.Y.; Quan, C.; Artis, D.R.; Rawson, T.; Blackburn, B.; Struble, M.; Fitzgerald, G.; Chan, K.; Mullins, S.; Burnier, J.P.; et al. Potent α4β1 peptide antagonists as potential anti-inflammatory agents. *J. Med. Chem.* **1997**, *40*, 3359–3368. [CrossRef] [PubMed]
45. Fotouhi, N.; Joshi, P.; Tilley, J.W.; Rowan, K.; Schwinge, V.; Wolitzky, B. Cyclic thioether peptide mimetics as VCAM–VLA-4 antagonists. *Bioorg. Med. Chem. Lett.* **2000**, *10*, 1167–1169. [CrossRef]
46. Tolomelli, A.; Baiula, M.; Viola, A.; Ferrazzano, L.; Gentilucci, L.; Dattoli, S.D.; Spampinato, S.; Juaristi, E.; Escudero, M. Dehydro-β-proline Containing α4β1 Integrin Antagonists: Stereochemical recognition in ligand–receptor interplay. *ACS Med. Chem. Lett.* **2015**, *6*, 701–706. [CrossRef] [PubMed]
47. Irie, A.; Kamata, T.; Puzon-McLaughlin, W.; Takada, Y. Critical amino acid residues for ligand binding are clustered in a predicted β-turn of the third N-terminal repeat in the integrin alpha 4 and alpha 5 subunits. *EMBO J.* **1995**, *14*, 5550–5556. [CrossRef] [PubMed]
48. Qasem, A.R.; Bucolo, C.; Baiula, M.; Spartà, A.; Govoni, P.; Bedini, A.; Fascì, D.; Spampinato, S. Contribution of α4β1 integrin to the antiallergic effect of levocabastine. *Biochem. Pharmacol.* **2008**, *76*, 751–762. [CrossRef]
49. Baiula, M.; Galletti, P.; Martelli, G.; Soldati, R.; Belvisi, L.; Civera, M.; Dattoli, S.D.; Spampinato, S.M.; Giacomini, D. New β-lactam derivatives modulate cell adhesion and signaling mediated by RGD-binding and leukocyte integrins. *J. Med. Chem.* **2016**, *59*, 9221–9742. [CrossRef]
50. De Marco, R.; Greco, A.; Calonghi, N.; Dattoli, S.D.; Baiula, M.; Spampinato, S.; Picchetti, P.; De Cola, L.; Anselmi, M.; Cipriani, F.; et al. Selective detection of α4β1 integrin (VLA-4)-expressing cells using peptide-functionalized nanostructured materials mimicking endothelial surfaces adjacent to inflammatory sites. *Pept. Sci.* **2018**, *110*, e23081. [CrossRef]
51. Martelli, G.; Baiula, M.; Caligiana, A.; Galletti, P.; Gentilucci, L.; Artali, R.; Spampinato, S.M.; Giacomini, D. Could dissecting the molecular framework of β-lactam integrin ligands enhance selectivity? *J. Med. Chem.* **2019**, *62*, 10156–10166. [CrossRef] [PubMed]
52. You, T.J.; Maxwell, D.S.; Kogan, T.P.; Chen, Q.; Li, J.; Kassir, J.; Holland, G.W.; Dixon, R.A. A 3D Structure model of integrin α4β1 complex: I. construction of a homology model of β1 and ligand binding analysis. *Biophys. J.* **2002**, *82*, 447–457. [CrossRef]
53. Singh, J.; Abraham, W.M.; Adams, S.P.; Van Vlijmen, H.; Liao, Y.; Lee, W.C.; Cornebise, M.; Harris, M.; Shu, I.H.; Gill, A.; et al. Identification of potent and novel α4β1 antagonists using in silico screening. *J. Med. Chem.* **2002**, *45*, 2988–2993. [CrossRef]
54. Thangapandian, S.; John, S.; Sakkiah, S.D.; Lee, K.W. Discovery of potential integrin VLA-4 antagonists using pharmacophore modeling, virtual screening and molecular docking studies. *Chem. Biol. Drug Des.* **2011**, *78*, 289–300. [CrossRef]
55. Hutt, O.E.; Saubern, S.; Winkler, D.A. Modeling the molecular basis for α4β1 integrin antagonism. *Bioorg. Med. Chem.* **2011**, *19*, 5903–5911. [CrossRef]
56. Jorgensen, W.L.; Schyman, P. Treatment of Halogen Bonding in the OPLS-AA Force Field: Application to Potent Anti-HIV Agents. *J. Chem. Theory Comput.* **2012**, *8*, 3895–3901. [CrossRef]

Article

Hybrid Multivalent Jack Bean α-Mannosidase Inhibitors: The First Example of Gold Nanoparticles Decorated with Deoxynojirimycin Inhitopes

Costanza Vanni [1], Anne Bodlenner [2,*], Marco Marradi [1], Jérémy P. Schneider [2], Maria de los Angeles Ramirez [3], Sergio Moya [3], Andrea Goti [1,4], Francesca Cardona [1,4], Philippe Compain [2] and Camilla Matassini [1,4,*]

1. Dipartimento di Chimica "Ugo Schiff", Università di Firenze, Via della Lastruccia 3-13, 50019 Sesto Fiorentino, Italy; costanza.vanni@unifi.it (C.V.); marco.marradi@unifi.it (M.M.); andrea.goti@unifi.it (A.G.); francesca.cardona@unifi.it (F.C.)
2. Laboratoire d'Innovation Moléculaire et Applications (LIMA), University of Strasbourg | University of Haute-Alsace | CNRS (UMR 7042), Equipe de Synthèse Organique et Molécules Bioactives (SYBIO), ECPM, 25 Rue Becquerel, 67000 Strasbourg, France; jeremy.schneider@live.fr (J.P.S.); philippe.compain@unistra.fr (P.C.)
3. Soft Matter Nanotechnology Lab, CIC biomaGUNE, Basque Research and Technology Alliance (BRTA), Paseo Miramón 182, 20014 Donostia-San Sebastián, Gipuzkoa, Spain; aangie.ramirez@gmail.com (M.d.l.A.R.); smoya@cicbiomagune.es (S.M.)
4. Associated with LENS, Via N. Carrara 1, 50019 Sesto Fiorentino, Italy
* Correspondence: annebod@unistra.fr (A.B.); camilla.matassini@unifi.it (C.M.); Tel.: +33-36-8852684 (A.B.); +39-055-457-3536 (C.M.)

Citation: Vanni, C.; Bodlenner, A.; Marradi, M.; Schneider, J.P.; Ramirez, M.d.l.A.; Moya, S.; Goti, A.; Cardona, F.; Compain, P.; Matassini, C. Hybrid Multivalent Jack Bean α-Mannosidase Inhibitors: The First Example of Gold Nanoparticles Decorated with Deoxynojirimycin Inhitopes. *Molecules* **2021**, *26*, 5864. https://doi.org/10.3390/molecules26195864

Academic Editor: Keykavous Parang

Received: 2 September 2021
Accepted: 23 September 2021
Published: 27 September 2021

Publisher's Note: MDPI stays neutral with regard to jurisdictional claims in published maps and institutional affiliations.

Copyright: © 2021 by the authors. Licensee MDPI, Basel, Switzerland. This article is an open access article distributed under the terms and conditions of the Creative Commons Attribution (CC BY) license (https://creativecommons.org/licenses/by/4.0/).

Abstract: Among carbohydrate-processing enzymes, Jack bean α-mannosidase (JBα-man) is the glycosidase with the best responsiveness to the multivalent presentation of iminosugar inhitopes. We report, in this work, the preparation of water dispersible gold nanoparticles simultaneously coated with the iminosugar deoxynojirimycin (DNJ) inhitope and simple monosaccharides (β-D-gluco- or α-D-mannosides). The display of DNJ at the gold surface has been modulated (i) by using an amphiphilic linker longer than the aliphatic chain used for the monosaccharides and (ii) by presenting the inhitope, not only in monomeric form, but also in a trimeric fashion through combination of a dendron approach with glyconanotechnology. The latter strategy resulted in a strong enhancement of the inhibitory activity towards JBα-man, with a K_i in the nanomolar range (K_i = 84 nM), i.e., more than three orders of magnitude higher than the monovalent reference compound.

Keywords: multivalency; gold nanoparticles; iminosugars; enzyme inhibition; Jack bean α-mannosidase

1. Introduction

More than ten years have passed since the first example of a trivalent deoxynojirimycin (DNJ) derivative, that displayed a small, but quantifiable, inhibitory multivalent effect for Jack bean α-mannosidase (JBα-man), was reported [1]. Until then, the multivalency concept was considered an exclusive prerogative of lectin-carbohydrate interactions, and its application to carbohydrate-processing enzymes was almost completely unexplored, being considered extremely challenging, from a practical standpoint, and theoretically arguable. The refutation of such speculations was apparent in 2010, when a fullerene-based 12-valent DNJ compound showed binding enhancements towards JBα-man up to three orders of magnitude over the monovalent counterpart (inhibition constant K_i = 0.15 µM vs. K_i = 188 µM) [2].

After that, a plethora of different scaffolds have been employed for the multimerization of DNJ, ranging from small polyols, β-cyclodextrins, porphyrins, calixarenes, and cyclopeptoids to more complex micellar self-assembled glycopeptides and polymeric dextrans [3–9].

The largest binding enhancement ever reported for an enzyme inhibitor, so far, was obtained with a cyclopeptoid core, decorated with 36 DNJ units [10], which showed a nanomolar inhibition value (K_i = 0.0011 µM). More importantly, the outstanding multivalent effect observed with this 36-valent cluster has been fully rationalized thanks to the recent achievement of the first high resolution crystal structure of its complex with JBα-man [11].

JBα-man enzyme (220 kDa) is a homodimer (LH)$_2$ bearing two active sites. The crystal structure revealed the formation of a 2:1 JBα-man:inhibitor complex, in which four DNJ inhitopes bind the four active sites of two homodimers [11]. The reported X-ray data of the JBα-man:inhibitor complex did not show any binding to secondary binding sites. Conversely, the observed multivalent effect was rationalized by invoking a chelate binding mode [11]. Through TEM studies, a different homodimer aggregation featuring an S-shape arrangement was proposed for JBα-man in the presence of a tetravalent pyrrolidine iminosugar 1,4-dideoxy-1,4-imino-D-arabinitol (DAB-1) inhibitor [12]. More generally, these data revealed that the binding modes of JBα-man with iminosugar-based multivalent inhibitors strongly depend not only on the bioactive inhitope but also on the size and shape of the prepared multivalent architectures. Surprisingly, glyco-coated multivalent clusters also displayed some inhibition power on glycosidases instead of being hydrolyzed, as shown by the 12-valent mannosylated fullerene (K_i of 320 µM for JBα-man) [13], glyconanodiamonds (K_i between 222 and 517 µM against JBα-man) [14], and perglucosylated or permannosylated cyclodextrins (IC_{50} of 32–132 µM against *Saccharomyces cerevisiae* α-glucosidase) [15] probably arising from a lectin-like behavior in those clusters.

In this context, we envisaged that glyco-gold nanoparticles (AuGNPs), decorated with different loading and spatial presentation of DNJ inhitopes, could represent a versatile and useful molecular tool for further mechanistic studies.

AuGNPs, as prepared in our laboratory, are water-dispersible and biocompatible gold nanoparticles coated with a 3D polyvalent carbohydrate shell, having a globular shape, chemically defined composition, and exceptionally small core size (about 2 nm) [16]. It is possible to simultaneously attach more than one kind of ligand onto the gold core (multifunctionality) and to modulate the ligand presentation on the metal surface in order to obtain multifunctional materials for application in nanomedicine (targeting, drug delivery, pathway inhibition, etc.) [17].

Despite the large number of different scaffolds employed for the DNJ multimerization, only three examples of nanoparticles (two inorganic [18,19] and one peptide-based [20]), decorated with this iminosugar, have been reported for different applications, to the best of our knowledge.

Some of us recently found that iminosugar-decorated AuGNPs retain inhibitory activity towards commercially available glycosidases [21]. We also reported that AuGNPs, decorated with DAB-1 iminosugar, were able to efficiently inhibit a therapeutically relevant enzyme [22]. Therein, a clickable iminosugar dendron approach [23] was combined with glyco-gold nanotechnology in order to obtain multivalent DAB-1 nanosystems with a denser iminosugar shell.

In the present work, we report our results, obtained applying this methodology to the preparation and characterization of AuGNPs decorated with simple monosaccharides, heterovalent DNJ-based ligands, either linear or dendritic (Figure 1), and their biological evaluation towards JBα-man.

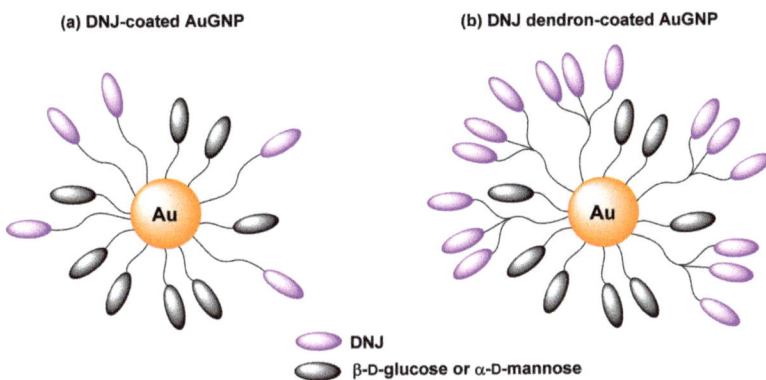

Figure 1. (a) Multivalent presentation of DNJ, using glyco-gold nanoparticles as scaffold and (b) the combination of glyco-gold nanotechnology with the iminosugar dendron approach.

2. Results and Discussion

2.1. Preparation of Ligands and Gold Nanoparticles

The gold nanoparticles (**1–5**) prepared in this work (Figure 2) are constituted by an approximately 2 nm gold core coated with two components: simple monosaccharides (D-glucose or D-mannose) and DNJ derivatives. We played on the nature of the inner monosaccharide and varied the density of DNJ heads on the periphery of the nanoparticles. Nanoparticles **6** and **7**, in Figure 2, bearing 100% monosaccharides with a short and linear five carbons (C_5) aliphatic linker, were prepared as control systems. For AuGNPs **1–5**, the carbohydrate moieties guarantee their water dispersibility, while the water soluble and flexible amphiphilic linker was designed to protrude the DNJ heads from the glyco-coated nanoparticle core. In addition, the N-C_6 aliphatic linker of the active component DNJ is taking part in the interaction with the catalytic site entry [11,24]. This strategy of ligand presentation was based on our previous experience with similar constructs [21,22,25].

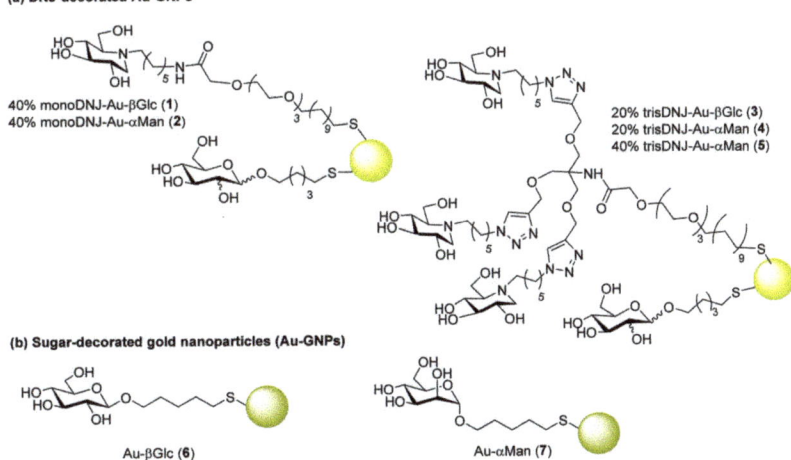

Figure 2. (a) Schematic representation of the multivalent glyco-gold nanoparticles coated with the iminosugar DNJ derivatives and monosaccharides (β-D-glucose or α-D-mannose derivatives) prepared in this work. (b) Nanoparticles coated with the monosaccharides (control systems).

In particular, AuGNPs **6** and **7** were prepared employing, as exclusive components, 5-mercaptopent-1-yl β-D-glucopyranoside (βGlcC$_5$SH, **13**) or 5-mercaptopent-1-yl α-D-mannopyranoside (αManC$_5$SH, **14**), following well established protocols [25]. AuGNPs **1** and **2** were prepared by employing the monovalent DNJ derivative **12** as the active component and **13** or **14**, respectively, as the inner component (Scheme 1).

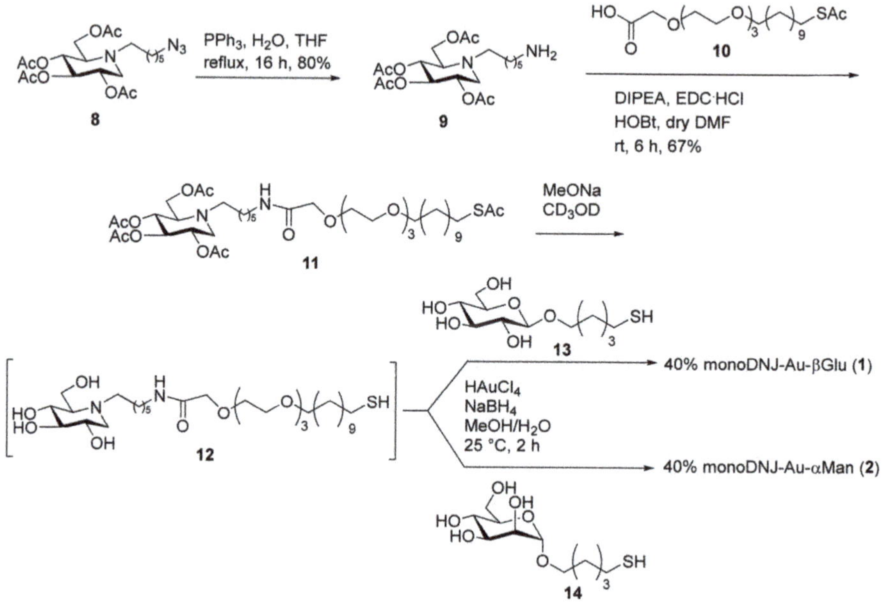

Scheme 1. Synthesis of the glyco-gold nanoparticles **1** and **2** decorated with the DNJ monovalent ligand **12**.

The preparation of the monovalent DNJ derivative **12** is described in Scheme 1. The peracetylated azido-functionalized DNJ **8** [26] was reduced to the corresponding amino derivative **9** through a standard Staudinger reaction (PPh$_3$, H$_2$O, THF, reflux for 16 h) in 80% yield. The same reaction performed with polymer-bound PPh$_3$ [22,27], and refluxing in THF/H$_2$O, required longer times and provided **9** in lower yields (48%).

The aminoalkyl DNJ **9** was then coupled to the thiol-protected carboxylic acid **10**, prepared, as reported in the literature [21], using 1-ethyl-3-(3-dimethylaminopropyl) carbodiimide (EDC) and 1-hydroxybenzotriazole (HOBt) in dry DMF. The amide **11** was obtained in 67% yield after purification by flash column chromatography (FCC). Removal of the acetyl groups of **11** was performed under Zemplén conditions (MeONa) but using CD$_3$OD, instead of MeOH, as solvent to follow the reaction via ^{1}H NMR. Once compound **12** was formed, it was directly used for the next step, avoiding its oxidation to disulfide. AuGNPs **1** and **2** were prepared by mixing **12** with βGlcC$_5$SH (**13**) or αManC$_5$SH (**14**), respectively, in the appropriate ratio (40/60) and then adding an Au(III) solution (the total amount of thiols being 3 equiv. with respect to gold) and the reducing agent NaBH$_4$ in excess (Scheme 1). The proportion of the ligands on the gold surface was evaluated by integrating diagnostic signals in the ^{1}H NMR spectrum of the initial mixture and in the ^{1}H NMR spectrum of the supernatant after AuGNPs formation (see Supplementary Materials for ^{1}H NMR spectra). The 40/60 active/inner component ratio allowed to preserve nanoparticle dispersibility in water [22]. After shaking for 2 h at room temperature, the supernatant was removed, and the nanoparticles were washed with methanol. The residue was dispersed in milliQ water, purified by dialysis, and freeze-dried. These nanoparticles were re-dispersed in water after the freeze-drying process, and no flocculation was observed even after several months. Full characterization with different techniques (see below) was carried out before enzymatic evaluation.

To further increase the DNJ loading onto AuGNPs, we combined this strategy with the dendron approach by preparing the trivalent DNJ derivative **18** (Scheme 2). The peracetylated azido-functionalized DNJ **8** was first reacted with the trialkyne **15** [28] to afford the dendron **16** in 85% yield through copper (I) catalyzed azide alkyne cycloaddition (CuAAC) reaction [29–31]. Subsequently, introduction of the linker was achieved by coupling the carboxylic acid **10** with **16** in the presence of 3-(diethoxyphosphoryloxy)-1,2,3-benzotriazin-4(3H)-one (DEPBT) as coupling agent affording the final ligand **17** in an excellent 94% yield. In analogy with the incorporation of the monovalent ligand **12** onto AuGNPs, trivalent thiol **18** was generated in situ, and its density on the nanoparticles was modulated with βGlcC5SH (**13**) (to yield AuGNPs **3**) or αManC5SH (**14**) (to yield AuGNPs **4** and **5**), as described above. The mannose-based AuGNPs **4** and **5** shared the same inner component but in different percentages (80% in **4**, 60% in **5**).

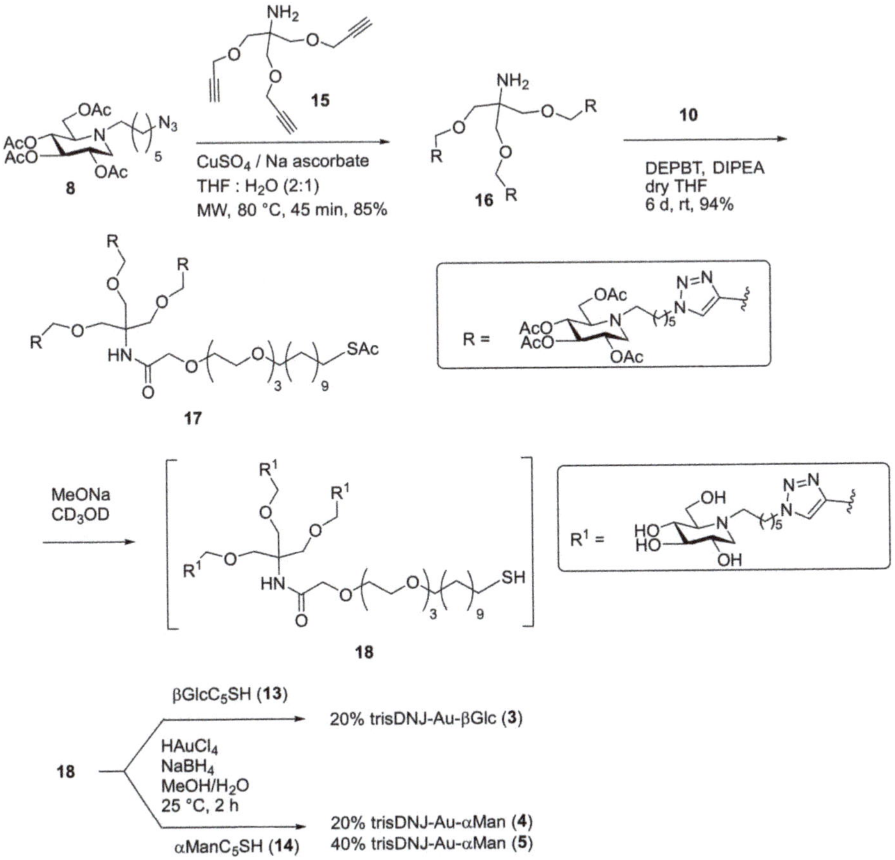

Scheme 2. Synthesis of the trivalent ligand **18** and preparation of AuGNPs **3–5**.

AuGNPs **1–5** were subjected to quantitative ^{1}H NMR (qNMR) analysis in D_2O with 3-(trimethylsilyl)propionic-2,2,3,3-d_4 acid (TSP-d_4) as an internal standard, following established protocols [21,32,33]. As an example, Figure 3a shows the qNMR for AuGNP **3**. In this way, it was possible to quantify the DNJ amount on the nanoparticle by integration of a diagnostic signal (see Supplementary Materials and Table S1) with respect to the internal standard. An average gold core size in 1.6–2.1 nm range was determined by TEM imaging for the AuGNPs (Figure 3c,d) [25,34]. This size was also confirmed by UV-Vis spectra, which did not show an absorption maximum at around 520 nm, typical of gold nanoparticles with a bigger core diameter [35] (Figure 3b).

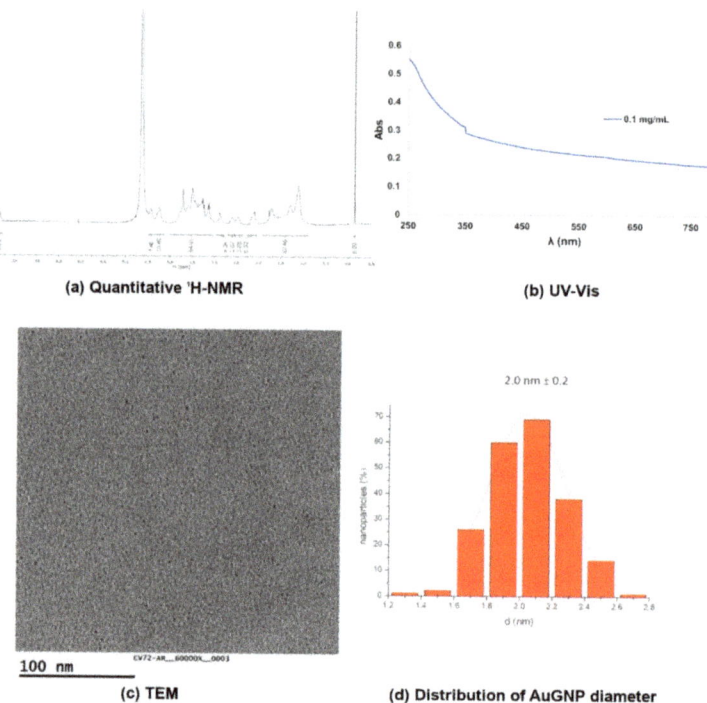

Figure 3. Characterization of AuGNP **3** (20% trisDNJ-Au-βGlc): (**a**) quantitative ^{1}H NMR (D$_2$O) spectrum (400 MHz); (**b**) UV/Vis spectrum; (**c**) TEM micrograph in H$_2$O; (**d**) size-distribution histogram, obtained by measuring 300 nanoparticles (average diameter 2.0 ± 0.2 nm).

2.2. Biological Evaluation

The DNJ-coated AuGNPs **1–5** were evaluated against Jack bean α-mannosidase (JBα-man), the glycosidase which showed the largest response to multivalent inhibitor presentation to date [3–9]. The inhibition mode and constants (K_i) were obtained from Dixon or Lineweaver Burk plots for a scale of inhibitor concentrations reflecting the DNJ concentration directly. The inhibition potency, relative to the corresponding monovalent reference **19** [36] (Figure 4), directly gives the relative potency per inhitope (*rpn*) (Table 1).

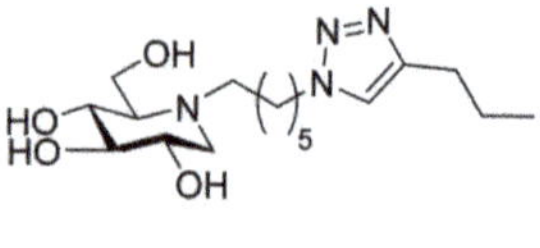

19
(K_i = 322 μM)[35]

Figure 4. DNJ derivative chosen as the monovalent reference to evaluate the relative potency of the AuGNPs synthesized in this work.

Table 1. Inhibitory activity of AuGNPs 1–7 towards Jack bean α-mannosidase (JBα-man).

AuGNP	DNJ Conc. (µM) in 2 mg mL^{-1} AuGNP [1]	K_i (µM) [2]	rpn [3]
1	413	16 ± 2	20
2	467	8 ± 2	40
3	567	0.198 ± 0.060	1626
4	450	0.175 ± 0.171	1840
5	503	0.084 ± 0.066	3833
6	0	n.i. [4]	-
7	0	n.i. [4]	-

[1] Estimated by qNMR. [2] The inhibition constant unit directly reflects the concentration of DNJ and not the concentration of nanoparticles. [3] Relative potency per DNJ unit, with respect to the monovalent reference 19; rpn = K_i (monovalent reference 19)/K_i (DNJ in AuGNP 1–5). [4] n.i. = no inhibition at 0.5 mg mL^{-1}.

The whole set of newly prepared heterovalent AuGNPs 1–5 showed good to excellent inhibitory activity towards JBα-man, with K_i values in the low micromolar range to the nanomolar one. All behaved as competitive inhibitors (see Supplementary Materials). Conversely, β-D-gluco- and α-D-mannoside decorated AuGNPs 6 and 7 did not inhibit the enzyme at 0.5 mg mL^{-1} (This concentration was a good compromise to allow spectrophotometric measurements due to the brown color of AuGNP solutions. The 12-valent mannosylated fullerene from Vincent and Nierengarten et al. (see ref [13]) was active at the hundreds of µM). However, trimerization of DNJ at the gold surface had a major impact: it strongly enhances the affinity towards JBα-man (3 vs. 1, 4 vs. 2, Table 1), with the affinity enhancement per inhitope increasing up to 1840 for AuGNP 4. A further enhancement of the inhibitory properties was reached upon increasing the loading of the trivalent DNJ derivative 18 onto the gold surface from 20% to 40%, with a K_i at 84 nM and an rpn value of 3833 observed for AuGNP 5 (Table 1).

Noticeably, the dendron strategy is a double-edged sword. On the one hand, the intrinsic hindrance of the iminosugar tripod seems to prevent loading as high as linear arms (It is worth noting that for a same percentage loading of D-mannose of 60%, the DNJ concentration on AuGNP 5 is not 3 times higher than that of AuGNP 2, suggesting that the total loading on the particle is lower than for AuGNP 2. However, the ratio between iminosugar DNJ derivatives and monosaccharides still reflects the composition of the preparation mixture, as attested by ^{1}H-NMR spectra recorded before and after (supernatant) the formation of the AuGNPs (see Supplementary Materials)). On the second hand, although the DNJ concentration is almost identical (503 vs. 467 for AuGNP 2 and 5), the inhibition constant of AuGNP 5 is more than 100 times better (Table 1 and Table S1). Despite this lower coating, the multivalent effect, as judged by the rpn, is remarkably increased by two orders of magnitude. The spatial distribution of DNJ at the AuGNP 5, as obtained through the dendron strategy, is highly beneficial to the enzymatic affinity. AuGNPs thus represent an opportunity to compare multivalent systems, having similar concentrations in DNJ inhitopes, but with different local densities. In DNJ dendron-coated AuGNPs, the higher local concentration may favor the bind-and-recapture of the reversible inhibitor heads.

3. Materials and Methods

3.1. General Experimental Procedures for the Syntheses

Commercial reagents were used as received. All reactions were carried out under magnetic stirring and monitored by TLC on 0.25 mm silica gel plates (Merck F254). Column chromatographies were carried out on Silica Gel 60 (32–63 µm) or on silica gel (230–400 mesh, Merck, Darmstadt, Germany). Yields refer to spectroscopically and analytically pure compounds unless otherwise stated. ^{1}H NMR spectra were recorded on a Varian Gemini 200 MHz, a Varian Mercury 400 MHz, or on a Varian INOVA 400 MHz instrument at 25 °C. Additionally, ^{13}C NMR spectra were recorded on a Varian Gemini 200 MHz or on a Varian Mercury 400 MHz instrument. Chemical shifts are reported relative to CDCl$_3$ (^{13}C:

δ = 77.0 ppm), or to CD$_3$OD (^{13}C: δ = 49.0 ppm). The following abbreviations were used to designate multiplicities: s = singlet, d = doublet, t = triplet, m = multiplet, br = broad, and dd = double-doublet. Integrals are in accordance with assignments, coupling constants are given in Hz. For detailed peak assignments, 2D spectra were measured (COSY, HSQC). IR spectra were recorded with a IRAffinity-1S SHIMADZU system spectrophotometer. ESI-MS spectra were recorded with a Thermo Scientific™ LCQ fleet ion trap mass spectrometer. Elemental analyses were performed with a Thermo Finnigan FLASH EA 1112 CHN/S analyzer. Optical rotation measurements were performed on a JASCO DIP-370 polarimeter. TEM analysis was performed with a LaB6-TEM of type JEOL JEM-1400PLUS (40 kV–120 kV, HC pole piece) equipped with a GATAN US1000 CCD camera (2 k × 2 k).

3.1.1. Synthesis of Monovalent DNJ Derivative 9

To a solution of **8** (36 mg, 79 µmol) in 1.5 mL of THF 4.6 mL and 3 µL of H$_2$O (145 µmol), PPh$_3$ (25 mg, 87 µmol) was added. The reaction mixture was then refluxed for 16 h, until a TLC analysis (CH$_2$Cl$_2$/MeOH 10: 1) showed the disappereance of the starting material (Rf = 0.91) and the formation of a new product (Rf = 0.10). The solvent was removed under reduced pressure and the crude was purified by FCC (from CH$_2$Cl$_2$/MeOH 10:1 to CH$_2$Cl$_2$/MeOH/NH$_3$ 5:1:0.2) affording pure **9** (25 mg, 58.1 µmol, Rf = 0.10 in CH$_2$Cl$_2$/MeOH 10:1) as a colourless oil in 80% yield. $[\alpha]_D^{22}$ = 4.75 (c = 0.59 in CHCl$_3$); ^{1}H-NMR (400 MHz, CDCl$_3$): δ = 5.05 (dt, J = 17.1, 9.2 Hz, 2H, H-3, H-4), 4.94 (td, J = 10.0, 5.0 Hz, 1H, H-2), 4.19–4.09 (m, 2H, H-6), 3.18 (dd, J = 11.5, 5.1 Hz, 1H, Ha-1), 2.75–2.68 (m, 1H, Ha-1′), 2.62 (dt, J = 11.6, 2.5 Hz, 1H, H-5), 2.57–2.50 (m, 1H, Hb-1′), 2.43 (bs, 2H, H-6′), 2.30 (t, J = 10.5 Hz, 1H, Hb-1), 2.06 (s, 3H, OCH$_3$), 2.01 (s, 6H, OCH$_3$), 2.00 (s, 3H, OCH$_3$), 1.48–1.22 (m, 8H, H-2′, H-3′, H-4′, H-5′) ppm; ^{13}C-NMR (100 MHz, CDCl$_3$): δ = 171.0–169.9 (4C, OCOCH$_3$), 74.8 (C-2), 69.7 (C-4), 69.6 (C-3), 61.7 (C-5), 59.7 (C-6), 53.0 (C-1), 51.8 (C-1′), 27.1–26.8 (4C, from C2′ to C5′), 24.9 (C-6′), 21.0–20.8 (4C, COCH$_3$) ppm; MS (ESI): m/z 431.04 (100, H$^+$). IR (CDCl$_3$): $\tilde{\nu}$ = 2935, 2860, 2255, 1744, 1664, 1601, 1508, 1369, 1234, 1061, 1032 cm^{-1}. Elemental analysis (%) for C$_{20}$H$_{34}$N$_2$O$_8$ (430.49): calcd C, 55.80; H, 7.96; N, 6.51; found: C, 56.34; H, 7.57; N, 6.90.

3.1.2. Synthesis of Monovalent DNJ Based Ligand 11

A solution of EDC·HCl (1-Ethyl-3-(3-dimethylaminopropyl) carbodimide hydrochloride (18.2 mg, 95 µmol), 1-hydroxybenzotriazole (HOBt, 11.7 mg, 87 µmol), and **10** (25.9 mg, 59 µmol) in dry DMF (0.2 mL) was left stirring for 10 min. and then added to a solution of DNJ derivative **9** (17.0 mg, 40 µmol) and N,N-diisopropylethylamine (19 µL, 107 µmol) in DMF (0.9 mL). The reaction mixture was left stirring at room temperature, under nitrogen atmosphere, for 6 h, then diluted with AcOEt (10 mL) and washed with H$_2$O (2 × 3 mL). The organic layer was then washed with a saturated solution of NaHCO$_3$ (1 × 3 mL) and brine (1 × 4 mL), dried over anhydrous Na$_2$SO$_4$ and concentrated under vacuum. The crude was purified by column chromatography (DCM/MeOH from 30:1 to 10:1) affording 22 mg of **11** (26 µmol, 67% yield). Rf = 0.32 (CH$_2$Cl$_2$/MeOH 10:1). $[\alpha]_D^{26}$ = 3.33 (c = 0.69 in CHCl$_3$). ^{1}H-NMR (400 MHz, CDCl$_3$): δ = 5.06 (dt, J = 9.2, 2.8 Hz, 2H, H-3, H-4), 4.99–4.92 (m, 1H, H-2), 4.17–4.13 (m, 2H, H-6), 3.98 (s, 2H, CH$_2$ = CO), 3.69–3.56 (m, 12H), 3.44 (t, J = 6.8 Hz, 2H), 3.29–3.16 (m, 3H, H-6′, Ha-1), 2.85 (t, J = 3.6 Hz, 2H, CH$_2$SAc), 2.77–2.69 (m, 1H, Ha-1′), 2.64–2.61 (m, 1H, H-5), 2.58–2.52 (m, 1H, Hb-1′), 2.32 (bs, 4H, SCOCH$_3$, Hb-1), 2.08 (s, 3H, OCH$_3$), 2.02 (s, 6H, OCH$_3$), 2.01 (s, 3H, OCH$_3$), 1.59–1.45 (m, 4H), 1.36–1.25 (m, 22H) ppm; ^{13}C-NMR (50 MHz, CDCl$_3$): δ = 196.0 (SCOCH$_3$), 170.8-170.0 (4C, –OCOCH$_3$), 169.7 (CONH), 74.7 (C-2), 71.6–70.0 (8C), 69.5 (C-4), 69.4 (d, C-3), 61.5 (C-5), 59.5 (C-6), 52.9 (C-1), 51.7 (C-1′), 38.7 (C-6′), 30.6 (SCOCH$_3$), 29.6–28.8 (11C), 26.9–26.1 (3C, C-2′, C-3′, C-4′), 20.8–20.6 (4C, COCH$_3$) ppm; MS (ESI): m/z 871.35 (100, M+Na$^+$). IR (CDCl$_3$): $\tilde{\nu}$ = 3005, 2929, 2856, 1744, 1673, 1540, 1466, 1370, 1237, 1107, 1033 cm^{-1}; Elemental analysis (%) for C$_{41}$H$_{72}$N$_2$O$_{14}$S (849.08): calcd C, 58.00; H, 8.55; N, 3.30; found: C, 56.44; H, 7.90; N, 2.39.

3.1.3. Synthesis of Trivalent DNJ Derivative 16

To a solution of **8** (63 mg, 138 µmol) in 3 mL of THF/H$_2$O = 2:1 CuSO$_4$ (2.1 mg, 13 µmol), sodium ascorbate (5 mg, 25 µmol) and **15** [28] (9.9 mg, 42 µmol) were added. The reaction mixture was stirred in microwave at 80 °C for 45 min, until a TLC analysis (CH$_2$Cl$_2$/MeOH 10: 1) showed the disappearance of the starting material (Rf = 0.46) and the formation of a new product (Rf = 0.00). The solvent was removed under reduced pressure and the crude was purified by FCC (CH$_2$Cl$_2$/MeOH/NH$_3$ from 30:1 to 5: 1: 0.1) affording pure **16** (57 mg, 35.4 µmol, Rf = 0.33 in CH$_2$Cl$_2$/MeOH/NH$_3$ 10:1:0.1) as a pale yellow oil, in 85% yield. $[\alpha]_D^{22}$ = 7.47 (c = 0.91 in CHCl$_3$). ^{1}H-NMR (400 MHz, CDCl$_3$): δ = 7.56 (br s, 3H, triazole), 5.09–5.01 (m, 6H, H-3, H-4), 4.97–4.91 (m, 3H, H-2), 4.60 (s, 6H, H-7′), 4.34 (t, J = 7.2 Hz, 6H, H-6′), 4.18–4.10 (m, 6H, H-6), 3.44 (s, 6H, H-8′), 3.18 (dd, J = 11.4, 5.0 Hz, 3 H, Ha-1), 2.76-2.69 (m, 3H, Ha-1′), 2.63–2.60 (m, 3H, H-5), 2.56–2.50 (m, 3H, Hb-1′), 2.29 (t, J = 10.8 Hz, 3H, Hb-1), 2.07 (s, 9H, OCH$_3$), 2.02–2.01 (m, 27H, OCH$_3$), 1.94–1.84 (m, 6H, H-5′), 1.45–1.25 (m, 18H, H-2′, H-3′, H-4′) ppm; ^{13}C-NMR (100 MHz, CDCl$_3$): δ = 170.9-169.9 (12C, OCOCH$_3$), 145.2 (3C, triazole), 122.5 (3C, triazole), 74.8 (3C, C-2), 72.4 (3C, C-8′), 69.7 (3C, C-4), 69.6 (3C, C-3), 65.2 (3C, C-7′), 61.9 (3C, C-5), 59.8 (3C, C-6), 56.3 (NH$_2$C(CH$_2$O-)$_3$), 53.0 (3C, C-1), NH$_2$C(CH$_2$O-)$_3$), 51.7 (3C, C-1′), 50.3 (3C, C-6′), 30.4 (3C, C-5′), 26.8 (3C, C-2′), 26.6 (3C, C-3′), 25.1 (3C, C-4′), 21.0–20.8 (12C, COCH$_3$) ppm; MS (ESI): m/z 1626.58 (100, Na$^+$); IR (CDCl3): $\tilde{\nu}$ = 3690, 3606, 2937, 2862, 2257, 1745, 1602, 1438, 1370, 1234, 1097, 1052, 1032 cm^{-1}. Elemental analysis (%) for C$_{73}$H$_{113}$N$_{13}$O$_{27}$ (1604.75): calcd C, 54.64; H, 7.10; N, 11.35; found: C, 54.60; H, 7.35; N, 10.37.

3.1.4. Synthesis of Trivalent DNJ Based Ligand 17

To a solution of compound **16** (56.8 mg, 35.4 µmol) in dry THF (570 µL) at 0 °C, 3-(diethoxyphosphoryloxy)-1,2,3-benzotriazin-4(3H)-one (DEPBT 21.2 mg, 70.8 µmol) and N,N-diisopropylethylamine (DIPEA 12.0 µL, 70.8 µmol) were added. The mixture was stirred at 25 °C for 15 min under a nitrogen atmosphere, then a solution of **10** (30.0 mg, 21.6 µmol) in dry THF (225 µL) was added and the reaction mixture was stirred at room temperature for 6 days, until a TLC analysis (CH$_2$Cl$_2$/MeOH 6:1) showed the formation of a new product (Rf = 0.7). The reaction mixture was diluted with AcOEt (5 mL), washed with NH$_4$Cl (2 × 3 mL), NaHCO$_3$ (2 × 3 mL) and H$_2$O (2 × 3 mL), dried over anhydrous Na$_2$SO$_4$ and concentrated under vacuum. Purification through gradient column chromatography (CH$_2$Cl$_2$/MeOH from 20:1 to 10:1) afforded pure **17** (61 mg, 30.1 µmol) as a yellow oil, in 94% yield. $[\alpha]_D^{18}$ = 5.29 (c = 0.87 in CHCl$_3$). ^{1}H-NMR (400 MHz, CDCl$_3$): δ = 7.54 (br s, 3H, triazole), 6.81 (s, 1H, CONH), 5.06–4.98 (m, 6H, H-3, H-4), 4.94–4.88 (m, 3H, H-2), 4.57 (s, 6H, H-7′), 4.31 (t, J = 7.2 Hz, 6H, H-6′), 4.15–4.07 (m, 6H, H-6), 3.84 (s, 2H), 3.77 (s, 6H, H-8′), 3.63–3.51 (m, 12 H), 3.39 (t, J = 6.8 Hz, 2H), 3.15 (dd, J = 11.5, 5.1 Hz, 3H, Ha-1), 2.82 (t, J = 7.4 Hz, 2H), 2.74–2.67 (m, 3H, Ha-1′), 2.60–2.58 (m, 3H, H-5), 2.53–2.47 (m, 3H, Hb-1′), 2.28–2.24 (m, 6H, SCOCH$_3$, Hb-1), 2.03 (s, 9H, OCH$_3$), 1.99 (s, 18H, OCH$_3$), 1.98 (s, 9H, OCH$_3$), 1.91–1.84 (m, 6H, H-5′), 1.56–1.48 (m, 4H), 1.33–1.22 (m, 32H) ppm; ^{13}C-NMR (50 MHz, CDCl$_3$): δ = 196.0 (SCOCH$_3$), 170.9–169.7 (12C, OCOCH$_3$), 169.7 (CONH), 144.9 (3C, triazole), 122.6 (3C, triazole), 74.8 (3C, C-2), 71.6–70.1 (8C), 69.6 (3C, C-4), 69.5 (3C, C-3), 68.9 (3C, C-8′), 65.1 (3C, C-7′), 61.8 (3C, C-5), 59.6 (4C, C-6, NHC(CH$_2$O-)$_3$), 53.0 (3C, C-1), 51.6 (3C, C-1′), 50.2 (3C, C-6′), 30.4 (CH$_2$SCOCH$_3$), 29.6–25.0 (22C), 20.9, 20.7 (12C, COCH$_3$) ppm; MS (ESI): m/z 1033.92 (100, (m/z) + 2Na$^+$); IR (CDCl3): $\tilde{\nu}$ = 3022, 2932, 2858, 1744, 1675, 1525, 1369, 1220, 1098, 1032 cm^{-1}. Elemental analysis (%) for C$_{94}$H$_{151}$N$_{13}$O$_{33}$S (2023.34): calcd C, 55.80; H, 7.52; N, 9.00; found: C, 55.76; H, 7.78; N, 8.22.

3.1.5. General Procedure for the In Situ Preparation of Ligands 12 and 18

To a solution of **11** or **17** in CD$_3$OD (10 mg/mL), 30 equivalents of NaOMe were added, and the reaction mixture was left stirring for 2 h at 25 °C under nitrogen atmosphere. The complete disappearance of the starting material was attested via ^{1}H NMR, and the crude, containing compounds **12** or **18**, respectively, was directly used for the preparation of AuGNPs.

3.1.6. General Procedure for the Preparation of AuGNPs 1–5

An aqueous solution of $HAuCl_4$ (25 mM, 1 equiv.) was added to a 12 mM methanolic solution of a suitable mixture of thiol-ending monosaccharide and DNJ ligands (3 equiv. overall). An aqueous solution of $NaBH_4$ (1 M, 27 equiv.) was then added in four portions, with vigorous shaking. The black suspension formed was shaken for 2 h at 25 °C. The residue was washed several times with MeOH. In order to effectively separate the nanoparticles from the supernatant, centrifugation (12,000 rpm, 2 min) was performed. The residue was dissolved in a minimal volume of HPLC gradient grade water and purified by dialysis (SnakeSkin® Pleated Dialysis Tubing, 10,000 MWCO and Slide-A-Lyzer® 10K Dialysis Cassettes, 10,000 MWCO). DNJ-coated AuGNPs were obtained as a dark-brown powder after freeze-drying and characterized via ^1H NMR, UV-Vis spectroscopy and TEM analysis (see Supplementary Materials). For the analysis of the ratio between the active component (DNJ-based ligand) and the inner component (monosaccharide ligand), ^1H NMR spectra of the initial mixture and the supernatant, after AuGNP formation, were recorded. The DNJ loading on the AuGNPs was evaluated by quantitative NMR (qNMR) using 3-(trimethylsilyl)propionic-2,2,3,3-d_4 acid, sodium salt (TSP-d_4) as an internal standard in the D_2O solution of the AuGNPs. The prepared AuGNPs can be stored at 4 °C for months while maintaining their biophysical properties.

Preparation of 40% monoDNJ-Au-βGlc 1

A 1:1.5 mixture of thiol-ending 12 (4.5 mg, 2.65 μmol) and βGlcC$_5$SH 13 (1.1 mg, 3.89 μmol) in CD_3OD (1.1 mL) was used to obtain 0.35 mg of AuGNP 1. TEM (average diameter): 1.8 ± 0.4 nm. Quantitative ^{1}H-NMR (400 MHz, D_2O containing 0.05 wt.% of 3-(trimethylsilyl)propionic-2,2,3,3-d_4 acid, sodium salt as an internal standard): 0.15 mg of 1 were dissolved in 200 μL of D_2O and 15 μL of D_2O, containing 0.05 wt.% TSP, were added and 31 nmoles of DNJ conjugate were found. In the quantitative NMR (qNMR) a mediated value of the multiplet corresponding to Ha-1' proton signal (δ = 2.84 ppm, 1H) of DNJ conjugate and the multiplet corresponding to H-6 (δ = 2.60-2.45 ppm, 2H) was selected for integration as it falls in a spectral region free of other signals. UV-Vis (H_2O, 0.1 mg/mL): absence of a maximum band at around 520 nm.

Preparation of 40% monoDNJ-Au-αMan 2

A 1:1.5 mixture of thiol-ending 12 (3.8 mg, 5.89 μmol) and αManC$_5$SH 14 (2.5 mg, 8.84 μmol) in CD_3OD (1.2 mL) was used to obtain 1.33 mg of AuGNPs 2. TEM (average diameter): 2.1 ± 0.6 nm. Quantitative ^{1}H-NMR (400 MHz, D_2O containing 0.05 wt.% of 3-(trimethylsilyl)propionic-2,2,3,3-d_4 acid, sodium salt as an internal standard): 0.60 mg of 2 were dissolved in 200 μL of D_2O and 40 μL of D_2O containing 0.05 wt.% TSP were added and 140 nmoles of DNJ conjugate were found. Significant peaks: δ = 2.85–2.95 (br signal, 1H, from DNJ conjugate) ppm. UV-Vis (H_2O, 0.1 mg/mL): absence of a maximum band at around 520 nm.

Preparation of 20% trisDNJ-Au-βGlc 3

A 1:4 mixture of thiol-ending 18 (10.9 mg, 7.41 μmol) and βGlcC$_5$SH 13 (8.4 mg, 29.64 μmol) in CD_3OD (3.1 mL) was used to obtain 4.5 mg of AuGNPs 3. TEM (average diameter): 2.1 ± 0.5 nm. Quantitative ^{1}H-NMR (400 MHz, D_2O containing 0.05 wt.% of 3-(trimethylsilyl)propionic-2,2,3,3-d_4 acid, sodium salt as an internal standard): 0.60 mg of 3 were dissolved in 180 μL of D_2O and 40 μL of D_2O containing 0.05 wt.% TSP were added and 170 nmoles of DNJ conjugate were found. Significant peaks: δ = 7.83 (br s, 3H, triazole from DNJ derivative) ppm. UV-Vis (H_2O, 0.1 mg/mL): absence of a maximum band at around 520 nm.

Preparation of 20% trisDNJ-Au-αMan 4

A 1:4 mixture of thiol-ending 18 (10.9 mg, 7.41 μmol) and αManC$_5$SH 14 (8.3 mg, 29.64 μmol) in CD_3OD (3.1 mL) was used to obtain 3.7 mg of AuGNP 4. TEM (average diameter): 2.0 ± 0.4 nm. Quantitative ^{1}H-NMR (400 MHz, D$_2$O containing 0.05 wt.% of 3-(trimethylsilyl)propionic-2,2,3,3-d_4 acid, sodium salt as an internal standard): 0.60 mg of **4** were dissolved in 180 µL of D$_2$O and 40 µL of D$_2$O containing 0.05 wt.% TSP were added and 135 nmoles of DNJ conjugate were found. Significant peaks: δ = 7.84 (br s, 3H, triazole from DNJ derivative) ppm. UV-Vis (H$_2$O, 0.1 mg/mL): absence of a maximum band at around 520 nm.

Preparation of 40% trisDNJ-Au-αMan **5**

A 1:1.8 mixture of thiol-ending **18** (6.9 mg, 4.70 µmol) and αManC$_5$SH **14** (2.4 mg, 8.37 µmol) in CD$_3$OD (0.97 mL) was used to obtain 1.7 mg of AuGNP **5**. TEM (average diameter): 2.1 ± 0.5 nm. Quantitative ^{1}H-NMR (400 MHz, D$_2$O containing 0.05 wt.% of 3-(trimethylsilyl)propionic-2,2,3,3-d_4 acid, sodium salt as an internal standard): 0.60 mg of **5** were dissolved in 180 µL of D$_2$O and 40 µL of D$_2$O containing 0.05 wt.% TSP were added and 151 nmoles of DNJ conjugate were found. Significant peaks: δ = 7.81 (br s, 3H, triazole from DNJ derivative), 2.57-2.44 (m, 6H, from DNJ conjugate) ppm. UV-Vis (H$_2$O, 0.1 mg/mL): absence of a maximum band at around 520 nm.

3.2. Biological Evaluation

p-Nitrophenyl-α-D-mannopyranoside and α-mannosidase (EC 3.2.1.24, from Jack Bean, K_m = 2.0 mM pH 5.5) were purchased from Merck (Darmstadt, Germany). Inhibition constants were determined by spectrophotometrically measuring the residual hydrolytic activities of the mannosidase against *p*-nitrophenyl-α-D-mannopyranoside in the presence and absence of the inhibitor with a VersaMax Microplate Reader. All kinetics were performed at 25 °C and started by substrate addition (20 µL) in a 100 µL assay medium (acetate buffer, 0.2 M, pH = 5) containing α-mannosidase (0.015 U/mL), in presence or absence of various concentrations of inhibitor and substrate (concentrations from km/8 to 2 km). After 15–30 min incubation, the reaction was quenched by the addition of 1M Na$_2$CO$_3$ 100 µL). The absorbance of the resulting solution was determined at 405 nm. Under these conditions, the *p*-nitrophenolate released led to optical densities linear with both reaction time and concentration of the enzyme. K_i values were determined, in duplicate or triplicate, using the Dixon or Lineweaver-Burk graphical methods with Microsoft Excel [37].

4. Conclusions

In conclusion, we report, in this work, the first example of gold nanoparticles decorated with the iminosugar deoxynojirimycin (DNJ). Jack bean α-mannosidase (JBα-man) was chosen as the target enzyme for the biological evaluation due to its well-known responsiveness to the multivalent presentation of iminosugar inhitopes. Monovalent and trivalent DNJ-derivative functionalized with a thiol were synthesized and used in mixture with a glucose or mannose thiol-ending inner component for the in-situ preparation of heterovalent AuGNPs. AuGNPs, bearing only the sugar component, were also prepared as control systems. Biological assays towards JBα-man revealed a competitive inhibition for the whole set of AuGNPs **1–5**, while glyco-coated AuGNPs **6–7** were inactive, thus demonstrating the fundamental role played by the iminosugar moiety for the inhibition. Comparison with the monovalent reference **19** highlighted affinity enhancements, per inhitope higher than 1600, for the AuGNPs **3–5** decorated with the trivalent DNJ-derivative, with the best result obtained for AuGNP **5** (K_i = 84 nM, *rpn* = 3833). Interestingly, AuGNPs **2** and **5** share similar concentration in DNJ but vary by their local inhitope density, which makes them powerful tools to measure the impact of DNJ distribution onto the affinity enhancements. The significant gain of affinity, observed between the DNJ-coated and DNJ dendron-coated AuGNPs, highlights the importance of the bind-and-recapture effect in the complex, interconnected mechanisms underlying the inhibitory multivalent effects. Further investigation with these new nanosystems is currently ongoing in our laboratories.

Supplementary Materials: The following are available online. The ^{1}H and ^{13}C-NMR spectra of DNJ-based ligands **9**, **11**, **16** and **17**; the characterization of AuGNPs **1–5** (^{1}H-NMR spectra, TEM graphs, UV-Vis spectra); the Dixon and Lineweaver-Burk plots for determination of inhibition constants of AuGNPs **1–5**; Table S1 which contains a resume of DNJ-based AuGNPs 1–5 characterization (average size diameter, calculated amount of iminosugar).

Author Contributions: C.M., F.C. and A.G. planned the synthetic strategy and wrote the manuscript; M.M., M.d.l.A.R. and S.M. characterized the nanosystems; C.V. and J.P.S. made the syntheses; A.B. performed the biological tests, interpreted the results and analyzed them with P.C.; P.C. raised funds. All authors contributed to the writing of the manuscript. All authors have read and agreed to the published version of the manuscript.

Funding: MIUR-Italy ("Progetto Dipartimenti di Eccellenza 2018−2022") allocated to the Department of Chemistry "Ugo Schiff", Università di Firenze and Regione Toscana (Bando Salute 2018) for the project: "Late onset Lysosomal Storage Disorders (LSDs) in the differential diagnosis of neurodegenerative diseases: development of new diagnostic procedures and focus on potential pharma-cological chaperones (PCs), Acronym: Lysolate). This work was also funded by the CNRS (UMR 7509), the University of Strasbourg and the Fondation pour la Recherche en Chimie (icFRC Strasbourg).

Acknowledgments: We thank Patrizia Andreozzi for fruitful discussion on experimental data.

Conflicts of Interest: The authors declare no conflict of interest.

Sample Availability: Samples of the compounds reported in the manuscript are not available from the authors.

References

1. Diot, J.; García-Moreno, M.I.; Gouin, S.G.; Ortiz Mellet, C.; Haupt, K.; Kovensky, J. Multivalent iminosugars to modulate affinity and selectivity for glycosidases. *Org. Biomol. Chem.* **2009**, *7*, 357–363. [CrossRef]
2. Compain, P.; Decroocq, C.; Iehl, J.; Holler, M.; Hazelard, D.; Mena Barragán, T.; Ortiz Mellet, C.; Nierengarten, J.-F. Glycosidase Inhibition with Fullerene Iminosugar Balls: A Dramatic Multivalent Effect. *Angew. Chem. Int. Ed.* **2010**, *49*, 5753–5756. [CrossRef] [PubMed]
3. Compain, P.; Bodlenner, A. The multivalent effect in glycosidase inhibition: A new, rapidly emerging topic in glycoscience. *ChemBioChem* **2014**, *15*, 1239–1251. [CrossRef] [PubMed]
4. Gouin, S.G. Multivalent inhibitors for carbohydrate-processing enzymes: Beyond the "lock-and-key" concept. *Chem. Eur. J.* **2014**, *20*, 11616–11628. [CrossRef] [PubMed]
5. Zelli, R.; Longevial, J.-F.; Dumy, P.; Marra, A. Synthesis and biological properties of multivalent iminosugars. *New J. Chem.* **2015**, *30*, 5050–5074. [CrossRef]
6. Kanfar, N.; Bartolami, E.; Zelli, R.; Marra, A.; Winum, J.-Y.; Ulrich, S.; Dumy, P. Emerging trends in enzyme inhibition by multivalent nanoconstructs. *Org. Biomol. Chem.* **2015**, *13*, 9894–9906. [CrossRef]
7. Matassini, C.; Parmeggiani, C.; Cardona, F.; Goti, A. Are enzymes sensitive to the multivalent effect? Emerging evidence with glycosidases. *Tetrahedron Lett.* **2016**, *57*, 5407–5415. [CrossRef]
8. Compain, P. Multivalent effect in glycosidase inhibition: The end of the beginning. *Chem. Rec.* **2020**, *20*, 10–22. [CrossRef]
9. González-Cuesta, M.; Ortiz Mellet, C.; García Fernández, J.M. Carbohydrate supramolecular chemistry: Beyond the multivalent effect. *Chem. Commun.* **2020**, *56*, 5207–5222. [CrossRef]
10. Lepage, M.L.; Schneider, J.P.; Bodlenner, A.; Meli, A.; De Riccardis, F.; Schmitt, M.; Tarnus, C.; Nguyen-Huynh, N.-T.; Francois, Y.-N.; Leize-Wagner, E.; et al. Iminosugar-Cyclopeptoid Conjugates Raise Multivalent Effect in Glycosidase Inhibition at Unprecedented High Levels. *Chem. Eur. J.* **2016**, *22*, 5151–5155. [CrossRef]
11. Howard, E.; Cousido-Siah, A.; Lepage, M.L.; Schneider, J.P.; Bodlenner, A.; Mitschler, A.; Meli, A.; Izzo, I.; Alvarez, A.; Podjarny, A.; et al. Structural Basis of Outstanding Multivalent Effects in Jack Bean α-Mannosidase Inhibition. *Angew. Chem. Int. Ed.* **2018**, *57*, 8002–8006. [CrossRef]
12. Mirabella, S.; D'Adamio, G.; Matassini, C.; Goti, A.; Delgado, S.; Gimeno, A.; Robina, I.; Moreno-Vargas, A.J.; Šesták, S.; Jiménez-Barbero, J.; et al. Mechanistic Insight into the Binding of Multivalent Pyrrolidines to α-Mannosidases. *Chem. Eur. J.* **2017**, *23*, 14585–14596. [CrossRef] [PubMed]
13. Abellán Flos, M.; García Moreno, M.I.; Ortiz Mellet, C.; García Fernández, J.M.; Nierengarten, J.-F.; Vincent, S.P. Potent Glycosidase Inhibition with Heterovalent Full-erenes: Unveiling the Binding Modes Triggering Multivalent Inhibition. *Chem. Eur. J.* **2016**, *22*, 11450–11460. [CrossRef] [PubMed]
14. Siriwardena, A.; Khanal, M.; Barras, A.; Bande, O.; Mena-Barragán, T.; Ortiz Mellet, C.; García Fernández, J.M.; Boukherroub, R.; Szunerits, S. Unprecedented Inhibition of Glyco-sidase-Catalyzed Substrate Hydrolysis by Nanodiamond-Grafted O-Glycosides. *RSC Adv.* **2015**, *5*, 100568–100578. [CrossRef]

15. Alali, U.; Vallin, A.; Bil, A.; Khanchouche, T.; Mathiron, D.; Przybylski, C.; Beaulieu, R.; Kovensky, J.; Benazza, M.; Bonnet, V. The Uncommon Strong Inhibition of α-Glucosidase by Multivalent Glycoclusters Based on Cyclodextrin Scaffolds. *Org. Biomol. Chem.* **2019**, *17*, 7228–7237. [CrossRef]
16. Marradi, M.; Chiodo, F.; García, I.; Penadés, S. Glyconanoparticles as multifunctional and multimodal carbohydrate systems. *Chem. Soc. Rev.* **2013**, *42*, 4728–4745. [CrossRef]
17. Marradi, M.; García, I.; Penadés, S. Carbohydrate-based nanoparticles for potential applications in medicine. In *Nanoparticles in Translational Science and Medicine*; Villaverde, A., Ed.; Elsevier: Amsterdam, The Netherlands, 2011; Volume 104, pp. 141–173, ISBN 9780124160200.
18. Alvarez-Dorta, D.; Brissonnet, Y.; Saumonneau, A.; Deniaud, D.; Bernard, J.; Yan, X.; Tellier, C.; Daligault, F.; Gouin, S.G. Magnetic Nanoparticles Coated with Thiomannosides or Iminosugars to Switch and Recycle Galactosidase Activity. *ChemistrySelect* **2017**, *2*, 9552–9556. [CrossRef]
19. Kleps, I.; Ignat, T.; Miu, M.; Craciunoiu, F.; Trif, M.; Simion, M.; Bragaru, A.; Dinescu, A. Nanostructured Silicon Particles for Medical Applications. *J. Nanosci. Nanotechnol.* **2010**, *10*, 2694–2700. [CrossRef]
20. Bonduelle, C.; Huang, J.; Mena-Barragán, T.; Ortiz Mellet, C.; Decrooq, C.; Etamé, E.; Heise, A.; Compain, P.; Lecommandoux, S. Iminosugar-based glycopolypeptides: Glycosidase inhibition with bioinspired glycoprotein analogue micellar self-assemblies. *Chem. Commun.* **2014**, *50*, 3350–3352. [CrossRef] [PubMed]
21. Matassini, C.; Marradi, M.; Cardona, F.; Parmeggiani, C.; Robina, I.; Moreno-Vargas, A.J.; Penadés, S.; Goti, A. Gold nanoparticles are suitable cores for building tunable iminosugar multivalency. *RSC Adv.* **2015**, *5*, 95817–95822. [CrossRef]
22. Matassini, C.; Vanni, C.; Goti, A.; Morrone, A.; Marradi, M.; Cardona, F. Multimerization of DAB-1 onto Au GNPs affords new potent and selective N-acetylgalactosamine-6-sulfatase (GALNS) inhibitors. *Org. Biomol. Chem.* **2018**, *16*, 8604–8612. [CrossRef]
23. Joosten, A.; Schneider, J.P.; Lepage, M.L.; Tarnus, C.; Bodlenner, A.; Compain, P. A Convergent Strategy for the Synthesis of Second-Generation Iminosugar Clusters Using "Clickable" Trivalent Dendrons. *Eur. J. Org. Chem.* **2014**, 1866–1872. [CrossRef]
24. Schneider, J.P.; Tommasone, S.; Della Sala, P.; Gaeta, C.; Talotta, C.; Tarnus, C.; Neri, P.; Bodlenner, A.; Compain, P. Synthesis and Glycosidase Inhibition Properties of Ca-lix[8]Arene-Based Iminosugar Click Clusters. *Pharmaceuticals* **2020**, *13*, 366–386. [CrossRef] [PubMed]
25. Martínez-Ávila, O.; Hijazi, K.; Marradi, M.; Clavel, C.; Campion, C.; Kelly, C.; Penadés, S. Gold Manno-Glyconanoparticles: Multivalent Systems to Block HIV-1 gp120 Binding to the Lectin DC-SIGN. *Chem. Eur. J.* **2009**, *15*, 9874–9888. [CrossRef]
26. Decrooq, C.; Rodríguez-Lucena, D.; Russo, V.; Mena Barragán, T.; Ortiz Mellet, C.; Compain, P. The Multivalent Effect in Glycosidase Inhibition: Probing the Influence of Architectural Parameters with Cyclodextrin-based Iminosugar Click Clusters. *Chem. Eur. J.* **2011**, *17*, 13825–13831. [CrossRef]
27. Ayesa, S.; Samuelsson, B.; Classon, B. A One-Pot, Solid-Phase Synthesis of Secondary Amines from Reactive Alkyl Halides and an Alkyl Azide. *Synlett* **2008**, 97–99.
28. Chabre, Y.M.; Contino-Pépin, C.; Placide, V.; Shiao, T.C.; Roy, R. Expeditive synthesis of glycodendrimer scaffolds based on versatile TRIS and mannoside derivatives. *J. Org. Chem.* **2008**, *73*, 5602–5605. [CrossRef]
29. Kolb, H.C.; Finn, M.G.; Sharpless, K.B. Click Chemistry: Diverse Chemical Function from a Few Good Reactions. *Angew. Chem. Int. Ed.* **2001**, *40*, 2004–2021. [CrossRef]
30. Rostovtsev, V.C.; Green, L.G.; Fokin, V.V.; Sharpless, K.B. A Stepwise Huisgen Cycloaddition Process: Copper(I)-Catalyzed Regioselective Ligation of Azides and Terminal Alkynes. *Angew. Chem. Int. Ed.* **2002**, *41*, 2596–2599. [CrossRef]
31. Tornøe, C.W.; Christensen, C.; Meldal, M. Peptidotriazoles on Solid Phase: [1,2,3]-Triazoles by Regiospecific Copper(I)-Catalyzed 1,3-Dipolar Cycloadditions of Terminal Alkynes to Azides. *J. Org. Chem.* **2002**, *67*, 3057–3064. [CrossRef]
32. Manea, F.; Bindoli, C.; Fallarini, S.; Lombardi, G.; Polito, L.; Lay, L.; Bonomi, R.; Mancin, F.; Scrimin, P. Multivalent, Saccharide-Functionalized Gold Nanoparticles as Fully Synthetic Analogs of Type A Neisseria meningitidis Antigens. *Adv. Mater.* **2008**, *20*, 4348–4352. [CrossRef]
33. Chiodo, F.; Enríquez-Navas, P.M.; Angulo, J.; Marradi, M.; Penadés, S. Assembling different antennas of the gp120 high mannose-type glycans on gold nanoparticles provides superior binding to the anti-HIV antibody 2G12 than the individual antennas. *Carbohydr. Res.* **2015**, *405*, 102–109. [CrossRef] [PubMed]
34. Hostetler, M.J.; Wingate, J.E.; Zhong, C.-J.; Harris, J.E.; Vachet, R.W.; Clark, M.R.; Londono, J.D.; Green, S.J.; Stokes, J.J.; Wignall, G.D.; et al. Alkanethiolate Gold Cluster Molecules with Core Diameters from 1.5 to 5.2 nm: Core and Monolayer Properties as a Function of Core Size. *Langmuir* **1998**, *14*, 17–30. [CrossRef]
35. Zhou, M.; Zeng, C.; Chen, Y.; Zhao, S.; Sfeir, M.Y.; Zhu, M.; Jin, R. Evolution from the plasmon to exciton state in ligand-protected atomically precise gold nanoparticles. *Nat. Commun.* **2016**, *7*, 13640. [CrossRef]
36. Decrooq, C.; Joosten, A.; Sergent, R.; Mena Barragan, T.; Ortiz Mellet, C.; Compain, P. The Multivalent Effect in Glycosidase Inhibition: Probing the Influence of Valency, Peripheral Ligand Structure, and Topology with Cyclodextrin-Based Iminosugar ClickClusters. *ChemBioChem* **2013**, *14*, 2038–2049. [CrossRef] [PubMed]
37. Segel, I.H. *Enzyme Kinetics: Behavior and Analysis of Rapid Equilibrium and Steady-State Enzyme Systems*; John Wiley & Sons: New York, NY, USA, 1975.

Article

α,α-Difluorophosphonohydroxamic Acid Derivatives among the Best Antibacterial Fosmidomycin Analogues

Aurore Dreneau, Fanny S. Krebs, Mathilde Munier, Chheng Ngov, Denis Tritsch, Didier Lièvremont, Michel Rohmer and Catherine Grosdemange-Billiard *

Laboratoire de Chimie et Biochimie de Molécules Bioactives, Université de Strasbourg/CNRS, UMR 7177, Institut Le Bel, 4 Rue Blaise Pascal, 67081 Strasbourg, France; aurore.dreneau@gmail.com (A.D.); krebs.fanny@live.fr (F.S.K.); mathilde.munier01@hotmail.fr (M.M.); chheng.ngov@alsace.cnrs.fr (C.N.); tritsch.denis@neuf.fr (D.T.); didier.lievremont@unistra.fr (D.L.); mirohmer@unistra.fr (M.R.)
* Correspondence: grosdemange@unistra.fr; Tel.: +33-368-851-349

Citation: Dreneau, A.; Krebs, F.S.; Munier, M.; Ngov, C.; Tritsch, D.; Lièvremont, D.; Rohmer, M.; Grosdemange-Billiard, C. α,α-Difluorophosphonohydroxamic Acid Derivatives among the Best Antibacterial Fosmidomycin Analogues. *Molecules* **2021**, *26*, 5111. https://doi.org/10.3390/molecules 26165111

Academic Editors: Camilla Parmeggiani, Camilla Matassini and Francesca Cardona

Received: 28 July 2021
Accepted: 22 August 2021
Published: 23 August 2021

Publisher's Note: MDPI stays neutral with regard to jurisdictional claims in published maps and institutional affiliations.

Copyright: © 2021 by the authors. Licensee MDPI, Basel, Switzerland. This article is an open access article distributed under the terms and conditions of the Creative Commons Attribution (CC BY) license (https://creativecommons.org/licenses/by/4.0/).

Abstract: Three α,α-difluorophosphonate derivatives of fosmidomycin were synthesized from diethyl 1,1-difluorobut-3-enylphosphonate and were evaluated on *Escherichia coli*. Two of them are among the best 1-deoxy-D-xylulose 5-phosphate reductoisomerase inhibitors, with IC_{50} in the nM range, much better than fosmidomycin, the reference compound. They also showed an enhanced antimicrobial activity against *E. coli* on Petri dishes in comparison with the corresponding phosphates and the non-fluorinated phosphonate.

Keywords: α,α-difluorophosphonate; deoxyxylulose phosphate reductoisomerase; 1-deoxy-D-xylulose 5-phosphate reductoisomerase (DXR); antimicrobial; fosmidomycin; isoprenoid biosynthesis; 2-C-methyl-D-erythritol 4-phosphate (MEP) pathway

1. Introduction

Antimicrobial resistance affecting anyone in any country is rising to dangerously high levels in all parts of the world and has been recognized as a global health crisis by the United Nations and the World Health Organization (WHO). As a result, the antibiotic treatment of a growing number of infections, e.g., tuberculosis, pneumonia, gonorrhea, and salmonellosis, are becoming less and less effective. In 2017, WHO reported a list of twelve priority pathogens, mostly Gram-negative bacteria belonging to e.g., the *Enterobacteriaceae* or to other groups (e.g., *Acinetobacter baumannii*, *Pseudomonas aeruginosa*...) for which it is urgent to find new treatments [1]. Recently, a study of the European Centre for Disease Prevention and Control (ECDPC) estimated that about 33,000 people died each year from an infection due to antimicrobial-resistant bacteria, frequently while receiving health care i.e., from nosocomial infections [2]. It is therefore crucial and urgent to identify new targets in order to elaborate and develop new drugs. In this respect, the biosynthesis of isopentenyl diphosphate (IPP) and dimethylallyl diphosphate (DMAPP), the two precursors of all isoprenoids, via the 2-C-methyl-D-erythritol 4-phosphate (MEP) pathway is an attractive prospect. In fact, this pathway is essential and present in many Gram-negative and Gram-positive bacteria as well as protozoans, e.g., *Plasmodium* species responsible for malaria [3]. As this pathway is absent in humans, each enzyme is a potential target to elaborate new antimicrobial compounds with expected minimal side effects for the patient. Part of our work on the design of new antimicrobials is based on the inhibition of the second enzyme of the MEP pathway, the 1-deoxy-D-xylulose 5-phosphate reductoisomerase (DXR), which catalyzes the conversion of 1-deoxy-D-xylulose 5-phosphate (DXP) **1** into MEP **2** in the presence of a divalent metal ion (Mg^{2+} or Mn^{2+}) and NADPH as cofactors (Scheme 1).

Indeed, fosmidomycin **3a** and its *N*-acetyl homologue, FR-900098 **3b**, two natural retrohydroxamate phosphonic acids isolated from *Streptomyces* spp., are selective inhibitors of DXR [4,5]. However, their use in antibiotherapy is limited due to their fast clearance and the rapid emergence of resistance as observed with fosmidomycin [6,7]. In an attempt to improve the efficiency of such inhibitors, numerous analogues of fosmidomycin have been synthesized. From these results, it is clear that neither the retro-hydroxamate chelating moiety nor the phosphonate anchoring group can be replaced without a drastic loss of activity except for the inversion of the retrohydroxamate into a hydroxamate (compounds **4a,b**, Scheme 1) [8] and for the replacement of a phosphonate with a phosphate group (compounds **5a,b** and **6a,b**, Scheme 1) as we previously reported [9].

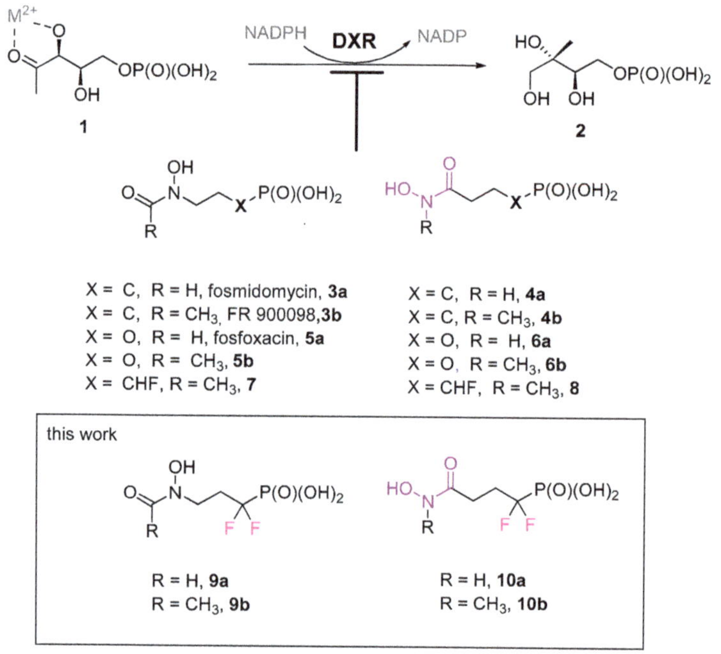

Scheme 1. The 1-deoxyxylulose 5-phosphate reductoisomerase (DXR) and its inhibitors.

Even if the phosphate derivatives (compounds **5a,b**, Scheme 1) were previously shown to be more potent inhibitors of the *Synechocystis* DXR than their phosphonate analogues [10], this conclusion is not valid for all DXRs, as we have shown that they are less efficient against the *E. coli* and *Mycobacterium smegmatis* DXRs [9]. Although a phosphonate group is considered to be isosteric to a phosphate group, some differences such as the pKa values and the C-O-P/C-C-P bond angles might impact their binding in the enzyme active site. Introduction of fluorine atoms in α position of the phosphonate moiety results in significant changes in biological properties and in metabolic stability as compared with the non-fluorinated compounds [11–13]. Van Calenbergh et al. reported the synthesis of FR-900098 and its hydroxamic derivative in which the phosphonate has been replaced with a phosphatase-stable α-monofluoromethylenephosphonate **7** and **8** [14]. The racemic mixtures of these synthetic compounds were evaluated in vitro and in vivo for their antimalarial potentials and were shown to be more effective than the reference compounds **3a,b**. The presence of an electron-withdrawing substituent, in particular fluorine, in the α position of the phosphonate moiety, resulted in the decrease of the pKa_2 of such compounds (from ca. 7.5 to ca. 6.4), almost identical to that of a phosphate group, which is in the dianionic form at the pH of the enzymatic assay, whereas the phosphonate is predominantly in a singly ionized form. By comparison with the α-monofluoromethylene group, which is isoacidic of a phosphate group [15], the α,α-difluoromethylene is an isopolar mimic of the oxygen component of the P-O-C linkage in phosphate [16] and has been used to prepare non-hydrolyzable phosphate analogues of nucleotides [17], phosphatidylinositol [18], glycerol 3-phosphate [11]. In fact, the presence of fluorine atoms able to form fluorine-hydrogen bonds in the DXR active site could affect the binding properties of the parent compounds and also increase their bioavailability [19]. Moreover, the presence of two fluorine atoms might increase the lipophilic properties of the compounds allowing a better cellular uptake [20]. In this context, the synthesis of the protected gem-difluoro FR-900098 derivative **16b** and N-H phosphonohydroxamic acid **10a** (Scheme 2) have been recently reported for a herbicide application [21].

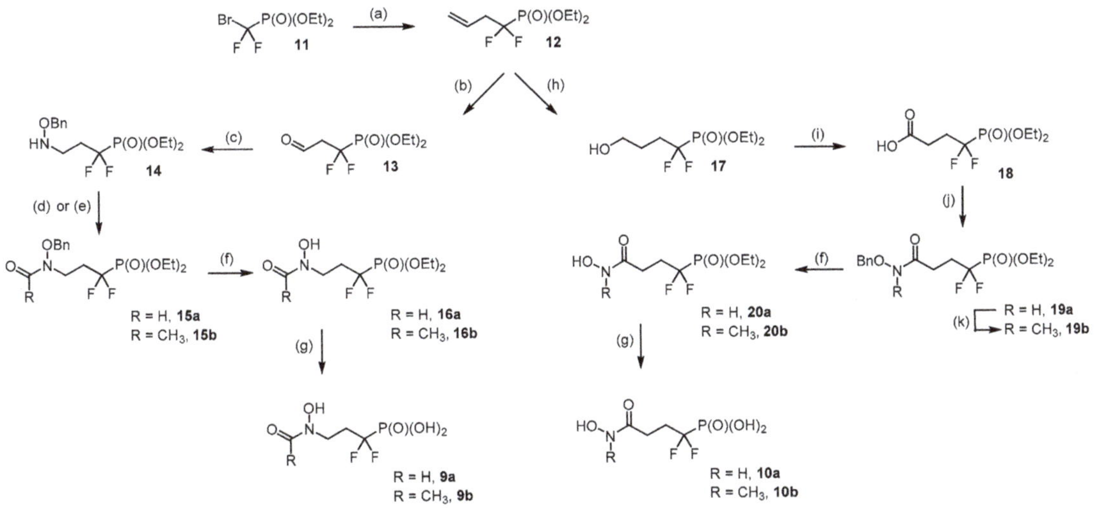

Scheme 2. Synthesis of the α,α-difluorophosphonated fosmidomycin derivatives **9** and **10**. Reagents and conditions: (a) Zn, CuBr, then allyl bromide, DMF(89%); (b) (i) O_3, MeOH, CH_2Cl_2, −78 °C; (ii) Me_2S, (quantitative); (c) $BnONH_2$·HCl, $NaBH_3CN$, HCl conc, MeOH (35%); (d) HCOOH/Ac_2O, THF, rt, **15a**, (83%); (d) Ac_2O, pyridine, rt, **15b**, (80%); (f) H_2, Pd/C, MeOH, rt, **16a** (74 %), **16b** (80 %), **20a** (41%), **20b** (65%); (g) TMSBr, DCM, 0 °C, then H_2O, **9b** (quantitative), **10a** (quantitative), **10b** (quantitative); (h,i) BH_3-THF complex (ii) 3M NaOH, H_2O_2 (39%); (i) TEMPO, BAIB, MeCN, H_2O, (86%); (j) CDI, DCM, 1 h, rt, then $BnONH_2$·HCl, Et_3N, DCM, (77%); (k) MeI, K_2CO_3, acetone, 30 min reflux (73%).

Our investigations are presently orientated toward the synthesis of α,α-difluorophosphonate fosmidomycin derivatives **9a** and **9b** and their analogues **10a** and **10b** to evaluate and determine their effect against *E. coli* DXR in order to develop more potent antimicrobials.

2. Results

2.1. Chemistry

For the introduction of the fluorine atoms into fosmidomycin **3a** and its analogues **3b** and **4**, we followed the procedure of Shibuya [22] to synthesize diethyl 1,1-difluorobut-3-enylphosphonate **12** [22,23], the parent precursor for all described compounds **9** and **10** (Scheme 2). The key precursor **12** has been prepared by a copper(I) catalyzed coupling reaction of [(diethoxyphosphinyl)difluoromethyl]zinc, formed in situ from the commercially available diethyl bromodifluorophosphonate **11** and Zn dust, with allyl bromide. The synthesis of the diethyl α,α-difluorophosphonate **14** was achieved by previously reported methods [21]. Formylation with the mixed acetyl/formyl anhydride generated in situ from a formic acid and acetic anhydride mixture led to the N-formylated compound **15a**, which was obtained as a mixture of conformers due to the restricted rotation around the C-N bond [9,24–26] and the large dipole moment of the C-F bond [27]. Acetylation with a mixture of acetic anhydride and pyridine gave the N-acetylated analogue **15b** as previously described [21].

The protective benzyl group was removed by catalytic hydrogenolysis with palladium over charcoal at atmospheric pressure and room temperature in methanol giving the deprotected hydroxylamines **16a** and **16b** as a mixture of conformers. Deprotection of the phosphonate group **16** using 10 equivalents of bromotrimethylsilane in DCM following by hydrolysis at room temperature led **9b** as a mixture of conformers. In these conditions, **9a** could not be obtained but led to a deformylated by-product as previously observed [21].

The key step of the synthesis of the α,α-difluorophosphonates **10** was the coupling reaction of the commercially available hydroxylamine hydrochloride with the carboxylic acid **18**. The latter compound was obtained in two steps. Hydroboration-oxidation of the parent precursor **12** in presence of THF complex of BH_3 and alkaline hydrogen peroxide gave a mixture of the primary and secondary alcohols in a 7/3 ratio, respectively. After purification by flash chromatography, the primary alcohol **17** was oxidized into the carboxylic acid **18** with 2,2,6,6-tetramethyl-1-piperidinyloxyl (TEMPO) in catalytic proportion in presence of [bis(acetoxy)iodo]benzene [28]. Treatment of the carboxylic acid with O-benzylhydroxylamine hydrochloride in the presence of 1-(3-dimethyl-aminopropyl)-3-ethylcarbodiimide and 1-hydroxybenzotriazole hydrochloride in DCM gave the hydroxamic acid derivative **19a** [21] as a mixture of conformers. The methyl group was introduced by reaction of **19a** with K_2CO_3 in acetone under reflux followed by addition of methyl iodide to give **19b** as a single conformer. Removal of the protective benzyl groups of **19a** and **19b** was achieved by catalytic hydrogenolysis with palladium over charcoal at atmospheric pressure and room temperature leading to **20a** and **20b** as a mixture of conformers. TMSBr-mediated deprotection of the α,α-difluorophosphonate phosphonohydroxamic acid analogues **20** in the same conditions as described for **9b** provided the free phosphonate **10**. Compounds **10a** and **10b** were obtained without further purification as a mixture of conformers.

All α,α-difluorophosphonated compounds **9b** and **10** were tested against His-tagged DXR enzyme of *E. coli* and for growth inhibition against a wild type *E. coli* and fosmidomycin-resistant *E. coli* strain (FosR *E. coli*) as described previously [9].

2.2. Biological Activity

2.2.1. Inhibition of *E. coli* H6-DXR with compounds **9b** and **10**

The inhibition potency of α,α-difluorophosphonohydroxamic acid derivatives was characterized by their IC_{50} value that was determined as previously described [8]. Postulating that the α,α-difluorophosphonated analogues act as slow binding inhibitors like fosmidomycin, they were pre-incubated with DXR during 2 min in the presence of NADPH. Residual activity was measured after initiating the enzymatic reaction by addition of DXP. The IC_{50} values are reported in Table 1.

Table 1. Inhibition of *E. coli* H6-DXR.

Compounds	IC_{50} (nM)
fosmidomycin, **3a**	42
FR-900098, **3b**	4
4a	180 **
4b	48 **
fosfoxacin, **5a**	342 *
5b	77 *
6a	2600 *
6b	46 *
9a	-
9b	9
10a	4600
10b	17

Mean from at least 2 different assays. Errors were <5%. * and ** values obtained from references [9] and [8] respectively.

The inhibitory concentration was determined by measuring the phosphorus content of the solution by the method of Lowry and Lopez [29]. N-methylated α,α-difluorophosphonates **9b** and **10b** show activity on *E. coli* DXR in the nanomolar concentration range and appear to be 2.5 to 5 times more efficient inhibitors than the parent compound fosmidomycin **3a** (IC_{50} = 9 nM and 17 nM respectively vs. IC_{50} = 42 nM) and slightly less potent inhibitors than FR-900098 **3b** (IC_{50} = 4 nM). The presence of two fluorine atoms in α,α position of the phosphonate group has clearly a positive effect on the affinity of the enzyme for these compounds.

As we have previously reported, except for fosmidomycin **3a**, non-N-methylated derivatives are weaker inhibitors than the N-methylated homologues [9]. In fact, the N-H α,α-difluorophosphonate **10a** is 280-fold less efficient (IC_{50} = 4600 nM) than its N-methylated analogue **10b** and therefore, the poorest inhibitor among all compounds of the non-methylated series. Those results indicated that the replacement of the methylene group or the oxygen atom by a difluoromethylene group enhances the inhibition potency of the N-methylated hydroxamic acid derivatives (**4b**, IC_{50} = 48 nM and **6b**, IC_{50} = 46 nM vs. **10b**, IC_{50} = 17 nM) but significantly decreased the inhibition by the N-H analogue **10a**.

2.2.2. Growth Inhibition of a Wild Type *E. coli* and Fosmidomycin Resistant *E. coli* FosR Strain by Compounds **9b**, **10a,b**

The antimicrobial activity of α,α-difluorophosphonate **9a** and **10** was determined using the paper disc diffusion method and was compared with the antimicrobial activity of the non-fluorinated phosphonate and phosphate compounds **3**, **4** et **6**. The diameters of the inhibition zone are given with respect to the amount of inhibitor deposited on the disc (Table 2).

Table 2. *E. coli* XL1 Blue growth inhibition on LB solid medium.

Compounds	nmoles/Disc	Growth Inhibition Zone (mm)
fosmidomycin, **3a**	2	35
FR-900098, **3b**	2 *	32 *
4a	400 **	12 **
4b	80 **	30 **
6a	400 *	<10 *
6b	160 *	20 *
9b	2	15
10a	3	-
10b	1.5	20

* and ** values obtained from references [9] and [8] respectively.

Fosmidomycin, the most efficient growth inhibitors of *E. coli*, was used as a positive control reference. Except for the N-H α,α-difluorophosphonate **10a** where no inhibition was observed, the N-methylated derivatives **9b** and **10b** were shown to be quite effective to inhibit bacterial growth (Figure 1A). Similar amounts of **9b** and **10b** had to be added to observe the same growth inhibition zones as those observed for fosmidomycin. Not only the N-Me α,α-difluorophosphonates **9b** and **10b** are able to inhibit the DXR in vitro but they are also potent *E. coli* growth inhibitors. Clearly, the presence of the two fluorine atoms in α position of the phosphonate enhances the antimicrobial efficiency of the N-methylated phosphonohydroxamic acid. Moreover, we observed in the fosmidomycin **3a** inhibition zone, colonies of tolerant bacteria, which did not appear with the difluoro compounds **9b** and **10b**. These persistent bacteria are known to be able to adapt rapidly to the antibiotic stress, although precise mechanisms are not fully understood [30]. Interestingly, α,α-difluoro compounds **9b** and **10b** eliminated this survival ability of the *E. coli* population.

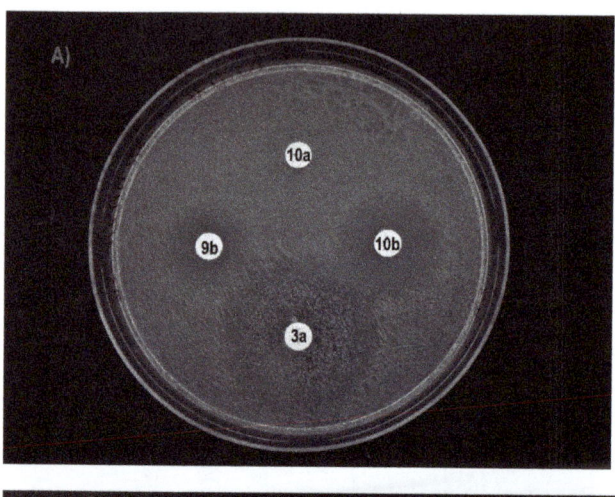

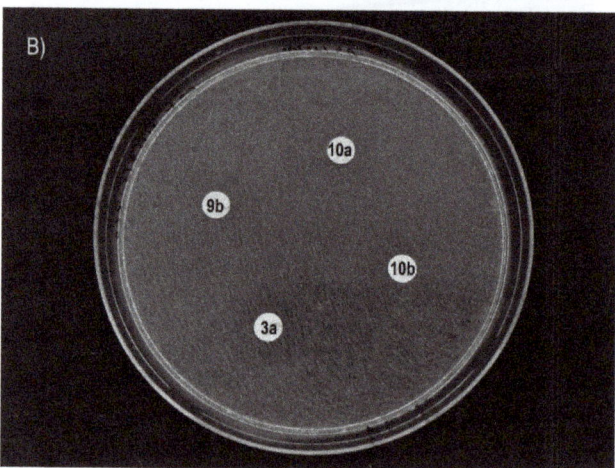

Figure 1. Antimicrobial activity of hydroxamic acids (**9b**, 2 nM), (**10a**, 3 nM) and (**10b**, 1.5 nM) compared to fosmidomycin (**3a**, 2 nM): (**A**) against *E. coli* XL1 Blue; (**B**) against fosmidomycin-resistant strain *E. coli* FosR.

All DXR α,α-difluorophosphonate inhibitors were tested on a fosmidomycin resistant strain of *E. coli* (FosR), but none of the compounds was able to affect bacterial growth (Figure 1B).

3. Discussion

Among a large variety of synthetic analogues of natural phosphono- and phosphoretrohydroxamic acids, e.g., fosmidomycin **3a**, FR-900098 **3b** and fosfoxacin **5a**, only the *N*-H gem-difluoro **10a** have been reported and evaluated as herbicide [21]. However, Van Calenbergh and co-workers observed that the racemic monofluoro analogues **7** and **8** were more active than the parent compounds fosmidomycin and **4b** against intraerythrocytic forms of *Plasmodium falciparum* (K1 strain). Interestingly, none of them have been tested against bacteria [14]. Such promising results prompted us to evaluate the efficiency of α,α-difluorophosphonate analogues of FR-900098 and its phosphonohydroxamic acid derivatives **10a** and **10b** against *E. coli*. Except for the N-H α,α-difluorophosphononate **10a**, the *N*-methylated CF_2-phosphonates **9b** and **10b** exhibited stronger inhibition activity than that of the reference compound fosmidomycin against *E. coli* DXR. With an IC_{50} value of 17 nM, the *N*-methyl α,α-difluorophosphono hydroxamic acid **10b** represents the most powerful inhibitor compared to the non-fluorinated **4b** (IC_{50} = 48 nM) and phosphate **6b** (IC_{50} = 46 nM) analogues. Even if the activity of the CF_2-FR-900098 derivative **9b** (IC_{50} = 9 nM) is two-fold less active than the parent compound **3b** (IC_{50} = 4 nM), it remains a better inhibitor of the *E. coli* DXR than its phosphate derivative **5b** (IC_{50} = 77 nM).

The introduction of the two electron-withdrawing fluorine atoms on the α-methylene group of the phosphonate significantly decreases the pKa2 from 7.6 for the phosphonate to ca. 5.4. The α,α-difluorophosphonates should therefore be in the dianionic form at the pH of the enzyme assay much like a phosphate (pKa2 = 6.4), thereby favoring a more efficient binding than the phosphonate, which is mostly in the singly ionized form. In addition, the dihedral C-CF_2-P angle (116.1°), wider than the C-CH_2-P (112.1°), closely resembles that of the phosphate C-O-P angle of 118.7° [12]. The α,α-difluorophosphonates were, however, shown to be better inhibitors than the phosphate analogues, resulting in a better setting in the *E. coli* DXR active site. The performance of the α,α-difluorophosphonate analogues cannot thus be attributed to ionization or geometry and could be mostly due to favorable modifications of the electrostatic and van der Waals interactions, leading to an increase of the affinity for those inhibitors.

Compared to the *N*-methylated phosphonate and phosphate, the CF_2-phosphonates **9b** and **10b** inhibited efficiently the growth of *E. coli* at doses similar to those of fosmidomycin and FR-900098, making them powerful promising antimicrobials. It is generally accepted that introduction of fluorine atoms enhances the lipophilicity of the compounds, which might facilitate a passive diffusion across the cell membrane of the α,α-difluorophosphonate inhibitors. However, we recently reported that, except for the phosphonate **4b**, all phosphate compounds **5** and **6** penetrated into the bacteria via the same transporters as those involved in the transport of e.g., the glycerol 3-phosphate and the hexose 6-phosphate like fosmidomycin and FR-900098 [9]. No growth inhibition by the α,α-difluoro phosphonates **9b** and **10** against fosmidomycin resistant strain *E. coli* (FosR), in which the GlpT/UhpT transporters are therefore inoperative, was observed implying that these inhibitors penetrate into the bacteria via these transporters (Figure 1B).

In summary, three α,α-difluorophosphonate derivatives of fosmidomycin **3a** were synthesized and were shown, except for the N-H difluoro compound **10a**, to be powerful inhibitors against *E. coli* DXR. Among the series of hydroxamic acids derivatives, the inhibitor **10b** surpasses the phosphonate and phosphate analogues in the inhibition of DXR enzyme as well as in the antimicrobial activity. For *N*-Me difluorophosphonate **9b** and **10b**, there is a direct relation between the capacity to inhibit the DXR and the bacterial growth. An important outcome of this study is that the introduction of two fluorine atoms on the α-methylene group, favors the inhibition on the DXR and enhances the antimicrobial activity in comparison with the phosphates and the non-fluorinated phosphonates.

4. Materials and Methods

4.1. Chemistry

4.1.1. General Methods

All non-aqueous reactions were run in oven-dried glassware under an argon atmosphere, using dry solvents. Commercial grade reagents were purchased from Sigma-Aldrich, Acros Organics or Thermo Fischer Scientific and used without further purification. Petroleum ether (PE) 40–60 °C (Sigma-Aldrich, St-Louis, MO, USA) was used for chromatography. Flash chromatography was performed on silica gel 60 230–400 mesh with the solvent system as indicated. TLC plates were revealed under UV light (254 nm) and/or by spraying with an ethanolic solution of phosphomolybdic acid (20%) or an ethanolic solution of potassium permanganate followed by heating. The NMR spectra (Supplementary Materials) were recorded on a Bruker Avance 300 (^{1}H-NMR: 300 MHz; ^{13}C-NMR, 75.5 MHz; ^{31}P-NMR 121.5 MHz; ^{19}F-NMR 282.4 MHz), a Bruker Avance 400 (^{1}H NMR: 400 MHz; ^{13}C NMR: 100 MHz; ^{31}P NMR: 162 MHz) or a Bruker Avance 500 (^{1}H-NMR:

500 MHz; ^{13}C-NMR, 125.8 MHz). ^{1}H-NMR experiments were performed in CDCl$_3$ with CHCl$_3$ (δ = 7.26 ppm) or CD$_3$OD with CD$_2$HOD (δ = 3.31 ppm) as internal references. ^{13}C-NMR experiments were performed in CDCl$_3$ with CDCl$_3$ (δ = 77.23 ppm) or in CD$_3$OD with CD$_2$HOD (δ = 49.0 ppm) as internal references. For ^{31}P-NMR and ^{19}F-NMR references, the spectrometer had external references, corresponding to 80% phosphoric acid in D$_2$O (δ = 0 ppm) and to 0.05% α,α,α-trifluorotoluene in CDCl$_3$ (δ = $-$62.75 ppm). Chemical shifts are expressed in ppm and signal multiplicities are described using the following abbreviations: s for singlet, d for doublet, t for triplet, q for quartet, p for quintet and m for multiplet. In the presence of conformers, signals were differentiated by a * sign added to the assignments. Negative or positive-mode electrospray MS were performed on a Bruker Daltonics microTOF spectrometer (Bruker Daltonik GmbH, Bremen, Germany) equipped with an orthogonal electrospray (ESI) interface. Calibration was performed using a solution of 10 mM sodium formate. Sample solutions were introduced into the spectrometer source with a syringe pump (Harvard type 55 1111: Harvard Apparatus Inc., South Natick, MA, USA) with a flow rate of 5 µL min^{-1}.

4.1.2. Synthesis of α,α-Difluorophosphonate Derivatives

Synthesis of the intermediate diethyl 1,1-difluorobut-3-enylphosphonate (**12**) [31]. A solution of BrZnCF$_2$P(O)(OEt)$_2$ was prepared from diethyl (bromodifluoromethyl)phosphonate (2.67 g, 10 mmol) and Zn dust (0.65 g, 10 mmol) in DMF (18 mL). The reaction was stirred for 3 h at room temperature. CuBr (1.43 g, 10 mmol) was added, and the reaction mixture was stirred for 30 min before the addition of allyl bromide (0.43 mL, 5 mmol). The reaction mixture was stirred overnight, quenched with a 10% aqueous solution of HCl (10 mL), filtered through celite and extracted with Et$_2$O (3 × 15 mL). The organic layers were combined, washed with a saturated solution of NaHCO$_3$ then with brine, dried over anhydrous Na$_2$SO$_4$ and evaporated to dryness. The product is obtained after purification by column chromatography (EtOAc/PE 3:7 to 1:1) as a colorless oil (1.02 g, 89% yield). R_f = 0.54 (EtOAc/PE 1:1); ^{1}H NMR (300 MHz, CDCl$_3$): δ = 1.37 (6H, t, J = 7.1 Hz), 2.73–2.93 (2H, m), 4.26 (4H, dq, J_{HH} = J_{PH} = 7.1 Hz), 5.22–5.31 (2H, m), 5.73–5.92 (1H, m); ^{13}C NMR (75.5 MHz, CDCl$_3$): δ = 16.5 (CH$_3$ × 2, d, J_{CP} = 5.6 Hz), 38.8 (CH$_2$, dt, J_{CF} = 21.5 Hz, J_{CP} = 15.1 Hz), 64.5 (CH$_2$ × 2, d, J_{CP} = 6.8 Hz), 119.7 (CF$_2$, dt, J_{CF} = 260.0 Hz, J_{CP} = 215.1 Hz), 121.4 (CH$_2$), 127.1 (CH, dt, J_{CF} = 11.4 Hz, J_{CP} = 5.6 Hz); ^{19}F NMR (282 MHz, CDCl$_3$): δ = $-$111.26 (d, J_{FP} = 107.5 Hz); ^{31}P NMR (121 MHz, CDCl$_3$): δ = 6.9 (t, J_{PF} = 107.5 Hz).

4.1.3. Synthesis of the Target Compounds (**9**)

Diethyl (1,1-difluoro-3-oxopropyl)phosphonate (**13**) [32]. Ozone was bubbled through a solution of diethyl 1,1-difluorobut-3-enylphosphonate **12** (2.28 g, 10 mmol) in DCM/MeOH (50 mL, 4:1) at $-$78 °C until the solution turned blue (5 min). Nitrogen was bubbled through the solution until the blue color disappeared (removal of excess ozone). Dimethylsulfide (1.8 mL, 24.5 mmol) was added at $-$78 °C and the solution was allowed to warm up to room temperature. Solvents and excess of Me$_2$S were removed under reduced pressure, giving **13** as a colorless oil, which was immediately used without further purification for the next reaction. R_f = 0.46 (PE/EtOAc 1:1); ^{1}H NMR (300 MHz, CDCl$_3$): δ = 1.36 (6H, td, J_{HH} = 7.1 Hz, J_{HP} = 0.6 Hz), 3.06 (2H, tdd, J_{HF} = 19.1 Hz, J_{HP} = 7.0 Hz, J_{HH} = 2.5 Hz), 4.27 (4H, dq, J_{HP} = 8.2 Hz, J_{HH} = 7.1 Hz), 9.77 (1H, t, J = 2.5 Hz).

Diethyl (3-((benzyloxy)amino)-1,1-difluoropropyl)phosphonate (**14**) [21]. Diethyl (1,1-difluoro-3-oxopropyl)phosphonate **13** (2.3 g, 10 mmol) and N-hydroxybenzylamine hydrochloride (1.6 g, 10 mmol) in 8 mL MeOH were stirred for 1 h at room temperature. The solution was diluted with more MeOH (142 mL), and sodium cyanoborohydride (1.89 g, 30 mmol) was added portion wise over 30 min. The reaction mixture was cooled to 0 °C before the dropwise addition of HCl 37% in water (10 mL, 100 mmol) over 40 min. Sodium cyanoborohydride (0.44 g) was then added at room temperature and the mixture left to stir for 2 h. The solution was evaporated and treated with aqueous KOH (10%) to obtain a basic pH. The product was extracted with EtOAc (3 × 70 mL), and the combined organic layers were dried over Na$_2$SO$_4$ and evaporated to dryness under reduced pressure. **14** (1.19 g, 35% yield) was obtained as a transparent oil after purification by chromatography column on silica gel with PE/EtOAc (100:0 to 50:50) as eluent. R$_f$ = 0.54 (EtOAc/PE 7:3); ^{1}H NMR (500 MHz, CDCl$_3$): δ = 1.37 (6H, t, *J* = 7.1 Hz), 2.36 (2H, tq, *J*$_{HF}$ = 20.2 Hz, *J*$_{HH}$ = *J*$_{HP}$ = 6.6 Hz), 3.18–3.25 (2H, t, *J* = 7.1 Hz), 4.22–4.31 (4H, m), 4.70 (2H, s), 5.76 (1H, br s), 7.27–7.39 (5H, m); ^{13}C NMR (125 MHz, CDCl$_3$): δ = 16.4 (CH$_3$ × 2, d, *J*$_{CP}$ = 5.1 Hz), 32.2 (CH$_2$CF$_2$, td, *J*$_{CF}$ = 20.4 Hz, *J*$_{CP}$ = 14.4 Hz), 44.4 (CH$_2$N, q, *J*$_{CF}$ = 5.1 Hz), 64.6 (CH$_2$ × 2, d, *J*$_{CP}$ = 6.7 Hz), 76.2 (CH$_2$Ph), 120.5 (CF$_2$, td, *J*$_{CF}$ = 257.8 Hz, *J*$_{CP}$ = 214.7 Hz), 127.9 (CH$_{ar}$), 128.4 (CH$_{ar}$ × 2), 128.4 (CH$_{ar}$ × 2), 137.7 (C$_{ar}$); ^{19}F NMR (282 MHz, CDCl$_3$): δ = −111.08 (d, *J*$_{FP}$ = 107.9 Hz); ^{31}P NMR (121 MHz, CDCl$_3$): δ = 7.00 (t, *J*$_{PF}$ = 107.3 Hz).

Diethyl (3-(N-(benzyloxy)formamido)-1,1-difluoropropyl)phosphonate (**15a**). Formic acid (2.8 mL, 74.1 mmol) and acetic anhydride (1.4 mL, 14.8 mmol) were stirred for 30 min at room temperature. The solution was cooled to 0 °C then diethyl (3-((benzyloxy)amino)-1,1-difluoropropyl)phosphonate **14** (500 mg, 1.5 mmol) dissolved in anhydrous THF (1.5 mL) was added dropwise. The reaction mixture was stirred for 10 min at 0 °C and overnight at room temperature. EtOAc (15 mL) was added and the resulting organic layer washed with water (2 × 10 mL) and aqueous KOH (0.1 M, 10 mL), dried over Na$_2$SO$_4$ and evaporated to dryness under reduced pressure. A purification by flash chromatography on silica gel with EtOAc/PE 7:3 as eluent gave **15a** as a light yellow oil (450 mg, 83% yield) and as a mixture of two conformers in a 2:8 ratio, respectively. R$_f$ = 0.54 (EtOAc/PE 7:3); ^{1}H NMR (300 MHz, CDCl$_3$): δ = 1.37 (6H, t, *J* = 7.2 Hz), 2.29–2.52 (2H, m), 3.48–3.70 (2/10 of 2H, s), 3.76–3.94 (8/10 of 2H, s), 4.20–4.33 (4H, m), 4.86 (2H, s), 7.27–7.39 (5H, m), 8.16 (1H, s); ^{13}C NMR (125 MHz, CDCl$_3$): δ = 16.4 (CH$_3$ × 2, d, *J*$_{CP}$ = 5.4 Hz), 30.6–31.4 (CH$_2$CF$_2$, m), 37.2 (CH$_2$N), 41.8 (CH$_2$N*), 64.72 (CH$_2$ × 2, d, *J*$_{CP}$ = 6.7 Hz), 77.8 (CH$_2$Ph), 119.6 (CF$_2$, td, *J*$_{CF}$ = 258.6 Hz, *J*$_{CP}$ = 215.1 Hz), 128.8 (CH$_{ar}$), 129.2 (CH$_{ar}$*), 129.5 (CH$_{ar}$), 134.1 (C$_{ar}$), 134.6 (C$_{ar}$*), 158.2 (CO*), 163.3 (CO); ^{19}F NMR (282 MHz, CDCl$_3$): δ = −113.0 (d, *J*$_{FP}$ = 121.2 Hz), −113.4 (d, *J*$_{FP}$ = 106.2 Hz); ^{31}P NMR (121 MHz, CDCl$_3$): δ = 6.38 (t, *J*$_{PF}$ = 105.8 Hz); HRMS (ESI$^+$) *m/z* calcd for C$_{15}$H$_{22}$F$_2$NNaO$_5$P [M + Na]$^+$ 388.1096, found 388.1099.

Diethyl (3-(N-(benzyloxy)acetamido)-1,1-difluoropropyl)phosphonate (**15b**) [21]. Anhydrous pyridine (0.4 mL, 4.4 mmol) was added dropwise to a solution of diethyl (3-((benzyloxy)amino)-1,1-difluoropropyl)phosphonate **14** (500 mg, 1.5 mmol) in acetic anhydride (6 mL, 63.5 mmol). The reaction mixture was stirred overnight at room temperature then evaporated to dryness under reduced pressure. A purification by flash chromatography on silica gel with EtOAc/PE 7:3 as eluent gave **15b** as a transparent oil (450 mg, 80% yield) and as a single conformer. R$_f$ = 0.55 (EtOAc/PE 7:3); ^{1}H NMR (300 MHz, CDCl$_3$): δ = 1.37 (6H, t, *J* = 7.6 H), 2.07 (3H, s), 2.29–2.53 (2H, m), 3.87–3.95 (2H, m), 4.26 (4H, pseudo p, *J*$_{HH}$ = *J*$_{HP}$ = 7.2 Hz), 4.83 (2H, s), 7.30–7.43 (5H, m); ^{13}C NMR (125 MHz, CDCl$_3$): δ = 16.4 (CH$_3$ × 2, d, *J*$_{CP}$ = 5.4 Hz), 20.6 (CH$_3$CO), 30.5–31.1 (CH$_2$CF$_2$, m), 38.5 (CH$_2$N), 64.7 (CH$_2$ × 2, d, *J*$_{CP}$ = 6.8 Hz), 76.5 (OCH$_2$Ph), 119.8 (CF$_2$, td, *J*$_{CF}$ = 259.9 Hz, *J*$_{CP}$ = 216.3 Hz), 128.8 (CH$_{ar}$ × 2), 129.1 (CH$_{ar}$), 129.3 (CH$_{ar}$ × 2), 134.2 (C$_{ar}$), 172.9 (CO); ^{19}F NMR (282 MHz, CDCl$_3$): δ = −112.33 (d, *J*$_{FP}$ = 106.7 Hz); ^{31}P NMR (121 MHz, CDCl$_3$): δ = 6.38 (t, *J*$_{PF}$ = 106.7 Hz).

Diethyl (1,1-difluoro-3-(N-hydroxyformamido)propyl)phosphonate (**16a**). Diethyl (3-(N-(benzyloxy)formamido)-1,1-difluoropropyl)phosphonate **15a** (438 mg, 1.2 mmol) and palladium on charcoal (45 mg, 10%) were introduced in a 2-neck-flask with a three-way tap under a nitrogen atmosphere. After the addition of HPLC grade MeOH (30 mL), the reaction mixture was degassed three times then left to stir for 24 h under hydrogen at atmospheric pressure. The solution was filtered over celite and evaporated to dryness under reduced pressure. A purification by chromatography column with EtOAc as eluent gave **16a** as an orange oil (245 mg, 74% yield) and as a mixture of two conformers in a 1:1 ratio. R$_f$ = 0.39 (EtOAc); ^{1}H NMR (500 MHz, CDCl$_3$): δ = 1.38 (6H, t, *J* = 7.1 Hz), 2.37–2.58 (2H, m), 3.80–3.88 (2H, m), 4.28 (4H, pseudo p, *J*$_{HH}$ = *J*$_{HP}$ = 7.4 Hz), 7.91 (1/2 of 1H, s), 8.35 (1/2 of 1H, s); ^{13}C NMR (125 MHz, CDCl$_3$): δ = 16.3 (CH$_3$ × 2, d, *J*$_{CP}$ = 5.4 Hz), 16.4 (CH$_3$ × 2, d, *J* = 5.4 Hz), 31.6 (CH$_2$CF$_2$, dt, *J*$_{CF}$ = 20.7 Hz, *J*$_{CP}$ = 14.4 Hz), 32.7 (CH$_2$CF$_2$*, dt,

J_{CF} = 20.7 Hz, J_{CP} = 14.4 Hz), 41.6 (CH$_2$N, q, J_{CF} = 6.1 Hz), 42.5 (CH$_2$N*, q, J_{CF} = 6.1 Hz), 65.0 (CH$_2$O × 2, d, J_{CP} = 7.0 Hz), 65.6 (CH$_2$O* × 2, d, J_{CP} = 7.0 Hz), 119.6 (CF$_2$, td, J_{CF} = 259.3 Hz, J_{CP} = 211.7 Hz), 119.9 (CF$_2$*, td, J_{CF} = 259.3 Hz, J_{CP} = 211.7 Hz), 156.3 (CO*), 163.1 (CO); ^{19}F NMR (282 MHz, CDCl$_3$): δ = −112.70 (d, J_{FP} = 105.0 Hz), −109.12 (d, J_{FP} = 107.7 Hz); ^{31}P NMR (121 MHz, CDCl$_3$) δ 6.14 (t, J_{PF} = 104.7 Hz), 7.39 (t, J_{PF} = 107.4 Hz); HRMS (ESI$^+$) m/z calcd for C$_8$H$_{16}$F$_2$NNaO$_5$P [M + Na]$^+$ 298.0626, found 298.0625.

Diethyl (1,1-difluoro-3-(N-hydroxyacetamido)propyl)phosphonate (**16b**) [21]. Diethyl (3-(N-(benzyloxy)acetamido)-1,1-difluoropropyl)phosphonate **15b** (450 mg, 1.2 mmol) and palladium on charcoal (45 mg, 10%) were introduced in a 2-neck-flask with a three-way tap under a nitrogen atmosphere. After the addition of HPLC grade MeOH (30 mL), the reaction mixture was degassed three times then left to stir for 24 h under a hydrogen atmosphere. The solution was filtered over a pad of celite and evaporated to dryness under reduced pressure. A purification by chromatography column with EtOAc as eluent gave **16b** as a light yellow oil (280 mg, 80% yield) and as a mixture of two conformers in a 2:8 ratio respectively. R_f = 0.41 (EtOAc); ^{1}H NMR (500 MHz, CDCl$_3$): δ = 1.38 (6H, t, J = 7.1 Hz), 2.13 (3H, s), 2.37–2.55 (2H, m), 3.82–3.90 (8/10 of 2H, t, J = 5.9 Hz), 3.91–3.99 (2/10 of 2H, m), 4.27 (4H, pseudo p, J_{HH} = J_{HP} = 6.8 Hz); ^{13}C NMR (125 MHz, CDCl$_3$): δ = 16.3 (CH$_3$ × 2, d, J_{CP} = 5.5 Hz), 18.3 (CH$_3$CO*), 20.4 (CH$_3$CO), 31.2–31.7 (CH$_2$CF$_2$*, m), 33.3 (CH$_2$CF$_2$, dt, J_{CF} = 22.0 Hz, J_{CP} = 14.7 Hz), 41.9–42.2 (CH$_2$N*, m), 43.3–43.6 (CH$_2$N, m), 64.7–65.0 (CH$_2$O* × 2, m), 65.50 (CH$_2$O × 2, d, J_{CP} = 7.3 Hz), 120.2 (CF$_2$, td, J_{CF} = 258.7 Hz, J_{CP} = 210.7 Hz), 164.8 (CO*), 172.6 (CO); ^{19}F NMR (282 MHz, CDCl$_3$): δ = −113.49 (d, J_{FP} = 105.1 Hz), −108.2 (d, J_{FP} = 108.5 Hz); ^{31}P NMR (121 MHz, CDCl$_3$): δ = 6.17 (t, J_{PF} = 106.1 Hz), 7.57 (t, J_{PF} = 108.1 Hz).

(1,1-Difluoro-3-(N-hydroxyacetamido)propyl)phosphonic acid (**9b**). A solution of diethyl (1,1-difluoro-3-(N-hydroxyacetamido)propyl)phosphonate **16b** (25 mg, 86 µmol) dissolved in DCM (0.32 mL) was cooled to 0 °C. TMSBr (0.11 mL, 0.9 mmol) was added dropwise at 0 °C then the reaction mixture was stirred overnight in the dark at room temperature. DCM and excess of TMSBr were evaporated under reduced pressure and the intermediate orange oil was treated with water (0.5 mL, 27.8 mmol) for 1.5 h. Removal of water under vacuum afforded **9b** as an orange oil (20 mg, quant. yield) and as a mixture of conformers. R_f = 0.14 (EtOAc); ^{1}H NMR (400 MHz, CD$_3$OD): δ = 2.02 (5/10 of 3H, s), 2.10 (5/10 of 3H, s), 2.34–2.46 (4/10 of 2H, m), 2.52–2.64 (6/10 of 2H, m), 3.53 (5/10 of 2H, t, J = 7.6 Hz), 3.89 (4/10 of 2H, t, J = 7.6 Hz), 4.10–4.26 (1/10 of 2H, m); ^{13}C NMR (125 MHz, CD$_3$OD): δ = 20.2 (CH$_3$CO*), 20.6 (CH$_3$CO), 29.6 (CH$_2$CF$_2$, td, J_{CF} = 22.3 Hz, J_{CP} = 15.7 Hz), 32.0 (CH$_2$CF$_2$*, td, J_{CF} = 20.2 Hz, J_{CP} = 14.5 Hz), 42.2–42.5 (CH$_2$N*, m), 45.6–45.8 (CH$_2$N, m), 120.6 (CF$_2$, td, J_{CF} = 258.9 Hz, J_{CP} = 210.2 Hz), 173.4 (CO), 173.9 (CO*); ^{19}F NMR (282 MHz, CD$_3$OD): δ = −115.5 (d, J_{FP} = 100.3 Hz), −115.7 (d, J_{FP} = 99.4 Hz), −116.3 (d, J_{FP} = 103.7 Hz); ^{31}P NMR (162 MHz, CD$_3$OD): δ = 5.00 (t, J_{PF} = 103.8 Hz), 3.96 (t, J_{PF} = 100.1 Hz), 3.78 (t, J_{PF} = 99.6 Hz); HRMS (ESI$^-$) m/z calcd for C$_5$H$_{10}$F$_2$NO$_5$P [M − H]$^-$ 232.0192, found 232.0181.

4.1.4. Synthesis of the Target Compounds (**10**)

Diethyl 1,1-difluorobutan-4-ol phosphonate (**17**). A 1 M solution of borane in THF (32 mL, 32 mmol) was added dropwise to a stirred solution of diethyl 1,1-difluorobut-3-enylphosphonate **12** (1.84 g, 8 mmol) in THF (20 mL) at 0 °C. After 30 min, the reaction mixture was allowed to warm up at room temperature for 4 h. More borane solution (16 mL) was added dropwise at 0 °C and the mixture was stirred overnight at room temperature. Then, methanol (10 mL), a 3 M solution of NaOH in water (4 mL) and a 30% aqueous solution of H$_2$O$_2$ (4 mL) were successively added. The mixture was heated at 50 °C for 1 h, quenched with brine (30 mL) and extracted with chloroform (3 × 20 mL). The organic layers were combined, dried over anhydrous Na$_2$SO$_4$ and evaporated to dryness. Purification by column chromatography (PE/EtOAc 3:2 to 100% EtOAc) yielded **17** as a colorless oil (770 mg, 39% yield). R_f = 0.29 (EtOAc/PE 7:3); ^{1}H NMR (300 MHz, CDCl$_3$): δ = 1.37 (6H, td, J_{HH} = 7.1 Hz, J_{PH} = 0.6 Hz), 1.62 (1H, br s, OH), 1.78–1.87 (2H, m), 2.06–2.26 (2H, m), 3.67

(2H, t, J = 6.2 Hz), 4.26 (4H, dq, J_{PH} = 7.9 Hz, J_{HH} = 7.1 Hz); ^{13}C NMR (125 MHz, CDCl$_3$): δ = 16.4 (CH$_3$, d, J_{CP} = 5.5 Hz), 24.0 (CH$_2$, dt, J_{CF} = 4.9 Hz, J_{CP} = 4.4 Hz), 30.5 (CH$_2$, td, J_{CF} = 21.1 Hz, J_{CP} = 14.9 Hz), 61.9 (CH$_2$OH), 64.5 (CH$_2$ × 2, d, J_{CP} = 6.8 Hz), 121.0 (CF$_2$, td, J_{CF} = 259.3 Hz, J_{CP} = 215.7 Hz); ^{19}F NMR (282 MHz, CDCl$_3$): δ = 112.7 (d, J_{FP} = 109.3 Hz); ^{31}P NMR (121 MHz, CDCl$_3$): δ = 7.49 (t, J_{PF} = 109.3 Hz).

Difluoro-4-(diethyl phosphonate) butanoic acid (18) [21]. To a stirred solution of **17** (400 mg, 1.6 mmol) in MeCN (4 mL) were successively added TEMPO (51 mg, 0.3 mmol), BAIB (1.151 g, 3.6 mmol) and water (4 mL). The reaction mixture was stirred overnight then concentrated under reduced pressure. An aqueous saturated solution of NaHCO$_3$ was added to reach a pH > 8. The solution was extracted with DCM (3 × 15 mL) to remove excess of starting materials. A 10% HCl solution was added to the aqueous layer until pH < 5. Then the solution was extracted with DCM (3 × 15mL), the organic layer was dried over anhydrous Na$_2$SO$_4$ and evaporated to dryness. The product was not further purified and obtained as a light yellow oil (420 mg, 99% yield). R$_f$ = 0.26 (EtOAc/PE 7:3); ^{1}H NMR (500 MHz, CDCl$_3$): δ = 1.38 (6H, t, J = 7.1 Hz), 2.35–2.51 (2H, m), 2.63–2.70 (2H, m), 4.28–4.31 (4H, pseudo p, J_{HH} = J_{PH} = 7.2 Hz), 9.61 (1H, br s); ^{13}C NMR (125 MHz, CDCl$_3$): δ = 16.5 (CH$_3$ × 2, d, J_{CP} = 5.5 Hz), 25.8 (CH$_2$CF$_2$, q, J_{CP} = 5.4 Hz), 29.1 (CH$_2$CO, dt, J_{CF} = 21.2 Hz, J_{CP} = 15.9 Hz), 64.8 (CH$_2$ × 2, d, J_{CP} = 6.9 Hz), 119.8 (CF$_2$, dt, J_{CF} = 259.8 Hz, J_{CP} = 217.2 Hz), 176.5 (CO); ^{19}F NMR (282 MHz, CDCl$_3$): δ = −112.75 (d, J_{FP} = 107.6 Hz); ^{31}P NMR (121 MHz, CDCl$_3$): δ = 6.48 (t, J_{PF} = 107.5 Hz.

Diethyl 4-(N-benzyloxy)-amino-1,1-difluoro-4-oxobutyl phosphonate (19a) [21]. CDI (0.528 g, 3.2 mmol) and BnONH$_2$·HCl (0.567 g, 3.5 mmol) were added to a solution of 4,4-difluoro-4-(diethylphosphonate)butanoic acid **18** (0.770 g, 2.9 mmol) in DCM (60 mL). The mixture was stirred overnight at room temperature. The reaction was quenched with a saturated aqueous solution of NH$_4$Cl (60 mL), and the resulting mixture was extracted with DCM (3 × 20 mL). The organic layers were combined, washed with brine (60 mL), dried over anhydrous Na$_2$SO$_4$ and evaporated to dryness under reduced pressure. The product was purified by flash chromatography (EtOAc/PE 7:3) and obtained as a colorless oil (995 mg, 92% yield) as a mixture of two conformers in a 7:3 ratio respectively. R$_f$ = 0.35 (EtOAc/PE 7:3); ^{1}H NMR (300 MHz, CDCl$_3$): δ = 1.37 (6H, t, J = 7.1 Hz), 2.27–2.55 (7/10 of 4H, m), 2.59–2.76 (3/10 of 4H, m), 4.23–4.31 (4H, pseudo p, J_{HH} = J_{PH} = 7.2 Hz), 4.90 (2H, s), 7.35–7.39 (5H, m), 8.03 (3/10 of 1H, br s, NH), 8.73 (7/10 of 1H, br s, NH); ^{13}C NMR (125 MHz, CDCl$_3$): δ = 16.4 (CH$_3$ × 2, d, J_{CP} = 5.5 Hz), 23.3–23.7 (CH$_2$CF$_2$*, m), 24.7–25.3 (CH$_2$CF$_2$, m), 28.0–28.7 (CH$_2$CO*, m), 29.7 (CH$_2$CO, dt, J_{CF} = 19.5 Hz, J_{CP} = 14.4 Hz), 64.3–65.0 (CH$_2$ × 2, m), 78.2 (CH$_2$Ph), 79.4 (CH$_2$Ph*), 120.3 (CF$_2$, dt, J_{CF} = 260.3 Hz, J_{CP} = 214.6 Hz), 128.6 (CH$_{ar}$ × 2), 128.7 (CH$_{ar}$), 129.1 (CH$_{ar}$ × 2), 135.3 (C$_{ar}$), 168.9 (CO), 175.1 (CO*); ^{19}F NMR (282 MHz, CDCl$_3$): δ = −112.52 (d, J_{FP} = 109.0 Hz), −111.34 (d, J_{FP} = 107.4 Hz); ^{31}P NMR (121 MHz, CDCl$_3$): δ = 6.37 (t, J_{PF} = 107.1 Hz), 6.65 (t, J_{PF} = 106.5 Hz.

Diethyl 4-(N,N-benzyloxy-methyl)-amino-1,1-difluoro-4-oxobutyl phosphonate (19b). To a stirred solution of diethyl 4-(N-benzyloxy)-amino-1,1-difluoro-4-oxobutyl phosphonate **18** (167 mg, 0.5 mmol) in anhydrous acetone (18 mL) was added anhydrous K$_2$CO$_3$ (82 mg, 0.6 mmol). The reaction was stirred under reflux for 30 min. Then, MeI (324 mg, 2.3 mmol) was added at room temperature and the solution was stirred overnight under reflux. The solution was filtered and evaporated to dryness under reduced pressure. The product was chromatographed on silica gel with gradient elution by EtOAc/PE 7:3 to 100% EtOAc to give **19b** as a yellow oil (165 mg, 94% yield) and as a single conformer. R$_f$ = 0.55 (EtOAc); ^{1}H NMR (500 MHz, CDCl$_3$): δ = 1.36 (6H, t, J = 7.1 Hz), 2.32–2.44 (2H, m), 2.67–2.70 (2H, m), 3.20 (3H, s), 4.23–4.31 (4H, m), 4.84 (2H, s), 7.36–7.40 (5H, m); ^{13}C NMR (125 MHz, CDCl$_3$): δ = 16.4 (CH$_3$ × 2, d, J_{CP} = 5.5 Hz), 24.0 (CH$_2$CO), 28.9 (CH$_2$CF$_2$, dt, J_{CF} = 20.9 Hz, J_{CP} = 15.0 Hz), 33.75 (CH$_3$N), 64.5 (CH$_2$ × 2, d, J_{CP} = 6.7 Hz), 76.4 (CH$_2$Ph), 120.4 (CF$_2$, dt, J_{CF} = 257.7 Hz, J_{CP} = 215.2 Hz), 128.8 (CH$_{ar}$ × 2), 129.1 (CH$_{ar}$), 129.4 (CH$_{ar}$ × 2), 134.2 (C$_{ar}$), 173.1 (CO); ^{19}F NMR (282 MHz, CDCl$_3$): δ = −113.39 (d, J_{FP} = 108.1 Hz); ^{31}P NMR (121 MHz, CDCl$_3$): δ = 7.07 (t, J_{PF} = 107.7 Hz); HRMS (ESI$^+$) m/z calcd for C$_{16}$H$_{25}$F$_2$NO$_5$P [M + H]$^+$ 380.1433, found 380.1445.

Diethyl 4-(N-hydroxyl)-amino-1,1-difluoro-4-oxobutyl phosphonate (**20a**) [21]. To a stirred solution of diethyl 4-(N-benzyloxy)-amino-1,1-difluoro-4-oxobutyl phosphonate **19a** (217 mg, 0.6 mmol) in MeOH (15 mL) was added 10% w/w Pd on activated carbon (22 mg, 0.02 mmol). The reaction flask was then connected to a balloon of H_2 at atmospheric pressure. Every 20 min, a slight vacuum was applied to the reaction flask, which was then backfilled with H_2. When all starting material was consumed, the reaction was filtered over a pad of celite and evaporated to dryness under reduced pressure. The product was obtained pure in a quantitative yield (165 mg) as a colorless oil and as a mixture of two conformers Z and E in a 55:45 ratio respectively. R_f = 0.23 (EtOAc); ^{1}H NMR (500 MHz, CDCl$_3$): δ = 1.37 (6H, t, J = 7.0 Hz), 2.28–2.88 (4H, m), 4.28 (4H, pseudo p, J_{HH} = J_{PH} = 7.1 Hz), 5.65 (45/100 of 1H, br s, OH), 5.78 (55/100 of 1H, br s, OH), 8.04 (45/100 of 1H, br s, NH), 9.30 (55/100 of 1H, br s, NH); ^{13}C NMR (125 MHz, CDCl$_3$): δ = 16.4 (CH$_3$ × 2, d, J_{CP} = 5.5 Hz), 24.7–24.9 (CH$_2$CF$_2$*, m), 27.2 (CH$_2$CF$_2$, q, J_{CF} = 4.8 Hz), 29.7 (CH$_2$CO, dt, J_{CF} = 21.3 Hz, J_{CP} = 16.2 Hz), 29.9 (CH$_2$CO*, dt, J_{CF} = 21.3 Hz, J_{CP} = 16.2 Hz), 64.7 (CH$_2$* × 2, d, J_{CP} = 6.9 Hz), 65.0 (CH$_2$ × 2, d, J_{CP} = 7.0 Hz), 120.2 (CF$_2$, J_{CF} = 257.5 Hz, J_{CP} = 216.0 Hz), 169.0 (CO*), 173.3 (CO); ^{19}F NMR (282 MHz, CDCl$_3$): δ = −111.22 (d, J_{FP} = 107.3 Hz), −112.69 (d, J_{FP} = 107.7 Hz); ^{31}P NMR (121 MHz, CDCl$_3$): δ = 6.31 (t, J_{PF} = 107.2 Hz), 6.52 (t, J_{PF} = 107.7 Hz).

Diethyl 4-(N,N-hydroxyl-methyl)-amino-1,1-difluoro-4-oxobutyl phosphonate (**20b**). To a stirred solution of diethyl 4-(N,N-benzyloxy-methyl)-amino-1,1-difluoro-4-oxobutyl phosphonate **19b** (120 mg, 0.3 mmol) in MeOH (8 mL) and water (2.4 mL) was added 10% w/w Pd on activated carbon (12 mg, 0.01 mmol). The reaction flask was then connected to a balloon of H_2 (1 atm). Every 20 min, a slight vacuum was applied to the reaction flask, which was then backfilled with H_2. When all starting material was consumed, the reaction was filtered over celite and evaporated to dryness under reduced pressure. The pure product was obtained as a light yellow oil (90 mg, quantitative yield) and as a mixture of conformers in a 6:4 ratio. R_f = 0.33 (EtOAc); ^{1}H NMR (500 MHz, CDCl$_3$): δ = 1.38 (6H, t, J = 7.0 Hz), 2.29–2.51 (2H, m), 2.57–2.70 (4/10 of 2H, m), 2.70–2.86 (6/10 of 2H, m), 3.25 (6/10 of 3H, br s), 3.36 (4/10 of 3H, br s), 4.19–4.35 (4H, pseudo p, J_{HH} = J_{PH} = 7.3 Hz); ^{13}C NMR (125 MHz, CDCl$_3$): δ = 16.4 (CH$_3$ × 2, d, J_{CP} = 4.7 Hz), 22.8 (CH$_2$CO*), 24.0 (CH$_2$CO), 27.8–28.1 (CH$_2$CF$_2$*, m), 29.0–30.3 (CH$_2$CF$_2$, m), 35.7 (CH$_3$N*), 36.1 (CH$_3$N), 64.7 (CH$_2$ × 2, d, J_{CP} = 6.7 Hz), 65.0 (CH$_2$* × 2, d, J_{CP} = 6.7 Hz), 119.9 (CF$_2$*, dt, J_{CF} = 261.0 Hz, J_{CP} = 220.3 Hz), 120.3 (CF$_2$, dt, J_{CF} = 261.0 Hz, J_{CP} = 220.3 Hz), 165.2 (CO*), 171.5 (CO*), 172.2 (CO), 173.6 (CO*); ^{19}F NMR (282 MHz, CDCl$_3$): δ = −110.9 (d, J_{FP} = 109.3 Hz), −112.4 (d, J_{FP} = 108.0 Hz), −113.4 (d, J_{FP} = 105.9 Hz), −113.6 (d, J_{FP} = 107.5 Hz); ^{31}P NMR (121 MHz, CDCl$_3$): δ = 6.61 (t, J_{PF} = 108.3 Hz), 6.75 (t, J_{PF} = 107.7 Hz); HRMS (ESI$^+$) m/z calcd for C$_9$H$_{18}$F$_2$NNaO$_5$P [M + Na]$^+$ 312.0783, found 312.0794.

(1,1-Difluoro-4-(hydroxyamino)-4-oxobutyl)phosphonic acid (**10a**). A solution of diethyl 4-(N-hydroxyl)-amino-1,1-difluoro-4-oxobutyl phosphonate **20a** (25 mg, 90.8 μmol) dissolved in DCM (0.34 mL) was cooled to 0 °C. TMSBr (0.12 mL, 0.9 mmol) was added dropwise at 0 °C then the reaction mixture was stirred overnight in the dark at room temperature. DCM and excess of TMSBr were evaporated under reduced pressure and the intermediate orange oil was treated with water (0.5 mL, 27.8 mmol) for 1.5 h. Removal of water under vacuum affords **10a** as a light orange solid (20 mg, quantitative yield) and as a mixture of conformers. R_f = 0.4 (EtOAc); ^{1}H NMR (400 MHz, CD$_3$OD): δ = 2.33–2.45 (2H, m), 2.50–2.53 (4/10 of 2H, m), 2.60 (6/10 of 2H, t, J = 7.5 Hz); ^{13}C NMR (125 MHz, CD$_3$OD): δ = 26.9–27.1 (CH$_2$CF$_2$, m), 28.0–28.1 (CH$_2$CF$_2$*, m), 30.3 (CH$_2$CO, dt, J_{CF} = 21.5 Hz, J_{CP} = 15.8 Hz), 30.6–31.1 (CH$_2$CO*, m), 121.5 (CF$_2$, dt, J_{CF} = 257.8 Hz, J_{CP} = 210.8 Hz), 174.2 (CO), 177.2 (CO*); ^{19}F NMR (282 MHz, CD$_3$OD): δ = −116.6 (d, J_{FP} = 104.6 Hz), −116.8 (d, J_{FP} = 104.4 Hz), −116.8 (d, J_{FP} = 104.1 Hz); ^{31}P NMR (162 MHz, CD$_3$OD): δ = 5.12 (t, J_{PF} = 104.0 Hz), 5.21 (t, J_{PF} = 104.4 Hz), 5.32 (t, J_{PF} = 104.9 Hz); HRMS (ESI$^+$) m/z calcd for C$_4$H$_9$F$_2$NO$_5$P [M + H]$^+$ 220.0181, found 220.0192.

(1,1-Difluoro-4-(hydroxy(methyl)amino)-4-oxobutyl)phosphonic acid (**10b**). A solution of diethyl 4-(N,N-hydroxyl-methyl)-amino-1,1-difluoro-4-oxobutyl phosphonate **20b** (20 mg, 69.2 μmol) dissolved in DCM (0.26 mL) was cooled to 0 °C. TMSBr (0.10 mL, 0.7 mmol)

was added dropwise at 0 °C then the reaction mixture was stirred overnight in the dark at room temperature. DCM and excess of TMSBr were evaporated under reduced pressure and the intermediate orange oil was treated with water (0.5 mL, 27.8 mmol) for 1.5 h. Removal of water under vacuum affords **10b** as an orange oil (16 mg, quant. yield) and as a mixture of conformers. R_f = 0.1 (EtOAc); ^{1}H NMR (400 MHz, CD$_3$OD): δ = 2.28–2.47 (2H, m), 2.56–2.63 (4/10 of 2H, m), 2.74–2.78 (6/10 of 2H, m), 2.95 (6/10 of 3H, s), 3.21 (4/10 of 3H, s); ^{13}C NMR (125 MHz, CD$_3$OD): δ = 25.0–25.2 (CH$_2$CF$_2$*, m), 26.8–27.0 (CH$_2$CF$_2$, m), 30.2 (CH$_2$CO*, dt, J_{CF} = 19.9 Hz, J_{CP} = 15.7 Hz), 30.3 (CH$_2$CO, dt, J_{CF} = 21.2 Hz, J_{CP} = 15.7 Hz), 36.4 (CH$_3$N*), 38.1 (CH$_3$N), 121.7 (CF$_2$, td, J_{CF} = 258.4 Hz, J_{CP} = 212.1 Hz), 174.0 (CO*), 174.2 (CO); ^{19}F NMR (282 MHz, CD$_3$OD): δ = −118.1 (d, J_{FP} = 104.1 Hz), −118.4 (d, J_{FP} = 104.1 Hz); ^{31}P NMR (162MHz, CD$_3$OD): δ = 5.21 (t, J_{PF} = 103.7 Hz), 5.33 (t, J_{PF} = 105.3 Hz), 5.48 (t, J_{PF} = 105.8 Hz); HRMS (ESI$^-$) m/z calcd for C$_5$H$_{10}$F$_2$NO$_5$P [M − H]$^-$ 232.0192, found 232.0206.

4.2. Biological Activity

4.2.1. His-Tagged DXR Activity

The assays were performed at 37 °C in a 50 mM Tris/HCl buffer pH 7.5 containing 3 mM MgCl$_2$ and 2 mM DTT. The concentrations of DXP and NADPH were 480 µM and 160 µM respectively. The decrease of absorbance at 340 nm due to NADPH oxidation was monitored to determine the initial rates. The retained values were the average of at least two measurements. The relative average deviation must be lower than 4%.

4.2.2. Inhibition of His-Tagged DXR

Fosmidomycin was purchased from Fujisawa Pharmaceutical. Inhibitor concentrations in the stock solutions were verified by spectrophotometric phosphorus determination [28]. The study compounds **9b** and **10a**, **10b** were tested against *E. coli* DXR using a photometric assay that was described earlier [8]. H-DXR was pre-incubated during 2 min in the presence of the inhibitors **9b**, **10a** and **10b** at different concentrations and NADPH. DXP was then added to measure the residual activity. The inhibitory potential of the tested compounds was quantified by determining the IC$_{50}$ values. They were obtained by plotting the percentage of residual activity versus the Log of inhibitor concentration.

4.2.3. Bacterial Growth Inhibition

The antimicrobial activity of hydroxamic acids **9b**, **10a** and **10b** against *E. coli* XL1 Blue [8] and fosmidomycin-resistant strain *E. coli* FosR [7], was determined using the paper disc diffusion method. LB agar plates (9 cm diameter) were inoculated with a suspension of bacteria (200 µL, mid-exponential phase). Paper discs (Durieux no. 268, diameter 6 mm) impregnated with a volume ≤8 µL of fosmidomycin derivatives were placed on Petri dishes. Growth inhibition was examined after 24 h incubation at 37 °C.

Supplementary Materials: The following are available online.^{1}H-NMR, ^{13}C-NMR, ^{19}F-NMR and ^{31}P-NMR spectra of compounds **5–20** are included.

Author Contributions: A.D., F.S.K., C.N. and M.M. are first coauthors contributing equally to the experiments. D.T. performed biological assays; Methodology, D.L.; Conceptualization, M.R., C.G.-B.; writing—original draft preparation, C.G.-B.; supervision, C.G.-B.; writing—review and editing, D.L., M.R., C.G.-B. All authors have read and agreed to the published version of the manuscript.

Funding: This research received no external funding.

Institutional Review Board Statement: Not applicable.

Informed Consent Statement: Not applicable.

Data Availability Statement: The data presented in this study are contained within the article and are also available in the Supplementary Materials.

Acknowledgments: M.M. and F.K. acknowledge financial support from the 'Ministère de la Recherche' and M.M. in addition a partial support from the Foundation "Frontier Research in Chemistry" (Strasbourg, France). The authors express their gratitude to L. Allouche, B. Vincent and M. Coppe for NMR measurements.

Conflicts of Interest: The authors declare no conflict of interest.

Sample Availability: Not available.

References

1. Word Health Organisation. News Released 27 February 2017. Available online: https://www.who.int/news-room/detail/27-02-2017-who-publishes-list-of-bacteria-for-which-new-antibiotics-are-urgently-needed (accessed on 10 May 2020).
2. European Centre for Disease Prevention and Control. 15 November 2018. Available online: https://www.ecdc.europa.eu/en/publications-data/infographic-antibiotic-resistance-increasing-threat-human-health (accessed on 10 May 2020).
3. Rohmer, M.; Grosdemange-Billiard, C.; Seemann, M.; Tritsch, D. Isoprenoid biosynthesis as a novel target for antibacterial and antiparasitic drugs. *Curr. Opin. Investig. Drugs* **2004**, *5*, 154–162.
4. Okuhara, M.; Kuroda, Y.; Goto, T.; Okamoto, M.; Terano, H.; Kohsaka, M.; Aoki, H.; Imanaka, H. Studies on new phosphonic acid antibiotics. I. FR-900098, isolation and characterization. *J. Antibiot.* **1980**, *33*, 13–17. [CrossRef] [PubMed]
5. Okuhara, M.; Kuroda, Y.; Goto, T.; Okamoto, M.; Terano, H.; Kohsaka, M.; Aoki, H.; Imanaka, H. Studies on new phosphonic acid antibiotics. III., isolation and characterization of FR-31564, FR-32863 and FR-33289. *J. Antibiot.* **1980**, *33*, 24–28. [CrossRef]
6. Lanaspa, M.; Moraleda, C.; Machevo, S.; González, R.; Serrano, B.; Macete, E.; Cisteró, P.; Mayor, A.; Hutchinson, D.; Kremsner, P.G.; et al. Inadequate efficacy of a new formulation of fosmidomycin-clindamycin combination in Mozambican children less than three years old with uncomplicated *Plasmodium falciparum* malaria. *Antimicrob. Agents Chemother.* **2012**, *56*, 2923–2928. [CrossRef]
7. Hemmerlin, A.; Tritsch, D.; Hammann, P.; Rohmer, M.; Bach, T.J. Profiling of defense responses in *Escherichia coli* treated with fosmidomycin. *Biochimie* **2014**, *99*, 54–62. [CrossRef] [PubMed]
8. Kuntz, L.; Tritsch, D.; Grosdemange-Billiard, C.; Hemmerlin, A.; Willem, A.; Bach, T.J.; Rohmer, M. Isoprenoid biosynthesis as a target for antibacterial and antiparasitic drugs: Phosphonohydroxamic acids as inhibitors of deoxyxylulose phosphate reducto-isomerase. *Biochem. J.* **2005**, *386*, 127–135. [CrossRef]
9. Munier, M.; Tritsch, D.; Krebs, F.; Esque, J.; Hemmerlin, A.; Rohmer, M.; Stote, R.H.; Grosdemange-Billiard, C. Synthesis and biological evaluation of phosphate isosters of fosmidomycin and analogs as inhibitors of *Escherichia coli* and *Mycobacterium smegmatis* 1-deoxyxylulose 5-phosphate reductoisomerases. *Bioorg. Med. Chem.* **2017**, *25*, 684–689. [CrossRef]
10. Woo, Y.-H.; Fernandes, R.P.M.; Proteau, P.J. Evaluation of fosmidomycin analogs as inhibitors of the *Synechocystis* sp. PCC6803 1-deoxy-D-xylulose 5-phosphate reductoisomerase. *Bioorg. Med. Chem.* **2006**, *14*, 2375–2385. [CrossRef] [PubMed]
11. Nieschalk, J.; O'Hagan, D. Monofluorophosphonates as phosphate mimics in bioorganic chemistry: A comparative study of CH_2-, CHF- and CF_2-phosphonate analogues of *sn*-glycerol-3-phosphate as substrates for *sn*-glycerol-3-phosphate dehydrogenase. *J. Chem. Soc. Chem. Commun.* **1995**, *7*, 719–720. [CrossRef]
12. Nieschalk, J.; Batsanov, A.S.; O'Hagan, D.; Howard, J. Synthesis of monofluoro- and difluoro-methylenephosphonate analogues of *sn*-glycerol-3-phosphate as substrates for glycerol-3-phosphate dehydrogenase and the X-ray structure of the fluoromethylenephosphonate moiety. *Tetrahedron* **1996**, *52*, 165–176. [CrossRef]
13. Romanenko, V.D.; Kukhar, V.P. Fluorinated phosphonates: Synthesis and biomedical application. *Chem. Rev.* **2006**, *106*, 3868–3935. [CrossRef] [PubMed]
14. Verbrugghen, T.; Cos, P.; Maes, L.; Van Calenbergh, S. Synthesis and evaluation of -halogenated analogues of 3-(acetylhydroxyamino)propylphosphonic acid (FR-900098) as antimalarials. *J. Med. Chem.* **2010**, *53*, 5342–5346. [CrossRef]
15. Berkowitz, D.B.; Bose, M. (α-Monofluoroalkyl)phosphonates: A class of isoacidic and "tunable" mimics of biological phosphates. *J. Fluor. Chem.* **2001**, *112*, 13–33. [CrossRef]
16. Blackburn, G.M.; England, D.A.; Kolkmann, F. Monofluoro- and difluoro-methylenebisphosphonic acids: Isopolar analogues of pyrophosphoric acid. *J. Chem. Soc. Chem. Commun.* **1981**, *17*, 930–932. [CrossRef]
17. Blackburn, G.M.; Kolkmann, F. The synthesis and metal binding characteristics of novel, isopolar phosphonate analogues of nucleotides. *J. Chem. Soc. Perkin Trans.* **1984**, *1*, 1119–1125. [CrossRef]
18. Berkowitz, D.B.; Eggen, M.J.; Shen, Q.; Shoemaker, R.K. Ready access to fluorinated phosphonate mimics of secondary phosphates. Synthesis of the (α,α-difluoroalkyl)phosphonate analogues of L-phosphoserine, L-phosphoallothreonine, and L-phosphothreonine. *J. Org. Chem.* **1996**, *61*, 4666–4675. [CrossRef]
19. Gillis, E.P.; Eastman, K.J.; Hill, M.D.; Donnelly, D.J.; Meanwell, N.A. Applications of fluorine in medicinal chemistry. *J. Med. Chem.* **2015**, *58*, 8315–8359. [CrossRef]
20. Shah, P.; Westwell, A.D. The role of fluorine in medicinal chemistry. *J. Enzyme Inhib. Med. Chem.* **2007**, *22*, 527–540. [CrossRef] [PubMed]
21. Volle, J.-N.; Midrier, C.; Blanchard, V.; Braun, R.; Haaf, K.; Willms, L.; Pirat, J.-L.; Virieux, D. Preparation of gem-difluorinated retrohydroxamic-fosmidomycin. *ARKIVOC-Online J. Org. Chem.* **2015**, *2015*, 117–126. [CrossRef]
22. Yokomatsu, T.; Suemune, K.; Murano, T.; Shibuya, S. Synthesis of (α,α-difluoroallyl)phosphonates from alkenyl halides or acetylenes. *J. Org. Chem.* **1996**, *61*, 7207–7211. [CrossRef]

23. Chambers, R.D.; Jaouharl, R.; O'Hagan, D. Fluorine in enzyme chemistry part 2. The preparation of difluoromethylenephosphonate analogues of glycolytic phosphates. Approaching an isosteric and isoelectronic phosphate mimic. *Tetrahedron* **1989**, *45*, 5101–5108. [CrossRef]
24. Brown, D.A.; Glass, W.K.; Mageswaran, R.; Girmay, B. *cis-trans* Isomerism in monoalkylhydroxamic acids by ^{1}H, ^{13}C and ^{15}N NMR spectroscopy. *Magn. Reson. Chem.* **1988**, *26*, 970–973. [CrossRef]
25. Brown, D.A.; Glass, W.K.; Mageswaran, R.; Mohammed, S.A. ^{1}H and ^{13}C NMR studies of isomerism in hydroxamic acids. *Magn. Reson. Chem.* **1991**, *29*, 40–45. [CrossRef]
26. O'Hagan, D. Understanding organofluorine chemistry. An introduction to the C-F bond. *Chem. Soc. Rev.* **2008**, *37*, 308–319. [CrossRef] [PubMed]
27. Zinglé, C.; Kuntz, L.; Tritsch, D.; Grosdemange-Billiard, C.; Rohmer, M. Isoprenoid biosynthesis via the methylerythritol phosphate pathway: Structural variations around phosphonate anchor and spacer of fosmidomycin, a potent inhibitor of deoxyxylulose phosphate reductoisomerase. *J. Org. Chem.* **2010**, *75*, 3203–3207. [CrossRef]
28. van den Bos, L.J.; Codée, J.D.C.; van der Toorn, J.C.; Boltje, T.J.; van Boom, J.H.; Overkleeft, H.S.; van der Marel, G. Thioglycuronides: Synthesis and application in the assembly of acidic oligosaccharides. *Org. Lett.* **2004**, *6*, 2165–2168. [CrossRef]
29. Lowry, O.H.; Lopez, J.A. The determination of inorganic phosphate in the presence of labile phosphate esters. *J. Biol. Chem.* **1946**, *162*, 421–428. [CrossRef]
30. Massey, R.C.; Buckling, A.; Peacock, S.J. Phenotypic switching of antibiotic resistance circumvents permanent costs in *Staphylococcus aureus*. *Curr. Biol.* **2001**, *11*, 1810–1814. [CrossRef]
31. Hikishima, S.; Hashimoto, M.; Magnowska, L.; Bzowska, A.; Yokomatsu, T. Structural-based design and synthesis of novel 9-deazaguanine derivatives having a phosphate mimic as multi-substrate analogue inhibitors for mammalian PNPs. *Bioorg. Med. Chem.* **2010**, *18*, 2275–2284. [CrossRef]
32. Chambers, R.D.; Jaouhari, R.; O'Hagan, D. Fluorine in enzyme chemistry: Part 1. Synthesis of difluoromethylene- phosphonate derivatives as phosphate mimics. *J. Fluor. Chem.* **1989**, *44*, 275–284. [CrossRef]

Article

Novel D-Annulated Pentacyclic Steroids: Regioselective Synthesis and Biological Evaluation in Breast Cancer Cells

Svetlana K. Vorontsova [1], Anton V. Yadykov [1], Alexander M. Scherbakov [2], Mikhail E. Minyaev [1], Igor V. Zavarzin [1], Ekaterina I. Mikhaevich [2], Yulia A. Volkova [1,*] and Valerii Z. Shirinian [1]

[1] N.D. Zelinsky Institute of Organic Chemistry, Russian Academy of Sciences, Leninsky Prosp. 47, 119991 Moscow, Russia; vorontsova_s@mail.ru (S.K.V.); antonyadykov@gmail.com (A.V.Y.); mminyaev@mail.ru (M.E.M.); igorzavarzin@yandex.ru (I.V.Z.); svbegunt@mail.ru (V.Z.S.)
[2] N.N. Blokhin National Medical Research Center of Oncology, Kashirskoye Shosse 24, 115522 Moscow, Russia; alex.scherbakov@gmail.com (A.M.S.); k.mihaevich@gmail.com (E.I.M.)
* Correspondence: yavolkova@gmail.com

Academic Editors: Francesca Cardona, Camilla Parmeggiani and Camilla Matassini
Received: 26 June 2020; Accepted: 30 July 2020; Published: 31 July 2020

Abstract: The acid-catalyzed cyclization of benzylidenes based on 16-dehydropregnenolone acetate (16-DPA) was studied. It was found that these compounds readily undergo regioselective interrupted Nazarov cyclization with trapping chloride ion and an efficient method of the synthesis of D-annulated pentacyclic steroids based on this reaction was proposed. The structures of the synthesized pentacyclic steroids were determined by NMR and X-ray diffraction. It was found that the reaction affords a single diastereomer, but the latter can crystallize as two conformers depending on the structure. Antiproliferative activity of synthesized compounds was evaluated against two breast cancer cell lines: MCF-7 and MDA-MB-231. All tested compounds showed relatively high antiproliferative activity. The synthetic potential of the protocol developed was illustrated by the gram-scale experiment.

Keywords: interrupted Nazarov cyclization; pentacyclic steroids; antiproliferative activity; D-annulated steroids; Lewis acid

1. Introduction

Steroids are an important class of both natural and synthetic products exhibiting various biological activities [1–11]. In the past decades, great efforts were made to accomplish further modification to synthesize structurally new and biologically interesting compounds [12–17]. The chemical modification of steroids is among the most efficient and attractive approaches to the design of new biologically active compounds, including pharmaceuticals [1,2,18,19]. An important application of chemical modification of steroid molecules is to synthesize compounds containing an additional fused ring. Steroids bearing an additional fused ring are common skeletons involved in many natural products [20–24] and pharmaceuticals [25–27]. It should be noted that some of these compounds exhibit no significant hormonal activity [28–31]. Both heterocyclic analogs **I** [32–39] and carbocyclic analogs **II** [40,41] of pentacyclic steroids have attracted considerable attention (Scheme 1).

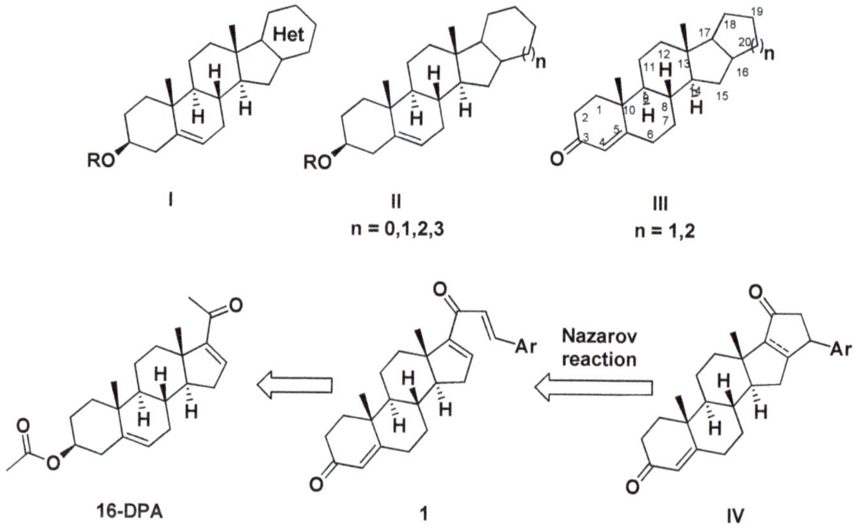

Scheme 1. Pentacyclic steroids.

Steroids bearing an additional carbocycle annulated at the 16 and 17 positions of the ring D (**III**) are characterized by unique biological activity [42–45]. Most of the known D-annulated pentacyclic steroids contain the cyclohexane or cyclopentane moiety as an additional ring. A number of synthetic strategies can be used to construct pentacyclic steroids, including Diels–Alder reactions and intermolecular condensation [40,41,46–49]. Commercially available 16-dehydropregnenolone acetate (16-DPA), which is a versatile building block for the preparation of various semi-synthetic steroidal drugs [50,51], is commonly utilized in the synthesis of D-annulated steroids. However, despite the availability of 16-DPA, most of the known methods for the construction of an additional carbocycle based on 16-DPA cannot be applied on a preparative scale and often suffer from drawbacks such as low yields, the lack of atom-economy or step efficiency, a narrow field of application, etc.

In a continuation of our research on the synthesis of semi-synthetic steroid derivatives and evaluation of their antitumor activity [36,37,52,53], we focused on the development of a convenient method for the synthesis of D-annulated pentacyclic steroid of the progesterone series **IV** bearing the cyclopentanone moiety as an additional ring. The Nazarov cyclization reaction represents one of the most effective methods for the construction of five-membered carbocyclic rings, and it has been applied in the total synthesis of many useful natural products. Due to high stereoselectivity, high yields, and readily available starting compounds, this reaction is widely used in organic synthesis for the preparation of diverse biologically active substances [54–72]. However, to the best of our knowledge, this reaction was not applied for the annulation of an additional cyclopentanone ring onto the steroid skeleton. Another important issue, which prompted us to perform this study, is that the starting benzylidenes **1** are readily available by the method proposed in our recent work [52,53]. Besides, the synthesized benzylidenes exhibited relatively high antiproliferative activity. Since the structure of the starting dienone does not undergo significant changes in the course of Nazarov cyclization, this modification would be expected to improve useful properties.

Despite the absence of data on the synthesis of pentacyclic steroids by the Nazarov reaction, several examples of the application of this transformation both in the synthesis and modification of steroids were reported [73–77]. However, since the pathway of the Nazarov reaction and the yields of the target products strongly depend on the nature of substituents and the catalyst, we analyzed the cyclization of compounds structurally similar to benzylidenes **1**, such as dienones, in which one vinyl moiety is involved in the carbocycle, while another vinyl moiety contains an aryl group at the β-position (compound **V**, Scheme 2A). Literature data analysis showed that there are only a few examples of cyclization of structurally related dienones (Scheme 2).

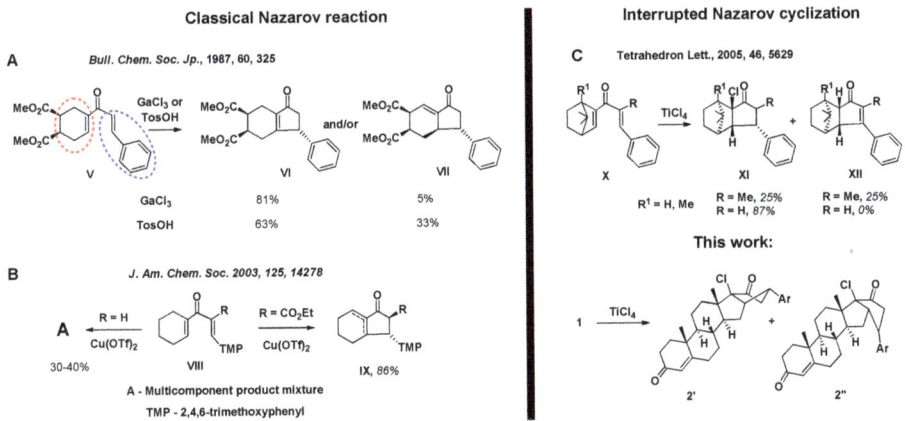

Scheme 2. Nazarov reaction of dienones in the presence of various catalysts (**A**—GaCl$_3$ or TosOH, **B**—Cu(OTf)$_2$, **C**—TiCl$_4$)

The cyclization of dienones in the presence of various Lewis acids (SnCl$_4$, TiCl$_4$, AlCl$_3$, GaCl$_3$, BF$_3$*Et$_2$O) and TsOH was studied [78]. It was found that despite the absence of the second α-substituent, the cyclization affords a classical Nazarov reaction product (Scheme 2A). The best yields were achieved using gallium chloride, which gave the minimum amount of the second isomer (5%). The use of p-toluenesulfonic acid as the catalyst leads to an increase in the amount of the minor isomer. It is worth noting that, despite the absence of the second α-substituent, the formation of interrupted Nazarov cyclization products was not observed. Dienone **VIII** (R = CO$_2$Et) also underwent classical Nazarov cyclization in the presence of copper (II) triflate (Scheme 2B) [79]. This reaction also produced two isomers. Under similar conditions, the cyclization of α-unsubstituted substrate **VIII** (R = H) affords a difficult-to-separate mixture. Interesting results were obtained in the study of the cyclization of camphor- and nopinone-derived dienones **X** [80]. The interrupted Nazarov cyclization via nucleophilic halide trapping was found to occur in the presence of halogen-contained Lewis acids (TiCl$_4$). The nucleophilic halide trapping was observed not only in the presence of titanium chloride but also in the presence of bulky bromide and iodide anions (TiBr$_4$ and TiI$_4$); the exception was the fluoride anion. The interrupted Nazarov cyclization proceeds with α-unsubstituted compounds. The introduction of the methyl group at this position facilitates the competitive classical Nazarov reaction, resulting in the formation of a mixture of products in moderate yields. It is worth noting that this is one of a few studies on the Nazarov reaction, in which halide trapping in the presence of Lewis acids took place [81,82].

Such transformations are also known for other cationic rearrangements [83–87]. Therefore, the analysis of the literature data shows that the pathway of the cyclization of such dienones strongly depends on both their structure and the nature of the catalyst. Since camphor-derived dienones **X** (R = Me) are structurally most similar to benzylidenes **1**, we decided to study the cyclization of these steroids under similar conditions.

2. Results and Discussion

2.1. Synthesis

We started our research by studying a series of Lewis acids (TiCl$_4$, SnCl$_4$, AlCl$_3$, FeCl$_3$) and hydrogen chloride (HCl). 4-Chlorophenyl-substituted dienone **1c** was used as a model substrate and dichloromethane, dichloroethane, and benzene as the solvents. In all cases, except for titanium (IV) chloride, the reaction was very slow or gave an unidentified mixture that is difficult to separate. Thus, the reaction in the presence of aluminum and iron chlorides was very slow at room temperature and gave a multicomponent mixture of products at elevated temperatures (the monitoring of the reaction in the presence of AlCl$_3$ is given in the SM, Figure S6). The reaction using tin (IV) chloride and hydrogen chloride affords a difficult-to-separate mixture of products even at 0 °C; a decrease in the temperature leads to a decrease in the reaction rate but does not improve the selectivity (see Table S1 in the Supplementary Materials (SM). Apparently, this is due to the low solubility of the steroid studied in dichloromethane at low temperatures (<−10 °C). Good results were obtained in the cyclization of benzylidenes in the presence of titanium (IV) chloride giving a single reaction product. The spectral analysis (mass spectrometry, ^{1}H and ^{13}C-NMR spectroscopy) of this product showed that it does not have a double bond but contains a chlorine atom. We suggested that, like the cyclization of camphor derived dienones [80], this reaction proceeds via interrupted Nazarov cyclization to form chloro-substituted pentacyclic steroid **2**. The optimization of the reaction conditions allowed us to synthesize products based on other benzylidenes in high yields (Scheme 3, Table 1). The reaction was performed in dichloromethane in the presence of freshly distilled titanium (IV) chloride as the catalyst under an inert atmosphere (argon) at a temperature from −5 °C to 20 °C (the catalyst was added at −5 to 0 °C). It should be noted that the use of freshly distilled TiCl$_4$ is a necessary condition to achieve high yields. Besides, the solvent and the substrate should be thoroughly dried because the presence of even traces of moisture leads to a significant decrease in the yield of the desired products.

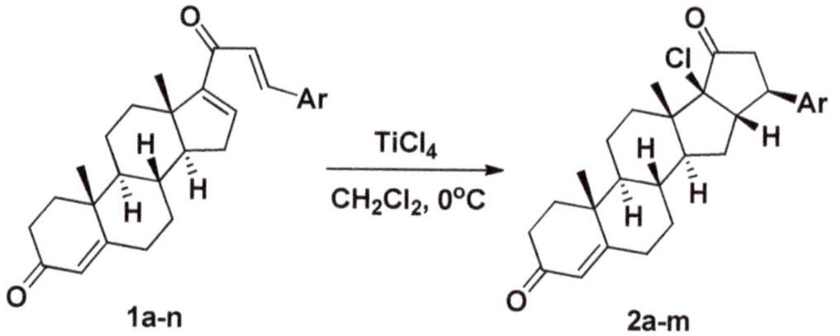

Scheme 3. Synthesis of pentacyclic steroids **2a–m** by the interrupted Nazarov cyclization.

Table 1. Yields of benzylidene cyclization products **2a–n**.

Entry	Codes	Ar	Yields	
			NMR *	Isolated
1	a	Ph	80	63
2	b	4-Cl-C$_6$H$_4$	95	78
3	c	4-Br-C$_6$H$_4$	95	70
4	d	3-Br-C$_6$H$_4$	88	60
5	e	2-F-C$_6$H$_4$	71	40
6	f	4-F-C$_6$H$_4$	80	41
7	g	2,4-Cl$_2$C$_6$H$_3$	92	73
8	h	2-Cl-6-F-C$_6$H$_3$	80	58
9	i	4-MeO-C$_6$H$_4$	78	55
10	j	3-MeO-C$_6$H$_4$	76	44
11	k	3,4-(MeO)$_2$-C$_6$H$_3$	91	43
12	l	3,4,5-(MeO)$_3$C$_6$H$_2$	70	47
13	m	2-Thienyl	92	64
14	n	2-Furyl	35	0

* TMS is used as internal standard.

As can be seen in Table 1, the cyclization of benzylidenes was efficient and gave products in high yields, as evidenced by ^{1}H-NMR analysis of the reaction mixture after the completion of the reaction (Table 1). The highest yields were obtained for halogen-substituted benzylidenes, the cyclization of which proceeds smoothly and with minimal side processes. The modest average yields of methoxy-substituted benzylidenes can be explained by the partial hydrolysis of methoxy groups promoted by a strong Lewis acid. Only the furan derivative was obtained in a low yield (35%) (Entry 14, Table 1). We failed to isolate this compound and characterize it as an individual product. Apparently, this is due to the high sensitivity of the furan derivative to Lewis acids, in particular, to titanium (IV) chloride, although the reaction of thiophenebenzylidene proved to be efficient and gave the target product in 64% yield. Despite high yields determined from the ^{1}H NMR data, the isolated yields were moderate (Table 1). Significant losses during the purification by column chromatography are apparently due to the sensitivity of labile functional groups (tertiary alkyl chlorides, carbonyl groups, and activated methylene moiety) to silica gel. It is worth noting that the utilization of neutral alumina instead of silica gel did not lead to a significant improvement in the outcome. The efficiency of this process and the absence of by-products were also demonstrated by NMR monitoring of the cyclization of benzylidene **1c** (Figure 1).

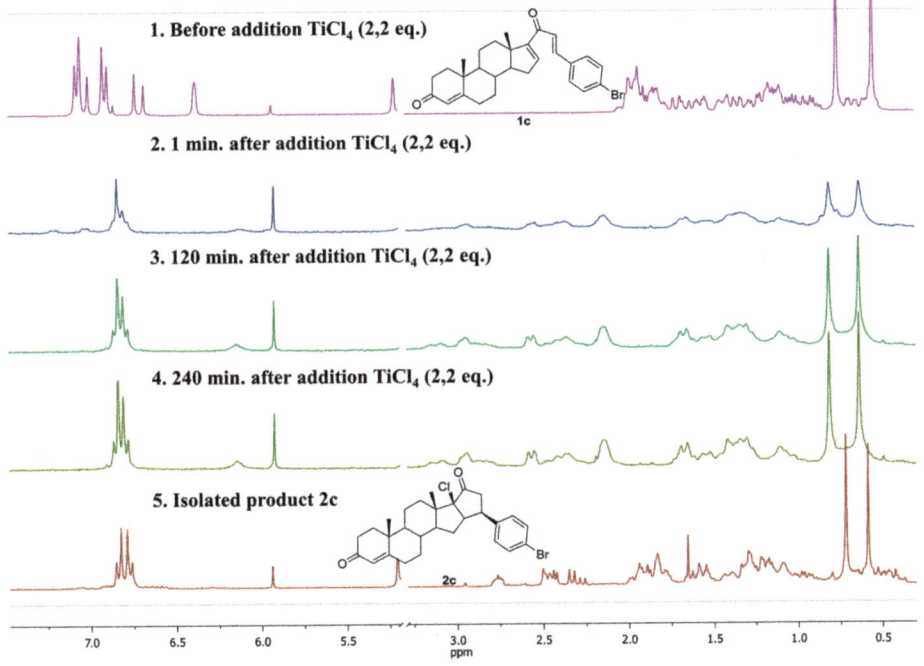

Figure 1. ^{1}H-NMR monitoring of the cyclization of benzylidene **1c**.

As can be seen in Figure 1, the reaction is not accompanied by side processes and gives a single isomer, although we initially expected that this reaction would afford two isomers structurally similar to the *exo* and *endo* isomers prepared by the cyclization of camphor-derived dienones [80]. (The absence of any other by-products is evidenced by the presence of a single pair of signals both in the aromatic region and for methyl groups; for more information see Figure S4 in the SM). An analysis of the ^{1}H and ^{13}C-NMR spectra showed that the second isomer was absent in all other reactions as well (see copies of the ^{1}H and ^{13}C-NMR spectra of the synthesized pentacyclic steroids in the SM). We performed additional NMR studies, which confirmed the formation of a single conformer in solution (the ^{1}H-NMR spectra were recorded in a wide temperature range from 20 to 100 °C, see Figure S5 in the SM). The heating of a solution of pentacyclic steroid **2m** in acetonitrile under reflux for 10 h also did not lead to any changes in the ^{1}H NMR spectrum. The formation of a single conformer is confirmed by the ^{1}H NMR spectrum of the crude mixture (see Figures S7 and S8 in the SM). The formation of one diastereomer can be explained by the rigidity of the steroid backbone, in which case the nucleophile always attacks with the formation of an *anti*-diastereomer. Similarly, the formation of a single diastereomer was observed in the case of bicyclic camphor and nopione structures [80].

2.2. Gram-Scale Synthesis of Pentacyclic Steroid **2g**

To demonstrate the scalability and application potential of the method for the synthesis of pentacyclic steroids, we performed a 5 mmol scale experiment for lead compound **2g**. The gram-scale synthetic protocol was carried out under the normal laboratory setup starting from commercially available 16-DPA under the optimized reaction conditions (Scheme 4).

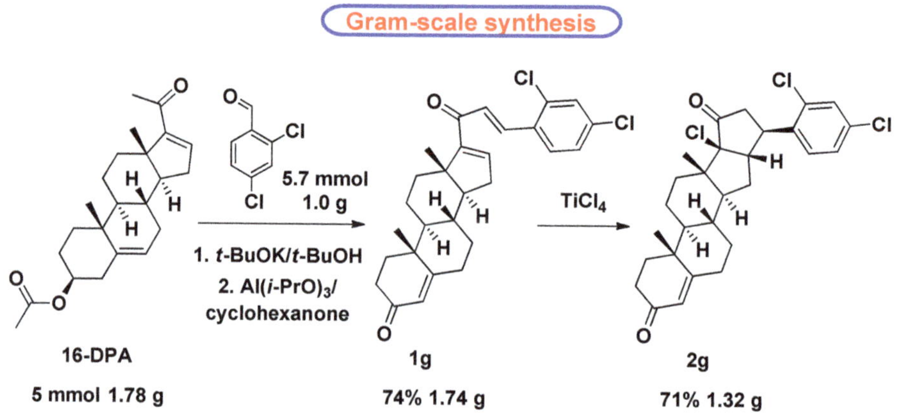

Scheme 4. Gram-scale synthesis of pentacyclic steroid **2g**.

To our delight, pentacyclic steroid **2g** was obtained without a significant decrease in yield (total 3 steps yield is 52%), which clearly demonstrates the practical applicability of our proposed method. (Benzylidene **1g** was obtained according to a previously developed method [53]).

2.3. X-ray Diffraction Study and Structure Optimization

The crystal structures of products **2** were determined by X-ray diffraction study of compounds **2b** and **2g** (Figure 2) [88]. Absolute configurations of studied molecules were established by anomalous X-ray scattering in both cases, which allowed us not only to confirm configurations of unchanged chiral centers but also to find configurations of two new chiral centers generated during cyclization. The unit cell of compound **2b** comprises two conformers (**2′b** and **2″b**) in a ratio of 1:1 (Figure 2), whereas compound **2g** crystallizes as a single conformer **2′g** (Figure 2).

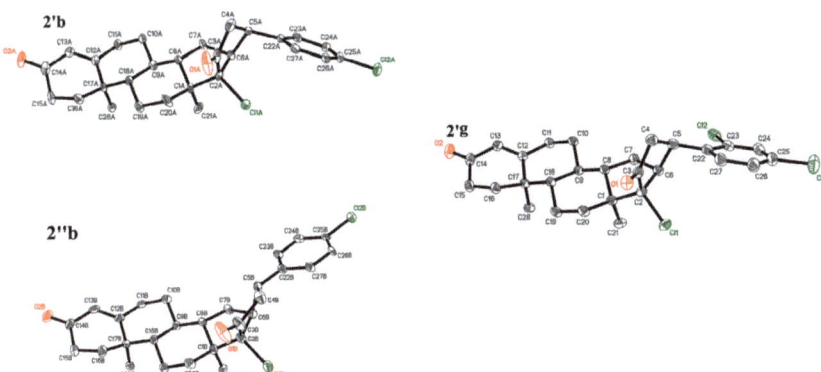

Figure 2. Crystal structures of **2b** (two conformers – **2′b** and **2″b**) and **2g** (a single conformer **2′g**). Two crystallographically independent molecules of **2′b** and molecule **2′g** are shown in a similar orientation, with thermal displacement ellipsoids drawn at the 50% probability level.

The new chiral centers at the 16 and 17 positions of the steroid system (the C2 and C6 atoms in Figure 2) adopt the same configuration in molecules **2′b**, **2″b**, and **2′g**. The geometry of conformer **2′g** is similar to that of **2′b**. An analysis of the geometry shows that the presence of two conformers of **2b** can be nearly entirely attributed to the conformational flexibility of the terminal five-membered ring. It should be noted that the five-membered ring D and the six-membered rings A–C of the steroid system in the two conformers (**2′b** and **2″b**) have a very similar geometry. In both conformers, the terminal five-membered ring adopts an envelope conformation. The C5A atom bound to the phenyl group deviates from the plane defined by the other four atoms of the ring in **2′b**, whereas the C4B atom located between the carbonyl group and the >CHAr group deviates from the corresponding plane in **2″b**. An analysis of the van der Waals contacts in the conformationally different moieties of molecules **2′b** and **2″b** shows that intramolecular repulsive forces in these molecules are similar in strength (see Table S6 in the SM). Hence, the conformers should have similar energies. It is worth noting that there are numerous intermolecular non-valence interactions (H···H, O···H, C_{Ph}···H_{Ph}, Cl···Cl), which can affect the molecular geometry in the crystal. Intramolecular van der Waals repulsion in molecule **2′g** is somewhat stronger compared to **2′b**. Besides, there are short contacts between hydrogen atoms and the *ortho*-chlorine atom of the phenyl group (H5···Cl2 and H6···Cl2, Table S6 in the SM) in **2′g**. In all three molecules (**2′b**, **2″b**, and **2′g**), the aryl substituents are in a similar orientation to the C4–C5 bond of the five-membered ring; the C4–C5–C22–C27 torsion angles vary in the range of 15.7°–16.7°. It should be noted that undetected conformer **2″g**, which is structurally similar to **2″b** should be much less stable than **2′g** due to intramolecular van der Waals contacts (see section IV in the SM).

The relative energies of **2′b** and **2″b** and the energy of the transition state between these structures were evaluated by DFT calculations. The geometry optimization and energy calculations for **2′b** and **2″b** were performed at the ωB97X-D/6-311++G(d,p)//ωB97X-D/6-31+G(d,p) level of theory, according to the literature data [89]. The energy difference between the conformers is rather small (0.2 kcal/mol) and the energy barrier for conformational transitions is ca. 1.1 kcal/mol (see section V in the SM). Therefore, a low activation barrier makes it impossible to detect each conformer in solution by NMR methods even at low temperatures because of fast exchange on the NMR time scale.

Calculated frontier orbitals for conformers **2′b** and **2″b** are shown in Figure 3. It might be noted that both HOMO and LUMO orbitals are localized, to some extent, at carbon atoms of the newly formed 5-membered ring and also at chlorine (Cl1), oxygen (O1) atoms bound to atoms C2, C3 of the ring.

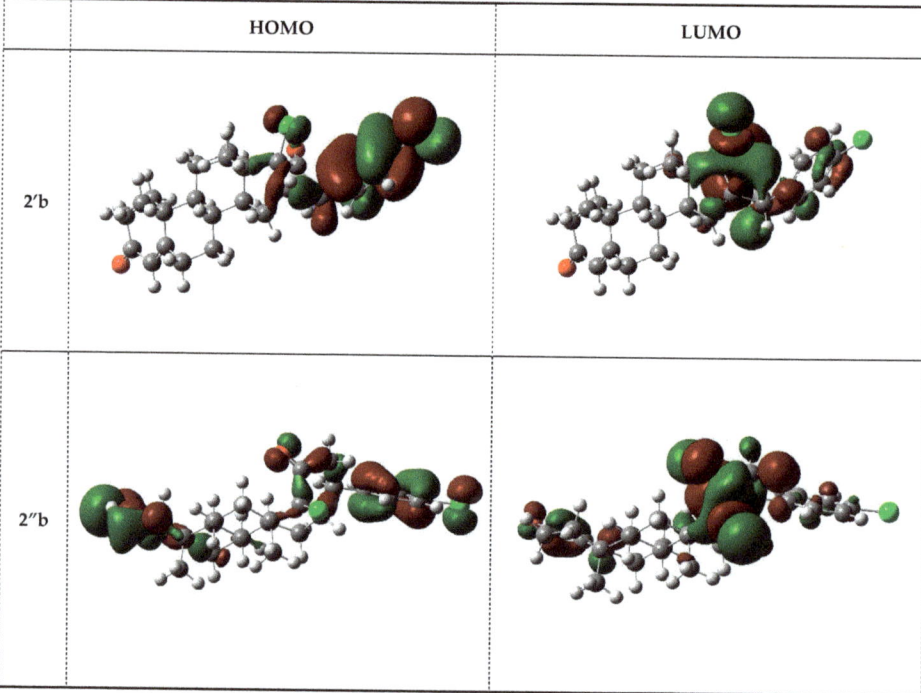

Figure 3. Localization of HOMO (**left**) and LUMO (**right**) for **2'b** (**top**) and **2"b** (**bottom**), calculated by the ωB97X-D functional for the gas phase.

Therefore, different parts of this terminal 5-membered ring should be considered as reactive centers that should be susceptible to an attack by either nucleophilic or, to a lesser extent, electrophilic reagents. This conclusion will be tested in our further works.

2.4. In Vitro Antiproliferative Activity

Biological assays of the synthesized pentacyclic steroids were performed in the following two breast cancer cell lines: the most commonly used estrogen-dependent breast cancer cell line MCF-7 [90] and the triple-negative breast cancer cell line MDA-MB-231. Triple-negative breast cancer (TNBC) does not express estrogen and progesterone receptors and HER2/neu. Hence, the development of effective therapeutic agents for the treatment of TNBC is a challenge. Triple-negative breast cancer is generally characterized by rapid progression and does not respond to hormonal therapy [91]. We evaluated the antiproliferative activity of 13 synthesized compounds using standard incubation of the cells with the tested compounds for 72 h (Table 2).

Table 2. Antiproliferative activity of pentacyclic steroids **2a–m** and cisplatin (cell growth for 72 h).

		IC$_{50}$ [a], µM	
Entry	Compounds	MCF-7	MDA-MB-231
1	2a	10.0 ± 1.5	8.8 ± 0.9
2	2b	6.0 ± 0.8	9.4 ± 1.1
3	2c	8.1 ± 0.9	>25
4	2d	7.5 ± 0.9	>25
5	2e	12.6 ± 1.6	19.6 ± 2.1
6	2f	7.1 ± 0.9	12.2 ± 1.6
7	2g	8.2 ± 0.9	6.3 ± 0.7
8	2h	17.7 ± 2.1	15.2 ± 1.6
9	2i	9.5 ± 1.2	13.0 ± 1.5
10	2j	7.2 ± 0.8	>25
11	2k	25 ± 2.2	24.8 ± 2.5
12	2l	10.6 ± 1.3	21.5 ± 2.4
13	2m	24.5 ± 2.5	>25
14	Cisplatin(CDDP)	6.3 ± 0.8	13.2 ± 1.5

IC$_{50}$ [a] is the concentration that causes 50% inhibition of cell growth.

All steroids caused 50% inhibition of proliferation of MCF-7 cells at concentrations lower than 25 µM. Compounds **2b**, **2d**, **2f**, and **2j** exhibited the highest activity with IC$_{50}$ values below 8 µM. Antiproliferative effects of these compounds are comparable with that of cisplatin used as the reference compound. Four other pentacyclic steroids (**2a, c, g, i**) also showed high activity. For these compounds, the IC$_{50}$ against MCF-7 cells was lower than 10 µM. The other steroids were less active and showed IC$_{50}$ values ranging from 10 to 25 µM. Steroid **2b** exhibited the highest activity in the breast cancer cell line MCF-7. Hence, the compound containing the 4-chlorophenyl moiety (4-Cl-C$_6$H$_4$) as the aryl substituent is the most promising for further assays as an agent against hormone-dependent cancers.

The activity of the tested compounds against triple-negative breast cancer varied in a wide range. Four pentacyclic steroids (**2c,d,j,m**) were inactive and caused 50% inhibition of MDA-MB-231 cell growth at concentrations below 25 µM. On the contrary, compounds **2a,b,g** exhibited high activity against TNBC with IC$_{50}$ lower than 10 µM. These steroids were more active than the reference compound cisplatin. Compounds **2f** and **2i** showed activity similar to that of cisplatin. Interestingly, 4-chlorophenyl-substituted steroid **2b**, which exhibited activity against the hormone-dependent cancer cell line MCF-7, showed relatively high activity against MDA-MB-231, but its effects were much weaker compared to **2g** (Table 2). Steroid **2g** was selected as the lead compound against MDA-MB-231. High antiproliferative activity of chlorine-containing steroids was also reported in the study [92]. The steroidal [17,16-*d*]pyrimidine derivative containing the 2-chlorophenyl moiety displayed a significant cytostatic effect in the cell lines HepG2 (human hepatocellular liver carcinoma cell line) and Huh7 (human hepatoma cell line) with IC$_{50}$ values lower than 6 µM. This steroid was less active in the cell line SGC-7901 (human gastric cancer cell line), where it showed an effect with IC$_{50}$ of ca. 10 µM. Probable mechanism of action of chlorine-containing steroids is considered in the study [93]. An androstanopyridine derivative bearing chlorine in the side chain was found to be an effective activator of the protein p53 in *in vitro* and *in vivo* experiments. The protein p53 is a key tumor suppressor and its directed activation can be considered as an approach to therapy using steroid derivatives, in particular pentacyclic steroids.

3. Materials and Methods

3.1. General Information

Proton nuclear magnetic resonance spectra (^{1}H-NMR) and carbon nuclear magnetic resonance spectra (^{13}C-NMR) were recorded in deuterated solvents on a spectrometers (Fourier 300 HD, Bruker BioSpin AG, Zürich, Switzerland) working at 300 MHz for ^{1}H, 75 MHz for ^{13}C. Both ^{1}H and ^{13}C NMR reported in parts per million (ppm) at 293 K. Data are represented as follows: chemical shift, multiplicity (s, singlet; d, doublet; m, multiplet; br, broad; q, quartet), coupling constant in hertz (Hz).

Melting points (mp) were recorded using an apparatus and not corrected. Mass spectra were obtained on a mass spectrometer (70 eV) with direct sample injection into the ion source. High resolution mass spectra were obtained from a TOF mass spectrometer with an SM source. All chemicals and anhydrous solvents were purchased from commercial sources and used without further purification. Silica column chromatography was performed using silica gel 60 (70–230 mesh); TLC analysis was conducted on silica gel 60 F254 plates. Benzylidines **1a–m** were prepared according to the previously reported method [53].

3.2. Synthesis and Characterization of Pentacyclic Steroids 2a–m

General method: To a stirring solution of benzylidine **1** (0.15 mmol) in abs methylene chloride (10 mL) at −5 °C the solution of freshly distilled titanium (IV) chloride (0.063 g, 0.33 mmol) in abs methylene chloride (5 mL) was added dropwise. The reaction mixture was stirred at the same temperature for 1.5 h and then it was kept at 20 °C until the reaction completion (monitoring by TLC). Then the reaction mixture was poured into the ice-water mixture (100 mL) and extracted with CH_2Cl_2 (3 × 20 mL). The combined organic phases were washed with water (2 × 50 mL), dried with sodium sulfate, and evaporated in a vacuum. The residue was purified by column chromatography by petroleum ester/ethyl acetate 3:1 and recrystallized from the same mixture.

3.2.1. (6a*R*,6b*S*,8a*S*,8b*S*,11*R*,11a*S*,12a*S*,12b*R*)-8b-chloro-6a,8a-dimethyl-11-phenyl-1,6,6a,6b,7,8,8a,8b, 10,11,11a,12,12a,12b-tetradecahydropentaleno[2,1-*a*]phenanthrene-4,9(2*H*,5*H*)-dione (**2a**)

Yield = 63% (0.041 g), given as a white powder; mp 203–205 °C; ^{1}H-NMR (300 MHz, CDCl$_3$): δ = 0.88–1.00 (m, 1H), 1.07 (s, 3H), 1.09–1.15 (m, 1H), 1.19 (s, 3H), 1.34–1.51 (m, 1H), 1.62–1.94 (m, 8H), 2.01–2.10 (m, 2H), 2.25–2.45 (m, 4H), 2.76 (dd, *J* = 18.2, 8.8 Hz,1H), 2.97–3.09 (m, 2H), 3.13–3.21 (m, 1H), 5.74 (s, 1H), 7.23–7.37 (m, 5H). ^{13}C NMR (75 MHz, CDCl$_3$): δ = 17.4, 17.5, 20.3, 30.4, 31.7, 32.5, 33.9, 35.4, 35.7, 36.6, 38.5, 45.0, 46.4, 48.6, 50.5, 53.1, 56.7, 79.0, 124.1, 126.8, 127.2 (2C), 128.6 (2C), 143.9, 170.1, 199.2, 211.6. HRMS (ESI-TOF) *m/z*: [M + H]$^+$ Calcd for $C_{28}H_{33}ClO_2$: 437.2237; Found: 437.2242.

3.2.2. (6a*R*,6b*S*,8a*S*,8b*S*,11*R*,11a*S*,12a*S*,12b*R*)-11-(4-chlorophenyl)-8b-chloro-6a,8a-dimethyl-1,6,6a, 6b,7,8,8a,8b,10,11, 11a,12,12a,12b-tetradecahydropentaleno[2,1-*a*]phenanthrene-4,9(2*H*,5*H*)-dione (**2b**)

Yield = 78% (0.055 g), given as a light yellow powder; mp 125–127 °C; ^{1}H-NMR (300 MHz, CDCl$_3$): δ = 0.89–0.98 (m, 1H), 1.06 (s, 3H), 1.10–1.14 (m,1H), 1.20 (s, 3H), 1.38–1.49 (m, 1H), 1.65–1.91 (m, 8H), 2.00–2.19 (m, 2H), 2.27–2.42 (m, 4H), 2.78 (dd, *J* = 18.4, 9.0 Hz, 1H), 2.93–3.01 (m, 2H), 3.15–3.19 (m, 1H), 5.74 (s, 1H), 7.23 (d, *J* = 9 Hz, 2H), 7.27–7.31 (m, 2H). ^{13}C-NMR (75 MHz, CDCl$_3$): δ = 17.4, 17.4, 20.3, 30.3, 31.7, 32.5, 33.9, 35.4, 35.7, 36.6, 38.5, 44.8, 45.7, 48.8, 50.6, 53.1, 56.6, 78.9, 124.1, 128.6 (2C), 128.7 (2C), 132.6, 142.4, 169.9, 199.2, 211.3. HRMS (ESI-TOF) *m/z*: [M + H]$^+$ Calcd for $C_{28}H_{32}Cl_2O_2$: 471.1844; Found: 471.1852.

3.2.3. (6a*R*,6b*S*,8a*S*,8b*S*,11*R*,11a*S*,12a*S*,12b*R*)-11-(4-bromophenyl)-8b-chloro-6a,8a-dimethyl-1,6,6a,6b, 7,8,8a,8b,10,11,11a,12,12a,12b-tetradecahydropentaleno[2,1-*a*]phenanthrene-4,9(2*H*,5*H*)-dione (**2c**)

Yield = 70% (0.046 g), given as a white powder; mp 123–125 °C; ^{1}H-NMR (300 MHz, CDCl$_3$): δ = 0.87–1.02 (m, 1H), 1.07 (s, 3H), 1.09–1.15 (m, 1H), 1.19 (s, 3H), 1.32–1.49 (m, 1H), 1.65–1.90 (m, 8H), 2.01–2.08 (m, 2H), 2.27–2.49 (m, 4H), 2.76 (dd, *J* = 18.5, 8.9 Hz, 1H), 2.99–3.07 (m, 2H), 3.10–3.20 (m, 1H), 5.74 (s, 1H), 7.18 (d, *J* = 8.3 Hz, 2H), 7.46 (d, *J* = 8.3 Hz, 2H). ^{13}C-NMR (75 MHz, CDCl$_3$): δ = 17.4, 17.4, 20.3, 30.3, 31.7, 32.5, 33.9, 35.4, 35.7, 36.6, 38.5, 44.7, 45.7, 48.8, 50.6, 53.1, 56.6, 78.9, 120.6, 124.1, 129.0 (2C), 131.7 (2C), 142.9, 170.0, 199.2, 211.4. HRMS (ESI-TOF) *m/z*: [M + H]$^+$ Calcd for $C_{28}H_{33}ClO_2$: 437.2237; Found: 437.2242.

3.2.4. (6a*R*,6b*S*,8a*S*,8b*S*,11*R*,11a*S*,12a*S*,12b*R*)-11-(3-bromophenyl)-8b-chloro-6a,8a-dimethyl-1,6,6a, 6b,7,8,8a,8b,10,11,11a,12,12a,12b-tetradecahydropentaleno[2,1-*a*]phenanthrene-4,9(2*H*,5*H*)-dione (**2d**)

Yield = 60% (0.046 g), given as a white powder; mp 230–232 °C; ^{1}H-NMR (300 MHz, CDCl$_3$): δ = 0.89–1.01 (m, 1H), 1.07 (s, 3H), 1.09–1.12 (m, 1H), 1.20 (s, 3H), 1.26–1.49 (m, 1H), 1.65–1.94 (m, 8H), 2.01–2.07 (m, 2H), 2.28–2.50 (m, 4H), 2.76 (dd, *J* = 18.3, 8.7 Hz, 1H), 2.96–3.03 (m, 2H), 3.14–3.20 (m, 1H),

5.74 (s, 1H), 7.18–7.23 (m, 2H), 7.38–7.40 (m, 1H), 7.45(s, 1H). ^{13}C-NMR (75 MHz, CDCl$_3$): δ = 17.4, 17.5, 20.3, 30.3, 31.7, 32.5, 33.9, 35.4, 35.7, 36.6, 38.5, 44.7, 45.9, 48.7, 50.5, 53.1, 56.5, 78.8, 122.7, 124.1, 125.7, 129.9, 130.2, 130.5, 146.2, 169.9, 199.2, 211.1. HRMS (ESI-TOF) m/z: [M + H]$^+$ Calcd for C$_{28}$H$_{32}$BrClO$_2$: 515.1359; Found: 515.1347.

3.2.5. (6aR,6bS,8aS,8bS,11R,11aS,12aS,12bR)-11-(2-fluorophenyl)-8b-chloro-6a,8a-dimethyl-1,6,6a, 6b,7,8,8a,8b,10,11,11a,12,12a,12b-tetradecahydropentaleno[2,1-a]phenanthrene-4,9(2H,5H)-dione (**2e**)

Yield = 40% (0.027 g), given as a white powder; mp 211–213 °C; ^{1}H-NMR (300 MHz, CDCl$_3$): δ = 0.89–1.00 (m, 1H), 1.06 (s, 3H), 1.09–1.17 (m, 1H), 1.20 (s, 3H), 1.26–1.49 (m, 1H), 1.65–1.92 (m, 8H), 2.01–2.16 (m, 2H), 2.28–2.48 (m, 4H), 2.79 (dd, J = 18.8, 9.0 Hz, 1H), 2.95–3.03 (m, 2H), 3.54–3.57 (m, 1H), 5.75 (s, 1H), 7.03–7.15 (m, 2H), 7.21–7.31 (m, 2H). ^{13}C NMR (75 MHz, CDCl$_3$): δ = 17.3, 17.4, 20.3, 30.3, 31.7, 32.5, 33.9, 35.3, 35.7, 36.6, 38.6 (d, ^3J(C-F) = 3.1 Hz), 38.7, 43.5, 49.0, 50.7, 53.1, 55.3, 78.8, 115.3 (d, ^2J(C-F) = 21.9 Hz), 124.1, 124.1, 127.9 (d, ^3J(C-F) = 4.1 Hz), 128.4 (d, ^3J(C-F) = 8.4 Hz), 130.7 (d, ^2J(C-F) = 13.9 Hz), 160.6 (d, ^1J(C-F) = 245.3 Hz), 170.1, 199.2, 212.0. HRMS (ESI-TOF) m/z: [M + H]$^+$ Calcd for C$_{28}$H$_{32}$FClO$_2$: 455.2149; Found: 455.2148.

3.2.6. (6aR,6bS,8aS,8bS,11R,11aS,12aS,12bR)-8b-chloro-11-(4-fluorophenyl)-6a,8a-dimethyl-1,6,6a, 6b,7,8,8a,8b,10,11,11a,12,12a,12b-tetradecahydropentaleno[2,1-a]phenanthrene-4,9(2H,5H)-dione (**2f**)

Yield = 71% (0.048 g), given as a white powder; mp 181–183 °C; ^{1}H-NMR (300 MHz, CDCl$_3$): δ = 0.89–1.00 (m, 1H), 1.06 (s, 3H), 1.09–1.14 (m, 1H), 1.19 (s, 3H), 1.34–1.49 (m, 1H), 1.64–1.90 (m, 8H), 1.99–2.11 (m, 2H), 2.27–2.49 (m, 4H), 2.76 (dd, J = 18.4, 8.9 Hz, 1H), 2.93–3.01 (m, 2H), 3.13–3.23 (m, 1H), 5.74 (s, 1H), 6.99–7.05 (m, 2H), 7.23–7.28 (m, 2H). ^{13}C NMR (75 MHz, CDCl$_3$): δ = 17.4, 17.4, 20.3, 30.3, 31.7, 32.5, 33.9, 35.4, 35.6, 36.6, 38.5, 45.0, 45.5, 48.8, 50.5, 53.1, 56.8, 78.9, 115.4 (d, ^2J(C-F) = 21.4 Hz, 2C), 124.1, 128.7 (d, ^3J(C-F) = 8.0 Hz, 2C), 139.6 (d, ^4J(C-F) = 3.0 Hz), 161.6 (d, ^1J(C-F) = 245.3 Hz), 170.1, 199.4, 211.7. HRMS (ESI-TOF) m/z: [M + H]$^+$ Calcd for C$_{28}$H$_{32}$FClO$_2$: 455.2155; Found: 455.2148.

3.2.7. (6aR,6bS,8aS,8bS,11R,11aS,12aS,12bR)-8b-chloro-11-(2,4-dichlorophenyl)-6a,8a-dimethyl-1,6, 6a,6b,7,8,8a,8b,10,11,11a,12,12a,12b-tetradecahydropentaleno[2,1-a]phenanthrene-4,9(2H,5H)-dione (**2g**)

Yield = 73% (0.055 g), given as a light yellow powder; mp 185–187 °C; ^{1}H-NMR (300 MHz, CDCl$_3$): δ = 0.89–0.98 (m, 1H), 1.05 (s, 3H), 1.10–1.14 (m, 1H), 1.20 (s, 3H), 1.39–1.49 (m, 1H), 1.65–1.89 (m, 8H), 2.03–2.21 (m, 2H), 2.29–2.43 (m, 4H), 2.89–2.96 (m, 3H), 3.68–3.74 (m, 1H), 5.75 (s, 1H, CH), 7.19–7.25 (m, 2H), 7.42 (s, 1H). ^{13}C-NMR (75 MHz, CDCl$_3$): δ = 17.1, 17.4, 20.2, 30.3, 31.7, 32.5, 33.9, 35.3, 35.7, 36.7, 38.5, 41.9, 43.8, 49.2, 50.8, 53.1, 55.0, 78.7, 124.2, 127.1, 128.7, 129.3, 133.1, 134.4, 139.8, 170.0, 199.2, 212.3. HRMS (ESI-TOF) m/z: [M + H]$^+$ Calcd for C$_{28}$H$_{31}$Cl$_3$O$_2$: 505.1462; Found: 505.1462.

3.2.8. (6aR,6bS,8aS,8bS,11R,11aS,12aS,12bR)-8b-chloro-11-(2-chloro-6-fluorophenyl)-6a,8a-dimethyl-1,6,6a,6b,7,8,8a,8b,10,11,11a,12,12a,12b-tetradecahydropentaleno[2,1-a]phenanthrene-4,9(2H,5H)-dione (**2h**)

Yield = 58% (0.042 g), given as a light yellow powder; mp 164–166 °C; ^{1}H-NMR (300 MHz, CDCl$_3$): δ = 0.93–0.99 (m, 1H), 1.09 (s, 3H), 1.12–1.17 (m, 1H), 1.20 (s, 3H), 1.39–1.51 (m, 1H), 1.63–1.98 (m, 9H), 2.02–2.09 (m, 2H), 2.28–2.45 (m, 4H), 2.57 (ddd, J = 17.8, 8.9, 3.4 Hz, 1H), 3.03–3.08 (m, 1H), 3.38 (dd, J = 17.8, 8.8 Hz, 1H), 3.60–3.67 (m,1H), 5.75 (s, 1H), 6.97–7.05 (m, 1H), 7.15–7.25 (m, 2H). ^{13}C NMR (75 MHz, CDCl$_3$): δ = 17.4, 17.6, 20.4, 30.9, 31.7, 32.5, 33.9, 34.5, 35.6, 36.4, 38.5, 40.4, 42.1 (d, ^3J(C-F) = 7.3 Hz), 48.5, 50.7, 53.0, 55.3, 79.0, 115.3 (d, ^2J(C-F) = 23.7 Hz), 124.0, 125.7 (d, ^4J(C-F) = 2.9 Hz), 128.0 (d, ^2J(C-F) = 14.3 Hz), 128.7 (d, ^3J(C-F) = 10.0 Hz), 134.9 (d, ^3J(C-F) = 7.0 Hz), 161.5 (d, ^1J(C-F) = 249.5 Hz), 170.0, 199.3, 210.4. HRMS (ESI-TOF) m/z: [M + H]$^+$ Calcd for C$_{28}$H$_{31}$Cl$_2$FO$_2$: 489.1751; Found: 489.1758.

3.2.9. (6a*R*,6b*S*,8a*S*,8b*S*,11*R*,11a*S*,12a*S*,12b*R*)-8b-chloro-11-(4-methoxyphenyl)-6a,8a-dimethyl-1,6, 6a,6b,7,8,8a,8b,10,11,11a,12,12a,12b-tetradecahydropentaleno[2,1-*a*]phenanthrene-4,9(2*H*,5*H*)-dione (**2i**)

Yield = 55% (0.038 g), given as an yellow powder; mp 180–182 °C; ^{1}H-NMR (300 MHz, CDCl$_3$): δ = 1.02 (s, 3H), 1.08–1.14 (m, 1H), 1.22 (s, 3H), 1.32–1.54 (m, 2H), 1.64–1.93 (m, 8H), 2.02–2.09 (m, 1H), 2.25–2.45 (m, 5H), 2.80–2.85 (m, 1H), 2.95–2.99 (m, 1H), 3.04–3.12 (m, 2H), 3.82 (s, 3H), 5.74 (s, 1H), 6.90 (d, *J* = 8.2 Hz, 2H), 7.19 (d, *J* = 8.3 Hz, 2H). ^{13}C NMR (75 MHz, CDCl$_3$): δ = 16.3, 17.5, 20.3, 31.8, 32.6, 33.4, 33.9, 34.0, 35.5, 35.7, 38.5, 46.5, 47.6, 48.6, 52.8, 52.9, 55.3, 61.4, 83.1, 114.3 (2C), 124.1, 127.6 (2C), 134.3, 158.7, 170.3, 199.2, 210.3. HRMS (ESI-TOF) *m/z*: [M + H]$^+$ Calcd for C$_{29}$H$_{35}$Cl$_2$O$_3$ 467.2345; Found: 467.2347.

3.2.10. (6a*R*,6b*S*,8a*S*,8b*S*,11*R*,11a*S*,12a*S*,12b*R*)-8b-chloro-11-(3-methoxyphenyl)-6a,8a-dimethyl-1,6, 6a,6b,7,8,8a,8b,10,11,11a,12,12a,12b-tetradecahydropentaleno[2,1-*a*]phenanthrene-4,9(2*H*,5*H*)-dione (**2j**)

Yield = 44% (0.030 g), given as an yellow powder; mp 126–128 °C; ^{1}H NMR (300 MHz, CDCl$_3$): δ = 0.90–1.00 (m, 1H), 1.08 (s, 3H), 1.12–1.16 (m, 1H), 1.21 (s, 3H), 1.36–1.50 (m, 1H), 1.63–1.95 (m, 8H), 2.01–2.10 (m, 2H), 2.29–2.45 (m, 4H), 2.75 (dd, *J* = 18.2, 8.7 Hz, 1H), 2.98–3.08 (m, 2H), 3.11–3.18 (m, 1H), 3.83 (s, 3H), 5.75 (s, 1H), 6.78–6.82 (m, 1H), 6.86–6.91 (m, 2H), 7.24–7.29 (m, 1H). ^{13}C-NMR (75 MHz, CDCl$_3$): δ = 17.4, 17.5, 20.3, 30.4, 31.7, 32.5, 33.9, 35.5, 35.7, 36.6, 38.5, 45.0, 46.4, 48.6, 50.5, 53.1, 55.2, 56.6, 79.1, 111.9, 113.2, 119.5, 124.1, 129.7, 145.6, 159.8, 170.1, 199.3, 211.6. HRMS (ESI-TOF) *m/z*: [M + H]$^+$ Calcd for C$_{29}$H$_{35}$ClO$_3$: 467.2333; Found: 467.2347.

3.2.11. (6a*R*,6b*S*,8a*S*,8b*S*,11*R*,11a*S*,12a*S*,12b*R*)-8b-chloro-11-(3,4-dimethoxyphenyl)-6a,8a-dimethyl-1, 6,6a,6b,7,8,8a,8b,10,11,11a,12,12a,12b-tetradecahydropentaleno[2,1-*a*]phenanthrene-4,9(2*H*,5*H*)-dione (**2k**)

Yield = 43% (0.032 g), given as an yellow powder; mp 182–184 °C; ^{1}H-NMR (300 MHz, CDCl$_3$): δ = 0.89–0.98 (m, 1H), 1.07 (s, 3H), 1.12–1,16 (m,1H), 1.20 (s, 3H), 1.35–1.49 (m, 1H), 1.65–1.91 (m, 8H), 2.00–2.09 (m, 2H), 2.27–2.49 (m, 4H), 2.77 (dd, *J* = 18.6, 9.0 Hz, 1H), 2.93–3.00 (m, 2H), 3.11–3.18 (m, 1H), 3.87 (s, 3H), 3.89 (s, 3H), 5.74 (s, 1H), 6.80–6.85 (m, 3H). ^{13}C NMR (75 MHz, CDCl$_3$): δ = 17.4, 17.5, 20.3, 30.4, 31.7, 32.5, 33.9, 35.6, 35.7, 36.0, 36.6, 38.5, 45.4, 46.1, 48.8, 50.5, 53.1, 55.9 (2C), 56.7, 79.2, 110.5, 111.1, 119.2, 124.1, 136.7, 147.9, 149.1, 170.0, 199.2, 212.0. HRMS (ESI-TOF) *m/z*: [M + H]$^+$ Calcd for C$_{30}$H$_{37}$ClO$_4$: 497.2447; Found: 497.2453.

3.2.12. (6a*R*,6b*S*,8a*S*,8b*S*,11*R*,11a*S*,12a*S*,12b*R*)-8b-chloro-11-(3,4,5-trimethoxyphenyl)-6a,8a-dimethyl-1,6,6a,6b,7,8,8a,8b,10,11,11a,12,12a,12b-tetradecahydropentaleno[2,1-*a*]phenanthrene-4,9(2*H*,5*H*)-dione (**2l**)

Yield = 47% (0.037 g), given as an yellow powder; mp 207–209 °C; ^{1}H-NMR (300 MHz, CDCl$_3$): δ = 0.89–1.00 (m, 1H), 1.09 (s, 3H), 1.12–1.16 (m, 1H), 1.20 (s, 3H), 1.40–1.48 (m, 1H), 1.65–1.91 (m, 8H), 2.02–2.10 (m, 2H), 2.27–2.44 (m, 4H), 2.79 (dd, *J* = 18.6, 9.1 Hz, 1H), 2.92–3.01 (m, 2H), 3.10–3.16 (m, 1H), 3.84 (s, 3H), 3.87 (s, 6H), 5.74 (s, 1H), 6.52 (s, 2H). ^{13}C-NMR (75 MHz, CDCl$_3$): δ = 17.5, 17.6, 20.4, 30.5, 31.8, 32.6, 33.9, 35.7, 36.0, 36.7, 38.6, 45.6, 47.0, 48.9, 50.6, 53.2, 56.2 (2C), 56.6, 60.9, 79.4, 104.5 (2C), 124.2, 136.9, 140.1, 153.3(2C) 170.1, 199.3, 212.1. HRMS (ESI-TOF) *m/z*: [M + H]$^+$ Calcd for C$_{31}$H$_{39}$ClO$_5$: 527.2551; Found: 527.2559.

3.2.13. (6a*R*,6b*S*,8a*S*,8b*S*,11*R*,11a*S*,12a*S*,12b*R*)-8b-chloro-6a,8a-dimethyl-11-(thiophen-2-yl)-1,6,6a, 6b,7,8,8a,8b,10,11,11a,12,12a,12b-tetradecahydropentaleno[2,1-*a*]phenanthrene-4,9(2*H*,5*H*)-dione (**2m**)

Yield = 64% (0.042 g), given as a light yellow powder; mp 193–195 °C; ^{1}H-NMR (300 MHz, CDCl$_3$): δ = 0.89–0.98 (m, 1H), 1.05–1.15 (m, 4H), 1.20 (s, 3H), 1.37–1.48 (m, 1H), 1.60–1.93 (m, 8H), 2.02–2.16 (m, 2H), 2.25–2.47 (m, 4H), 2.83 (dd, *J* = 17.6, 7.4 Hz, 1H), 2.99–3.17 (m, 2H), 3.42–3.54 (m, 1H), 5.74 (s, 1H), 6.86–6.98 (m, 2H), 7.17–7.24 (m, 1H). ^{13}C NMR (75 MHz, CDCl$_3$): δ = 17.4, 17.5, 20.4, 30.4, 31.8, 32.6, 33.9, 34.9, 35.7, 36.7, 38.6, 41.3, 45.6, 49.0, 50.8, 53.2, 56.9, 78.5, 123.9, 124.0, 124.2,126.9, 147.8, 170.1, 199.3, 211.1. HRMS (ESI-TOF) *m/z*: [M + H]$^+$ Calcd for C$_{26}$H$_{31}$ClO$_2$S 443.1801; Found: 443.1806.

4. Conclusions

In summary, we studied the cyclization of 16-DPA-based benzylidenes in the presence of Lewis or Bronsted acids, in particular, aluminum, iron (III), tin, titanium (IV) chlorides and dry hydrogen chloride. It was found that these compounds readily undergo highly regioselective interrupted Nazarov cyclization with trapping chloride ion. An efficient method of the synthesis of D-annulated pentacyclic steroids was developed. The structures of the synthesized pentacyclic steroids were determined by NMR and X-ray diffraction. It was found that the reaction affords a single diastereomer, but it can crystallize as two conformers depending on the structure. The formation of one diastereomer can be explained by the rigidity of the steroid backbone, in which case the nucleophile always attacks with the formation of an antidiastereomer. Similarly, the formation of a single diastereomer was observed in the case of bicyclic camphor and nopione structures. The synthetic potential of this protocol has been illustrated by the gram-scale experiment. Antiproliferative activity of the synthesized compounds was evaluated against two breast cancer cell lines: MCF-7 and MDA-MB-231. All the tested compounds exhibit relatively high antiproliferative activity. The activity of a number of steroids is comparable to that of cisplatin used as the reference compound. Besides, we expected that the steroid compounds with high antiproliferative activity would be less toxic to normal tissues compared to cisplatin, which displays serious side effects significantly complicating the treatment process. A comparison of the antiproliferative activity of the newly synthesized pentacyclic steroids with that of the starting benzylidenes (see [51]) shows that the Nazarov cyclization of the latter does not lead to a loss of activity. Further preclinical and clinical trials are required to identify intracellular targets for the lead compounds (**2b** and **2g**). In particular, it remains to determine whether the selected molecules affect steroid hormone receptors and steroid metabolism enzymes in luminal breast cancer cells and to analyze whether p53 is involved in the cytotoxic effect caused by new-class steroids.

Supplementary Materials: The following are available online biological assays, X-ray diffraction studies, DFT calculations, transition state, thermodynamic calculations, ^{1}H NMR monitoring, copies of ^{1}H and ^{13}C NMR spectra, copies of HRMS spectra.

Author Contributions: Conceptualization, V.Z.S.; software, M.E.M. (X-ray); investigation, S.K.V., synthesis; A.V.Y., NMR study; A.M.S. and E.I.M., biology; writing—original draft preparation, Y.A.V., (manuscript, SM); A.V.Y. (Experimental section); writing—review and editing, V.Z.S.; supervision, I.V.Z. All authors have read and agreed to the published version of the manuscript.

Funding: This research received no external funding.

Acknowledgments: Crystal structure determination was performed in the Department of Structural Studies of the Zelinsky Institute of Organic Chemistry, Russian Academy of Sciences, Moscow.

Conflicts of Interest: The authors declare no conflict of interest.

References and Note

1. Ke, S. Recent Progress of novel steroid derivatives and their potential biological properties. *Mini-Rev. Med. Chem.* **2018**, *18*, 745–775. [CrossRef] [PubMed]
2. Hanson, J.R. Steroids: Partial synthesis in medicinal chemistry. *Nat. Prod. Rep.* **2010**, *27*, 887–899. [CrossRef] [PubMed]
3. Heasley, B. Chemical synthesis of the cardiotonic steroid glycosides and related natural products. *Chem. Eur. J.* **2012**, *18*, 3092–3120. [CrossRef] [PubMed]
4. Salvador, J.A.R.; Carvalho, J.F.S.; Neves, M.A.C.; Silvestre, S.M.; Leitao, A.J.; Silva, M.M.C.; SáeMelo, M.L. Anticancer steroids: Linking natural and semi-synthetic compounds. *Nat. Prod. Rep.* **2013**, *30*, 324–374. [CrossRef] [PubMed]
5. Shagufta, A.I.; Panda, G. Quest for steroidomimetics: Amino acids derived steroidal and nonsteroidal architectures. *Eur. J. Med. Chem.* **2017**, *133*, 139–151. [CrossRef]
6. Khan, M.O.F.; Lee, H.J. Synthesis and pharmacology of anti-Inflammatory steroidal antedrugs. *Chem. Rev.* **2008**, *108*, 5131–5145. [CrossRef]
7. Chung, S.-K.; Ryoo, C.H.; Yang, H.W.; Shim, J.-Y.; Kang, M.G.; Lee, K.W.; Kang, H.I. Synthesis and bioactivities of steroid derivatives as antifungal agents. *Tetrahedron* **1998**, *54*, 15899–15914. [CrossRef]

8. Sieghart, W.; Savic, M.M. International union of basic and clinical pharmacology. CVI: GABAA receptor subtype- and function-selective ligands: Key issues in translation to humans. *Pharmacolog. Rev.* **2018**, *70*, 836–878. [CrossRef]
9. Zhang, Y.-L.; Li, Y.-F.; Wang, J.-W.; Yu, B.; Shi, Y.-K.; Liu, H.-M. Multicomponent assembly of novel antiproliferative steroidal dihydropyridinyl spirooxindoles. *Steroids* **2016**, *109*, 22–28. [CrossRef]
10. Zorumski, C.F.; Mennerick, S.; Isenberg, K.E.; Covey, D.F. Potential clinical uses of neuroactive steroids. *Curr. Opin. Investig. Drugs* **2000**, *1*, 360–369.
11. Blanco, M.-J.; La, D.; Coughlin, Q.; Newman, C.A.; Griffin, A.M.; Harrison, B.L.; Salituro, F.G. Breakthroughs in neuroactive steroid drug discovery. *Bioorg. Med. Chem. Lett.* **2018**, *28*, 61–70. [CrossRef] [PubMed]
12. Nongthombam, G.S.; Boruah, R.C. Divergent synthesis of steroid analogs from steroidal β-formylenamides, conjugated enones and β-formylvinyl halides. *Heterocycles* **2019**, *98*, 19–62.
13. Ibrahim-Ouali, M. Total synthesis of steroids and heterosteroids from BISTRO. *Steroids* **2015**, *98*, 9–28. [CrossRef]
14. Ibrahim-Ouali, M.; Romero, E.; Hamze, K. Stereoselective synthesis of pentacyclic steroids functionalized at C-11. *Steroids* **2012**, *77*, 1092–1100. [CrossRef]
15. Yu, B.; Shi, X.-J.; Zheng, Y.-F.; Fang, Y.; Zhang, E.; Yu, D.-Q.; Liu, H.-M. A novel [1,2,4]triazolo[1,5-a] pyrimidine-based phenyl-linked steroid dimer: Synthesis and its cytotoxic activity. *Eur. J. Med. Chem.* **2013**, *69*, 323–330. [CrossRef] [PubMed]
16. Yu, B.; Sun, X.N.; Shi, X.J.; Qi, P.P.; Fang, Y.; Zhang, E.; Yu, D.Q.; Liu, H.M. Stereoselective synthesis of novel antiproliferative steroidal (E, E) dienamides through a cascade aldol/cyclization process. *Steroids* **2013**, *78*, 1134–1140. [CrossRef] [PubMed]
17. Yu, B.; Wang, S.; Liu, H.M. Recent advances on the synthesis and antitumor evaluation of exonuclear heterosteroids. *Chin. J. Org. Chem.* **2017**, *37*, 1952–1962. [CrossRef]
18. Bansal, R.; Achary, P.C. Man-made cytotoxic steroids: Exemplary agents for cancer therapy. *Chem. Rev.* **2014**, *114*, 6986–7005. [CrossRef]
19. Singh, H.; Jindal, D.P.; Yadav, M.R.; Kumar, M. Heterosteroids and drug research. *Progr. Med. Chem.* **1991**, *28*, 233–300.
20. Zhang, Z.; Giampa, G.M.; Draghici, C.; Huang, Q.; Brewer, M. Synthesis of demissidine by a ring fragmentation 1,3-dipolar cycloaddition approach. *Org. Lett.* **2013**, *15*, 2100–2103. [CrossRef]
21. Wang, Z.; Li, M.; Liu, X.; Yu, B. Synthesis of steroidal saponins bearing an aromatic E ring. *Tetrahedron Lett.* **2007**, *48*, 7323–7326. [CrossRef]
22. Arthan, D.; Svasti, J.; Kittakoop, P.; Pittayakhachonwut, D.; Tanticharoen, M.; Thebtaranonth, Y. Antiviral isoflavonoid sulfate and steroidal glycosides from the fruits of Solanum torvum. *Phytochemistry* **2002**, *59*, 459–463. [CrossRef]
23. Mewshaw, R.E.; Taylor, M.D.; Smith, A.B., III. Indole diterpene synthetic studies. 2. First-generation total synthesis of (-)-paspaline. *J. Org. Chem.* **1989**, *54*, 3449–3462. [CrossRef]
24. Kobayashi, J.; Shinonaga, H.; Shigemori, H.; Ymeyama, A.; Shoji, N.; Arihara, Sh. Xestobergsterol C, a new pentacyclic steroid from the okinawan marine sponge IRCINIA sp. and absolute stereochemistry of xestobergsterol A. *J. Nat. Prod.* **1995**, *58*, 312–318. [CrossRef]
25. Tietze, L.F.; Modi, A. Multicomponent domino reactions for the synthesis of biologically active natural products and drugs. *Med. Res. Rev.* **2000**, *20*, 304–322. [CrossRef]
26. Lambert, J.J.; Belelli, D.; Harney, S.C.; Peters, J.A.; Frenguelli, B.G. Modulation of native and recombinant GABAA receptors by endogenous and synthetic neuroactive steroids. *Brain Res. Rev.* **2001**, *37*, 68–80. [CrossRef]
27. Jiang, X.; Manion, B.D.; Benz, A.; Rath, N.P.; Evers, A.S.; Zorumski, C.F.; Mennerick, S.; Covey, D.F. Neurosteroid analogues. 9. Conformationally constrained pregnanes: Structure-activity studies of 13,24-cyclo-18,21-dinorcholane analogues of the GABA modulatory and anesthetic steroids (3r,5r)- and (3r,5â)-3-hydroxypregnan-20-one. *J. Med. Chem.* **2003**, *46*, 5334–5348. [CrossRef]
28. Kamernitzky, A.V.; Levina, I.S. Pregna-D'-pentaranes, Progestins and Antiprogestins: I. Separation of Biological Functions of Steroid Hormones. *Rus. J. Bioorg. Chem.* **2005**, *31*, 105–118. [CrossRef]
29. Kamernitzky, A.V.; Levina, I.S. Pregna-D'-pentaranes, Progestins and Antiprogestins: II. 1 Pathways and Realization Mechanisms of Separate Biological Functions of Steroid Hormones. *Rus. J. Bioorg. Chem.* **2005**, *31*, 199–209. [CrossRef]

30. Shchelkunova, T.A.; Rubtsov, P.M.; Levina, I.S.; Kamerntsky, A.V.; Smirnov, A.N. Pregna-D-pentarane structure influences progesterone receptor affinity for DNA. *Steroids* **2002**, *67*, 323–332. [CrossRef]
31. Shoji, N.; Umeyama, A.; Shin, K.; Takeda, K.; Arihara, S.; Kobayashi, J.; Takei, M. Two unique pentacyclic steroids with cis C/D ring junction from Xestospongia bergquistia Fromont, powerful inhibitors of histamine release. *J. Org. Chem.* **1992**, *11*, 2996–2997. [CrossRef]
32. Singh, R.; Panda, G. An overview of synthetic approaches for heterocyclic steroids. *Tetrahedron* **2013**, *69*, 2853–2884. [CrossRef]
33. Baranovskii, A.V.; Litvinovskaya, R.P.; Khripach, V.A. Steroids with a side chain containing a heterocyclic fragment: Synthesis and transformations. *Russ. Chem. Rev.* **1993**, *62*, 661–682. [CrossRef]
34. Monier, M.; El-Mekabaty, A.; Abdel-Latif, D.; DogruMert, B.; Elattar, K.M. Heterocyclic steroids: Efficient routes for annulation of pentacyclic steroidal pyrimidines. *Steroids* **2020**, *154*, 108548–108549. [CrossRef] [PubMed]
35. Singh, R.; Thota, S.; Bansal, R. Studies on 16,17-pyrazoline substituted heterosteroids as anti-Alzheimer and anti-Parkinsonian agents using LPS induced neuroinflammation models of mice and rats. *Acs Chem. Neurosci.* **2018**, *9*, 272–283. [CrossRef] [PubMed]
36. Scherbakov, A.M.; Komkov, A.V.; Komendantova, A.S.; Yastrebova, M.A.; Andreeva, O.E.; Shirinian, V.Z.; Hajra, A.; Zavarzin, I.V.; Volkova, Y.A. Steroidal pyrimidines and dihydrotriazines as novel classes of anticancer agents against hormone-dependent breast cancer cells. *Front. Pharm.* **2018**, *8*, 979–992. [CrossRef]
37. Komendantova, A.S.; Scherbakov, A.M.; Komkov, A.V.; Chertkova, V.V.; Gudovanniy, A.O.; Chernoburova, E.I.; Sorokin, D.V.; Dzichenka, Ya.U.; Shirinian, V.Z.; Volkova, Yu.A.; et al. Novel steroidal 1,3,4-thiadiazines: Synthesis and biological evaluation in androgen receptor-positive prostate cancer 22Rv1 cells. *Bioorg. Chem.* **2019**, *91*, 103142. [CrossRef]
38. Zhang, B.-L.; Song, L.-X.; Li, Y.-F.; Li, Y.-L.; Guo, Y-Z.; Zhang, E.; Liu, H.-M. Synthesis and biological evaluation of dehydroepiandrosterone-fused thiazole, imidazo[2,1-*b*]thiazole, pyridine steroidal analogues. *Steroids* **2014**, *80*, 92–101. [CrossRef]
39. Pilgrim, B.S.; Gatland, A.E.; Esteves, C.H.A.; McTernan, C.T.; Jones, G.R.; Tatton, M.R.; Procopiou, P.A.; Donohoe, T.J. Palladium-catalyzed enolatearylation as a key C–C bond-forming reaction for the synthesis of isoquinolines. *Org. Biomol. Chem.* **2016**, *14*, 1065–1090. [CrossRef]
40. Ibrahim-Ouali, M. Synthesis of pentacyclic steroids. *Steroids* **2008**, *73*, 775–797. [CrossRef]
41. Zavarzin, I.V.; Chertkova, V.V.; Levina, I.S.; Chernoburova, E.I. Steroids fused to heterocycles at positions 16, 17 of the D-ring. *Russ. Chem. Rev.* **2011**, *80*, 661–682. [CrossRef]
42. Lopes, S.M.M.; Gomes, C.S.B.; Pinho e Meló, T.M.V.D. Reactivity of steroidal 1-azadienes toward carbonyl Compounds under Enamine Catalysis: Chiral Penta- and hexacyclic steroids. *Org. Lett.* **2018**, *20*, 4332–4336. [CrossRef] [PubMed]
43. Jung, M.E.; Johnson, T.W. Synthesis of 7-deoxyxestobergestrol A, a novel pentacyclic steroid of the Xestobergestrol class. *J. Am. Chem. Soc.* **1997**, *119*, 12412–12413. [CrossRef]
44. Sharma, U.; Ahmed, Sh.; Boruah, R.C. A facile synthesis of annelated pyridines from β-formyl enamides under microwave irradiation. *Tetrahedron Lett.* **2000**, *41*, 3493–3495. [CrossRef]
45. Dutta, M.; Saikia, P.; Gogoi, S.; Boruah, R.C. Microwave-promoted and Lewis acid catalysed synthesis of steroidal A- and D-ring fused 4,6-diarylpyridines. *Steroids* **2013**, *78*, 387–395. [CrossRef]
46. Schomburg, D.; Thielmann, M.; Winterfeldt, E. Dienes as chiral templates. *Tetrahedron Lett.* **1986**, *27*, 5833–5834. [CrossRef]
47. LeBideau, F.; Dagorne, S. Synthesis of transition-metal steroid derivatives. *Chem. Rev.* **2013**, *113*, 7793–7850. [CrossRef]
48. Gupta, A.; Sathish, B.; Negi, A.S. Current status on development of steroids as anticancer agents. *J. Steroid Biochem. Mol. Biol.* **2013**, *137*, 242–270. [CrossRef]
49. Krafft, M.E.; Dasse, O.A.; Fu, Z. Synthesis of the C/D/E and A/B rings of Xestobergsterol-(A). *J. Org. Chem.* **1999**, *64*, 2475–2485. [CrossRef]
50. Chowdhury, P.; Borah, J.M.; Bordoloi, M.; Goswami, P.K.; Goswami, A.; Barua, N.C.; Rao, P.G. A simple efficient process for the synthesis of 16-dehydropregnenolone acetate (16-DPA)—A key steroid drug intermediate from diosgenin. *J. Chem. Eng. Process Technol.* **2011**, *2*, 117–124. [CrossRef]
51. Kumar, M.; Rawat, P.; Khan, M.; Rawat, A.K.; Srivastava, A.K.; Maurya, R. Aza-annulation on the 16-dehydropregnenolone, via tandem intermolecular Aldol process and intramolecular Michael addition. *Bioorganic Med. Chem. Lett.* **2011**, *21*, 2232–2237. [CrossRef] [PubMed]

52. Volkova, Yu.A.; Kozlov, A.S.; Kolokolova, M.K.; Uvarov, D.Y.; Gorbatov, S.A.; Andreeva, O.E.; Scherbakov, A.M.; Zavarzin, I.V. Steroidal N-sulfonylimidates: Synthesis and biological evaluation in breast cancer cells. *Eur. J. Med. Chem.* **2019**, *179*, 694–706. [CrossRef] [PubMed]
53. Scherbakov, A.M.; Zavarzin, I.V.; Vorontsova, S.K.; Hajra, A.; Andreeva, O.E.; Yadykov, A.V.; Levina, I.S.; Volkova, Y.A.; Shirinian, V.Z. Synthesis and evaluation of the antiproliferative activity of benzylidenes of 16-dehydroprogesterone series. *Steroids* **2018**, *138*, 91–101. [CrossRef]
54. Yadykov, A.V.; Shirinian, V.Z. Recent advances in the interrupted Nazarov reaction. *Adv. Synth. Catal.* **2020**, *4*, 702–723. [CrossRef]
55. Grant, T.N.; Riedera, C.J.; West, F.G. Interrupting the Nazarov reaction: Domino and cascade processes utilizing cyclopentenyl cations. *Chem. Commun.* **2009**, *38*, 5676–5688. [CrossRef]
56. Pellissier, H. Recent developments in the Nazarov process. *Tetrahedron* **2005**, *61*, 6479–6517. [CrossRef]
57. Frontier, A.J.; Collison, C. The Nazarov cyclization in organic synthesis. *Recent Adv. Tetrahedron* **2005**, *61*, 7577–7606. [CrossRef]
58. Vaidya, T.; Eisenberg, R.; Frontier, A.J. Catalytic Nazarov cyclization: The state of the art. *Chem. Cat. Chem.* **2011**, *3*, 1531–1548. [CrossRef]
59. Denmark, S.E.; Jones, T.K. Silicon-directed Nazarov cyclization. *J. Am. Chem. Soc.* **1982**, *104*, 2642–2645. [CrossRef]
60. Aggarwal, V.K.; Belfield, A.J. Catalytic asymmetric Nazarov reactions promoted by chiral Lewis acid complexes. *Org. Lett.* **2003**, *5*, 5075–5078. [CrossRef]
61. Janka, M.; He, A.; Frontier, A.J. Efficient catalysis of Nazarov cyclization using a cationic iridium complex possessing adjacent labile coordination sites. *J. Am. Chem. Soc.* **2004**, *126*, 6864–6865. [CrossRef] [PubMed]
62. Liang, G.; Trauner, D. Enantioselective Nazarov reactions through catalytic asymmetric proton transfer. *J. Am. Chem. Soc.* **2004**, *126*, 9544–9545. [CrossRef]
63. Shirinian, V.Z.; Lvov, A.G.; Yadykov, A.V.; Yaminova, L.V.; Kachala, V.V.; Markosyan, A.I. Triaryl-substituted divinyl ketones cyclization: Nazarov reaction versus Friedel–Crafts electrophilic substitution. *Org. Lett.* **2016**, *18*, 6260–6263. [CrossRef]
64. Janka, M.; He, W.; Haedicke, I.E.; Fronczek, F.R.; Frontier, A.J.; Eisenberg, R. Tandem Nazarov cyclization–Michael addition sequence catalyzed by an Ir(III) complex. *J. Am. Chem. Soc.* **2006**, *128*, 5312–5313. [CrossRef] [PubMed]
65. Rieder, C.J.; Winberg, K.J.; West, F.G. Cyclization of cross-conjugated trienes: The vinylogous Nazarov reaction. *J. Am. Chem. Soc.* **2009**, *131*, 7504–7505. [CrossRef] [PubMed]
66. Dhoro, F.; Tius, M.A. Interrupted Nazarov cyclization on silica gel. *J. Am. Chem. Soc.* **2005**, *127*, 12472–12473. [CrossRef] [PubMed]
67. Fujiwara, M.; Kawatsura, M.; Hayase, S.; Nanjo, M.; Itoh, T. Iron(III) Salt-catalyzed Nazarov cyclization/Michael addition of pyrrole derivatives. *Adv. Synth. Catal.* **2009**, *351*, 123–128. [CrossRef]
68. Basak, A.K.; Shimada, N.; Bow, W.F.; Vicic, D.A.; Tius, M.A. An Organocatalytic asymmetric Nazarov cyclization. *J. Am. Chem. Soc.* **2010**, *132*, 8266–8267. [CrossRef]
69. Murugan, K.; Srimurugan, S.; Chen, C.P. A mild, catalytic and efficient Nazarov cyclization mediated by phosphomolybdic acid. *Chem. Commun.* **2010**, *46*, 1127–1129. [CrossRef]
70. Shirinian, V.Z.; Lvov, A.G.; Yanina, A.M.; Kachala, V.V.; Krayushkin, M. Synthesis of new photochromic diarylethenes of cyclopentenone series by Nazarov reaction. *Chem. Heterocycl. Comp.* **2015**, *51*, 234–241. [CrossRef]
71. Vaidya, T.; Manbeck, G.F.; Chen, S.; Frontier, A.J.; Eisenberg, R. Divergent reaction pathways of a cationic intermediate: Rearrangement and cyclization of 2-substituted furyl and benzofurylenones catalyzed by iridium(III). *J. Am. Chem. Soc.* **2011**, *133*, 3300–3303. [CrossRef] [PubMed]
72. Huang, J.; Leboeuf, D.; Frontier, A.J. Understanding the fate of the oxyallyl cation following Nazarov electrocyclization: Sequential Wagner–Meerwein migrations and the synthesis of spirocycliccyclopentenones. *J. Am. Chem. Soc.* **2011**, *133*, 6307–6317. [CrossRef] [PubMed]
73. Bender, J.A.; Arif, A.M.; West, F.G. Nazarov-initiated diastereoselective cascade polycyclization of aryltrienones. *J. Am. Chem. Soc.* **1999**, *121*, 7443–7444. [CrossRef]
74. Saxena, H.O.; Faridi, U.; Kumar, J.K.; Luqman, S.; Darokar, M.P.; Shanker, K.; Chanotiya, C.S.; Gupta, M.M.; Negi, A.S. Synthesis of chalcone derivatives on steroidal framework and their anticancer activities. *Steroids* **2007**, *72*, 892–900. [CrossRef]

75. Singh, R.; Panda, G. Application of Nazarov type electrocyclization to access [6,5,6] and [6,5,5] core embedded new polycycles: An easy entry to tetrahydrofluorene scaffolds related to Taiwaniaquinoids and C-nor-D homosteroids. *Org. Biomol. Chem.* **2011**, *9*, 4782–4790. [CrossRef]
76. Mousavizadeh, F.; Meyer, D.; Giannis, A. Synthesis of C-nor-D-homo-steroidal alkaloids and their derivatives. *Synthesis* **2018**, *50*, 1587–1600.
77. Krieger, J.; Smeilus, T.; Schackow, O.; Giannis, A. Lewis Acid mediated Nazarov cyclization as a convergent and enantioselective entry to C-nor-D-homo-Steroids. *Chem. Eur. J.* **2017**, *23*, 5000–5004. [CrossRef]
78. Vada, E.; Fujiwara, I.; Kanemasa, S.; Tsuge, O. Synthesis of hexaindenones through a Diels-Alder cycloaddition and Nazarov cyclization sequence of triene alcohols. *Bull. Chem. Soc. Jpn.* **1987**, *60*, 325–334.
79. He, W.; Sun, X.; Frontier, A.J. Polarizing the Nazarov cyclization: efficient catalysis under mild conditions. *J. Am. Chem. Soc.* **2003**, *125*, 14278–14279. [CrossRef]
80. White, T.D.; West, F.G. Halide trapping of the Nazarov intermediate in strained polycyclic systems: A new interrupted Nazarov reaction. *Tetrahedron Lett.* **2005**, *46*, 5629–5632. [CrossRef]
81. Marx, V.M.; Cameron, T.S.; Burnell, D.J. Formation of halogenated cyclopent-2-enone derivatives by interrupted Nazarov cyclizations. *Tetrahedron Lett.* **2009**, *50*, 7213–7216. [CrossRef]
82. Schatz, D.J.; Kwon, Y.; Scully, T.W.; West, F.G. Interrupting the Nazarov cyclization with bromine. *J. Org. Chem.* **2016**, *81*, 12494–12498. [CrossRef] [PubMed]
83. Davis, C.E.; Coates, R.M. Stereoselective Prinscyclizations of δ,ε-unsaturated Ketones to cis-3-chlorocyclohexanols with $TiCl_4$. *Angew. Chem. Int. Ed.* **2002**, *41*, 491–493. [CrossRef]
84. Willmore, N.D.; Goodman, R.; Lee, H.-H.; Kennedy, R.M. A short synthesis of (.+-.)-.beta.-isocomene. *J. Org. Chem.* **1992**, *57*, 1216–1219. [CrossRef]
85. Liu, H.-J.; Sun, D. Polyene cyclization promoted by the cross conjugated α-carbalkoxyenone system. Application to the total synthesis of ($\pm$)-Dehydrochamaecynenol. *Tetrahedron Lett.* **1997**, *38*, 6159–6162. [CrossRef]
86. Browder, C.C.; West, F.G. Formation of hydrindans and tricyclo[4.3.0.03,8]nonanes via 6-endo trapping of the Nazarovoxyallyl intermediate. *Synlett* **1999**, *1999*, 1363–1366. [CrossRef]
87. Harmata, M.; Elomari, S.; Barnes, C.L. Intramolecular 4 + 3 cycloadditions. Cycloaddition reactions of cyclic alkoxyallylic and oxyallylic cations. *J. Am. Chem. Soc.* **1996**, *118*, 2860–2871. [CrossRef]
88. Crystallographic data was deposited with the Cambridge Crystallographic Data Centre (CCDC No. 1990621 and 1990622. For details, see section IV in the SM).
89. Li, Z.; Boyd, R.J.; Burnell, D.J. Computational examination of (4 + 3) versus (3 + 2) Cycloaddition in the Interception of Nazarov reactions of allenyl vinyl ketones by dienes. *J. Org. Chem.* **2015**, *80*, 12535–12544. [CrossRef]
90. Levenson, A.S.; Jordan, V.C. MCF-7: The first hormone-responsive breast cancer cell line. *Cancer Res.* **1997**, *57*, 3071–3078.
91. Yu, K.; Rohr, J.; Liu, Y.; Li, M.; Xu, J.; Wang, K.; Chai, J.; Zhao, D.; Liu, Y.; Ma, J.; et al. Progress in triple negative breast carcinoma pathophysiology: Potential therapeutic targets. *Pathol. Res. Pr.* **2020**, *216*, 152874. [CrossRef]
92. Ke, S.; Shi, L.; Zhang, Z.; Yang, Z. Steroidal[17,16-*d*]pyrimidines derived from dehydroepiandrosterone: A convenient synthesis, antiproliferation activity, structure-activity relationships, and role of heterocyclic moiety. *Sci. Rep.* **2017**, *7*, 44439. [CrossRef] [PubMed]
93. Hussein, M.M.M.; Amr, A.E.-G.E.; Abdalla, M.M.; Al-Omar, M.A.; Safwat, H.M.; Elgamal, M.H. Synthesis of androstanopyridine and pyrimidine compounds as novel activators of the tumor suppressor protein p53. *Z. Nat. C* **2015**, *70*, 205–216. [CrossRef] [PubMed]

Sample Availability: Samples of the compounds **2a–m** are available from the authors.

© 2020 by the authors. Licensee MDPI, Basel, Switzerland. This article is an open access article distributed under the terms and conditions of the Creative Commons Attribution (CC BY) license (http://creativecommons.org/licenses/by/4.0/).

Review

Strigolactones, from Plants to Human Health: Achievements and Challenges

Valentina Dell'Oste [1],*, Francesca Spyrakis [2],* and Cristina Prandi [3],*

1. Department of Public Health and Pediatric Science, University of Turin, via Santena 9, 10126 Turin, Italy
2. Department of Drug Science and Technology, University of Turin, via P. Giuria 9, 10125 Turin, Italy
3. Department of Chemistry, University of Turin, via P. Giuria 7, 10125 Turin, Italy
* Correspondence: valentina.delloste@unito.it (V.D.); francesca.spyrakis@unito.it (F.S.); cristina.prandi@unito.it (C.P.)

Abstract: Strigolactones (SLs) are a class of sesquiterpenoid plant hormones that play a role in the response of plants to various biotic and abiotic stresses. When released into the rhizosphere, they are perceived by both beneficial symbiotic mycorrhizal fungi and parasitic plants. Due to their multiple roles, SLs are potentially interesting agricultural targets. Indeed, the use of SLs as agrochemicals can favor sustainable agriculture via multiple mechanisms, including shaping root architecture, promoting ideal branching, stimulating nutrient assimilation, controlling parasitic weeds, mitigating drought and enhancing mycorrhization. Moreover, over the last few years, a number of studies have shed light onto the effects exerted by SLs on human cells and on their possible applications in medicine. For example, SLs have been demonstrated to play a key role in the control of pathways related to apoptosis and inflammation. The elucidation of the molecular mechanisms behind their action has inspired further investigations into their effects on human cells and their possible uses as anti-cancer and antimicrobial agents.

Keywords: strigolactones; Strigol; anti-cancer; antimicrobials; sustainable agriculture

1. Introduction

Strigolactones (SLs) are carotenoid-derived sesquiterpene lactones whose structure is characterized by a four-ring system that is generally identified as an ABC tricyclic core linked to a fourth ring, named the D-ring, by means of an enol-ether bridge (Figure 1). The partial elucidation of their biosynthesis in several plant species has identified the involvement of the following genes: DWARF27 (D27; β-carotene isomerase), Carotenoid Cleavage Dioxygenase 7 and 8 (CCD7 and CCD8), and MAX1 homologs (cytochrome P450s) [1]. The first SL, Strigol, was isolated from cotton-root exudate in 1966 [2]; it took over 40 years for its activity as a hyphal branching inducer to be uncovered [3], and for the role of SLs as a new class of phytohormones to be assessed [4,5]. Since then, the boom in interest in the use of these challenging molecules in sustainable agricultural practices indicates that there will be promising forthcoming developments [6–8]. Anti-cancer activity has been reported for multiple classes of plant hormones, including cytokinins, methyl jasmonate and brassinosteroids [9], and the first report on the antiproliferative activity of SLs was published in 2012 [10]. An ever-increasing number of references on the exploitation of SLs in the biomedical field have subsequently appeared in the literature. This review highlights the prospects of these future opportunities by outlining the accumulated knowledge on SLs and their potential applications in human health.

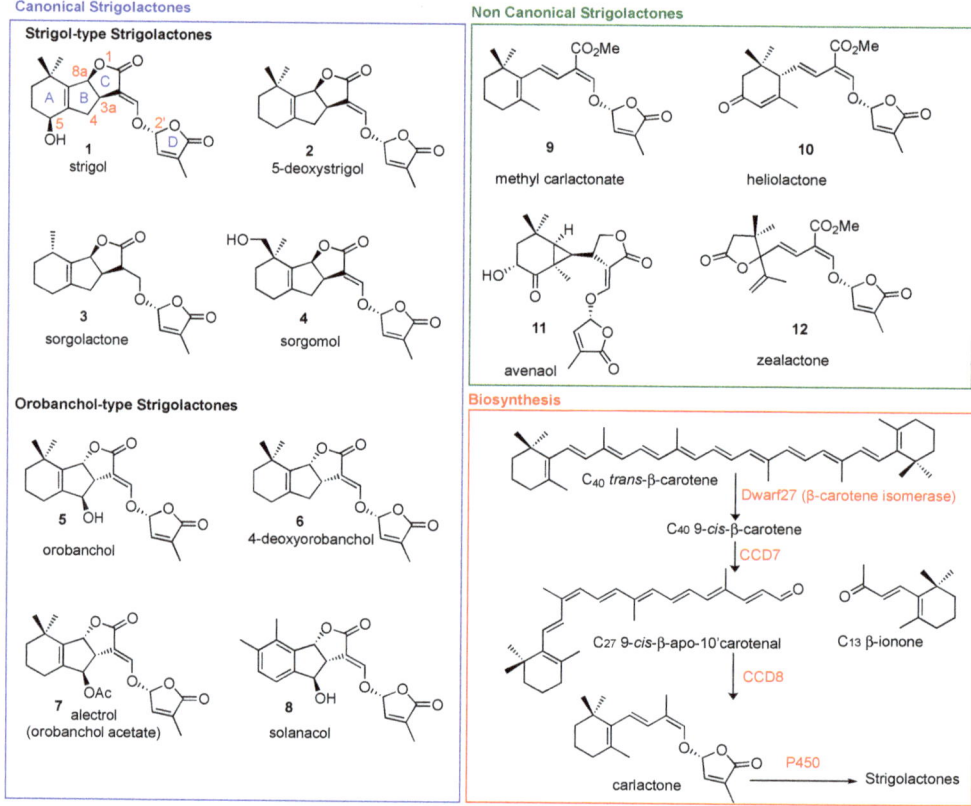

Figure 1. Structural diversity of natural canonical (blue box) and non-canonical SLs (green box). Biosynthesis in red box.

2. Structure and Synthesis of SLs

2.1. Naturally Occurring SLs

According to recent reviews, more than 25 SLs have been identified across the plant kingdom, with different plant species usually exuding different blends of several SLs [11].

Natural SLs are classified into two main classes: canonical and non-canonical SLs (Figure 1), according to the presence or absence, respectively, of the complete ABC-ring system [12,13]. The D-ring and the enol-ether bridge, which acts as a connection to the ABC core of the molecule, are a conserved feature in both canonical and non-canonical SLs. The structural variations in SLs are reflected in their functional diversity [14]. Stereochemistry plays a crucial role in the fine tuning of the biological properties ascribed to SLs [14,15]. Naturally occurring SLs can be divided into two families, strigol- (3aR,8aS in Strigol, Figure 1) and orobanchol-type SLs, (3aS,8aR in Orobanchol, Figure 1), depending on different orientations of the B/C junction, while the D-ring is always R configured (Figure 1). In biosynthetic pathways, the AB-rings can be modified via demethylation, hydroxylation, epoxidation and acetoxylation [16], giving rise to the structural diversification present in natural SLs.

2.2. SL Analogs and Mimics

Once the potential applications of SLs in agriculture and biomedicine became striking, synthetic SLs turned out to be an important tool with which to elucidate the functions of these signaling molecules and, at the same time, foster research in the field. Chemical synthesis involves either a total synthesis of the entire SL structure, or the synthesis of analogues with simplified structures that retain SL bio-properties [6,17,18].

The synthesis of SL analogues is based on the identification of the bioactiphore in SLs. This is the D-ring and the enol-ether bridge connecting C and D-ring (see Strigol 1 in Figure 1), which are required for activity, apparently as a Michael acceptor. Stereochemistry at the D-ring often plays a crucial role, with the most active derivatives showing the same configuration as natural SLs (2'R). A selection of the huge number of synthetic SLs that have been produced so far is provided in Figure 2 and includes GR24 [6], Nijmegen-1 [19], as well as indole derivatives EGO10, TH-EGO and EDOT [20]. Reports have shown that the structural modification of the D-ring into a c-lactam functional group may provide insight into the variations in SL-binding interactions with their receptor [21,22]. Other important analogues are fluorescent SLs, which can be used to track SL perception and trafficking, and include the fluorescence turn-on probe Yoshimulactone Green [23–27]. All these synthetic SLs have greatly contributed to improving our understanding of the biological role of SLs.

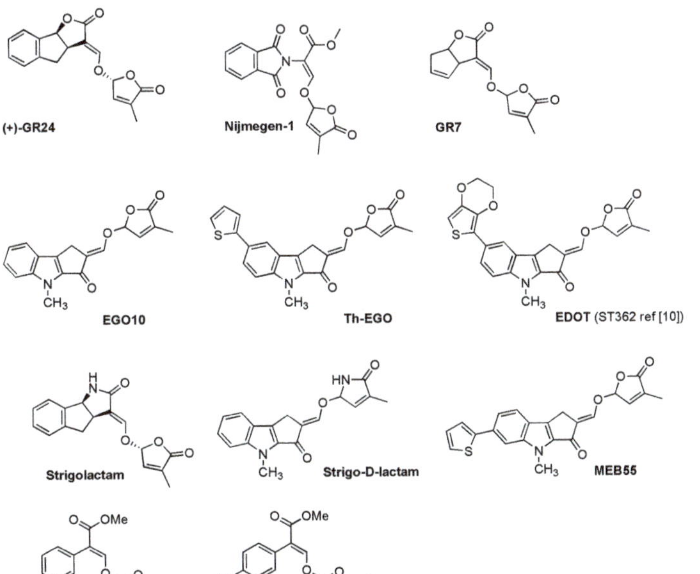

Figure 2. Panel of synthetic strigolactones. (+)-GR24 [6], Nijmegen-1 [19], GR7 [18], EGO10 [20], TH-EGO [20], EDOT [20], Strigolactam [21], Strigo-D-lactam [22], MEB 55 [10], TIT3 and TIT7 [28].

3. Roles of SLs in Plant Biology

After the first SL was isolated from the root parasite plant *Striga lutea* (witchweed) for use as a germination stimulant, many other SLs, with similar functions, were identified in *Striga* spp., broomrapes, *Alectrs* spp., and other host and non-host plants [2,29–31].

All SLs derive from carlactone (CL), which is synthesized in plasmids from all-trans-β-carotene by three different enzymes D27 [32,33], CCD7 and CCD8 [34]. In particular D27, an iron-binding enzyme, catalyzes the isomerization of all-trans-β-carotene to 9-cys-β-carotene, CCD7 converts that to 9-cys-β-apo-10′-carotenal and CCD8 then converts the carotenal into (Z)-(R)-carlactone (CL). The latter is then oxidized by cytochrome P450 monooxygenase MAX1, or other homologous enzymes, to generate the different SLs (Figure 1) [4,5,35]. Interestingly, the genes responsible for SL biosynthesis have been identified in several plants, algae and bryophytes, which suggests that SLs are fundamental molecules that have been maintained by evolution for a very long time [35]. When produced, SLs accumulate in the roots, the main storage organ [36], and then leave them by exudation to reach the rhizosphere, where they can exert their signaling activity [37]. This transport and exudation are regulated by the PhPDR1 transporter, whose mutants have shown a highly reduced level of SLs in root exudate [38].

The biological receptor through which SLs exert their action in plants has been identified as the α/β hydrolase receptor DWARF14 (D14), which is responsible for both the perception and deactivation of hormone signals [39,40]. D14 was first identified in a rice SL-insensitive mutant, but orthologs were soon found in Arabidopsis, petunia and pea [41–43]. This enzyme possesses a typical hydrolase catalytic triad, i.e., Ser, His, Asp, and cleaves SLs into the ABC- and the D-ring by performing a nucleophilic attack. The subject has been widely debated [42,44] but, recently, Seto et al., have demonstrated that the active signaling is activated by intact SLs. Upon SL binding, the receptor undergoes a transient conformational adjustment by which it becomes catalytically inactive, but able to interact with D53/SMXLs and D3/MAX2 signaling partners. When the latter are degraded by ubiquitination, the catalytic triad is reconstructed, SLs are hydrolyzed and the signal transduction is interrupted. Thus, the signaling process is triggered by intact SLs and not by hydrolysis or intermediate products [40].

SLs play a number of different roles, which will be explained in more detail hereafter: (i) they control the architecture of above-ground and underground plant organs [4,45–47]; (ii) they induce germination in root parasitic plants in genera such as *Striga*, *Orobanche*, *Alectra* and *Phelipanche* spp. [48], which have limited seed reserves, no photosynthetic activity and represent a real threat for agriculture thanks to SL-mediated activation; (iii) they regulate the symbiosis between plants and arbuscular mycorrhizal fungi (AMF); and (iv) they maintain plant life under hostile ecological conditions [35].

SLs are fundamental to controlling plant growth and architecture. Indeed, they induce root growth and the elongation of root hair, but inhibit secondary shoot branching [49]. They also participate, with auxins, in regulating leaf senescence, stem growth and seed germination [50–52]. Other phytohormones, such as abscisic acid (ABA), seem to positively regulate SL biosynthesis [53], while SLs behave antagonistically with respect to cytokinins [54], thus underlining the complex interplay of phytohormones that guarantees proper plant behavior. This integrative pathway also sustains the plant response to stress conditions [55]. SL production is, in fact, also regulated by nutrient starvation, such as salt stress, water stress, temperature and nutrient stress conditions. In the case of water stress, for instance, SLs inhibit shoot growth, but stimulate lateral root growth to increase water uptake from the soil. In the case of nutrient stress, the higher amount of produced SLs leads to shoot-branching suppression and stimulates symbiosis with AMF [4]. This latter can, in fact, guarantee the necessary water, phosphate and nitrogen supply, through hyphal extensions. Interestingly, phosphate starvation, as well as AMF colonization, GR24 treatment and naphthylacetic acid, induces the expression of the PhPDR1 transporter [37].

Similarly, in nitrogen-limited conditions, the expression of SL biosynthesis genes is boosted [56]. Moreover, SLs have been proven to respond to biotic stress [57] caused, in particular, by *Rhodococcus fascians*, *Pectobacterium carotovorum* and *Pseudomonas syringae* [58], whose infection induces the upregulation of genes associated to SL production, such as max1, max3 and max4 [59]. However, no alteration was detected in infections caused by bacteria such as *Pythium irregulare* and *Fusarium oxysporum*, thus suggesting that SLs only take part in plant immune response when stimulated by specific bacteria and fungi [35].

4. SLs for Sustainable Agriculture: The First Translation

The application of SLs in sustainable agriculture is a challenging goal, but one that is supported by the widespread use in agriculture of technologies that are based on plant hormones to control crop development [60].

Promising results have been achieved with SLs and SL analogs in agriculture, both when they are used as agrochemicals and via developing crop varieties modified for SL production and signaling [61,62].

The core components of sustainable agriculture strategies in which SLs display their main application domains are: (i) the control of parasitic weeds; (ii) drought mitigation; (iii) the efficiency of nutrient assimilation and crop development (Figure 3).

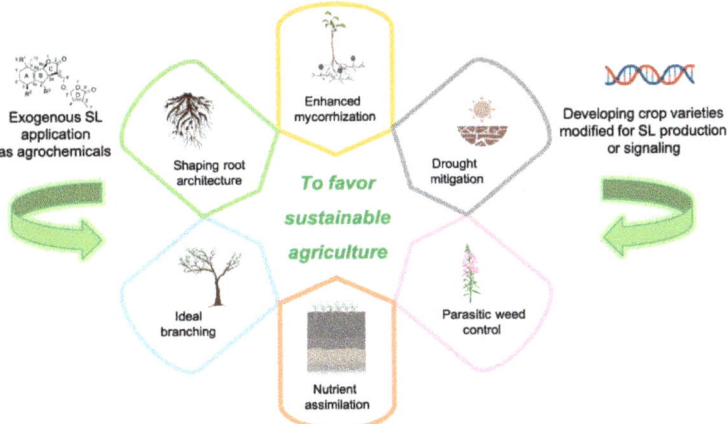

Figure 3. Agricultural applications of SLs. Exogenous applications of SLs as agrochemicals and the development of crops with modified SL production or signaling have the potential to favor sustainable agriculture via a number of mechanisms: shaping root architecture, promoting ideal branching, stimulating nutrient assimilation, controlling parasitic weeds, mitigating drought and enhancing mycorrhization. Created with BioRender.com.

4.1. SLs in the Control of Parasitic Weeds

One of the most thoroughly investigated applications of SLs is the control of the dangerous parasitic weeds species *Striga* (witchweeds) and *Orobanche* (broomrapes), which are estimated to infest upwards of 60 million hectares of farmland worldwide, resulting in severe yield losses every year [63].

Indeed, crops that produce significantly fewer SLs are more resistant to *Striga* and/or *Orobanche* infection than other cultivars. This has been observed for different species, including rice [64], tomato (*Solanum lycopersicum*) [65], the faba bean (*Vicia faba*) [66,67], and pea (*Pisum sativum*) [68]. However, the complete loss of SL exudation is not desirable since it can affect some symbiotic mycorrhizal associations, which are particularly needed in soils that are profoundly affected by *Striga* infestations.

The finding that different SLs have different properties towards mycorrhizae and parasitic weeds has allowed *Striga*-resistant varieties with normal mycorrhization to be obtained. For example, sorghum species mutated at the Low Germination Stimulant 1 (LGS1) locus are resistant to *Striga hermonthica* and *Striga asiatica*, and this resistance can be attributed to a change in profile from strigol-type to orobanchol-type SLs [69]. In field trials, a yield increase in sorghum [70], and maize [71], has been observed in farms across sub-Saharan Africa, where *Striga*-resistant crops were combined with other control measures, such as fertilization and the procedure of non-host trap crops.

When used as agrochemicals, SL analogs have proven themselves to be a realistic opportunity for controlling *Striga* and *Orobanche* via suicidal seed germination. This approach entails the application of SL analogs to soil, followed by the induced germination of the parasites, which cannot survive without the host, thereby depleting the seed bank in the soil. For example, the carbamate SL mimic T-010 reduced *S. hermonthica* emergence by 94–100% in pots and by 33% in sorghum, and is associated with 187–241% increases in sorghum dry weight [72]. Similarly, the SL analogs Nijmegen-1 and Nijmegen-1 Me were effective in controlling *Orobanche ramosa* in tobacco (*Nicotiana tabacum*) crops [73], while a novel class of SLs analogues, derived from dihydroflavonoids, exhibited higher potential in the suicidal germination of the Broomrapes, even compared to the control GR24 [74]. Another approach for the prevention of parasitic seed germination is to antagonize SL responses using SL receptor inhibitors, such as triazole ureas, as agrochemicals [75].

4.2. SLs in Drought Mitigation

Another challenging application of SLs is in the improvement of drought tolerance and decreasing yield losses that are caused by adverse climate conditions that lead to low water availability and high salinity [55,76].

It has been observed that water deprivation increase the expression of SL biosynthesis genes in *Arabidopsis* leaves [55], tomato shoots [76], and rice [77]. Interestingly, rice root extracts exhibited increased SL content under water deprivation, and the expression of the genes involved in SL biosynthesis was increased in both the roots and shoots of different species, such as in the crown of tall fescue (*Festuca arundinacea*) [78,79]. By contrast, osmotic stress represses SL biosynthesis in tomato [76], and *Lotus japonicus* roots [80].

A number of observations have highlighted that the foliar application of GR24, a synthetic SL analog, in SL mutants of *Arabidopsis thaliana* or grape, can lessen the effects of drought [55,81].

The underlying molecular mechanism is still to be understood, but data are available about the capability of SLs to promote stomata closure to reduce transpiration-associated water loss by interacting with ABA [78,82,83].

Another possibility, in addition to the examples of SL agrochemicals for drought, is the development of drought-tolerant crop varieties via the upregulation of SL signaling. For example, transgenic rice that overexpresses the OsNAC14 transcription factor was observed to upregulate SL biosynthesis genes as well as other genes involved in plant defense, stress response and DNA-damage repair. These transgenic plants had a better survival rate and chlorophyll fluorescence under drought conditions than non-transgenic controls [84].

4.3. SLs in the Promotion of Nutrient Assimilation and Crop Development

SLs can also be optimized to improve nutrient assimilation, thus favoring crop enhancement [85,86].

For example, there is evidence to demonstrate that both natural and synthetic SLs (e.g., GR24) endorse plant growth by positively influencing root vigor in different species [47,49].

One interesting strategy is the shaping of the root microbiome by recruiting specific beneficial microorganisms, such as arbuscular mycorrhizal (AM) fungi that promote hyphal branching [87], spore germination, mitochondrial biogenesis and respiration [88], and the exudation of oligosaccharide and protein signals required for AM recognition by the host [89–91]. Interestingly, plants mutated for the petunia hybrida ABC transporter (PDR1), a cellular SL exporter with a key role in regulating the development of AM and axillary branches, displayed reduced symbiotic interactions at the root level, indicating that SLs are critical for the establishment of an appropriate root microbiome [38]. A plant's genetic background influences the degree of mycorrhization and is a key factor in crop success in low-phosphate soils, as confirmed by experiments on SL transporter overexpression, which led to faster mycorrhization in *M. truncatula* [92].

This is a critical point as it supports the idea that SLs play a role in the adaptation of root architecture to variable nutrient accessibility in the soil, mainly nitrogen or phosphorus.

Results obtained from field trials have shown that SL analogs increased the capability of maize and sunflower to efficiently uptake nitrogen, when few fertilizers and pesticides are added [93], and of zucchini squash (*Cucurbita pepo*) [94] and "Hamlin" sweet oranges (*Citrus sinensis*) [95] to do the same under normal growth conditions.

SLs are also involved in legume nodulation processes, and thereby play an important role in nitrogen acquisition. For instance, the application of GR24 increased nodulation in alfalfa (*Medicago sativa*) [96], pea [97], and soybean (*Glycine max*) [98]; conversely, fewer nodules have been observed in SL-biosynthesis mutants than in wild-type plants in *L. japonicus*, pea and soybean [97,99–101].

SLs are increased by nutrient stress, such as low phosphate, nitrogen and sulfur conditions [102,103]. For example, the reduction of phosphate levels induces SLs in different families, including cereals, legumes and nightshades [102].

From a different perspective, crop yield can also be increased upon the reduction of SL biosynthesis or perception. For example, genetic approaches applied to modify a rice allele in order to alter SL signaling lead to an improvement in rice architecture. In particular, the analysis of 147 rice accessions identified the CCD7 gene as causing the partial loss of SL-biosynthesis function. Interestingly, CCD7 is widely co-selected with gibberellin deficiency in rice and contributed to improving grain yields during the green revolution [104]. This observation was further confirmed by the detection of an inverse correlation between different levels of tillering across commercial rice cultivars and SL levels [105], as well as by increased tillering in the Nipponbare background upon the silencing of the CCD7 gene by CRISPR/Cas9 [106].

Even though several "proofs of principle" on the potential application of SL in agriculture are now available, future and continued investments will be crucial for their routine and successful application in agriculture. More data about SL bioavailability and stability in plants and soil are certainly required, and the levels of uptake following application under field conditions should be determined. The fast degradation of natural and synthetic SLs in soil [107], and the limited information on early chemical uptake into seeds are major limits [108]. Finally, more clarity is required in the legislation on the production, commercialization and use of potential future SL-based technologies [7].

The optimization of these aspects could potentially allow SLs to be applied at very low quantities, in a range of 1–10 grams/hectare, with undoubtedly positive implications in terms of costs, environment and human safety [61].

5. Potential of SLs in Human Health

In recent years, a few studies have begun to shed light on the effects of plant hormones on human health. Molecules such as ABA, salicylic acid, indole-3-acetic acid (the best-known auxin) and cytokinins, which have been extensively studied as plant regulators, are also produced by and elicit biological activities in human cells and animal models [9,109].

Interestingly, several phytohormones can also be produced by human gut microbes, in addition to dietary intake, and likely influence many physiological pathways, such as glucose homeostasis, inflammatory responses and other cellular processes [110,111].

Some phytohormones affect human diseases, such as diabetes, inflammatory bowel disease and cancers, which are also modulated by the gut microbiota [112]. For instance, previous findings have revealed the beneficial effects of ABA against inflammation-related diseases such as type 2 diabetes (T2D), colitis, atherosclerosis, glioma and depression [111]. Salicylates, on the other hand, have long been appreciated as pharmacological agents [113].

Considering these effects, the use of phytohormones as multifunctional nutraceuticals against inflammation-associated diseases, in particular metabolic syndrome and its diverse comorbid symptoms, has been proposed [112]. Overall, the optimal formulation and dosage for phytohormone supplements are still to be established, although the ABA extract of fig fruit has recently been proposed for sugar control against T2D [114].

In this context, the value of SLs in the medical field is only emerging recently. The following sections outline the main discoveries in the applications of SLs for human health.

5.1. Modulation of Inflammation

Apart from being involved in the regulation of plant physiology, phytohormones have also been reported to affect human processes including, among others, cell division, glucose metabolism and inflammation [115]. More than ten years ago, it was observed that specific brassinosteroids improve oral glucose tolerance in mice by decreasing the expression of the gluconeogenic enzymes PEPCK and G6Pase and increasing ACT phosphorylation in the liver and muscles [116]. ABA also improves insulin resistance and has a positive effect on neuroinflammation [117], while gibberellic acid (GA) inhibits the release of proinflammatory interleukins, indicating that a GA-enriched diet may alleviate inflammatory disorders [109].

The representative SL GR24 has also been studied for its possible effects on glucose metabolism, and it was found to upregulate and activate SIRT1, a NAD^+-dependent deacetylase that plays a key role in glucose homeostasis and energy metabolism, and to enhance insulin signaling, glucose uptake, GLUT4 translocation and mitochondrial biogenesis. It is thus a possible new treatment for insulin resistance in skeletal muscle [118].

Recent studies have reported interesting anti-inflammatory activity for GR24 when tested in vitro and in vivo in RAW263.7 cells and zebrafish larvae, respectively [119]. Two GR24 isomers, in particular, were observed to significantly inhibit the release of the pro-inflammatory mediator NO in lipopolysaccharide (LPS)-stimulated cells, as well as the levels of TNF-α and IL-6, compared to the glucocorticoid dexamethasone. Similarly, the levels of phosphorylated NF-κB p65, IκBα, ERK1/2 and p38 MAPK significantly decreased upon treatment with GR24 isomers in a concentration-dependent manner. Indeed, the suppression of NF-κB and MAPK cascades directly resulted in decreased NO, TNF-α and IL-6 production. Important outcomes in the migration of neutrophils and primitive macrophages in zebrafish injuries were also observed. Apart from widening the many possible roles played by SLs, these results also confirmed the importance of the absolute SL configuration and the unsaturated D-ring, whose absence significantly reduced the aforementioned effects [119].

More recently, the role of SLs in neuroinflammation was studied in more detail when phenotypic screenings were performed on SIM-A9 microglial cell lines treated with a GR24 racemic mixture [120]. Again, a reduction in LPS-induced NO production was observed, and this reduction is comparable to that exerted by 1400W, which is a selective irreversible inhibitor of inducible nitric oxide synthase (iNOS). Both mRNA and iNOS levels, generally elevated in neurodegenerative disorders [121,122], were significantly reduced in a dose-dependent manner. ELISA and Western blot again confirmed the downregulation of the TNF-α gene and the consequent inhibition of TNF-α, known to be involved in the activation of α- and β-secretases, which, in turn, stimulate Aβ deposition and the consequent microglial cytokine storm. The suppression of IL-1β production was been registered. These observations support the potential anti-neuroinflammatory and neuroprotective effects of GR24, and reasonably those of SLs in general, against neurodegenerative disorders and the early events of Alzheimer disease (AD). It was also found that GR24 is able to provide the strong dose-dependent downregulation of COX-2, which is responsible for the production of prostaglandins in inflammatory processes [120]. It has to be noted that a clear correlation exists between COX-2 expression and dementia severity in patients affected by dementia, AD and Parkinson disease [123,124]. Interestingly, the nuclear deposition of LPS-induced NF-κB also decreased 3-fold, while PPARγ protein expression, suppressed by LPS treatment, was restored almost completely. Indeed, it has been reported that PPARγ activation can treat and prevent neurodegenerative diseases, and that PPARγ agonists prevent LPS-induced neuronal death [125]. GR24 has also been proven to increase the accumulation of Nrf2, which is the main transcription factor that controls the expression of several cytoprotective enzymes, in microglia cells. In fact, GR24 treatment induced the increased expression of NADPH quinone dehydrogenase-1 (NQO1) and heme oxygenase-1 (HO-1). GR24 seems to have positive efficacy on BBB endothelial cell permeabilization in reducing the negative effects provided by LPS. In particular, treatment with 20 μM GR24 reduced Evans Blue dye extravasation and increased the expression of tight junction proteins, such as occludins. Overall, Kurt et al., have soundly demonstrated that GR24 promotes the downregulation of proinflammatory genes/proteins and the upregulation of cytoprotective ones in microglia and BBB endothelial cells, thus making it an interesting candidate for the development of new treatments for neurodegenerative and neuroinflammatory diseases [120].

Similar effects were previously reported by the same authors in the treatment of murine RAW macrophages and hepatic Hepa1c1c7 cell lines with GR24 [115]. Having confirmed the potent inhibition exerted by the compound on LPS-induced NO production, molecular docking simulations were performed towards the iNOS enzyme, and confirmed the hypothesis to some extent; better interactions and score values were obtained for GR24 enantiomers compared to the positive control 1400 W [115]. As mentioned above, GR24 has an effect on Nrf2 expression. Nrf2 signaling is regulated by the repressor Kelch-like ECH-associated protein 1 (Keap1), which promotes Nrf2 ubiquitination [126]. The disruption of the Nrf2-Keap1 association allows Nrf2 to translocate within the nucleus and induce the expression of phase II detoxification enzymes. Docking simulations were therefore also performed in the Keap1 crevice bound by the Nrf2 peptide. Again, better poses and interaction energies were obtained compared to the control compounds sulforaphane and curcumin. All these data strongly suggest that there exists a link between the activation of Nrf2 and the increased expression of HO-1 and NQO1 cytoprotective enzymes. Indeed, other phytochemicals, such as resveratrol, carnosol, oroxylin A and epigallocatechin-3-gallate, have been demonstrated to exert their protective role through Nrf2 activation in numerous chronic inflammatory diseases, T2D, neurodegenerative disorders, cancer and cardiovascular diseases [127–129].

5.2. SLs as Anti-Cancer Agents

Several plant-derived compounds have shown anti-cancer activity. The most famous of these include curcumin, which is able to suppress NF-κB and cause apoptosis, vinblastine, an alkaloid that targets microtubule, and paclitaxel (Taxol), which also acts on microtubules [130]. More recently, phytohormones enlarged this category with brassinosteroids, which cause G1 arrest and apoptosis [131], methyl jasmonate, which depletes ATP in cancer cells through mitochondrial perturbation [9,132] and cytokinins. SL analogues have also demonstrated anti-cancer activity in vitro and in vivo.

The first anti-tumoral effect of SLs in breast cancer cells was reported by Pollock et al., who found that these natural compounds can specifically inhibit proliferation and induce apoptosis in cancer cells, while sparing non-cancer cells [10]. The effect of GR24 was first evaluated on ER+ tumorigenic, ER- metastatic and normal non-neoplastic fibroblasts. Significant growth reduction was observed at 2.5–5 ppm concentrations in both cancer lines, while no significant effect was observed on fibroblasts. GR24 was also observed to inhibit the growth and reduce the viability of MCF-7 tumorigenic cells that propagated as mammospheres in non-adherent growing conditions. Similar positive results were obtained when five synthetic SL analogues were tested. In particular, ST362 and MEB55 (later renamed TH-EGO and EDOT, respectively), which are characterized by an indolyl-based structure with an enol-ether bridge connecting the C and D ring, were found to be the most potent and to exert a non-reversible reduction in cell viability after only four hours. This effect is likely associated to the inhibition of the phosphorylation of p38 MAPK and JNK1/2, which are stress-activated kinases that play a key role in a stress-signaling cascade and are associated with cell-cycle arrest and apoptosis [133,134]. Indeed, the mechanism by which SLs exerts their anti-tumoral activity has been associated to the blockage of cell-cycle progression and the consequent induction of apoptosis. The authors only observed a dose-dependent increase of cells in the G2/M phase in tumorigenic cell lines, while normal fibroblasts did not show sensitivity to SLs in this context. This may be linked to the higher division rate of cancer cells and to the capacity of SLs to target rapidly dividing cells.

ST362 and MEB55 were also tested, alone and in combination with the breast cancer chemotherapy drug paclitaxel, in xenograft models [135]. The administration of MEB55 led to reductions in tumor volume and tumor-growth rate in mice implanted with MDA-MB-231 xenografts. ST632 also showed promising results, comparable to those of paclitaxel. The co-treatment of cancer cell lines with MEB55 and paclitaxel showed a two-fold decrease in MEB55 IC_{50}, thus suggesting that the two molecules could have an additive effect. However, fewer promising results were obtained on xenografts, as the tumor volume reduction obtained by the co-administration was not significant with respect to treatment with MEB55 alone. The effect of SLs on microtubule bundling has also been studied, and it was found that the phytohormones might affect microtubule network integrity and, consequently, inhibit the migration of the most invasive breast cancer cell lines [136,137]. This might also have an effect on tumor metastatic character, which is strictly related to cell-migration capability. It is interesting to note that paclitaxel also mainly targets microtubules to exert its potent action.

Having demonstrated the inhibition exerted by SLs in breast cancer cells and breast cancer stem cells, the same authors widened the study to other solid and non-solid cancer cell lines, including prostate, colon, lung, melanoma, osteosarcoma and leukemic cells [138]. They found that SLs, in particular the analogues EG5, EG9c, ST357, ST362 and MEB55, were able to inhibit the growth of the cell lines and to induce a cellular-stress response that turned into cell-cycle arrest and apoptosis in all cases, except fibroblasts. In particular, the authors again reported that SLs were able to arrest the cell cycle at the G2 state. This arrest was primarily associated to the down-regulation of cyclin B1, the Cdc25C protein and mRNA levels, and to the activation of stress signaling, such as the induction of multiple heat shock proteins (HSP) and cytokine [139]. The activation of stress signaling exerted by SLs also affects the stress-induced transcription factor FOXO4, p38 MAPK and JNK1/2, which are again involved in the signaling for cell-cycle arrest [133], and apoptosis [134,140]. Moreover, SLs induce the expression of several pro-apoptotic genes and inhibit the expression of survival factors such as ALDH1, which is a key regulator of stem cell viability and self-renewal.

The two most potent molecules were, again, MEB55 and ST362, which were able to induce apoptosis in all tested cell lines and to specifically reduce the viability of prostate tumor conditionally reprogrammed cells (CRC), in which a significant reduction in cyclin B level and a pronounced stress response (pp38 induction) were observed. Similarly, a stronger apoptotic response was observed in CRC tumor cells than in normal cells. These findings support the potential of SL analogues to induce a significant non-reversible apoptotic response in transformed cells and in patient-derived tumor cells, while having significantly lower toxic effects in normal cells [138]. The two compounds were also proven to induce DNA double-strand breaks (DSBs) and consequently activate DNA damage response in osteosarcoma cells [141]. However, at the same time, SLs downregulate the DNA repair protein RAD51 via ubiquitination and, consequently, also the homology-directed repair (HDR) system, which are possibly associated to resistance towards DNA-damaging chemotherapy and radiotherapy. It follows that RAD51 downregulation may be a useful strategy for restoring and enhancing the effectiveness of cancer chemotherapy [141,142]. Importantly, no DSB or cell death was detected in non-transformed fibroblasts, which once again highlights the potential clinical relevance of these molecules.

SLs analogues were also tested in cell lines of hepatocellular carcinoma (HCC), which is the predominant form of liver cancer and the fifth most common type of cancer in men [143]. Two of the tested compounds, namely TIT3 and TIT7 (Figure 2), showed antiproliferative effects (cell viability reduction) on HepG2 cells, but had a lower effect on hamster kidney cells (BHK cells). The two compounds were also tested on PC3 prostate cancer and T-cell acute lymphoblastic leukemia cell lines, with dose-dependent cell-viability inhibition being shown. This indicates that the two compounds have a capability to inhibit cell proliferation in both solid and hematological tumors. Interestingly, the compounds showed a minimal inhibitory effect on healthy cells compared to cancer cells. The authors also performed a wound-healing assay on HepG2 cells to check cell migration, which was effectively inhibited by both TIT3 and TIT7. A possible mechanism, as already suggested for SLs, could involve the compounds interfering with the microtubular network, but the exact targets and a detailed mechanism of action that explains compound selectivity is currently difficult to define [141].

Taken together, these results clearly support the anti-cancer effects of SLs, which are emerging as a new possible treatment for advanced prostate cancer and other types of tumors.

5.3. SLs with Antimicrobial Activity

Despite the multifaceted roles of SLs in plant biology and their promising features as drug candidates for different kinds of cancers, their antimicrobial and antiviral activity is still unexplored (Figure 4).

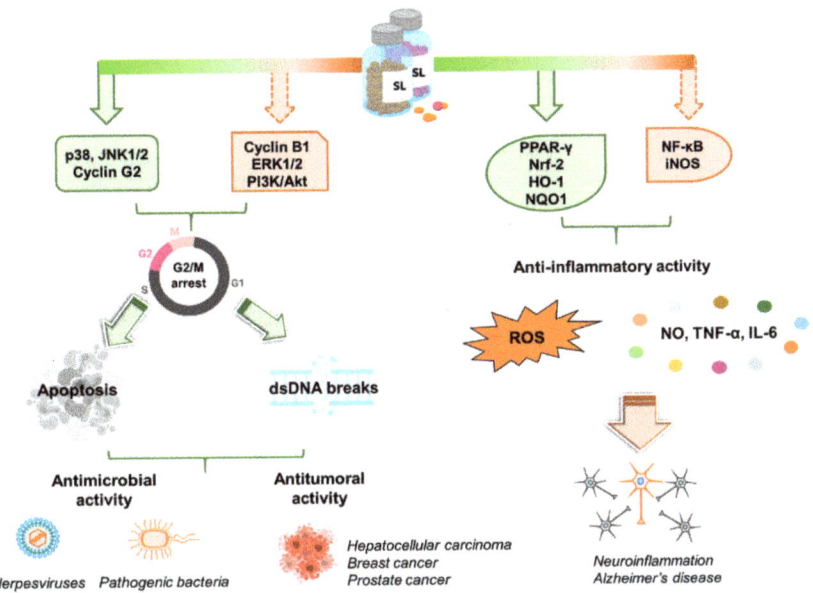

Figure 4. SL activity in human cells and their potential in medicine. *Left panel.* Synthetic analogs of SLs control multiple pathways leading to the arrest of the cell cycle in the G2/M phase. Apoptosis and DSBs are then induced. These properties grant SLs antimicrobial as well as anti-tumoral activity. *Right panel.* SLs exert anti-inflammatory effects by inhibiting the release of inflammatory molecules (e.g., NO, TNF-α, IL-6, ROS). This makes SLs promising scaffolds for the development of novel anti-Alzheimer's disease candidates. Created with BioRender.com.

The lessons learned from phytopathogenic fungi, in which the SL analog GR24 impairs the growth of root pathogens (e.g., *Fusarium oxysporum f. sp. melonis*, *Fusarium solani f. sp. mango*, *Sclerotinia sclerotiorum* and *Macrophomina phaseolina*), and the foliar pathogens *Alternaria alternata*, *Colletotrichum acutatum* and *Botrytis cinerea* [144], led to the hypothesis that SL antimicrobial activity could be extended to human pathogens.

The possibility of using SLs as antibiotics has been explored for the novel SL analog TIT3 against different pathogenic bacteria. Promising results were obtained for *Staphylococcus aureus*, *Salmonella typhimurium*, *Escherichia coli*, *Klebsiella pneumonia* and *Bacillus subtilis*, indicating that SLs may be a viable alternative for the treatment of different strains of bacteria that are resistant to conventional antibiotics [28].

Recently, our group has demonstrated, for the first time, the efficacy of a group of SL analogues as antivirals against members of the *Herpesviridae* family, in particular human cytomegalovirus (HCMV) [145]. HCMV is a widespread pathogen that can cause severe disease in immunocompromised individuals [146]. In addition, HCMV infection is the most frequent cause of congenital malformation in developed countries [147]. Although nucleoside analogues have been successfully used against HCMV, their use is hampered by the occurrence of serious side effects, the rapid emergence of resistance and the fact that their efficacy is limited to alleviating symptoms, without eradicating the latent infection [148,149]. There is, therefore, an urgent clinical need for new antiviral drugs that can overcome these limitations. Of the different SL analogs screened, there are two compounds that significantly inhibit HCMV replication in vitro, i.e., TH-EGO and EDOT-EGO. These results are challenging in the field of antiviral research, since, besides inhibiting the late phases of the viral cycle, apoptosis has been shown to be a novel strategy that SLs rely on to exert their inhibitory role against viral replication. These results have been confirmed by in-silico molecular docking simulations, which predict a stable protein-ligand complex between the SL analogs and the modeled structure of the putative target IE1, which is employed by HCMV to escape apoptosis [145].

In this context, further investigations on physiologically relevant targets for HCMV infection, such as endothelial and epithelial cells, and on cells that do not progress to a lytic infection, such as monocytes, will be crucial to corroborating and expanding the data obtained on HFFs. Furthermore, it will be essential to extend the analysis to other HCMV proteins, such as other antiapoptotic HCMV proteins (vMIA, cICA, UL38 and IE2), as well as to other members of the *Herpesviridae* family and emerging viruses, for which medical demand is an absolute priority at this time.

6. Conclusions

SLs are versatile and challenging molecules. In this review, we have demonstrated how the blooming and interdisciplinary research on SLs continuously unveils exciting, new biological functions and properties for these molecules. The exploitation of these properties is not without challenges: (i) lead compounds with unbiased activity and uncontroversial benefits should be identified; (ii) the synthesis of SLs is complicated, and designing the proper structure to emphasize specific activity is a difficult task; (iii) it is necessary to find the right balance between the stability and reactivity of SLs and effective formulations must be set up. We can foresee that the deep and full understanding of the molecular mechanisms, the elucidation of the transduction signal pathways and the development of their synthetic chemistry will pave the way for a variety of potential applications in agriculture and medicine.

7. Patents

Patent "Strigolattoni per uso nella prevenzione e/o trattamento di infezioni da virus della famiglia Herpesviridae" (No: 102018000010142, PCT/IB2019/059611, E7527/19-EW, University of Turin, Italy).

Author Contributions: F.S., V.D., C.P.: writing—review and editing. All authors have read and agreed to the published version of the manuscript.

Funding: The authors are actively involved in research on the topic of the Review, funded by: Proof of Concept (PoC)-TOINPROVE/2020, the "Cassa di Risparmio" Foundation of Turin (V.D.), the Italian Ministry of Education, University and Research-MIUR (PRIN 2015RMNSTA) (V.D.), the University of Turin (RILO2020) (V.D.) (C.P.) (F.S.), and the National Agency for the evaluation of Universities and Research Institutes (ANVUR, Basic Research 2017 to P.C.).

Conflicts of Interest: The authors declare no conflict of interest.

References

1. Lopez-Obando, M.; Ligerot, Y.; Bonhomme, S.; Boyer, F.D.; Rameau, C. Strigolactone biosynthesis and signaling in plant development. *Development* **2015**, *142*, 3615–3619. [CrossRef]
2. Cook, C.E.; Whichard, L.P.; Turner, B.; Wall, M.E.; Egley, G.H. Germination of witchweed (striga lutea lour.): Isolation and properties of a potent stimulant. *Science* **1966**, *154*, 1189–1190. [CrossRef]
3. Akiyama, K.; Ogasawara, S.; Ito, S.; Hayashi, H. Structural requirements of strigolactones for hyphal branching in AM fungi. *Plant Cell Physiol.* **2010**, *51*, 1104–1117. [CrossRef]
4. Gomez-Roldan, V.; Fermas, S.; Brewer, P.B.; Puech-Pagès, V.; Dun, E.A.; Pillot, J.P.; Letisse, F.; Matusova, R.; Danoun, S.; Portais, J.C.; et al. Strigolactone inhibition of shoot branching. *Nature* **2008**, *455*, 189–194. [CrossRef]

5. Umehara, M.; Hanada, A.; Yoshida, S.; Akiyama, K.; Arite, T.; Takeda-Kamiya, N.; Magome, H.; Kamiya, Y.; Shirasu, K.; Yoneyama, K.; et al. Inhibition of shoot branching by new terpenoid plant hormones. *Nature* **2008**, *455*, 195–200. [CrossRef] [PubMed]
6. Zwanenburg, B.; Blanco-Ania, D. Strigolactones: New plant hormones in the spotlight. *J. Exp. Bot.* **2018**, *69*, 2205–2218. [CrossRef] [PubMed]
7. Vurro, M.; Prandi, C.; Baroccio, F. Strigolactones: How far is their commercial use for agricultural purposes? *Pest Manag. Sci.* **2016**, *72*, 2026–2034. [CrossRef] [PubMed]
8. Aliche, E.B.; Screpanti, C.; De Mesmaeker, A.; Munnik, T.; Bouwmeester, H.J. Science and application of strigolactones. *New Phytol.* **2020**, *227*, 1001–1011. [CrossRef]
9. Lin, L.; Tan, R.X. Cross-kingdom actions of phytohormones: A functional scaffold exploration. *Chem. Rev.* **2011**, *111*, 2734–2760. [CrossRef]
10. Pollock, C.B.; Koltai, H.; Kapulnik, Y.; Prandi, C.; Yarden, R.I. Strigolactones: A novel class of phytohormones that inhibit the growth and survival of breast cancer cells and breast cancer stem-like enriched mammosphere cells. *Breast Cancer Res. Treat.* **2012**, *134*, 1041–1055. [CrossRef]
11. Bürger, M.; Chory, J. The Many Models of Strigolactone Signaling. *Trends Plant Sci.* **2020**, *25*, 395–405. [CrossRef]
12. Wang, Y.; Bouwmeester, H.J. Structural diversity in the strigolactones. *J. Exp. Bot.* **2018**, *69*, 2219–2230. [CrossRef] [PubMed]
13. Yoneyama, K.; Xie, X.; Yoneyama, K.; Kisugi, T.; Nomura, T.; Nakatani, Y.; Akiyama, K.; McErlean, C.S.P. Which are the major players, canonical or non-canonical strigolactones? *J. Exp. Bot.* **2018**, *69*, 2231–2239. [CrossRef] [PubMed]
14. Scaffidi, A.; Waters, M.T.; Sun, Y.K.; Skelton, B.W.; Dixon, K.W.; Ghisalberti, E.L.; Flematti, G.R.; Smith, S.M. Strigolactone hormones and their stereoisomers signal through two related receptor proteins to induce different physiological responses in arabidopsis. *Plant Physiol.* **2014**, *165*, 1221–1232. [CrossRef] [PubMed]
15. Flematti, G.R.; Scaffidi, A.; Waters, M.T.; Smith, S.M. Stereospecificity in strigolactone biosynthesis and perception. *Planta* **2016**, *243*, 1361–1373. [CrossRef]
16. Al-Babili, S.; Bouwmeester, H.J. Strigolactones, a novel carotenoid-derived plant hormone. *Annu. Rev. Plant Biol.* **2015**, *66*, 161–186. [CrossRef]
17. Prandi, C.; McErlean, C.S.P. The chemistry of strigolactones. In *Strigolactones—Biology and Applications*; Springer: Berlin/Heidelberg, Germany, 2019; pp. 163–198.
18. Zwanenburg, B.; Čavar Zeljković, S.; Pospíšil, T. Synthesis of strigolactones, a strategic account. *Pest Manag. Sci.* **2016**, *72*, 15–29. [CrossRef] [PubMed]
19. Nefkens, G.H.L.; Thuring, J.W.J.F.; Beenakkers, M.F.M.; Zwanenburg, B. Synthesis of a Phthaloylglycine-Derived Strigol Analogue and Its Germination Stimulatory Activity toward Seeds of the Parasitic Weeds Striga hermonthica and Orobanche crenata. *J. Agric. Food Chem.* **1997**, *45*, 2273–2277. [CrossRef]
20. Prandi, C.; Occhiato, E.G.; Tabasso, S.; Bonfante, P.; Novero, M.; Scarpi, D.; Bova, M.E.; Miletto, I. New potent fluorescent analogues of strigolactones: Synthesis and biological activity in parasitic weed germination and fungal branching. *Eur. J. Org. Chem.* **2011**, *2011*, 3781–3793. [CrossRef]
21. De Mesmaeker, A.; Screpanti, C.; Fonné-Pfister, R.; Lachia, M.; Lumbroso, A.; Bouwmeester, H. Design, synthesis and biological evaluation of strigolactone and strigolactam derivatives for potential crop enhancement applications in modern agriculture. *Chimia* **2019**, *73*, 549–560. [CrossRef]
22. Lombardi, C.; Artuso, E.; Grandi, E.; Lolli, M.; Spyrakis, F.; Priola, E.; Prandi, C. Recent advances in the synthesis of analogues of phytohormones strigolactones with ring-closing metathesis as a key step. *Org. Biomol. Chem.* **2017**, *15*, 8218–8231. [CrossRef]
23. Tsuchiya, Y.; Yoshimura, M.; Sato, Y.; Kuwata, K.; Toh, S.; Holbrook-Smith, D.; Zhang, H.; McCourt, P.; Itami, K.; Kinoshita, T.; et al. Probing strigolactone receptors in Striga hermonthica with fluorescence. *Science* **2015**, *349*, 864–868. [CrossRef] [PubMed]
24. Prandi, C.; Ghigo, G.; Occhiato, E.G.; Scarpi, D.; Begliomini, S.; Lace, B.; Alberto, G.; Artuso, E.; Blangetti, M. Tailoring fluorescent strigolactones for in vivo investigations: A computational and experimental study. *Org. Biomol. Chem.* **2014**, *12*, 2960–2968. [CrossRef] [PubMed]
25. Rasmussen, A.; Heugebaert, T.; Matthys, C.; Van Deun, R.; Boyer, F.D.; Goormachtig, S.; Stevens, C.; Geelen, D. A fluorescent alternative to the synthetic strigolactone GR24. *Mol. Plant* **2013**, *6*, 100–112. [CrossRef] [PubMed]
26. Fridlender, M.; Lace, B.; Wininger, S.; Dam, A.; Kumari, P.; Belausov, E.; Tsemach, H.; Kapulnik, Y.; Prandi, C.; Koltai, H. Influx and Efflux of Strigolactones Are Actively Regulated and Involve the Cell-Trafficking System. *Mol. Plant* **2015**, *8*, 1809–1812. [CrossRef]
27. Prandi, C.; Rosso, H.; Lace, B.; Occhiato, E.G.; Oppedisano, A.; Tabasso, S.; Alberto, G.; Blangetti, M. Strigolactone analogs as molecular probes in chasing the (SLs) receptor/s: Design and synthesis of fluorescent labeled molecules. *Mol. Plant* **2013**, *6*, 113–127. [CrossRef]
28. Al-Malki, A.L.; Huwait, E.A.; Moselhy, S.S. Synthesis, characterization and in vitro antibacterial activity of a novel strigolactones analogues TIT3. *J. Pure Appl. Microbiol.* **2020**, *14*, 2425–2430. [CrossRef]
29. Xie, X.; Yoneyama, K.; Yoneyama, K. The strigolactone story. *Annu. Rev. Phytopathol.* **2010**, *48*, 93–117. [CrossRef]
30. Mangnus, E.M.; Zwanenburg, B. Tentative Molecular Mechanism for Germination Stimulation of Striga and Orobanche Seeds by Strigol and Its Synthetic Analogues. *J. Agric. Food Chem.* **1992**, *40*, 1066–1070. [CrossRef]
31. Yokota, T.; Sakai, H.; Okuno, K.; Yoneyama, K.; Takeuchi, Y. Alectrol and orobanchol, germination stimulants for Orobanche minor, from its host red clover. *Phytochemistry* **1998**, *49*, 1967–1973. [CrossRef]

32. Wang, Y.; Li, J. Branching in rice. *Curr. Opin. Plant Biol.* **2011**, *14*, 94–99. [CrossRef]
33. Hao, L.; Renxiao, W.; Qian, Q.; Meixian, Y.; Xiangbing, M.; Zhiming, F.; Cunyu, Y.; Biao, J.; Zhen, S.; Jiayang, L.; et al. DWARF27,an iron-containing protein required for the biosynthesis of strigolactones, regulates rice tiller bud outgrowth. *Plant Cell* **2009**, *21*, 1512–1525. [CrossRef]
34. Alder, A.; Jamil, M.; Marzorati, M.; Bruno, M.; Vermathen, M.; Bigler, P.; Ghisla, S.; Bouwmeester, H.; Beyer, P.; Al-Babili, S. The path from β-carotene to carlactone, a strigolactone-like plant hormone. *Science* **2012**, *335*, 1348–1351. [CrossRef]
35. Mishra, S.; Upadhyay, S.; Shukla, R.K. The role of strigolactones and their potential cross-talk under hostile ecological conditions in plants. *Front. Physiol.* **2017**, *7*, 691. [CrossRef] [PubMed]
36. Yoneyama, K.; Xie, X.; Kusumoto, D.; Sekimoto, H.; Sugimoto, Y.; Takeuchi, Y.; Yoneyama, K. Nitrogen deficiency as well as phosphorus deficiency in sorghum promotes the production and exudation of 5-deoxystrigol, the host recognition signal for arbuscular mycorrhizal fungi and root parasites. *Planta* **2007**, *227*, 125–132. [CrossRef] [PubMed]
37. Ruyter-Spira, C.; Al-Babili, S.; van der Krol, S.; Bouwmeester, H. The biology of strigolactones. *Trends Plant Sci.* **2013**, *18*, 72–83. [CrossRef] [PubMed]
38. Kretzschmar, T.; Kohlen, W.; Sasse, J.; Borghi, L.; Schlegel, M.; Bachelier, J.B.; Reinhardt, D.; Bours, R.; Bouwmeester, H.J.; Martinoia, E. A petunia ABC protein controls strigolactone-dependent symbiotic signalling and branching. *Nature* **2012**, *483*, 341–344. [CrossRef]
39. Sanchez, E.; Artuso, E.; Lombardi, C.; Visentin, I.; Lace, B.; Saeed, W.; Lolli, M.L.; Kobauri, P.; Ali, Z.; Spyrakis, F.; et al. Structure–activity relationships of strigolactones via a novel, quantitative in planta bioassay. *J. Exp. Bot.* **2018**, *69*, 2333–2343. [CrossRef] [PubMed]
40. Seto, Y.; Yasui, R.; Kameoka, H.; Tamiru, M.; Cao, M.; Terauchi, R.; Sakurada, A.; Hirano, R.; Kisugi, T.; Hanada, A.; et al. Strigolactone perception and deactivation by a hydrolase receptor DWARF14. *Nat. Commun.* **2019**, *10*, 191. [CrossRef]
41. Hamiaux, C.; Drummond, R.S.M.; Janssen, B.J.; Ledger, S.E.; Cooney, J.M.; Newcomb, R.D.; Snowden, K.C. DAD2 is an α/β hydrolase likely to be involved in the perception of the plant branching hormone, strigolactone. *Curr. Biol.* **2012**, *22*, 2032–2036. [CrossRef]
42. De Saint Germain, A.; Clavé, G.; Badet-Denisot, M.A.; Pillot, J.P.; Cornu, D.; Le Caer, J.P.; Burger, M.; Pelissier, F.; Retailleau, P.; Turnbull, C.; et al. An histidine covalent receptor and butenolide complex mediates strigolactone perception. *Nat. Chem. Biol.* **2016**, *12*, 787–794. [CrossRef] [PubMed]
43. Waters, M.T.; Nelson, D.C.; Scaffidi, A.; Flematti, G.R.; Sun, Y.K.; Dixon, K.W.; Smith, S.M. Specialisation within the DWARF14 protein family confers distinct responses to karrikins and strigolactones in Arabidopsis. *Development* **2012**, *139*, 1285–1295. [CrossRef] [PubMed]
44. Yao, R.; Ming, Z.; Yan, L.; Li, S.; Wang, F.; Ma, S.; Yu, C.; Yang, M.; Chen, L.; Chen, L.; et al. DWARF14 is a non-canonical hormone receptor for strigolactone. *Nature* **2016**, *536*, 469–473. [CrossRef] [PubMed]
45. Kapulnik, Y.; Resnick, N.; Mayzlish-Gati, E.; Kaplan, Y.; Wininger, S.; Hershenhorn, J.; Koltai, H. Strigolactones interact with ethylene and auxin in regulating root-hair elongation in Arabidopsis. *J. Exp. Bot.* **2011**, *62*, 2915–2924. [CrossRef]
46. Kohlen, W.; Charnikhova, T.; Liu, Q.; Bours, R.; Domagalska, M.A.; Beguerie, S.; Verstappen, F.; Leyser, O.; Bouwmeester, H.; Ruyter-Spira, C. Strigolactones are transported through the xylem and play a key role in shoot architectural response to phosphate deficiency in nonarbuscular mycorrhizal host arabidopsis. *Plant Physiol.* **2011**, *155*, 974–987. [CrossRef]
47. Ruyter-Spira, C.; Kohlen, W.; Charnikhova, T.; van Zeijl, A.; van Bezouwen, L.; de Ruijter, N.; Cardoso, C.; Lopez-Raez, J.A.; Matusova, R.; Bours, R.; et al. Physiological effects of the synthetic strigolactone analog GR24 on root system architecture in arabidopsis: Another belowground role for strigolactones? *Plant Physiol.* **2011**, *155*, 721–734. [CrossRef]
48. Yoneyama, K.; Kisugi, T.; Xie, X.; Yoneyama, K. Chemistry of Strigolactones: Why and How do Plants Produce so Many Strigolactones? In *Molecular Microbial Ecology of the Rhizosphere*; John Wiley and Sons: Hoboken, NJ, USA, 2013; Volume 1, ISBN 9781118296172.
49. Koltai, H. Strigolactones are regulators of root development. *New Phytol.* **2011**, *190*, 545–549. [CrossRef]
50. Yamada, Y.; Furusawa, S.; Nagasaka, S.; Shimomura, K.; Yamaguchi, S.; Umehara, M. Strigolactone signaling regulates rice leaf senescence in response to a phosphate deficiency. *Planta* **2014**, *240*, 399–408. [CrossRef]
51. Agusti, J.; Herold, S.; Schwarz, M.; Sanchez, P.; Ljung, K.; Dun, E.A.; Brewer, P.B.; Beveridge, C.A.; Sieberer, T.; Sehr, E.M.; et al. Strigolactone signaling is required for auxin-dependent stimulation of secondary growth in plants. *Proc. Natl. Acad. Sci. USA* **2011**, *108*. [CrossRef]
52. de Saint Germain, A.; Ligerot, Y.; Dun, E.A.; Pillot, J.-P.; Ross, J.J.; Beveridge, C.A.; Rameau, C. Strigolactones Stimulate Internode Elongation Independently of Gibberellins. *Plant Physiol.* **2013**, *163*, 1012–1025. [CrossRef] [PubMed]
53. López-Ráez, J.A.; Kohlen, W.; Charnikhova, T.; Mulder, P.; Undas, A.K.; Sergeant, M.J.; Verstappen, F.; Bugg, T.D.H.; Thompson, A.J.; Ruyter-Spira, C.; et al. Does abscisic acid affect strigolactone biosynthesis? *New Phytol.* **2010**, *187*, 343–354. [CrossRef] [PubMed]
54. Dun, E.A.; de Saint Germain, A.; Rameau, C.; Beveridge, C.A. Antagonistic action of strigolactone and cytokinin in bud outgrowth control. *Plant Physiol.* **2012**, *158*, 487–498. [CrossRef] [PubMed]
55. Van Ha, C.; Leyva-Gonzalez, M.A.; Osakabe, Y.; Tran, U.T.; Nishiyama, R.; Watanabe, Y.; Tanaka, M.; Seki, M.; Yamaguchi, S.; Van Dong, N.; et al. Positive regulatory role of strigolactone in plant responses to drought and salt stress. *Proc. Natl. Acad. Sci. USA* **2014**, *111*, 851–856. [CrossRef] [PubMed]

56. Ito, S.; Ito, K.; Abeta, N.; Takahashi, R.; Sasaki, Y.; Yajima, S. Effects of strigolactone signaling on Arabidopsis growth under nitrogen deficient stress condition. *Plant Signal. Behav.* **2016**, *11*, e1126031. [CrossRef]
57. Marzec, M. Strigolactones as Part of the Plant Defence System. *Trends Plant Sci.* **2016**, *21*, 900–903. [CrossRef]
58. Piisilä, M.; Keceli, M.A.; Brader, G.; Jakobson, L.; Jõesaar, I.; Sipari, N.; Kollist, H.; Palva, E.T.; Kariola, T. The F-box protein MAX2 contributes to resistance to bacterial phytopathogens in Arabidopsis thaliana. *BMC Plant Biol.* **2015**, *15*, 53. [CrossRef]
59. Stes, E.; Depuydt, S.; De Keyser, A.; Matthys, C.; Audenaert, K.; Yoneyama, K.; Werbrouck, S.; Goormachtig, S.; Vereecke, D. Strigolactones as an auxiliary hormonal defence mechanism against leafy gall syndrome in Arabidopsis thaliana. *J. Exp. Bot.* **2015**, *66*, 5123–5134. [CrossRef] [PubMed]
60. Rademacher, W. Plant Growth Regulators: Backgrounds and Uses in Plant Production. *J. Plant Growth Regul.* **2015**, *34*, 845–872. [CrossRef]
61. Screpanti, C.; Fonné-Pfister, R.; Lumbroso, A.; Rendine, S.; Lachia, M.; De Mesmaeker, A. Strigolactone derivatives for potential crop enhancement applications. *Bioorg. Med. Chem. Lett.* **2016**, *26*, 2392–2400. [CrossRef] [PubMed]
62. Mostofa, M.G.; Li, W.; Nguyen, K.H.; Fujita, M.; Tran, L.S.P. Strigolactones in plant adaptation to abiotic stresses: An emerging avenue of plant research. *Plant Cell Environ.* **2018**, *41*, 2227–2243. [CrossRef]
63. Parker, C. Observations on the current status of orobanche and striga problems worldwide. *Pest Manag. Sci.* **2009**, *65*, 453–459. [CrossRef]
64. Jamil, M.; Rodenburg, J.; Charnikhova, T.; Bouwmeester, H.J. Pre-attachment Striga hermonthica resistance of New Rice for Africa (NERICA) cultivars based on low strigolactone production. *New Phytol.* **2011**, *192*, 964–975. [CrossRef]
65. Dor, E.; Yoneyama, K.; Wininger, S.; Kapulnik, Y.; Yoneyama, K.; Koltai, H.; Xie, X.; Hershenhorn, J. Strigolactone deficiency confers resistance in tomato line SL-ORT1 to the parasitic weeds *Phelipanche* and *Orobanche* spp. *Phytopathology* **2011**, *101*, 213–222. [CrossRef]
66. Fernández-Aparicio, M.; Kisugi, T.; Xie, X.; Rubiales, D.; Yoneyama, K. Low strigolactone root exudation: A novel mechanism of broomrape (*Orobanche* and *Phelipanche* spp.) resistance available for faba bean breeding. *J. Agric. Food Chem.* **2014**, *62*, 7063–7071. [CrossRef]
67. Trabelsi, I.; Yoneyama, K.; Abbes, Z.; Amri, M.; Xie, X.; Kisugi, T.; Kim, H.I.; Kharrat, M.; Yoneyama, K. Characterization of strigolactones produced by Orobanche foetida and Orobanche crenata resistant faba bean (*Vicia faba* L.) genotypes and effects of phosphorous, nitrogen, and potassium deficiencies on strigolactone production. *S. Afr. J. Bot.* **2017**, *108*, 15–22. [CrossRef]
68. Pavan, S.; Schiavulli, A.; Marcotrigiano, A.R.; Bardaro, N.; Bracuto, V.; Ricciardi, F.; Charnikhova, T.; Lotti, C.; Bouwmeester, H.; Ricciardi, L. Characterization of low-strigolactone germplasm in pea (*Pisum sativum* L.) resistant to crenate broomrape (Orobanche crenata Forsk.). *Mol. Plant-Microbe Interact.* **2016**, *29*, 743–749. [CrossRef]
69. Gobena, D.; Shimels, M.; Rich, P.J.; Ruyter-Spira, C.; Bouwmeester, H.; Kanuganti, S.; Mengiste, T.; Ejeta, G. Mutation in sorghum LOW GERMINATION STIMULANT 1 alters strigolactones and causes Striga resistance. *Proc. Natl. Acad. Sci. USA* **2017**, *114*, 4471–4476. [CrossRef]
70. Tesso, T.; Gutema, Z.; Deressa, A.; Ejeta, G. An integrated Striga management option offers effective control of Striga in Ethiopia. In *Integrating New Technologies for Striga Control: Towards Ending the Witch-Hunt*; World Scientific Publishing Co.: Singapore, 2007; pp. 199–212. ISBN 9789812771506.
71. Douthwaite, B.; Schulz, S.; Olanrewaju, A.S.; Ellis-Jones, J. Impact pathway evaluation of an integrated Striga hermonthica control project in Northern Nigeria. *Agric. Syst.* **2007**, *92*, 201–222. [CrossRef]
72. Samejima, H.; Babiker, A.G.; Takikawa, H.; Sasaki, M.; Sugimoto, Y. Practicality of the suicidal germination approach for controlling Striga hermonthica. *Pest Manag. Sci.* **2016**, *72*, 2035–2042. [CrossRef]
73. Zwanenburg, B.; Mwakaboko, A.S.; Kannan, C. Suicidal germination for parasitic weed control. *Pest Manag. Sci.* **2016**, *72*, 2016–2025. [CrossRef]
74. Jin, Z.; Xu, X.; Kang, Y.; Pang, Z.; Xu, N.; Chen, F. Strigolactone analogues derived from dihydroflavonoids as potent seed germinators for the broomrapes. *J. Agric. Food Chem.* **2020**, *68*, 11077–11087. [CrossRef]
75. Nakamura, H.; Hirabayashi, K.; Miyakawa, T.; Kikuzato, K.; Hu, W.; Xu, Y.; Jiang, K.; Takahashi, I.; Niiyama, R.; Dohmae, N.; et al. Triazole Ureas Covalently Bind to Strigolactone Receptor and Antagonize Strigolactone Responses. *Mol. Plant* **2019**, *12*, 44–58. [CrossRef]
76. Visentin, I.; Vitali, M.; Ferrero, M.; Zhang, Y.; Ruyter-Spira, C.; Novák, O.; Strnad, M.; Lovisolo, C.; Schubert, A.; Cardinale, F. Low levels of strigolactones in roots as a component of the systemic signal of drought stress in tomato. *New Phytol.* **2016**, *212*, 954–963. [CrossRef]
77. Du, H.; Huang, F.; Wu, N.; Li, X.; Hu, H.; Xiong, L. Integrative Regulation of Drought Escape through ABA-Dependent and -Independent Pathways in Rice. *Mol. Plant* **2018**, *11*, 584–597. [CrossRef]
78. Haider, I.; Andreo-Jimenez, B.; Bruno, M.; Bimbo, A.; Floková, K.; Abuauf, H.; Ntui, V.O.; Guo, X.; Charnikhova, T.; Al-Babili, S.; et al. The interaction of strigolactones with abscisic acid during the drought response in rice. *J. Exp. Bot.* **2018**, *69*, 2403–2414. [CrossRef] [PubMed]
79. Zhuang, L.; Wang, J.; Huang, B. Drought inhibition of tillering in Festuca arundinacea associated with axillary bud development and strigolactone signaling. *Environ. Exp. Bot.* **2017**, *142*, 15–23. [CrossRef]

80. Liu, J.; He, H.; Vitali, M.; Visentin, I.; Charnikhova, T.; Haider, I.; Schubert, A.; Ruyter-Spira, C.; Bouwmeester, H.J.; Lovisolo, C.; et al. Osmotic stress represses strigolactone biosynthesis in Lotus japonicus roots: Exploring the interaction between strigolactones and ABA under abiotic stress. *Planta* **2015**, *241*, 1435–1451. [CrossRef]
81. Min, Z.; Li, R.; Chen, L.; Zhang, Y.; Li, Z.; Liu, M.; Ju, Y.; Fang, Y. Alleviation of drought stress in grapevine by foliar-applied strigolactones. *Plant Physiol. Biochem.* **2019**, *135*, 99–110. [CrossRef] [PubMed]
82. Lv, S.; Zhang, Y.; Li, C.; Liu, Z.; Yang, N.; Pan, L.; Wu, J.; Wang, J.; Yang, J.; Lv, Y.; et al. Strigolactone-triggered stomatal closure requires hydrogen peroxide synthesis and nitric oxide production in an abscisic acid-independent manner. *New Phytol.* **2018**, *217*, 290–304. [CrossRef] [PubMed]
83. Zhang, Y.; Lv, S.; Wang, G. Strigolactones are common regulators in induction of stomatal closure in planta. *Plant Signal. Behav.* **2018**, *13*, e1444322. [CrossRef]
84. Shim, J.S.; Oh, N.; Chung, P.J.; Kim, Y.S.; Choi, Y.D.; Kim, J.K. Overexpression of OsNAC14 improves drought tolerance in rice. *Front. Plant Sci.* **2018**, *9*, 310. [CrossRef] [PubMed]
85. Chesterfield, R.J.; Vickers, C.E.; Beveridge, C.A. Translation of Strigolactones from Plant Hormone to Agriculture: Achievements, Future Perspectives, and Challenges. *Trends Plant Sci.* **2020**, *25*, 1087–1106. [CrossRef] [PubMed]
86. Bouwmeester, H.J.; Fonne-Pfister, R.; Screpanti, C.; De Mesmaeker, A. Strigolactones: Plant Hormones with Promising Features. *Angew. Chem. Int. Ed.* **2019**, *58*, 12778–12786. [CrossRef] [PubMed]
87. Akiyama, K.; Matsuzaki, K.I.; Hayashi, H. Plant sesquiterpenes induce hyphal branching in arbuscular mycorrhizal fungi. *Nature* **2005**, *435*, 824–827. [CrossRef]
88. Besserer, A.; Puech-Pagès, V.; Kiefer, P.; Gomez-Roldan, V.; Jauneau, A.; Roy, S.; Portais, J.C.; Roux, C.; Bécard, G.; Séjalon-Delmas, N. Strigolactones stimulate arbuscular mycorrhizal fungi by activating mitochondria. *PLoS Biol.* **2006**, *4*, 1239–1247. [CrossRef] [PubMed]
89. Schlemper, T.R.; Leite, M.F.A.; Lucheta, A.R.; Shimels, M.; Bouwmeester, H.J.; van Veen, J.A.; Kuramae, E.E. Rhizobacterial community structure differences among sorghum cultivars in different growth stages and soils. *FEMS Microbiol. Ecol.* **2017**, *93*. [CrossRef] [PubMed]
90. Sasse, J.; Martinoia, E.; Northen, T. Feed Your Friends: Do Plant Exudates Shape the Root Microbiome? *Trends Plant Sci.* **2018**, *23*, 25–41. [CrossRef] [PubMed]
91. Carvalhais, L.C.; Rincon-Florez, V.A.; Brewer, P.B.; Beveridge, C.A.; Dennis, P.G.; Schenk, P.M. The ability of plants to produce strigolactones affects rhizosphere community composition of fungi but not bacteria. *Rhizosphere* **2019**, *9*, 18–26. [CrossRef]
92. Banasiak, J.; Borghi, L.; Stec, N.; Martinoia, E.; Jasiński, M. The Full-Size ABCG Transporter of Medicago truncatula Is Involved in Strigolactone Secretion, Affecting Arbuscular Mycorrhiza. *Front. Plant Sci.* **2020**, *11*, 1. [CrossRef]
93. Liu, G.; Pfeifer, J.; de Brito Francisco, R.; Emonet, A.; Stirnemann, M.; Gübeli, C.; Hutter, O.; Sasse, J.; Mattheyer, C.; Stelzer, E.; et al. Changes in the allocation of endogenous strigolactone improve plant biomass production on phosphate-poor soils. *New Phytol.* **2018**, *217*, 784–798. [CrossRef]
94. Pokluda, R.; Shehata, S.M.; Kopta, T. Vegetativer chemischer Status und Produktivität von Gartenkürbispflanzen (Cucurbita pepo L.) in Reaktion auf die Blattanwendung von Pentakeep und Strigolactonen unter NPK-Raten. *Gesunde Pflanz.* **2018**, *70*, 21–29. [CrossRef]
95. Zheng, Y.; Kumar, N.; Gonzalez, P.; Etxeberria, E. Strigolactones restore vegetative and reproductive developments in Huanglongbing (HLB) affected, greenhouse-grown citrus trees by modulating carbohydrate distribution. *Sci. Hortic.* **2018**, *237*, 89–95. [CrossRef]
96. Soto, M.J.; Fernández-Aparicio, M.; Castellanos-Morales, V.; García-Garrido, J.M.; Ocampo, J.A.; Delgado, M.J.; Vierheilig, H. First indications for the involvement of strigolactones on nodule formation in alfalfa (*Medicago sativa*). *Soil Biol. Biochem.* **2010**, *42*, 383–385. [CrossRef]
97. Foo, E.; Davies, N.W. Strigolactones promote nodulation in pea. *Planta* **2011**, *234*, 1073–1081. [CrossRef]
98. Rehman, N.u.; Ali, M.; Ahmad, M.Z.; Liang, G.; Zhao, J. Strigolactones promote rhizobia interaction and increase nodulation in soybean (*Glycine max*). *Microb. Pathog.* **2018**, *114*, 420–430. [CrossRef] [PubMed]
99. Foo, E.; Yoneyama, K.; Hugill, C.J.; Quittenden, L.J.; Reid, J.B. Strigolactones and the regulation of pea symbioses in response to nitrate and phosphate deficiency. *Mol. Plant* **2013**, *6*, 76–87. [CrossRef]
100. Liu, J.; Novero, M.; Charnikhova, T.; Ferrandino, A.; Schubert, A.; Ruyter-Spira, C.; Bonfante, P.; Lovisolo, C.; Bouwmeester, H.J.; Cardinale, F. Carotenoid cleavage dioxygenase 7 modulates plant growth, reproduction, senescence, and determinate nodulation in the model legume *Lotus japonicus*. *J. Exp. Bot.* **2013**, *64*, 1967–1981. [CrossRef]
101. Haq, B.U.I.; Ahmad, M.Z.; Ur Rehman, N.; Wang, J.; Li, P.; Li, D.; Zhao, J. Functional characterization of soybean strigolactone biosynthesis and signaling genes in Arabidopsis MAX mutants and GmMAX3 in soybean nodulation. *BMC Plant Biol.* **2017**, *17*, 1–20. [CrossRef]
102. Yoneyama, K.K.; Xie, X.; Kim, H.I.; Kisugi, T.; Nomura, T.; Sekimoto, H.; Yokota, T.; Yoneyama, K.K. How do nitrogen and phosphorus deficiencies affect strigolactone production and exudation? *Planta* **2012**, *235*, 1197–1207. [CrossRef]
103. Shindo, M.; Shimomura, K.; Yamaguchi, S.; Umehara, M. Upregulation of DWARF27 is associated with increased strigolactone levels under sulfur deficiency in rice. *Plant Direct* **2018**, *2*, e00050. [CrossRef] [PubMed]
104. Wang, Y.; Shang, L.; Yu, H.; Zeng, L.; Hu, J.; Ni, S.; Rao, Y.; Li, S.; Chu, J.; Meng, X.; et al. A Strigolactone Biosynthesis Gene Contributed to the Green Revolution in Rice. *Mol. Plant* **2020**, *13*, 923–932. [CrossRef]

105. Jamil, M.; Charnikhova, T.; Houshyani, B.; van Ast, A.; Bouwmeester, H.J. Genetic variation in strigolactone production and tillering in rice and its effect on Striga hermonthica infection. *Planta* **2012**, *235*, 473–484. [CrossRef]
106. Butt, H.; Jamil, M.; Wang, J.Y.; Al-Babili, S.; Mahfouz, M. Engineering plant architecture via CRISPR/Cas9-mediated alteration of strigolactone biosynthesis. *BMC Plant Biol.* **2018**, *18*, 174. [CrossRef]
107. Lumbroso, A.; Villedieu-Percheron, E.; Zurwerra, D.; Screpanti, C.; Lachia, M.; Dakas, P.Y.; Castelli, L.; Paul, V.; Wolf, H.C.; Sayer, D.; et al. Simplified strigolactams as potent analogues of strigolactones for the seed germination induction of Orobanche cumana Wallr. *Pest Manag. Sci.* **2016**, *72*, 2054–2068. [CrossRef] [PubMed]
108. Lachia, M.; Fonne-Pfister, R.; Screpanti, C.; Rendine, S.; Renold, P.; Witmer, D.; Lumbroso, A.; Godineau, E.; Hueber, D.; De Mesmaeker, A. New and scalable access to karrikin (KAR1) and evaluation of its potential application on corn germination. *Helv. Chim. Acta* **2018**, *101*, e201800081. [CrossRef]
109. Chanclud, E.; Lacombe, B. Plant Hormones: Key Players in Gut Microbiota and Human Diseases? *Trends Plant Sci.* **2017**, *22*, 754–758. [CrossRef] [PubMed]
110. Karadeniz, A.; Topcuoğlu, Ş.F.; İnan, S. Auxin, Gibberellin, Cytokinin and Abscisic Acid Production in Some Bacteria. *World J. Microbiol. Biotechnol.* **2006**, *22*, 1061–1064. [CrossRef]
111. Lievens, L.; Pollier, J.; Goossens, A.; Beyaert, R.; Staal, J. Abscisic Acid as Pathogen Effector and Immune Regulator. *Front. Plant Sci.* **2017**, *8*, 587. [CrossRef]
112. Kim, S.W.; Goossens, A.; Libert, C.; Van Immerseel, F.; Staal, J.; Beyaert, R. Phytohormones: Multifunctional nutraceuticals against metabolic syndrome and comorbid diseases. *Biochem. Pharmacol.* **2020**, *175*, 113866. [CrossRef]
113. Klessig, D.F.; Tian, M.; Choi, H.W. Multiple Targets of Salicylic Acid and Its Derivatives in Plants and Animals. *Front. Immunol.* **2016**, *7*, 206. [CrossRef]
114. Atkinson, F.S.; Villar, A.; Mulà, A.; Zangara, A.; Risco, E.; Smidt, C.R.; Hontecillas, R.; Leber, A.; Bassaganya-Riera, J. Abscisic Acid Standardized Fig (Ficus carica) Extracts Ameliorate Postprandial Glycemic and Insulinemic Responses in Healthy Adults. *Nutrients* **2019**, *11*, 1757. [CrossRef]
115. Tumer, T.B.; Yılmaz, B.; Ozleyen, A.; Kurt, B.; Tok, T.T.; Taskin, K.M.; Kulabas, S.S. GR24, a synthetic analog of Strigolactones, alleviates inflammation and promotes Nrf2 cytoprotective response: In vitro and in silico evidences. *Comput. Biol. Chem.* **2018**, *76*, 179–190. [CrossRef] [PubMed]
116. Esposito, D.; Kizelsztein, P.; Komarnytsky, S.; Raskin, I. Hypoglycemic effects of brassinosteroid in diet-induced obese mice. *Am. J. Physiol. Endocrinol. Metab.* **2012**, *303*, E652. [CrossRef]
117. Sánchez-Sarasúa, S.; Moustafa, S.; García-Avilés, Á.; López-Climent, M.F.; Gómez-Cadenas, A.; Olucha-Bordonau, F.E.; Sánchez-Pérez, A.M. The effect of abscisic acid chronic treatment on neuroinflammatory markers and memory in a rat model of high-fat diet induced neuroinflammation. *Nutr. Metab.* **2016**, *13*, 1–11. [CrossRef]
118. Modi, S.; Yaluri, N.; Kokkola, T.; Laakso, M. Plant-derived compounds strigolactone GR24 and pinosylvin activate SIRT1 and enhance glucose uptake in rat skeletal muscle cells. *Sci. Rep.* **2017**, *7*, 17606. [CrossRef] [PubMed]
119. Zheng, J.X.; Han, Y.S.; Wang, J.C.; Yang, H.; Kong, H.; Liu, K.J.; Chen, S.Y.; Chen, Y.R.; Chang, Y.Q.; Chen, W.M.; et al. Strigolactones: A plant phytohormone as novel anti-inflammatory agents. *MedChemComm* **2018**, *9*, 181–188. [CrossRef]
120. Kurt, B.; Ozleyen, A.; Antika, G.; Yilmaz, Y.B.; Tumer, T.B. Multitarget Profiling of a Strigolactone Analogue for Early Events of Alzheimer's Disease: In Vitro Therapeutic Activities against Neuroinflammation. *ACS Chem. Neurosci.* **2020**, *11*, 501–507. [CrossRef]
121. Bagasra, O.; Michaels, F.H.; Zheng, Y.M.; Bobroski, L.E.; Spitsin, S.V.; Fu, Z.F.; Tawadros, R.; Koprowski, H. Activation of the inducible form of nitric oxide synthase in the brains of patients with multiple sclerosis. *Proc. Natl. Acad. Sci. USA* **1995**, *92*, 12041–12045. [CrossRef]
122. Brosnan, C.F.; Battistini, L.; Raine, C.S.; Dickson, D.W.; Casadevall, A.; Lee, S.C. Reactive nitrogen intermediates in human neuropathology: An overview. *Dev. Neurosci.* **1994**, *16*, 152–161. [CrossRef] [PubMed]
123. Minghetti, L. Role of COX-2 in inflammatory and degenerative brain diseases. *Subcell. Biochem.* **2007**, *42*, 127–141. [CrossRef]
124. Wang, P.; Guan, P.P.; Wang, T.; Yu, X.; Guo, J.J.; Wang, Z.Y. Aggravation of Alzheimer's disease due to the COX-2-mediated reciprocal regulation of IL-1β and Aβ between glial and neuron cells. *Aging Cell* **2014**, *13*, 605–615. [CrossRef] [PubMed]
125. Chen, Y.C.; Wu, J.S.; Tsai, H.D.; Huang, C.Y.; Chen, J.J.; Sun, G.Y.; Lin, T.N. Peroxisome proliferator-activated receptor gamma (PPAR-γ) and neurodegenerative disorders. *Mol. Neurobiol.* **2012**, *46*, 114–124. [CrossRef]
126. Wakabayashi, N.; Slocum, S.L.; Skoko, J.J.; Shin, S.; Kensler, T.W. When NRF2 talks, who's listening? *Antioxid. Redox Signal.* **2010**, *13*, 1649–1663. [CrossRef]
127. Houghton, C.A.; Fassett, R.G.; Coombes, J.S. Sulforaphane and Other Nutrigenomic Nrf2 Activators: Can the Clinician's Expectation Be Matched by the Reality? *Oxid. Med. Cell. Longev.* **2016**, *2016*, 7857186. [CrossRef] [PubMed]
128. Tumer, T.B.; Rojas-Silva, P.; Poulev, A.; Raskin, I.; Waterman, C. Direct and indirect antioxidant activity of polyphenol- and isothiocyanate-enriched fractions from moringa oleifera. *J. Agric. Food Chem.* **2015**, *63*, 1505–1513. [CrossRef] [PubMed]
129. Ye, M.; Wang, Q.; Zhang, W.; Li, Z.; Wang, Y.; Hu, R. Oroxylin A exerts anti-inflammatory activity on lipopolysaccharide-induced mouse macrophage via Nrf2/ARE activation. *Biochem. Cell Biol.* **2014**, *92*, 337–348. [CrossRef]
130. Hasan, M.N.; Razvi, S.S.I.; Kuerban, A.; Balamash, K.S.; Al-Bishri, W.M.; Abulnaja, K.O.; Choudhry, H.; Khan, J.A.; Moselhy, S.S.; Ma, Z.; et al. Strigolactones—A novel class of phytohormones as anti-cancer agents. *J. Pest. Sci.* **2018**, *43*, 168. [CrossRef]

131. Steigerová, J.; Oklestkova, J.; Levková, M.; Rárová, L.; Kolář, Z.; Strnad, M. Brassinosteroids cause cell cycle arrest and apoptosis of human breast cancer cells. *Chem. Biol. Interact.* **2010**, *188*, 487–496. [CrossRef]
132. Cohen, S.; Flescher, E. Methyl jasmonate: A plant stress hormone as an anti-cancer drug. *Phytochemistry* **2009**, *70*, 1600–1609. [CrossRef]
133. Corrèze, C.; Blondeau, J.P.; Pomérance, M. p38 mitogen-activated protein kinase contributes to cell cycle regulation by cAMP in FRTL-5 thyroid cells. *Eur. J. Endocrinol.* **2005**, *153*, 123–133. [CrossRef] [PubMed]
134. Iyoda, K.; Sasaki, Y.; Horimoto, M.; Toyama, T.; Yakushijin, T.; Sakakibara, M.; Takehara, T.; Fujimoto, J.; Hori, M.; Wands, J.R.; et al. Involvement of the p38 mitogen-activated protein kinase cascade in hepatocellular carcinoma. *Cancer* **2003**, *97*, 3017–3026. [CrossRef]
135. Mayzlish-Gati, E.; Laufer, D.; Grivas, C.F.; Shaknof, J.; Sananes, A.; Bier, A.; Ben-Harosh, S.; Belausov, E.; Johnson, M.D.; Artuso, E.; et al. Strigolactone analogs act as new anti-cancer agents in inhibition of breast cancer in xenograft model. *Cancer Biol. Ther.* **2015**, *16*, 1682–1688. [CrossRef]
136. Kaverina, I.; Straube, A. Regulation of cell migration by dynamic microtubules. *Semin. Cell Dev. Biol.* **2011**, *22*, 968–974. [CrossRef]
137. Schiff, P.B.; Horwitz, S.B. Taxol stabilizes microtubules in mouse fibroblast cells. *Proc. Natl. Acad. Sci. USA* **1980**, *77*, 1561–1565. [CrossRef]
138. Pollock, C.B.; McDonough, S.; Wang, V.S.; Lee, H.; Ringer, L.; Li, X.; Prandi, C.; Lee, R.J.; Feldman, A.S.; Koltai, H.; et al. Strigolactone analogues induce apoptosis through activation of p38 and the stress response pathway in cancer cell lines and in conditionally reprogrammed primary prostate cancer cells. *Oncotarget* **2014**, *5*, 1683–1698. [CrossRef]
139. Murphy, M.E. The HSP70 family and cancer. *Carcinogenesis* **2013**, *34*, 1181–1188. [CrossRef]
140. Chang, H.L.; Wu, Y.C.; Su, J.H.; Yeh, Y.T.; Yuan, S.S.F. Protoapigenone, a novel flavonoid, induces apoptosis in human prostate cancer cells through activation of p38 mitogen-activated protein kinase and c-Jun NH2-terminal kinase 1/2. *J. Pharmacol. Exp. Ther.* **2008**, *325*, 841–849. [CrossRef] [PubMed]
141. Croglio, M.P.; Haake, J.M.; Ryan, C.P.; Wang, V.S.; Lapier, J.; Schlarbaum, J.P.; Dayani, Y.; Artuso, E.; Prandi, C.; Koltai, H.; et al. Analogs of the novel phytohormone, strigolactone, trigger apoptosis and synergize with PARP inhibitors by inducing DNA damage and inhibiting DNA repair. *Oncotarget* **2016**, *7*, 13984–14001. [CrossRef]
142. Ward, A.; Khanna, K.K.; Wiegmans, A.P. Targeting homologous recombination, new pre-clinical and clinical therapeutic combinations inhibiting RAD51. *Cancer Treat. Rev.* **2015**, *41*, 35–45. [CrossRef]
143. Hasan, M.N.; Choudhry, H.; Razvi, S.S.; Moselhy, S.S.; Kumosani, T.A.; Zamzami, M.A.; Omran, Z.; Halwani, M.A.; Al-Babili, S.; Abualnaja, K.O.; et al. Synthetic strigolactone analogues reveal anti-cancer activities on hepatocellular carcinoma cells. *Bioorg. Med. Chem. Lett.* **2018**, *28*, 1077–1083. [CrossRef]
144. Dor, E.; Joel, D.M.; Kapulnik, Y.; Koltai, H.; Hershenhorn, J. The synthetic strigolactone GR24 influences the growth pattern of phytopathogenic fungi. *Planta* **2011**, *234*, 419–427. [CrossRef]
145. Biolatti, M.; Blangetti, M.; D'arrigo, G.; Spyrakis, F.; Cappello, P.; Albano, C.; Ravanini, P.; Landolfo, S.; De Andrea, M.; Prandi, C.; et al. Strigolactone analogs are promising antiviral agents for the treatment of human cytomegalovirus infection. *Microorganisms* **2020**, *8*, 703. [CrossRef]
146. Griffiths, P.; Reeves, M. Pathogenesis of human cytomegalovirus in the immunocompromised host. *Nat. Rev. Microbiol.* **2021**. [CrossRef]
147. Gugliesi, F.; Coscia, A.; Griffante, G.; Galitska, G.; Pasquero, S.; Albano, C.; Biolatti, M. Where do we Stand after Decades of Studying Human Cytomegalovirus? *Microorganisms* **2020**, *8*, 685. [CrossRef] [PubMed]
148. Perera, M.R.; Wills, M.R.; Sinclair, J.H. HCMV Antivirals and Strategies to Target the Latent Reservoir. *Viruses* **2021**, *13*, 817. [CrossRef]
149. Chen, S.-J.; Wang, S.-C.; Chen, Y.-C. Antiviral Agents as Therapeutic Strategies Against Cytomegalovirus Infections. *Viruses* **2019**, *12*, 21. [CrossRef] [PubMed]

Review

Molecular Targets of Cannabidiol in Experimental Models of Neurological Disease

Serena Silvestro †, Giovanni Schepici †, Placido Bramanti and Emanuela Mazzon *

Istituto di Ricovero e Cura a Carattere Scientifico (IRCCS) Centro Neurolesi "Bonino-Pulejo", Via Provinciale Palermo, Contrada Casazza, 98124 Messina, Italy; serena.silvestro@irccsme.it (S.S.); giovanni.schepici@irccsme.it (G.S.); placido.bramanti@irccsme.it (P.B.)
* Correspondence: emanuela.mazzon@irccsme.it; Tel.: +39-090-6012-8172
† These authors contributed equally to this work.

Academic Editor: Francesca Cardona
Received: 16 October 2020; Accepted: 5 November 2020; Published: 7 November 2020

Abstract: Cannabidiol (CBD) is a non-psychoactive phytocannabinoid known for its beneficial effects including antioxidant and anti-inflammatory properties. Moreover, CBD is a compound with antidepressant, anxiolytic, anticonvulsant and antipsychotic effects. Thanks to all these properties, the interest of the scientific community for it has grown. Indeed, CBD is a great candidate for the management of neurological diseases. The purpose of our review is to summarize the in vitro and in vivo studies published in the last 15 years that describe the biochemical and molecular mechanisms underlying the effects of CBD and its therapeutic application in neurological diseases. CBD exerts its neuroprotective effects through three G protein coupled-receptors (adenosine receptor subtype 2A, serotonin receptor subtype 1A and G protein-coupled receptor 55), one ligand-gated ion channel (transient receptor potential vanilloid channel-1) and one nuclear factor (peroxisome proliferator-activated receptor γ). Moreover, the therapeutical properties of CBD are also due to GABAergic modulation. In conclusion, CBD, through multi-target mechanisms, represents a valid therapeutic tool for the management of epilepsy, Alzheimer's disease, multiple sclerosis and Parkinson's disease.

Keywords: cannabidiol; molecular mechanisms; neurological diseases; neuroprotective effects

1. Introduction

Neurological diseases are complex conditions affecting millions of people around the world [1]. These disorders can have different etiologies; indeed, they can be caused by genetic or environmental factors [2–5]. Although for some of these pathologies treatment that can delay or control the clinical symptoms is available, they remain incurable diseases.

Phytocannabinoids, such as CBD, represent a new class of compounds characterized by beneficial effects in various neurodegenerative and psychiatric diseases [6,7].

CBD is extracted from *Cannabis sativa*, and, together with the psychoactive Δ^9-tetrahydro-cannabinol (Δ^9-THC), they represent the main neuroactive components of the plant. Unlike Δ^9-THC that induces psychotropic effects, CBD is the main non-psychotropic compound present in the plant [8,9].

CBD shows a relatively low toxicity and dependence profile [10]. For these reasons, the role of CBD as adjuvant therapy is being evaluated in those conditions in which the available treatments are not satisfying. Furthermore, CBD has a broad spectrum of therapeutic properties, such as anxiolytic [7,11], neuroprotective [7,12–15], antidepressant [16], anti-inflammatory [17–19] and immunomodulating activities [20,21]. The neuroprotective effects of CBD are due to its antioxidant and anti-inflammatory activities and the modulation of a large number of brain biological targets, such as receptors and channels, involved in the development and maintenance of neurodegenerative diseases [22]. CBD can exert its antioxidant action directly by the modulation of oxidative stress or indirectly through molecules targets associated with the redox system such as nuclear factor erythroid 2-related factor 2 (Nrf2), implicated in the transcription of genes that encodes for antioxidant proteins, such as superoxide dismutase (SOD) and glutathione (GSH) peroxidase [23,24]. CBD can increase the activity of GSH peroxidase and reductase, favoring the reduction of malondialdehyde and thus preventing oxidative stress [25]. Moreover, thanks to its ability to reduce reactive oxygen species (ROS), CBD maintains the correct GSH levels, necessary for the antioxidant activity of vitamins A, C and E [26]. CBD is a regulator of the expression of nitrotyrosine and inducible nitric oxide synthase (iNOS), thus promoting the reduction of the production of ROS [27]. Moreover, CBD exerts its anti-inflammatory action, modulating the release of proinflammatory cytokines such as interleukin-6 (IL-6) and interleukin 1-β (IL-1β) and interacting with transcription factors such as tumor necrosis factor α (TNF-α), nuclear factor κB (NF-κB) or peroxisome proliferator-activated receptor γ (PPARγ) [28–30]. CBD also performs its anti-inflammatory action by regulating the transient receptor potential (TRP) channels such as transient receptor potential vanilloid (TRPV) Type 1 and 2, a non-selective cationic channel whose activation allows the entry of Ca^{2+} [31]. Indeed, in the neuroinflammatory conditions, an increase of the density and sensitivity of TRPV1 was demonstrated. Conversely, the binding of the CBD with TRPV1 leads to the desensitization of these channel receptors and a consequent reduction of neuroinflammation, thus explaining the neuroprotective properties of CBD [32].

Therefore, in recent years, the scientific community has shown interest in this compound due to its neuroprotective effects in several neurological disorders, including Parkinson's disease [33,34], Alzheimer's diseases [35–37] and epilepsy [38]. Additionally, CBD shows other actions such as antidepressant, antipsychotic, antiepileptic and analgesic effects, as highlighted in preclinical studies and human clinical trials [7,39–41].

The purpose of this review is to describe the molecular mechanisms associated with the efficacy of CBD in neurological diseases. In the present review, the experimental studies highlighting the CBD's mechanisms of action in neurological disorders are summarized.

2. Methodology

In this review, the articles published from 2004 to 2020 are considered. Specifically, the bibliography research in PubMed was performed using the following keywords: "cannabidiol", "neurological disease", "adenosine receptors", "serotonin receptors", "transient receptor potential", "TRPV receptors", "GPR55 receptors", "peroxisome proliferator-activated receptor-γ", "PPARγ receptors" and "GABA receptors". In this way, 67 articles were found, as shown in the Prisma flow diagram (Figure 1). Articles that evaluate the biochemical and molecular mechanisms underlying the effects of CBD and its therapeutic application in neurological diseases are considered.

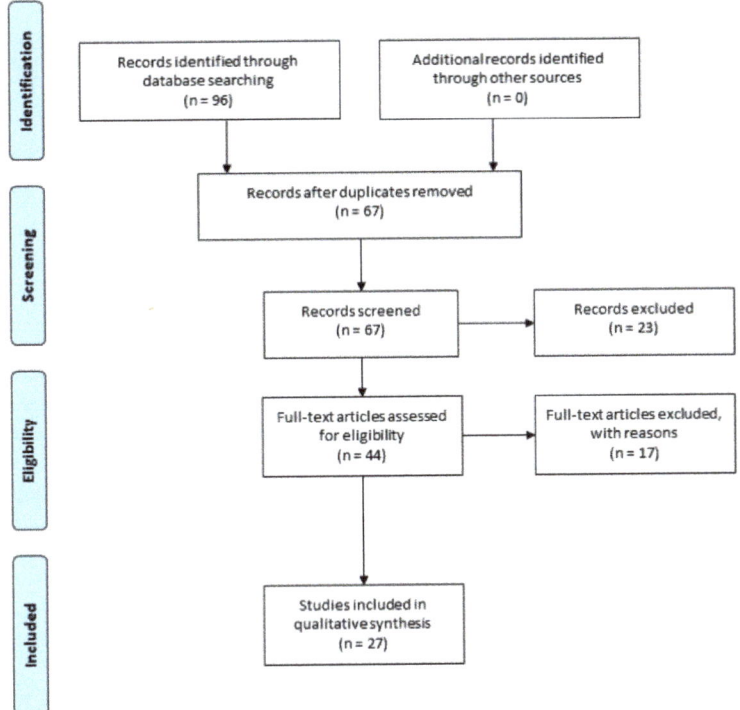

Figure 1. Prisma flow diagram illustrating the selection methodology of the preclinical studies used for the writing of the review. Duplicate articles were excluded from the total of the studies recorded. Instead, articles that evaluate the biochemical and molecular mechanisms underlying the effects of CBD and its therapeutic application in neurological diseases are considered (The PRISMA Statement is published in [42]).

3. Chemical Properties of Cannabidiol

CBD was first isolated in 1940 from Mexican marijuana by Roger Adams and from Indian charas by Alexander Todd [43,44]. However, its crystalline structure was determined in 1977 (Jones et al. 1977) [45]. CBD 2-[(1R,6R)-3-methyl-6-prop-1-en-2-ylcyclohex-2-en-1-yl]-5-pentylbenzene-1,3-diol is a terpenophenol containing 21 carbon atoms, with the formula $C_{21}H_{30}O_2$. CBD is a cyclohexene which is substituted in position 1 by a methyl group, by a 2,6-dihydroxy-4-pentylphenyl group at position 3 and with a prop-1-en-2-yl group at position 4 [46]. The aromatic ring and the terpene ring are almost perpendicular to each other. In the chemical nomenclature, its chemical numbering is determined by the terpene ring (Figure 2).

The reactivity of CBD is mainly due to the methyl group in position C-1 of the cyclohexane ring, to the hydroxyl groups present in the aromatic ring in position C-2' and C-6' and to the pentyl chain present in position C-4'. The hydroxyl groups are also capable of binding threonine, tyrosine and glutamic acid [47]. Moreover, CBD, thanks to its hydroxyl groups in the aromatic ring, can exert an antioxidant action by inactivating the free radicals [48].

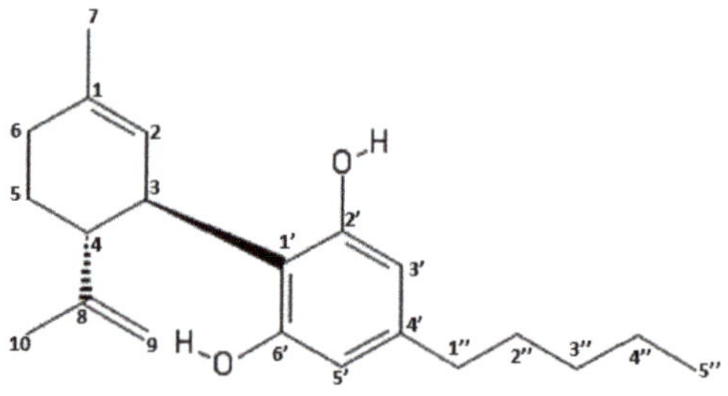

Figure 2. Chemical Structure and numbering system of Cannabidiol (CBD).

4. Cannabidiol Mechanism of Action

Unlike other cannabinoids, CBD has a poor affinity for cannabinoid receptor type 1 (CB1) and cannabinoid receptor type 2 (CB2); however, it acts as a non-competitive negative allosteric modulator of CB1 [49]. It has recently been shown that CBD is a CB1R-negative allosteric modulator of Δ^9-THC and the 2-arachidonoyl-glycerol (2-AG), this evidence may explain some of the in vivo effects of this nonpsychoactive phycompound [50]. Furthermore, CBD modulates the tone of endocannabinoids by inhibiting cellular uptake of the endocannabinoids arachidonoylethanolamide (AEA) [51]. AEA, CB1, CB2 and 2-AG constitute the endocannabinoid system (ECS) responsible for the synthesis and degradation of endocannabinoids [52]. Therefore, CBD can lead to increased levels of AEA, thus interacting with CB receptors [53]. AEA is involved in different biological processes such as mood regulation, pain sensation and appetite. The synthesis of AEA and 2-AG is regulated by increased of intracellular calcium (Ca^{2+}). Moreover, AEA and 2-AG are, respectively, metabolized by fatty acid amide hydrolase (FAAH) and monoglyceride lipase [54]. The increase of intracellular Ca^{2+} induces the production and release into the synaptic space of AEA and 2-AG that acts as retrograde synaptic messengers [55,56]. Indeed, these endocannabinoids acting on CB1 at a presynaptic level block the release by neuronal terminals of neurotransmitters such as γ-aminobutyric acid (GABA), dopamine, glutamate, serotonin (5-HT), norepinephrine and acetylcholine [57,58].

However, it was shown that CBD has more affinity for 5-HT receptors, non-endocannabinoid G protein-coupled receptors (GPCRs) and other targets such as enzymes and ion channels [53]. Indeed, some anti-inflammatory and immunosuppressive effects of CBD may be partly mediated by 5-HT and adenosine receptors (ARs) which are not considered part of the ECS. CBD acts as a $5-HT_{1A}$ agonist, as a partial agonist of $5-HT_{2A}$ and as a non-competitive antagonist of $5-HT_{3A}$ [59–61]. Furthermore, CBD is capable of activating the ARs [62]. Moreover, CBD acts as a GPR55 antagonist and as an agonist of TRPV1 and TRPV2 [63]. It has already been reported that CBD exerts its anti-inflammatory and neuroprotective effects also due to its ability to activate PPARγ [17,64]. Furthermore, it is a positive allosteric modulator of $GABA_A$ receptors, thereby exerting its anticonvulsant, analgesic and anxiolytic properties [65]. However, despite all this evidence, the molecular mechanisms underlying the effects of CBD remain complex.

5. Pharmacokinetic Properties of Cannabidiol

The pharmacokinetics and observed effects of CBD are related to the formulation and route of administration [66].

CBD can be administered orally, inhaled and vaporized [67]. CBD administered by inhalation is effectively absorbed into the lungs from the circulating blood, showing similar pharmacokinetics to the intravenous route [68]. Administered by inhalation, CBD reaches peak plasma concentrations in 5–10 min and shows bioavailability of 31%. However, the need for specialized equipment for these routes of administration limits the development of this mode of delivery [69]. On the contrary, oral administration shows a variable pharmacokinetic profile, probably due to the poor solubility of CBD in water [70]. Furthermore, the maximum plasma concentrations are lower than those reached when the drug is administered by inhalation. Indeed, CBD in human showed an oral bioavailability of

6% [71]. The administration of CBD by the oro-mucosal and sublingual route shows a less variable pharmacokinetic profile than the oral administration. Instead, when CBD is administered intravenously, it quickly passes the blood–brain barrier (BBB) to distribute in the brain, adipose tissue and other organs [68]. Moreover, thanks to its liposolubility, CDB forms aggregates which can be released slowly into adipose tissue [72].

CBD is metabolized in the liver by cytochrome P450 enzymes (CYPs) such as CYP2C19 and CYP3A4, CYP1A1, CYP1A2, CYP2C9 and CYP2D6 and is converted to 7-hydroxycannabidiol (7-OH-CBD) [73]. After hydroxylation, CBD is further metabolized in the liver and subsequently excreted, mainly in feces and urine. CBD exhibits a half-life of 18–32 h and a clearance of 57.6–93.6 L/h [69].

CBD is well tolerated and shows a good safety profile at therapeutic dosages. However, some studies have shown that CBD possesses strong inhibitory activity against CYP2C, CYP2D6 and CYP3A isoforms [74].

6. Molecular Targets of CBD for Application in Neurodegenerative Diseases

The neuroprotective effects of CBD based on its anti-inflammatory and antioxidant action are directed to the modulation of receptors and channels involved in neurodegenerative diseases [22]. It is known that CBD interacts with many non-endocannabinoid signaling systems such as G protein coupled-receptors, TRPV1 and PPARγ [75]. Additionally, CBD therapeutic potential possibly derives from its GABAergic modulation [65].

6.1. GPCRs

6.1.1. Adenosine Receptors

ARs (A_1R, $A_{2A}R$, $A_{2B}R$ and A_3R) are GPCRs stimulated by endogenous adenosine, involved in several physiological and pathological processes [76]. The stimulation of A_1R and A_3R by adenosine leads to the activation inhibiting G ($G_{i/o}$) proteins with the consequent inhibition of adenylate cyclase and intracellular reduction of cyclic adenosine monophosphate (cAMP). Instead, $A_{2A}R$ and $A_{2B}R$ promote the G protein activation and the subsequent increase of cAMP [77]. ARs activation involves the modulation of second messengers and additional signaling mechanisms such as phospholipase C; the protein kinase C dependent on Ca^{2+} involved in cell communication; phosphoinositide 3-kinases/protein kinase B (PI$_3$K/Akt) signaling involved in cell proliferation, growth and differentiation; and the activation of ion channels and regulation of Ca^{2+} [76,78]. ARs are located in immune cells, blood vessels, astrocytes, microglia, corpus striatum and spinal cord [79]. ARs can affect both in the central nervous system (CNS) and peripheral tissues, thus could represent a useful tool for the development of new neuroprotective strategies [79,80].

Although CBD exerts its beneficial action through several signaling pathways, its anti-inflammatory effects seem to involve $A_{2A}R$ [81]. The anti-inflammatory effect of CBD can be directly mediated by $A_{2A}R$ whose activation induces the regulation of the immune response, as well as a reduction of proinflammatory cytokines [82,83]. CBD can improve the adenosine signaling, leading to an increase of extracellular adenosine and a consequent reduction of the neuroinflammation [81]. Mecha et al. demonstrated the anti-inflammatory effects of CBD, through $A_{2A}R$ activation, in a viral model of multiple sclerosis. They showed in mice infected with Theiler's murine encephalomyelitis virus (TMEV) that CBD (5 mg/kg), administered via intraperitoneal (i.p.) daily, for seven days, led to a reduction of leukocyte migration in the blood and of the inflammatory response. Moreover, CBD treatment induces the downregulation of the levels of adhesion expression of vascular cells molecule-1 (VCAM-1), chemokine ligand 2 (CCL2) and chemokine ligand 5 (CCL5). Likewise, CBD reduced IL-1β and microglia activation, thus demonstrating its immunosuppressive and neuroprotective action. In addition, CBD has also improved motor deficits, especially in the chronic phase of the disease. To confirm the action of CBD on $A_{2A}R$ the animals were treated with ZM241385 (5 mg/kg), a selective $A_{2A}R$ antagonist, at the time of TMEV infection and 30 min before CBD treatment. ZM241385, especially in the early stages, attenuated some of the anti-inflammatory effects of CBD, such as the inhibition of the expression of VCAM-1, the infiltration of immune cells and the reduction of immunoreactivity. Moreover, it was also demonstrated that ZM241385 (5 µM) with a dose-dependent mechanism, antagonizes $A_{2A}R$ blocking completely the inhibitory action of CBD on the release of VCAM-1. Conversely, the single administration of ZM241385 did not affect TMEV mice. Therefore, the results suggest that the $A_{2A}R$ mediated anti-inflammatory effects of CBD could be useful for the management of inflammatory diseases such as multiple sclerosis [18].

The involvement of $A_{2A}R$ in neuroprotective effects of CBD was also demonstrated in vitro on hypoxic-ischemic immature brain of mice. In this study, Castillo et al. showed the effects of CBD in forebrain sections of C57BL6 mice incubated in absence of oxygen and glucose and treated for 15 min with CBD or vehicle. CBD treatment (100 µM) has significantly reduced acute brain damage and apoptosis, evaluated through the decrease in the efflux of lactate dehydrogenase. In the same way, CBD led to a reduction of glutamate and increase of caspase-9. CBD has also reduced the neuroinflammation leading to the reduction of IL-6, TNF-α, cyclooxygenase-2 (COX-2) and iNOS. Moreover, it was shown that the administration of SCH58261, an $A_{2A}R$ antagonist, or AM630, a CB2 antagonist, abolished the neuroprotective effects of CBD. The possible affinity of CBD to CB2 receptors is prompted by the effect of AM630 on CBD neuroprotection. Therefore, the study demonstrated how the neuroprotective effects of CBD can be mediated by $A_{2A}R$ and CB2 receptors [28]. The neuroprotective effects of CBD were also demonstrated by Martin-Moreno et al. in vitro and in vivo model of Alzheimer's disease. CBD inhibited ATP-induced intracellular Ca^{2+} increase in cultured N13 cells and primary microglial cells. The use of ZM241385, an A_{2A} receptor antagonist, reversed the effects of CBD on intracellular Ca^{2+} in N13 microglial cells and primary rat microglial. This result confirms the implication of $A_{2A}R$ in the action of CBD. In vivo, CBD at a dose of 20 mg/kg was able to prevent cognitive impairment induced by amyloid-β (Aβ). Indeed, after i.p. administration for three weeks, CBD reduced the expression of the gene encoding for IL-6, a proinflammatory cytokine. Therefore, in light of these results, CBD, interacting with $A_{2A}R$, could be a useful approach for Alzheimer's disease [84].

Instead, Magen et al. evaluated the $A_{2A}R$-mediated therapeutic effects of CBD in the experimental model of hepatic encephalopathy induced by bile duct ligation. After four weeks of treatment, CBD (5 mg/kg), administered i.p. daily, induced a reduction of expression of the TNF-α-receptor 1 gene in the hippocampus. Conversely, increased the expression of the brain-derived neurotrophic factor (BDNF) gene. CBD exerts these effects through an indirect modulation of the $A_{2A}Rs$. Indeed, the use of its antagonist ZM241385 (1 mg/kg) reversed the effects of CBD, confirming the involvement of the $A_{2A}Rs$. In this way, the chronic treatment with CBD, through the indirect activation of the $A_{2A}Rs$, improved the cognitive and motor function of the rats with hepatic encephalopathy. As demonstrated by these studies, CBD, probably inhibiting the reuptake of adenosine, exerts an indirect modulation of ARs [82].

In conclusion, the results of these studies show that CBD, through $A_{2A}R$ activation, exerts anti-inflammatory effects in models of multiple sclerosis, hypoxic-ischemic damage, Alzheimer's disease and hepatic encephalopathy (Table 1).

Table 1. Neuroprotective effects of CBD in different neurological diseases through the activation of the $A_{2A}Rs$.

In Vitro and in Vivo Models	CBD Dose	Treatments	Biological/Pharmacological Effect	Neurological Diseases	Ref.
Female SJL/J mice	5 mg/kg	Once-daily during Days 1–7 post-infection	CBD attenuated the activation of microglia downregulating the expression of VCAM-1, CCL2 and CCL5 and the proinflammatory cytokine IL-1β. Moreover, CBD improved motor deficits in the chronic phase of the disease	multiple sclerosis	[18]
Newborn C57BL6 mice	0.1–1000 μM	15 min. pre-incubation	CBD reduced acute brain damage and apoptosis. Moreover, it induced a reduction concentration of glutamate and IL-6 and decreased the expression of TNF-α, COX-2 and iNOS.	hypoxic-ischemic brain damage	[28]
Primary Rat Microglial and N13 Microglial Cells and C57Bl/6 mice	20 mg/kg	Once-daily during the first week, then 3 days/week for 2 weeks	CBD inhibited ATP-induced intracellular Ca^{2+} increase in cultured N13 and primary microglial cells and A_{2A} receptors may be involved in this mechanism. In vivo, CBD reduced the gene expression of proinflammatory cytokine IL-6 and prevented cognitive impairment induced by Aβ.	Alzheimer's disease	[84]
Female Sabra mice	5 mg/kg	Every day for 4 weeks	CBD reduced the expression of the TNF-α-receptor 1 gene in the hippocampus. Conversely, enhanced the expression of the BDNF gene. Moreover, CBD, through the indirect activation of the $A_{2A}R$, improved the cognitive and motor function of the rats with Hepatic Encephalopathy.	hepatic encephalopathy	[82]

CBD, cannabidiol; VCAM-1, vascular cell adhesion molecule-1; CCL-2, chemokine ligand 2; CCL-5, chemokine ligand 5; IL-6, interleukin-6; TNF-α, tumor necrosis factor α; COX-2, cyclooxygenase-2; iNOS, inducible nitric oxide synthase; $A_{2A}R$, adenosine 2A receptors; CB2, cannabinoid receptors type 2; Aβ, β-amyloid; BDNF, brain-derived neurotrophic factor.

6.1.2. 5-HT Receptors

5-HT receptors are activated by serotonin and involved in the release of neurotransmitters and hormones, thus regulating many of the processes that occur in the nervous system. The family of receptors 5-HT$_1$ is coupled to G$_{i/o}$ proteins and to adenylate cyclase which leads to the production of cAMP. In particular, the 5-HT$_{1A}$ receptor inhibits the Ca^{2+} channel and activates a ligand-dependent potassium (K$^+$) channel [85]. 5-HT$_{1A}$ receptors are associated to GPCR and they modulate neurotransmission through K$^+$ and Ca^{2+} channels. Moreover, the 5-HT$_1$ receptors were classified into five subtypes (5-HT$_{1A}$, 5-HT$_{1B}$, 5-HT$_{1D}$, 5-HT$_{1E}$ and 5-HT$_{1F}$), which are located in different areas of the brain at pre- and post-synaptic level [86]. It was demonstrated that the main neuroprotective effects of CBD are related to the 5-HT$_{1A}$ receptor. Russo et al. demonstrated that CBD while showing a low-affinity agonism towards the 5-HT$_{1A}$ receptor could enhance 5-HT$_{1A}$-mediated neurotransmission [59].

Mishima et al. explored the 5-HT$_{1A}$ receptors-mediate neuroprotective effects of CBD, in mice with middle cerebral artery (MCA) occlusion. Mice received 3 or 10 mg/kg of CBD immediately before and 3 h after occlusion. CBD, at a dose of 3 mg/kg, significantly decreased the infarct volume induced by MCA occlusion. However, treatment with CBD (3 mg/kg) plus WAY100135 (10 mg/kg; a 5-HT$_{1A}$ antagonist) inhibited the effects of CBD. This data suggested the involvement, at least in part, of 5-HT$_{1A}$ receptors in the neuroprotective effects of CBD against cerebral ischemia [87].

Gomes et al. evaluated the beneficial effects of CBD, through the facilitation of the 5-HT1A receptors, in motor-related striatal disorders, such as Parkinson's disease. For the study, catalepsy was induced in mice, using pharmacological mechanisms, in order to test the motor function disorders. Thirty minutes before receiving the drugs that induce catalepsy, mice were treated with CBD 5, 15, 30 or 60 mg/kg (i.p.). The pretreatment with CBD attenuated the cataleptic effects, in a dose-dependent manner. To explain the mechanism of action by which this phytocannabinoid exerts its anticataleptic action, mice were treated intraperitoneally with WAY100635 (0.1 mg/kg), a 5-HT$_{1A}$ receptor antagonist, 30 min before of the treatment with CBD (30 mg/kg). The administration of WAY100635 prevented the anticataleptic effect of CBD. Therefore, CBD exerting its anticataleptic action, through a mechanism that involves 5-HT$_{1A}$ receptors, could be a possible therapeutic tool in Parkinson's disease [88]. Moreover, Sonego et al. evaluated the 5-HT$_{1A}$ receptors-mediated anticataleptic effect of CBD. In this study, the researchers induced catalepsy, in male Swiss mice, with haloperidol (0.6 mg/kg). The pre-treatment with CBD (15–60 mg/kg) i.p., prevented the catalepsy. To understand the mechanism of action, it was demonstrated that administration i.p. of WAY100635 (0.1 mg/kg) reduced the anticataleptic effect of CBD and its action on the expression of c-FOS. Moreover, it was shown that the administration of bilateral injections of CBD (60 nmol) into the dorsal striatum, followed by treatment with haloperidol (0.6 mg/kg), reduced the catalepsy, in a similar way to systemic administration. These data suggest that this compound, via activation of 5-HT$_{1A}$ receptor, could represent a therapeutic opportunity for the treatment of striatal disorders such as Parkinson's disease [89].

Instead, Pelz et al. evaluated the role of the 5-HT$_{1A}$ receptors in the anticonvulsant effect of CBD. In this study, to induce the experimental model of generalized seizure, male Wistar Kyoto rats were given a single i.p. injection of 85 mg/kg pentylenetetrazole (PTZ), to induce seizures. Mice received CBD at a dose of 100 mg/kg 60 min before induction of seizures. The results show that CBD significantly reduced the proconvulsant activity induced by PTZ. In particular, CBD significantly reduced seizure severity and the number of animals exhibiting seizure activity and prevented the severe consequences of seizures. Serotonergic signaling is known to be involved in seizure susceptibility and CBD shows a binding affinity for both 5-HT$_{1A}$ and 5-HT$_{2A}$. To test the mechanism of action used by CBD to exert these anticonvulsant effects, mice were pretreated with WAY 100635 (1 mg/kg), a 5-HT$_{1A}$ antagonist, or MDL-100907 (0.3 mg/kg), a 5-HT$_{2A}$ antagonist. Contrary to what might be expected, pretreatment with WAY 100635 and MDL-100907 did not reduce the anticonvulsant effect of CBD. However, this study does not prove that CBD exerted its effects through the 5-HT$_{1A}$ or 5-HT$_{2A}$ receptors [90].

CBD has anxiolytic and analgesic effects and it is known that, at least in part, the anxiolytic effects of CBD depend on the activation of 5-HT$_{1A}$ mediated neurotransmission. De Gregorio et al. evaluated these effects using rats subjected to the spared nerve injury for 24 days to induce a neuropathic pain. The animals were treated acutely with increasing intravenous (i.v.) CBD doses (0.1–1.0 mg/kg). As CBD is poorly soluble in water, it was prepared in a vehicle of ethanol/Tween 80/0.9% saline (3:1:16). Acute CBD treatment reduced the activating activity of 5-HT neurons in the dorsal raphe nucleus. Additionally, to simulate the drug regimen used by patients using CBD to treat chronic neuropathic pain and anxiety, the animals were treated with subcutaneous injections of CBD (5 mg/kg) for seven days. Treatment with CBD for one week decreased mechanical allodynia and anxiety-like behavior, which increased following the spared nerve injury. Furthermore, CBD normalized the activity of 5-HT neurons in the dorsal raphe nucleus. To investigate the mechanism used by CBD, the animals were subjected to single injections of WAY 100635 (0.3 mg/kg; i.v.), 5-HT_{1A} antagonist, capsazepine (1 mg/kg; i.v.), a TRPV1 antagonist and AM 251 (1 mg/kg; i.v.), to CB1 receptor antagonist. Capsazepine treatment completely reduced the antiallodynic effect of CBD, while WAY100635 reduced this effect partially. Instead, the anxiolytic effect of CBD was blocked following treatment with WAY100635. Therefore, it is possible to conclude that treatment with CBD in low doses protects 5-HT neurotransmission, exerts antiallodynic effects through the activation of TRPV1 and anxiolytic properties through the activation of 5-HT_{1A} receptors. In this way, CBD results as a possible candidate for treating neuropathic pain and behavior disorders [91].

Instead, Magen et al. evaluated the 5-HT_{1A} receptors-mediated therapeutic effects of CBD on hepatic encephalopathy induced by bile duct ligation in a model of chronic liver disease. Mice subjected to bile-duct ligation were treated with CBD (5 mg/kg) administrated i.p. every day for 28 days. CBD treatment improved cognitive impairments and motor function. After four weeks of treatment, CBD induced a reduction of expression of the TNF-α-receptor 1 gene in the hippocampus. Conversely, increased the expression of the BDNF gene. To verify other mechanisms of action used by CBD to exert these beneficial effects, the animals were co-treated with WAY 100635. Co-administration of CBD and WAY 100635 reversed the effects of CBD, confirming the involvement of the 5-HT_{1A} receptors [83]. The same positive effects of CBD in the cognitive and locomotor deficits were observed in another model of hepatic encephalopathy induced by injection of thioacetamide. One day after thioacetamide-administration, the animals were given a single dose of 5 mg/kg i.p. Treatment with CBD improved neurological and motor function, assessed, respectively, two and three days after the induction of liver damage. While eight days after the induction of hepatic insufficiency, CBD significantly ameliorated cognitive deficits, impaired following thioacetamide. However, 12 days after treatment with thioacetamide, CBD normalized the 5-HT levels in the brain and induced an improvement in liver function. In this way, the chronic treatment with CBD, through the indirect activation of the 5-HT_{1A} receptors, improved the cognitive and motor function of the rats with hepatic encephalopathy [92].

In conclusion, the findings of these studies suggest that the neuroprotective effects of CBD are mediated, at least in part, by 5-HT_{1A} receptors. In this way, CBD protects against cerebral ischemia and ameliorates the motor-related striatal damage in experimental models. In the same way, it improved the cognitive and motor function in an in vivo model of hepatic encephalopathy. Moreover, CBD exhibited anxiolytic properties through the activation of 5-HT_{1A} receptors, in experimental models of anxiety-like behavior disorders (Table 2).

Table 2. Neuroprotective effects of CBD in different neurological diseases through the activation of the 5-HT$_{1A}$.

In Vivo Models	CBD Dose	Treatments	Biological/Pharmacological Effect	Neurological Diseases	Ref.
MCA occlusion male mice	3 or 10 mg/kg	Before and 3 h after damage	CBD, at dose of 3 mg/kg, significantly reduced the infarct volume induced by MCA occlusion, at least in part, through the 5-HT$_{1A}$ receptor.	cerebral ischemia	[87]
Male Swiss mice	5, 15, 30 or 60 mg/kg	30 min before receiving the drugs that induce catalepsy	Pretreatment with CBD reduced the cataleptic effects, in a dose-dependent manner, through the 5-HT$_{1A}$ receptor.	striatal disorders	[88]
Male Swiss mice	15–60 mg/kg or 60 nmol	30 min before or 2.5 h after receiving the drugs that induce catalepsy	Pretreatment with CBD reduced the cataleptic effects, in a dose-dependent manner, through the 5-HT$_{1A}$ receptor.	striatal disorders	[89]
Male Wistar Kyoto rats	100 mg/kg	60 min before induction of seizures	CBD significantly mitigated PTZ-induced seizure.	seizure disorders	[90]
Adult male Wistar rats	0.1–1.0 mg/kg and 5 mg/kg	Acute treatment with cumulative injections of CBD every 5 min and repeated treatment with 5 mg/kg/day for 7 days	CBD (5 mg/kg) protects nerve injury-induced deficits in dorsal raphe nucleus 5-HT neuronal activity. Moreover, CBD exerts antiallodynic effects through the activation of TRPV1 and anxiolytic properties through the activation of 5-HT$_{1A}$ receptors.	allodynia and anxiety-like behavior	[91]
Female Sabra mice	5 mg/kg	28 days	CBD, through the 5-HT$_{1A}$ receptor activation, improved cognition and motor function, which were impaired by bile-duct ligation. Moreover, in the animal model of hepatic encephalopathy, CBD also reduced neuroinflammation, increasing expression of the BDNF genes and reducing TNF-α receptor 1 gene expression.	hepatic encephalopathy	[83]
Female Sabra mice	5 mg/kg	Single dose	CBD ameliorated cognitive impairments and locomotor activity. Moreover, CBD restored the 5-HT levels in the brain and improved the liver function.	hepatic encephalopathy	[92]

CBD, cannabidiol; MCA, middle cerebral artery; 5-HT$_{1A}$, serotonin 5-hydroxytriptamine1A; PTZ, pentylenetetrazole; BDNF, brain-derived neurotrophic factor; TNF-α, tumor necrosis factor-α.

6.1.3. GPR55

The CBD also exhibits a high affinity towards the G-protein-coupled receptor 55 (GPR55), a class of receptors implicated in the synaptic transmission. GPR55 is a G protein-coupled receptor widely expressed in the immune and nervous systems. GPR55 is involved in the modulation of parameters and cell processes such as blood pressure and bone density modulation, cell migration and proliferation, inflammation, neuropathic pain, energy balance and antiepileptic action [93]. GPR55 is a seven-transmembrane receptor that acts as a G protein inducing the intracellular increase of Ca^{2+} and the phosphorylation of the extracellular receptor-activated kinases (ERK) protein, which in turn is involved in proliferation, differentiation and cytoskeletal modulation [93]. GPR55 was found at the post-synaptic level in endothelial cells and at the pre-synaptic level in the hippocampus where it appears to increase the release of vesicular glutamate [94,95]. The facilitating effect of GPR55 contrasts the action of CB1 receptors, inhibiting neurotransmitter release. On the contrary, CBD can suppress GPR55 activation, thus increasing the release of neurotransmitters [96]. This mechanism, at least in part, elucidates the anti-convulsive effect of cannabinoids towards epileptic pharmaco-resistant patients [97]. In this regard, Kaplan et al. evaluated the GPR55-mediated antiepileptic properties of CBD in a mouse model of genetically-induced Dravet syndrome (DS). CBD (100 mg/kg or 200 mg/kg) was administered intraperitoneally, twice daily for one week. The acute treatment of CBD, in a manner dependent on its concentration, reduced the thermally-induced seizures and significantly decreased the rate of spontaneous seizures. Additionally, CBD treatment ameliorated hyperactivity due to disease. Moreover, the mouse model of genetically-induced DS showed a decrease in the GABA inhibitory transmission. The CBD treatment restored the excitability of neurons inhibitors in the dentate gyrus of the hippocampus, which represents an important zone for convulsions propagation. To confirm the involvement of GPR55 in anticonvulsant effects of CBD, mice were treated with a CID16020046 (10 µM), a GPR55 antagonist. The treatment with CID16020046 abolished the beneficial effects of CBD in inhibitory. Thus, this result suggests that the therapeutic effects of CBD are mediated through GPR55. Therefore, GPR55 could be an important therapeutic target for the treatment of epilepsy [98].

GPR55 can be involved in motor function. Celorrio et al. studied the effects of CBD and abnormal-CBD on the modulation of GPR55, in an experimental model of Parkinson's disease. To induce Parkinson's model, adult male C57BL/6 mice were treated, for five weeks, with 1-methyl-4-phenyl-1,2,3,6-tetrahydropyridine (MPTP) (20 mg/kg) and probenecid (250 mg/kg), a drug capable of reducing the renal disposal capacity of MPTP and its metabolites. MPTP mice were treated with CBD (5 mg/kg) and abnormal-CBD (5 mg/kg), administered chronically for five weeks. Abnormal-CBD is a synthetic CBD isomer that owns a high affinity towards GPR55. The study demonstrated that abnormal-CBD prevented the motor deficits MPTP-induced, while CBD did not produce a significant effect in motor behavior tests. However, both CBD and abnormal-CBD induced morphological changes in the microglia, probably due to an anti-inflammatory response. The anti-parkinsonian effect of abnormal-CBD was also confirmed in cataleptic mice induced by i.p. injection of haloperidol (1 mg/kg). Instead, the administration of CBD (5 mg/kg) did not show anti-cataleptic effects but rather abolished the action of abnormal-CBD. To confirm the involvement of GPR55 in the results obtained, it was also shown that the treatment with PSB1216 (10 mg/kg), a GPR55 antagonist, abolished the effect of abnormal-CBD. Conversely, the treatment with GPR55 agonists such as CID1792197 and CID2440433, similarly to abnormal-CBD, showed anti-cataleptic effects. Therefore, the study demonstrated that compounds able to activate GPR55 could be beneficial in combating PD [99].

The anti-inflammatory effects of CBD mediated by GPR55 were also tested by González-García et al. in experimental autoimmune encephalomyelitis (EAE) mice, a model of multiple sclerosis. The experimental model was induced in female C57BL/6J mice through i.p. injection of encephalitogenic cells cultured with Myelin Oligodendrocyte Glycoprotein peptide 35–55 (25 µg/mL) and interleukin-12 (25 ng). After the induction of the model, animals were treated with CBD (5–10 mg/kg) or CBD (50 mg/kg) i.p. The study showed an improvement in the disease already at low doses of CBD (5 mg/kg). In addition, it was shown an improvement of the disease with no signs of toxicity also at the high doses of CBD (50 mg/kg). EAE induction caused a significant reduction in the levels of CB1 and the levels were not restored by CBD treatment. On the contrary, CB2 was found in scarce levels in healthy control mice, whereas its expression was significantly increased in the EAE mice. CBD induced a significant reduction of CB2 expression in the treated animals. In the same way, the high level of EAE-induced GPR55 was reduced after CBD treatment. Although the role of GPR55 in EAE is not fully understood, its function for the disease can be very important. Indeed, mice with genetic deletion of GPR55 showed a less severe form of EAE. Therefore, the action of CBD as an antagonist of the GPR55 receptor could be useful to counteract the disease [100].

In conclusion, via the antagonist action of the GPR55 receptor, CBD carries out its anti-inflammatory effects in experimental models of DS, Parkinson's disease and EAE disease (Table 3).

6.2. TRP

TRP is a family of ion channels mainly located on the plasma membrane of many animal cells. CBD could interact with TRP, thus modulating the inflammation [101]. Interestingly, CBD is a potent and selective agonist of TRPV1, a TRP channel from the vanilloid subfamily. TRPV1 is a non-selective cationic channel present on sensory tissues such as skin, lungs, heart and blood vessels. TRPV1 activation induces the release of neuropeptides involved in pain perception, neuroinflammation, and regulation of body temperature [102]. Indeed, this receptor coupled to G proteins is characterized by six transmembrane domains which can be activated by capsaicin and stimuli such as a low pH, heat (>43 °C) and phytocannabinoids [101]. TRPV1 receptors are mainly expressed in the ganglia of dorsal roots in the spinal cord, while in the CNS they located in the hypothalamus and hippocampus [103]. It is known that TRPV1 antagonists possess analgesic properties. However, the activation of TRPV1 receptors by some agonists, such as capsaicin, leads to the entry of Ca^{2+} and Na^+ with consequent desensitization of the channel [104]. Consequently, the increase of Ca^{2+} induced by capsaicin actives the calcineurin protein which dephosphorylates TRPV1 and other proteins in the voltage-gated Ca^{2+} channels also involved in the nociceptive transmission. Thus, the receptor desensitization in response to capsaicin can render the TRPV1 channel insensitive to further painful stimuli [105]. In the same way, CBD agonist action towards TRPV1 induces the desensitization of these channels [106,107]. Therefore, at least in part, the anti-nociceptive and antihyperalgesic actions of CBD appear to be mediated by activation, dephosphorization and strong desensitization of TRPV1 channels [108].

Costa et al. showed the effects of CBD and the mechanisms associated with antihyperalgesic action in a mouse model of acute inflammation. CBD (10 mg/kg) administered orally 2 h after the induction of the model abolished the thermal hyperalgesia induced by carrageenan (0.1 mL). The CBD co-administration with capsazepine (2 mg/kg), a synthetic capsaicin antagonist, suppressed the carrageenan-induced hyperalgesia. Capsazepine at the higher dose (10 mg/kg) inhibited the CBD-induced antihyperalgesic action. These data suggest how CBD can exert its antihyperalgesic effect by directly involving TRPV1. Therefore, CBD could represent a valid therapeutical tool in the treatment of pathological conditions such as neuropathy [31].

Moreover, CBD is also able to activate TRPV2, transient receptor potential ankyrin 1 (TRPA1) and antagonize also the transient receptor potential cation channel subfamily M member 8 (TRPM8) [101]. CBD through TRP channels, involved in the proliferation and release of proinflammatory cytokines, can regulate Ca^{2+} in immune and inflammatory cells [24].

Table 3. Neuroprotective effects of CBD in different neurological diseases through the antagonize activation of the GPR55.

In Vitro and in Vivo Models	CBD Dose	Treatments	Biological/Pharmacological Effect	Neurological Diseases	Ref.
Scn1a mutant mice	10, 20, 100 or 200 mg/kg	Twice daily for 1 week	Acute treatment of CBD decreased thermally-induced seizures and reduced the rate of spontaneous seizures. Moreover, the low doses of CBD ameliorated the autism-type social interaction deficits in the mouse model of genetically-induced DS. CBD also increased the GABA inhibitory transmission which was impaired in DS. These therapeutic effects of CBD are mediated through GPR55.	DS	[98]
Adult male C57BL/6 mice	5 mg/kg	5 days a week for 5 weeks	Abnormal-CBD, but not CBD, ameliorated MPTP-induced motor damage. Instead, both compounds significantly reduced the density of microglial cells in the cell body. In the haloperidol-induced catalepsy mouse model, abnormal-CBD also showed anti-cataleptic effects, through the GPR55-activation.	Parkinson's disease	[99]
Male and female C57BL/6 mice	5–10 and 50 mg/kg	Increasing doses from 5 to 10 mg/kg three times per week, or daily, at a dose of 50 mg/kg, for 23 days	CBD, both at low and high doses, ameliorated the EAE disease. Moreover, CBD treatment reduced the vitality of encephalitogenic cells, levels of IL-6, production of ROS with consequent decrease of the apoptosis process. Additionally, it decreased the levels of GPR55 receptors in the CNS.	EAE disease	[100]

CBD, cannabidiol; DS, Dravet syndrome; GABA, γ-aminobutyric acid; MPTP, 1-methyl-4-phenyl-1,2,3,6-tetrahydropyridine; EAE, experimental autoimmune encephalomyelitis; ROS, reactive oxygen species; CNS, central nervous system.

The effects of CBD and its affinity for the TRPV2 receptors were also demonstrated by Luo et al. in human brain endothelial cells forming the BBB. CBD promoted a long-lasting increase of intracellular Ca^{2+} level, especially at 15 μM dose. After 24 h of incubation, CBD treatment (0.1, 0.3, 1, 3 and 10 μM), in a dose-dependent manner, induced the cell growth of hCMEC/D3 cells, and, after 4 h, it significantly enhanced cell migration. Instead, after 7 or 24 h, CBD significantly increased the tubulogenesis of hCMEC/D3 cells. Additionally, after 72 h to seeding, treatment with 1 μM CBD increased trans-endothelial resistance in human Primary Brain Microvascular Endothelial Cells (hPBMECs) Monolayers. To demonstrate the possible involvement of TRPV2 in CBD-induced effect, cells were pretreated, 5 min before added CBD, with ruthenium red, a nonspecific TRP antagonist, or with tranilast, a selective TRPV2 inhibitor. Ruthenium red and tranilast suppressed the CBD-induced long-lasting increase of intracellular Ca^{2+} levels. The same results were also obtained in silencing cells with TRPV2 siRNA. These data highlight the role of TRPV2 in the CBD mechanism of action. Therefore, CBD, due to the high affinity of TRPV2, could be a potential pharmacological tool to regulate the BBB characteristic [109].

Nabissi et al. evaluated the role of CBD to contrast the proliferation of glioblastoma, through activation of TRPV2. In this study, the authors showed that CBD (10 μM) improved the action of cytotoxic agents able to contrast the proliferation of glioblastoma. CBD, through activation of TRPV2 and the consequent entry of Ca^{2+}, improved the action of chemotherapy drugs. To confirm the agonize effect of CBD on TRPV2, it was performed deletion of the TRPV2 poredomain. Since the poredomain of the TRP channels is important for Ca^{2+} entry, the authors demonstrated that the deletion of this region prevents CBD-induced influx of Ca^{2+} by reducing drug absorption and cytotoxic effects. Therefore, with this result, it has been shown that CBD co-administered together with chemotherapeutic agents, activating TRPV2, enhances drug absorption and cytotoxic activity in human glioma cells. Consequently, the administration of chemotherapy drugs together with CBD could improve the efficacy of therapy useful to counteracting the glioma cells [110].

Moreover, CBD, through the activation of the TRPV, is able to promote PI_3K/Akt signaling, which in turn inhibits the glycogen synthase kinase 3β (GSK-3β). The GSK-3β inhibition induces an increase of the Wnt/β-catenin pathway, thus exerting a neuroprotective action against the oxidative stress and neurotoxicity induces of Aβ in the Alzheimer's disease [54]. In line with this evidence in an in vitro model of Alzheimer's disease, CBD treatment suppressed the hyperphosphorylation of tau protein-mediated to β-catenin and GSK-3β, in Aβ-stimulated PC12 neuronal cells [111]. Moreover, CBD decreased Aβ levels in SH-SY5Y cells transfected with the amyloid precursor protein (SH-SY5Y^{APP+}) [64], and, in an Alzheimer's disease mouse model, CBD administration ameliorated cognitive impairment [112]. In this context, our research group, in a previous study, investigated the involvement of TRPV2 in the molecular CBD's mechanism of action, comparing the expression profiles of human gingival mesenchymal stem cells treated with CBD (5 μM) to those without treatment. The results of the transcriptomic analysis show that CBD decreased the expression of genes related to Alzheimer's disease. Conversely, CBD upregulated genes coding for the PI_3K subunits and for AKT1. Thus, CBD, through modulation of PI_3K/Akt signaling, is capable of regulating GSK3β activity and consequently improve hallmarks of Alzheimer's disease. To study how CBD modulated the PI_3K/Akt signaling, human gingival mesenchymal stem cells were treated with antagonists for CB1R (SR141716A), CB2R (AM630), or TRPV1 (capsazepine) receptors. Noteworthy, only the pretreatment with capsazepine reversed the CBD-mediated effects. Therefore, CBD, by TRPV1 activation, promoted the PI_3K/Akt pathway that inactivates GSK3β. Thus, CBD could reduce Alzheimer's hallmarks [37].

The results of the studies show that CBD activates and rapidly desensitize TRPV1, inducing antihyperalgesic effects. Moreover, TRPV1 activation induced the PI_3K/Akt pathway signaling, which can reduce Alzheimer's hallmarks. Instead, CBD, through activation of TRPV2, enhanced cell proliferation and improved the action of chemotherapy drugs (Table 4).

Table 4. Neuroprotective effects of CBD in different neurological diseases through the activation of the TRPV receptors.

In Vitro and in Vivo Models	CBD Dose	Treatments	Biological/Pharmacological Effect	Neurological Diseases	Ref.
Male Wistar rats	10 mg/kg	2 h after the induction of model	CBD inhibited the carrageenan-induced hyperalgesia through the desensitization of the TRPV1 receptor	Hyperalgesia	[31]
hPBMECs and hCMEC/D3 Cells	0.1, 0.3, 1, 3, 10 and 15 µM	7 or 24 h of incubation	CBD, in a dose-dependent manner, led a last-lasting increase in intracellular Ca^{2+} level, through activation of TRPV2. In this way, CBD, enhanced cell proliferation, cell migration and tubulogenesis in human brain endothelial cells.	-	[109]
U87MG glioma cell line	10 µM	Cells were treated with different doses of CBD for 1 day or co-treated with CBD 10 µM and chemotherapeutic drugs for 6 h.	CBD, through activation of TRPV2 and the consequent entry of Ca^{2+}, improved the action of chemotherapy drugs enhancing drug absorption and ameliorated cytotoxic activity in human glioma cells.	-	[110]
human Gingival Mesenchymal Stem Cells	5 µM	24 h of incubation	CBD, through TRPV1 desensitization, promoted the PI_3K/Akt pathway signaling, which can reduce Alzheimer's hallmarks.	Alzheimer's disease	[37]

CBD, cannabidiol; hPBMECs, human primary brain microvascular endothelial cell.

6.3. PPARγ Receptors

PPARγ represents a member of the nuclear receptor family and is also a transcription factor modulated by a ligand that regulates the expression levels of genes involved in inflammation, trophic factors production, redox equilibrium, metabolism of glucose and lipids [24,113,114]. PPARγ is also a ubiquitin E3 ligase consisting of several residues of lysine, including Lysine48 capable of generating polyubiquitin chains [115]. The polyubiquitin linked to Lysine48 of PPARγ is responsible for proteasomal degradation of p65, which in turn leads to the inhibition of the inflammatory pathway mediated by NF-κB. Indeed, the activation of PPARγ inhibits the transcription of proinflammatory genes, cytokines such as TNF-α, IL-1β and IL-6, thus preventing the NF-κB signaling pathway [35]. Therefore, PPARγ agonists such as CBD, inhibiting the transcription of downstream genes mediated by NF-κB can perform an anti-inflammatory action through a molecular mechanism regulated by GSK3β [116].

This mechanism of action was evaluated by Scuderi et al. in SH-SY5Y^{APP+} cells, an in vitro model of Alzheimer's disease. SH-SY5Y^{APP+} cells were treated with CBD (10^{-9}–10^{-6} M) for 24 h. Treatment with CBD reduced the expression of the APP protein, as well as its ubiquitination, thus leading to the reduction of Aβ and neuronal apoptosis. To demonstrate the selective involvement of PPARγ in mediating CBD activity in SH-SY5Y^{APP+} cells, CBD was co-administered with MK886 (3 μM) or GW9662 (9 nM), selective antagonists of PPARα and PPARγ, respectively. The results show that treatment with GW9662 (9 nM) led to the almost complete lack of efficacy of CBD, thus confirming that, the neuroprotective role of this phytocannabinoid, involved the PPARγ receptors [64]. The neuroprotective effect of CBD through the activation of PPARγ was observed in an experimental study of Alzheimer's disease. Esposito et al. showed both in vitro and in vivo the properties of PPARγ agonists and non-agonists on neurotoxicity induced by Aβ. Cultures primary of rat astrocytes were induced with Aβ (1 μg/mL). The treatment with CBD (10^{-9}–10^{-7} M), in a concentration-dependent manner, reduced the effect of Aβ mediated through the inhibition of NF-κB. On the contrary, the treatment with GW9662 (9 nM), a PPARγ antagonist, was shown to reverse the anti-inflammatory effect of CBD. To confirm the results obtained with CBD in reactive gliosis, Esposito et al. performed the study in vivo. Male Sprague-Dawley rats were induced with Aβ (10 μg/mL). After the induction of the Alzheimer's disease model, the animals were treated for 15 days with CBD (10 mg/kg; i.p.). The data obtained show that CBD preserved from the neuronal damage induced by Aβ and also led to a reduction of gliosis and glial fibrillary acidic protein. Conversely, the administration of GW9662 (10 mg/kg), the antagonist PPARγ, has completely reversed the neuroprotective effects of CBD. Therefore, the results obtained both in vitro and in vivo confirm the important role of PPARγ in mediating the neuroprotective actions of CBD in experimental models of Alzheimer's disease [17]. The role of PPARγ in neuroprotective effects of CBD was also demonstrated by Hughes et al. The hippocampal slices of C57/black 6 mice induced with soluble oligomeric Aβ$_{1-42}$ were treated with CBD 30 min before to the addition of Aβ$_{1-42}$. The treatment with CBD improved the synaptic transmission and the potentiation long-term in the hippocampus, thereby preserving it from cognitive deficits induced by Aβ$_{1-42}$. To understand the mechanism of action of CBD, WAY 100635, (300 nM; 5-HT$_{1A}$ antagonists), ZM241385 (100 nM; A$_{2A}$R antagonist), AM 251 (2 μM; CB1 inverse agonist) or GW9662 (2 μM; PPARγ antagonist) was added to the perfusate 30 min before of CBD. It was shown that only the treatment with GW9662 attenuated the neuroprotective effects of CBD. In addition, this study suggested that CBD, at least in part, through interaction with PPARγ, could be a therapeutic potential for the treatment of Alzheimer's disease [117].

The anti-inflammatory effect of CBD was also shown in tardive dyskinesia, a disease characterized by chronic use of drugs capable of reducing or blocking the dopaminergic neurotransmission, which in turn induces the abnormal and repetitive involuntary movements that mainly involve the orofacial region [118]. In this context, Sonego et al. investigated the PPARγ-mediated protective effect of CBD in vivo and in vitro models of dyskinesia haloperidol-induced. Swiss mice were received two daily i.p. injections of CBD (60 mg/kg) for 21 days and 30 min later were treated with haloperidol (2 and 3 mg/kg). The behavioral analysis showed that CBD treatment prevented dyskinesia induced by haloperidol and reduced the oxidative stress and activation of microglial and inflammatory cytokine (IL-1β and TNF-α) in the corpus striatum. Moreover, it was shown to increase the expression of peroxisome proliferator-activated receptor-γ coactivator 1-α, a co-activator of PPARγ, thereby confirming PPARγ as a molecular target of CBD. To confirm the involvement of PPARγ in the effect of CBD, animals received the injection of GW9662 (2 mg/kg), 30 min after CBD. The administration of GW9662, inhibited the positive effect of CBD on dyskinesia. Moreover, to confirm the involvement of PPARγ, the authors also performed an in vitro study. Primary microglial cells were pretreated with GW9662 (0.1, 1 and 10 μM), 30 min after treatment with CBD (10 μM) and 4 h after the cells were stimulated with lipopolysaccharide (10 ng/mL). The in vitro study confirmed the results obtained in vivo, showing the involvement of PPARγ in the neuroprotective effects of CBD [119].

Instead, Dos-Santos-Pereira et. evaluated the PPARγ-mediated effects of CBD on L-3,4-dihydroxyphenylalanine (L-DOPA)-induced dyskinesia, a mouse model of Parkinson's disease. Male C57/BL6 mice were treated with 6-hydroxydopamine (6-OHDA) neurotoxin, which induces hallmarks of Parkinson's disease. To induce dyskinesia, 6-OHDA-lesioned animals were treated with L-DOPA for 21 days. Subsequently, 15 min before L-DOPA administration, mice received CBD (15, 30 and 60 mg/kg) i.p. for three days. CBD alone was not able to prevent the L-DOPA-induced dyskinesia. To understand the CBD's mechanism of action, it was co-administered with capsazepine (1 or 5 mg/kg), a TRPV-1 antagonist, or with arachidonoyl-serotonin (AA-5-HT; 5 mg/kg), an enzyme responsible for inhibiting anandamide metabolism FAAH and TRPV-1. Co-administration with CBD and capsazepine, through an increase in AEA and an antagonism of TRPV1 receptors, leads to the reduction of dyskinesia. Additionally, co-treatment of CBD with AM 251 (1 mg/kg), a CB1 antagonist, or with GW9662 (4 mg/kg), an antagonist of PPARγ receptors, reversed the anti-dyskinetic effect of CBD and capsazepine. In conclusion, CBD together with capsazepine, through the interaction with CB1 and PPARγ receptors, could be a valid therapeutic strategy to prevent the L-DOPA-induced dyskinesia in patients with Parkinson's disease [120]. Hind et al. also demonstrated the role of PPARγ in mediating the neuroprotective effects of CBD in experimental models of ischemic damage. To induce ischemic damage, the cells were simulated using oxygen–glucose deprivation. CBD treatment (100 nM, 1 μM and 10 μM) was administered either before or immediately after the induction of ischemic damage. CBD (10 μM) reduced the increase of BBB permeability following the ischemic damage. To assess the CBD's mechanism of action, cells were treated with AM 251 (100 nM; CB1 receptors antagonists), AM630 (100 nM; CB2 receptors antagonist), capsazepine (1 μM; TRPV1 channels antagonists), GW9662 (100 nM; PPARγ antagonist), SCH58261 (100 nM; A_{2A}Rs antagonist) and WAY100135 (300 nM; $5-HT_{1A}$ receptors antagonist). The neuroprotective effect of CBD was abolished by the administration of GW9662 (100 nM) and partially reduced by WAY100135 (300 nM). To confirm the involvement of PPARγ and $5-HT_{1A}$ receptors underlying the neuroprotective effect of CBD, cells were also treated with pioglitazone (a PPARγ agonist) and 8-OH-DPAT (an agonist of $5-HT_{1A}$ receptors). The treatment with these receptors showed similar effects of CBD on the permeability induced by ischemic damage. Therefore, the study reported useful results on the neuroprotective effect of CBD in the permeability of the BBB, through the activation of PPARγ and $5-HT_{1A}$ [121].

In addition, in an experimental model of multiple sclerosis, PPARγ exerted an important role in mediating the effects of CBD. Giacoppo et al. showed the effects of CBD in C57BL/6 mice with EAE immunized with Myelin Oligodendrocyte Glycoprotein peptide 35–55 (300 μg). Fourteen days after the induction of the EAE model, the mice were treated daily with CBD (10 mg/kg) i.p. The results of the treatment demonstrate that CBD restored the $PI_3K/Akt/mTOR$ pathway, which was downregulated after EAE induction. CBD treatment has also led to the reduction of inflammatory cytokines interferon-γ (IFN-γ) and interleukin 17 (IL-17) and significantly increased the levels of PPARγ. These results suggest that, at least in part, the effects of CBD could be related to the increased level of PPARγ. In conclusion, CBD, through both enhanced PPARγ and modulation of the $PI_3K/Akt/mTOR$ pathway, could be an interesting therapeutic target for multiple sclerosis [122].

In conclusion, the data obtained from these studies underline the implication of PPARγ in the anti-inflammatory effects of CBD, in experimental models of Alzheimer's disease, Ischemic stroke, Parkinson's disease and EAE disease (Table 5).

Table 5. Neuroprotective effects of CBD in different neurological diseases through the activation of the PPARγ.

In Vitro and in Vivo Models	CBD Dose	Treatments	Biological/Pharmacological Effect	Neurological Diseases	Ref.
SH-SY5Y^{APP+}	10^{-9}–10^{-6} M	24 h	CBD reduced the expression of the APP protein, as well as its ubiquitination, thus leading to the reduction of Aβ and neuronal apoptosis. These CBD's effects were mediated by PPARγ activation.	Alzheimer's disease	[64]
Cultures primary of astrocytes rat and male Sprague-Dawley rats	10^{-9}–10^{-7} M for in vitro study; 10 mg/kg for in vivo study.	Daily for 15 days	In the in vitro study, CBD in a concentration-dependent manner reduced the effect of Aβ mediated through the inhibition of NF-κB. In addition, in vivo, CBD ameliorated neuronal damage induced by Aβ and led to a reduction of gliosis and glial fibrillary acidic protein. CBD exerts these effects through PPARγ activation.	Alzheimer's disease	[17]
Hippocampal slices from C57Bl/6 mice	10 μM	30 min before to the addition of Aβ	The treatment with CBD improved the synaptic transmission and the potentiation long-term in the hippocampus slice of C57/black 6 mice, thereby preserving it from cognitive deficits induced by Aβ$_{1-42}$. CBD exerts these effects, at least in part, through interaction with PPARγ.	Alzheimer's disease	[117]
Primary microglial cultures from brain of male and female newborn C57/BL6 mice and Swiss mice	60 mg/kg; for in vivo study; 10 μM for in vitro study	Two daily injections 30 min before received haloperidol for 21 days	In mice, CBD treatment prevented dyskinesia induced by haloperidol. Moreover, in the corpus striatum, CBD reduced oxidative stress, activation of microglial, inflammatory cytokine (such as IL-1β and TNF-α) and increased anti-inflammatory cytokine IL-10. It was demonstrated that PPARγ is a molecular target of CBD. In the same way, it was also confirmed the effect of CBD through PPARγ on lipopolysaccharide-stimulated microglial cells.	Tardive dyskinesia	[119]
Male adult C57/BL6 mice	15, 30 and 60 mg/kg	15 min before the L-DOPA administration for three days	CBD alone was not able to prevent the L-DOPA-induced dyskinesia. The co-treatment with CBD and capsazepine, through the interaction with CB1 and PPARγ receptors, ameliorate dyskinesia.	Parkinson's disease	[120]
Human brain microvascular endothelial cell and human astrocyte co-cultures modeled	100 nM, 1 and 10 μM	Either before or immediately after the induction of ischemic damage	CBD (10 μM) prevented the enhance of BBB permeability following the ischemic damage induced by oxygen-glucose deprivation, through the activation of PPARγ and 5-HT$_{1A}$ receptors.	Ischemic stroke	[121]
Male C57Bl/6 mice	10 mg/kg	Daily treated, approximately 14 days after disease induction, for 14 days	CBD treatment ameliorated the clinical evidence of disease in EAE mice. CBD restored the PI3K/Akt/mTOR pathway that was downregulated after EAE induction. Moreover, CBD reduced inflammatory cytokines IFN-γ and IL-17 significantly and increased the levels of PPARγ. Probably, the anti-inflammatory effects of CBD are linked to the increased of PPARγ.	EAE disease	[122]

CBD, cannabidiol; SH-SY5Y^{APP+}, SH-SY5Y cells transfected with the amyloid precursor protein; Aβ, amyloid-β; IL-1β, interleukin 1-β; IL-6, interleukin 6; TNF-α, tumor necrosis factor-α; L-DOPA, L-3,4-dihydroxyphenylalanine; BBB, blood-brain barrier; IFN-γ, interferon-γ; IL-17, interleukin-17; EAE, experimental autoimmune encephalomyelitis.

6.4. GABA Receptors

GABA is the principal inhibitory neurotransmitter of the CNS. This receptor binds three classes of type A receptor: $GABA_A$, $GABA_B$ and $GABA_C$. Among these, the $GABA_A$ receptor subfamily appears to be involved in many neurological diseases [123–125]. Deficits of this receptor are responsible for several neurological disorders such as Huntington's disease [126], cognitive alterations [3], epileptic disorders [127], drug addiction [128], chronic stress and anxiety [129]. For these reasons, $GABA_A$ receptor represents an interesting target for new compounds. CBD is known to potentiate $GABA_A$-mediated inhibitory currents by acting on the $GABA_A$ receptor [65]. In accordance with this evidence, our research team in a previous work suggested that CBD might be able to decrease neuronal excitability in NSC-34 motor neuron-like cells through enhancing the expression of genes linked in GABA release and increasing the $GABA_A$ receptor gene expression [130].

Recently, CBD has been shown to be a promising compound in the treatment of patients with drug-resistant DS [131–133]. In this context, Ruffolo et al. studied the effects of CBD on $GABA_A$ receptor-mediated neuronal transmission using human cortical tissue obtained from patients with DS and tuberous sclerosis complex (TSC). Cell membranes obtained from cortical tissue of patients were transplanted in *Xenopus oocytes* to perform experiments. To evaluate the effect of CBD on GABA currents, the cells were preincubated for 10 s with CBD (5 µM) before the co-application of GABA and CBD. CBD enhanced the amplitude of the GABA-evoked current, in cortical tissue of patients with DS. A similar effect was also obtained using cortical tissues of patients with TSC. These results highlight that CBD, increasing the average amplitude of the currents evoked from GABA, could be a new therapeutic discovery in drug-resistant DS [134]. In light of this evidence, in 2018, the US Food and Drug Administration, approved CBD to treat DS and Lennox–Gastaut syndrome, two drug-resistant epileptic syndromes [135].

In the treatment of these drug-resistant epileptic syndromes, CBD is often co-administered with common antiepileptic drugs, such as clobazam (CLB). However, several clinical studies have highlighted the existence of drug–drug interaction between CBD and CLB. In these studies, it was observed that CBD causes an increase in plasma concentrations of both CLB and its active metabolite, *N*-desmethylclobazam (N-CLB) [136–138]. Similar to CBD, CLB and N-CLB are positive allosteric modulators of $GABA_A$ receptors [139–141]. In view of these this data, Anderson et al., in vivo and in vitro, explored the pharmacodynamic and pharmacokinetic interactions of these compounds highlighted the involvement of $GABA_A$ receptors. For the study, the researchers used mice with heterozygous loss of function *SCN1A* (*Scn1a$^{+/-}$*), a genetic model of DS. CBD was administrated i.p. 45 min before to receive CLB (0.1–10 mg/kg). The combined treatment of CBD and CLB resulted in an increase of anticonvulsant effect against seizures compared to the administration of the single compound. In this regard, the researchers investigated the novel pharmacodynamic mechanism where CBD and CLB together enhanced inhibitory $GABA_A$ receptor activation using *Xenopus oocytes* expressing $GABA_A$ receptors. For the in vitro study, GABA treatment (15 µM) was performed three times with a washout period of 7–12 min between GABA applications. Subsequently, CBD (10 µM) was co-applied with GABA, for 60 s. The results show that CBD and CLB exert their anticonvulsant action by enhancing the activity of the $GABA_A$ receptor. This pharmacological interaction between CBD and GABAergic drugs, like CLB, could explain the anti-seizures effect of CBD [142].

In another study, Aso et al. showed that chronic co-administration of Δ^9-THC and CBD improves cognitive deficits in transgenic *AβPP/PS1* mice, a model of Alzheimer's disease. Interestingly, the positive effects of co-administration of these two compounds may be related to a reduction in glutamate ionotropic receptors AMPA Type Subunits 2/3 and an increase in $GABA_A$ receptors. Therefore, this study further confirms the involvement of cannabinoids in excitatory and inhibitory neural activity [143]. These studies demonstrated that CBD, enhancing inhibitory $GABA_A$ receptor activation, could be an interesting approach for conditions such as epilepsy (Table 6).

Table 6. Neuroprotective effects of CBD in different neurological diseases through positive allosteric modulation of GABA$_A$ receptors.

In Vitro and in Vivo Models	CBD Dose	Treatments	Biological/Pharmacological Effect	Neurological Diseases	Ref.
Surgical human DS and TSC cortical tissue in *Xenopus* oocytes	5 μM	Pre-incubation of cells of 10 s before the co-application of GABA and CBD	CBD, through positive modulation of GABA$_A$ receptors, enhanced the amplitude of the GABA-evoked current, in brain tissues of patients with DS and TSC.	DS and TSC	[134]
Male and female *Scn1a*$^{+/-}$ mice and *Xenopus* oocytes expressing GABA$_A$ receptors	12 mg/kg or 100 mg/kg for in vivo study; 10 μM for in vitro study	In in vivo study, CBD was administered i.p. 45 min before CLB; in in vitro study CBD (10 μM) was co-applied with GABA, for 60 s	CBD significantly increased the concentrations of CLB and its active metabolite N-CLB, both in the plasma and in the brain. Co-administration of both compounds significantly increased the anticonvulsant effect. CBD and CLB exert their anticonvulsant action by enhancing the activity of the GABA$_A$ receptor.	DS	[142]

CBD, cannabidiol; GABA, γ-aminobutyric acid; DS, Dravet syndrome; TSC, tuberous sclerosis complex; *Scn1a*$^{+/-}$, heterozygous loss of function *SCN1A*; CLB, clobazam; N-CLB, N-desmethylclobazam.

7. Conclusions

The neuroprotective properties of CBD are performed via several mechanisms of action. CBD mainly exerts these effects through multiple biological targets. Specifically, CBD, through $A_{2A}R$ activation, exerts anti-inflammatory effects in animal models of Alzheimer's disease and multiple sclerosis. Moreover, at least in part, it exerts neuroprotective effects by the activation of 5-HT_{1A} receptors. In experimental models of Parkinson's disease and DS, CBD performs its beneficial effects antagonizing the GPR55 receptor. Noteworthy, CBD through the TRPV1 activation could modulate PI_3K/Akt signaling, thus ameliorating the Alzheimer's hallmarks. Additionally, the neuroprotective properties of CBD in Alzheimer's and Parkinson's diseases are also mediated by the interaction with PPARγ. Instead, CBD, enhancing the inhibitor GABA transmission, could be an interesting tool to treat epilepsy condition. The preclinical evidence reviewed here, linked with the already reported safety profile of CBD in humans, highlights that CBD represents a new opportunity for the treatment of several neurological diseases. However, further studies are needed to elucidate the molecular mechanisms underlying the properties of CBD and identify new molecular targets.

Author Contributions: Conceptualization, E.M. and P.B.; writing—original draft preparation, S.S. and G.S.; and writing—review and editing, E.M. and P.B. All authors have read and agreed to the published version of the manuscript.

Funding: This study was supported by a Current Research Funds 2020, Ministry of Health, Italy.

Acknowledgments: The authors would like to thank the Ministry of Health, Italy.

Conflicts of Interest: The authors declare no conflict of interest.

Abbreviations

CBD	cannabidiol
Δ^9-THC	Δ^9-tetrahydro-cannabinol
Nrf2	nuclear factor erythroid 2 – related factor 2
SOD	superoxide dismutase
GSH	glutathione
ROS	reactive oxygen species
iNOS	inducible nitric oxide synthase
IL-6	interleukin-6
IL-1β	interleukin-1β
TNF-α	tumor necrosis factor α
NF-κB	nuclear factor κB
PPARγ	peroxisome proliferator-activated receptor γ
TRP	transient receptor potential
TRPV	transient receptor potential vanilloid
CB1	cannabinoid receptor type 1
CB2	cannabinoid receptor type 2
2-AG	2-arachidonoyl glycerol
AEA	arachidonoylethanolamide
ECS	endocannabinoid system
Ca^{2+}	calcium
FAAH	fatty acid amide hydrolase
GABA	γ-aminobutyric acid
5-HT	Serotonin
GPCRs	G protein-coupled receptors
ARs	adenosine receptors
BBB	blood–brain barrier
CYPs	cytochrome P450 enzymes
7-OH-CBD	7-hydroxycannabidiol
$G_{i/o}$	inhibiting G

cAMP	cyclic adenosine monophosphate
PI$_3$K/Akt	phosphoinositide 3-kinases/protein kinase B
CNS	central nervous system
TMEV	Theiler's murine encephalomyelitis virus
i.p.	intraperitoneal
VCAM-1	vascular cell adhesion molecule-1
CCL2	chemokine 2
CCL5	chemokine 5
COX-2	cyclooxygenase-2
Aβ	amyloid-β
BDNF	brain-derived neurotrophic factor
K$^+$	potassium
MCA	middle cerebral artery occlusion
PTZ	pentylenetetrazole
i.v.	intravenous
GPR55	G protein-coupled receptors 55
ERK	extracellular receptor-activated kinases
DS	Dravet syndrome
MPTP	1-methyl-4-phenyl-1,2,3,6-tetrahydropyridine
EAE	experimental autoimmune encephalomyelitis
TRPA1	transient receptor potential ankyrin 1
TRPM8	transient receptor potential cation channel subfamily M member 8
hPBMECs	human Primary Brain Microvascular Endothelial Cells
GSK-3β	glycogen synthase kinase 3β
SH-SY5Y^{APP+}	SH-SY5Y cells transfected with the amyloid precursor protein
L-DOPA	L-3,4-Dihydroxyphenylalanine
6-OHDA	6-hydroxydopamine
AA-5-HT	arachidonoyl-serotonin
IFN-γ	interferon-γ
IL-17	interleukin-17
TSC	tuberous sclerosis complex
CLB	clobazam
N-CLB	N-desmethylclobazam
Scn1a$^{+/-}$	heterozygous loss of function SCN1A

References

1. WHO. *WHO Methods and Data Sources for Global Burden of Disease Estimates 2000–2011*; Department of Health Statistics and Information Systems: Geneva, Switzerland, 2013; Available online: http://www.who.int/healthinfo/statistics/GlobalDALYmethods_2000_2011.pdf?ua=1 (accessed on 6 November 2020).
2. Roy, M.; Tapadia, M.G.; Joshi, S.; Koch, B. Molecular and genetic basis of depression. *J. Genet.* **2014**, *93*, 879–892. [CrossRef] [PubMed]
3. Lopizzo, N.; Bocchio Chiavetto, L.; Cattane, N.; Plazzotta, G.; Tarazi, F.I.; Pariante, C.M.; Riva, M.A.; Cattaneo, A. Gene–environment interaction in major depression: Focus on experience-dependent biological systems. *Front. Psychiatry* **2015**, *6*, 68. [CrossRef] [PubMed]
4. Nowicka, N.; Juranek, J.; Juranek, J.K.; Wojtkiewicz, J. Risk factors and emerging therapies in amyotrophic lateral sclerosis. *Int. J. Mol. Sci.* **2019**, *20*, 2616. [CrossRef] [PubMed]
5. Ascherio, A.; Munger, K. Epidemiology of multiple sclerosis: From risk factors to prevention, *Semin. Neurol.* **2016**, *36*, 103–114. [CrossRef] [PubMed]
6. Fernández-Ruiz, J.; Sagredo, O.; Pazos, M.R.; García, C.; Pertwee, R.; Mechoulam, R.; Martínez-Orgado, J. Cannabidiol for neurodegenerative disorders: Important new clinical applications for this phytocannabinoid? *Br. J. Clin. Pharm.* **2013**, *75*, 323–333. [CrossRef]
7. Campos, A.C.; Moreira, F.A.; Gomes, F.V.; Del Bel, E.A.; Guimaraes, F.S. Multiple mechanisms involved in the large-spectrum therapeutic potential of cannabidiol in psychiatric disorders. *Philos. Trans. R. Soc. B Biol. Sci.* **2012**, *367*, 3364–3378. [CrossRef]

8. Mechoulam, R.; Gaoni, Y. Hashish—IV: The isolation and structure of cannabinolic cannabidiolic and cannabigerolic acids. *Tetrahedron* **1965**, *21*, 1223–1229. [CrossRef]
9. Pertwee, R.G. Pharmacological actions of cannabinoids. In *Cannabinoids*; Springer: Heidelberg, Germany, 2005; pp. 1–51.
10. Fiani, B.; Sarhadi, K.J.; Soula, M.; Zafar, A.; Quadri, S.A. Current application of cannabidiol (CBD) in the management and treatment of neurological disorders. *Neurol. Sci. Off. J. Ital. Neurol. Soc. Ital. Soc. Clin. Neurophysiol.* **2020**, *41*, 3085–3098. [CrossRef] [PubMed]
11. Zuardi, A.W.; Cosme, R.; Graeff, F.; Guimarães, F. Effects of ipsapirone and cannabidiol on human experimental anxiety. *J. Psychopharmacol.* **1993**, *7*, 82–88. [CrossRef] [PubMed]
12. Campos, A.; Brant, F.; Miranda, A.; Machado, F.; Teixeira, A. Cannabidiol increases survival and promotes rescue of cognitive function in a murine model of cerebral malaria. *Neuroscience* **2015**, *289*, 166–180. [CrossRef]
13. Silveira, J.W.; Issy, A.C.; Castania, V.A.; Salmon, C.E.; Nogueira-Barbosa, M.H.; Guimarães, F.S.; Defino, H.L.; Del Bel, E. Protective effects of cannabidiol on lesion-induced intervertebral disc degeneration. *PLoS ONE* **2014**, *9*, e113161. [CrossRef] [PubMed]
14. Schiavon, A.P.; Soares, L.M.; Bonato, J.M.; Milani, H.; Guimaraes, F.S.; de Oliveira, R.M.W. Protective effects of cannabidiol against hippocampal cell death and cognitive impairment induced by bilateral common carotid artery occlusion in mice. *Neurotox. Res.* **2014**, *26*, 307–316. [CrossRef] [PubMed]
15. Kwiatkoski, M.; Guimaraes, F.S.; Del-Bel, E. Cannabidiol-treated rats exhibited higher motor score after cryogenic spinal cord injury. *Neurotox. Res.* **2012**, *21*, 271–280. [CrossRef] [PubMed]
16. Linge, R.; Jiménez-Sánchez, L.; Campa, L.; Pilar-Cuéllar, F.; Vidal, R.; Pazos, A.; Adell, A.; Díaz, Á. Cannabidiol induces rapid-acting antidepressant-like effects and enhances cortical 5-HT/glutamate neurotransmission: Role of 5-HT1A receptors. *Neuropharmacology* **2016**, *103*, 16–26. [CrossRef]
17. Esposito, G.; Scuderi, C.; Valenza, M.; Togna, G.I.; Latina, V.; De Filippis, D.; Cipriano, M.; Carratù, M.R.; Iuvone, T.; Steardo, L. Cannabidiol reduces Aβ-induced neuroinflammation and promotes hippocampal neurogenesis through PPARγ involvement. *PLoS ONE* **2011**, *6*, e28668. [CrossRef] [PubMed]
18. Mecha, M.; Feliú, A.; Iñigo, P.; Mestre, L.; Carrillo-Salinas, F.; Guaza, C. Cannabidiol provides long-lasting protection against the deleterious effects of inflammation in a viral model of multiple sclerosis: A role for A2A receptors. *Neurobiol. Dis.* **2013**, *59*, 141–150. [CrossRef] [PubMed]
19. Napimoga, M.H.; Benatti, B.B.; Lima, F.O.; Alves, P.M.; Campos, A.C.; Pena-dos-Santos, D.R.; Severino, F.P.; Cunha, F.Q.; Guimarães, F.S. Cannabidiol decreases bone resorption by inhibiting RANK/RANKL expression and pro-inflammatory cytokines during experimental periodontitis in rats. *Int. Immunopharmacol.* **2009**, *9*, 216–222. [CrossRef]
20. Malfait, A.; Gallily, R.; Sumariwalla, P.; Malik, A.; Andreakos, E.; Mechoulam, R.; Feldmann, M. The nonpsychoactive cannabis constituent cannabidiol is an oral anti-arthritic therapeutic in murine collagen-induced arthritis. *Proc. Natl. Acad. Sci. USA* **2000**, *97*, 9561–9566. [CrossRef]
21. Kozela, E.; Pietr, M.; Juknat, A.; Rimmerman, N.; Levy, R.; Vogel, Z. Cannabinoids Δ9-tetrahydrocannabinol and cannabidiol differentially inhibit the lipopolysaccharide-activated NF-κB and interferon-β/STAT proinflammatory pathways in BV-2 microglial cells. *J. Biol. Chem.* **2010**, *285*, 1616–1626. [CrossRef]
22. Mannucci, C.; Navarra, M.; Calapai, F.; Spagnolo, E.V.; Busardo, F.P.; Cas, R.D.; Ippolito, F.M.; Calapai, G. Neurological Aspects of Medical Use of Cannabidiol. *Cns. Neurol. Disord. Drug Targets* **2017**, *16*, 541–553. [CrossRef]
23. Peres, F.F.; Lima, A.C.; Hallak, J.E.C.; Crippa, J.A.; Silva, R.H.; Abilio, V.C. Cannabidiol as a Promising Strategy to Treat and Prevent Movement Disorders? *Front Pharm.* **2018**, *9*, 482. [CrossRef]
24. Atalay, S.; Jarocka-Karpowicz, I.; Skrzydlewska, E. Antioxidative and Anti-Inflammatory Properties of Cannabidiol. *Antioxidants* **2019**, *9*, 21. [CrossRef]
25. Costa, B.; Trovato, A.E.; Comelli, F.; Giagnoni, G.; Colleoni, M. The non-psychoactive cannabis constituent cannabidiol is an orally effective therapeutic agent in rat chronic inflammatory and neuropathic pain. *Eur. J. Pharm.* **2007**, *556*, 75–83. [CrossRef] [PubMed]
26. Pan, H.; Mukhopadhyay, P.; Rajesh, M.; Patel, V.; Mukhopadhyay, B.; Gao, B.; Hasko, G.; Pacher, P. Cannabidiol Attenuates Cisplatin-Induced Nephrotoxicity by Decreasing Oxidative/Nitrosative Stress, Inflammation, and Cell Death. *J. Pharm. Exp.* **2009**, *328*, 708–714. [CrossRef] [PubMed]
27. Esposito, G.; De Filippis, D.; Maiuri, M.C.; De Stefano, D.; Carnuccio, R.; Iuvone, T. Cannabidiol inhibits inducible nitric oxide synthase protein expression and nitric oxide production in β-amyloid stimulated PC12

neurons through p38 MAP kinase and NF-κB involvement. *Neurosci. Lett.* **2006**, *399*, 91–95. [CrossRef] [PubMed]
28. Castillo, A.; Tolon, M.R.; Fernandez-Ruiz, J.; Romero, J.; Martinez-Orgado, J. The neuroprotective effect of cannabidiol in an in vitro model of newborn hypoxic-ischemic brain damage in mice is mediated by CB2 and adenosine receptors. *Neurobiol. Dis.* **2010**, *37*, 434–440. [CrossRef] [PubMed]
29. Jean-Gilles, L.; Gran, B.; Constantinescu, C.S. Interaction between cytokines, cannabinoids and the nervous system. *Immunobiology* **2010**, *215*, 606–610. [CrossRef] [PubMed]
30. Jastrzab, A.; Gegotek, A.; Skrzydlewska, E. Cannabidiol Regulates the Expression of Keratinocyte Proteins Involved in the Inflammation Process through Transcriptional Regulation. *Cells-Basel* **2019**, *8*, 827. [CrossRef]
31. Costa, B.; Giagnoni, G.; Franke, C.; Trovato, A.E.; Colleoni, M. Vanilloid TRPV1 receptor mediates the antihyperalgesic effect of the nonpsychoactive cannabinoid, cannabidiol, in a rat model of acute inflammation. *Br. J. Pharm.* **2004**, *143*, 247–250. [CrossRef] [PubMed]
32. Rajan, T.S.; Giacoppo, S.; Iori, R.; De Nicola, G.R.; Grassi, G.; Pollastro, F.; Bramanti, P.; Mazzon, E. Anti-inflammatory and antioxidant effects of a combination of cannabidiol and moringin in LPS-stimulated macrophages. *Fitoterapia* **2016**, *112*, 104–115. [CrossRef] [PubMed]
33. Santos, N.A.G.; Martins, N.M.; Sisti, F.M.; Fernandes, L.S.; Ferreira, R.S.; Queiroz, R.H.C.; Santos, A.C. The neuroprotection of cannabidiol against MPP+-induced toxicity in PC12 cells involves trkA receptors, upregulation of axonal and synaptic proteins, neuritogenesis, and might be relevant to Parkinson's disease. *Toxicol. In Vitro* **2015**, *30*, 231–240. [CrossRef] [PubMed]
34. Chagas, M.H.N.; Zuardi, A.W.; Tumas, V.; Pena-Pereira, M.A.; Sobreira, E.T.; Bergamaschi, M.M.; dos Santos, A.C.; Teixeira, A.L.; Hallak, J.E.; Crippa, J.A.S. Effects of cannabidiol in the treatment of patients with Parkinson's disease: An exploratory double-blind trial. *J. Psychopharmacol.* **2014**, *28*, 1088–1098. [CrossRef] [PubMed]
35. Vallée, A.; Lecarpentier, Y.; Guillevin, R.; Vallée, J.-N. Effects of cannabidiol interactions with Wnt/β-catenin pathway and PPARγ on oxidative stress and neuroinflammation in Alzheimer's disease. *Acta Biochim. Et Biophys. Sin.* **2017**, *49*, 853–866. [CrossRef]
36. Watt, G.; Karl, T. In vivo evidence for therapeutic properties of cannabidiol (CBD) for Alzheimer's disease. *Front Pharm.* **2017**, *8*, 20. [CrossRef] [PubMed]
37. Libro, R.; Diomede, F.; Scionti, D.; Piattelli, A.; Grassi, G.; Pollastro, F.; Bramanti, P.; Mazzon, E.; Trubiani, O. Cannabidiol Modulates the Expression of Alzheimer's Disease-Related Genes in Mesenchymal Stem Cells. *Int. J. Mol. Sci.* **2017**, *18*, 26. [CrossRef]
38. Hussain, S.A.; Zhou, R.; Jacobson, C.; Weng, J.; Cheng, E.; Lay, J.; Hung, P.; Lerner, J.T.; Sankar, R. Perceived efficacy of cannabidiol-enriched cannabis extracts for treatment of pediatric epilepsy: A potential role for infantile spasms and Lennox–Gastaut syndrome. *Epilepsy Behav.* **2015**, *47*, 138–141. [CrossRef] [PubMed]
39. Devinsky, O.; Cilio, M.R.; Cross, H.; Fernandez-Ruiz, J.; French, J.; Hill, C.; Katz, R.; Di Marzo, V.; Jutras-Aswad, D.; Notcutt, W.G. Cannabidiol: Pharmacology and potential therapeutic role in epilepsy and other neuropsychiatric disorders. *Epilepsia* **2014**, *55*, 791–802. [CrossRef]
40. Bergamaschi, M.M.; Queiroz, R.H.C.; Chagas, M.H.N.; De Oliveira, D.C.G.; De Martinis, B.S.; Kapczinski, F.; Quevedo, J.; Roesler, R.; Schröder, N.; Nardi, A.E. Cannabidiol reduces the anxiety induced by simulated public speaking in treatment-naive social phobia patients. *Neuropsychopharmacology* **2011**, *36*, 1219–1226. [CrossRef]
41. Zhornitsky, S.; Potvin, S. Cannabidiol in humans-the quest for therapeutic targets. *Pharmaceuticals* **2012**, *5*, 529–552. [CrossRef]
42. Moher, D.; Liberati, A.; Tetzlaff, J.; Altman, D.G.; Group, P. Preferred reporting items for systematic reviews and meta-analyses: The PRISMA statement. *PLoS Med.* **2009**, *6*, e1000097. [CrossRef]
43. Adams, R. Marihuana: Harvey lecture, February 19, 1942. *Bull. New York Acad. Med.* **1942**, *18*, 705.
44. Todd, A. The chemistry of hashish. *Sci. J. R. Coll. Sci.* **1942**, *12*, 37.
45. Jones, P.G.; Falvello, L.; Kennard, O.; Sheldrick, G.; Mechoulam, R. Cannabidiol. *Acta Cryst. Sect. B Struct. Cryst. Cryst. Chem.* **1977**, *33*, 3211–3214. [CrossRef]
46. Morales, P.; Reggio, P.H.; Jagerovic, N. An Overview on Medicinal Chemistry of Synthetic and Natural Derivatives of Cannabidiol. *Front Pharm.* **2017**, *8*, 422. [CrossRef] [PubMed]
47. Elmes, M.W.; Kaczocha, M.; Berger, W.T.; Leung, K.; Ralph, B.P.; Wang, L.Q.; Sweeney, J.M.; Miyauchi, J.T.; Tsirka, S.E.; Ojima, I.; et al. Fatty Acid-binding Proteins (FABPs) Are Intracellular Carriers for Delta (9)-Tetrahydrocannabinol (THC) and Cannabidiol (CBD). *J. Biol. Chem.* **2015**, *290*, 8711–8721. [CrossRef]

48. Borges, R.S.; Batista, J.; Viana, R.B.; Baetas, A.C.; Orestes, E.; Andrade, M.A.; Honorio, K.M.; da Silva, A.B.F. Understanding the Molecular Aspects of Tetrahydrocannabinol and Cannabidiol as Antioxidants. *Molecules* **2013**, *18*, 12663–12674. [CrossRef] [PubMed]
49. Laprairie, R.B.; Bagher, A.M.; Kelly, M.E.M.; Denovan-Wright, E.M. Cannabidiol is a negative allosteric modulator of the cannabinoid CB1 receptor. *Br. J. Pharm.* **2015**, *172*, 4790–4805. [CrossRef]
50. Morales, P.; Goya, P.; Jagerovic, N.; Hernandez-Folgado, L. Allosteric modulators of the CB1 cannabinoid receptor: A structural update review. *Cannabis Cannabinoid Res.* **2016**, *1*, 22–30. [CrossRef]
51. Leweke, F.; Piomelli, D.; Pahlisch, F.; Muhl, D.; Gerth, C.; Hoyer, C.; Klosterkötter, J.; Hellmich, M.; Koethe, D. Cannabidiol enhances anandamide signaling and alleviates psychotic symptoms of schizophrenia. *Transl. Psychiatry* **2012**, *2*, e94. [CrossRef]
52. Campos, A.C.; Fogaca, M.V.; Sonego, A.B.; Guimaraes, F.S. Cannabidiol, neuroprotection and neuropsychiatric disorders. *Pharm. Res.* **2016**, *112*, 119–127. [CrossRef]
53. Ibeas Bih, C.; Chen, T.; Nunn, A.V.; Bazelot, M.; Dallas, M.; Whalley, B.J. Molecular Targets of Cannabidiol in Neurological Disorders. *Neurotherapeutics* **2015**, *12*, 699–730. [CrossRef] [PubMed]
54. Cassano, T.; Villani, R.; Pace, L.; Carbone, A.; Bukke, V.N.; Orkisz, S.; Avolio, C.; Serviddio, G. From Cannabis sativa to Cannabidiol: Promising Therapeutic Candidate for the Treatment of Neurodegenerative Diseases. *Front Pharm.* **2020**, *11*, 124. [CrossRef] [PubMed]
55. Kano, M.; Ohno-Shosaku, T.; Maejima, T. Retrograde signaling at central synapses via endogenous cannabinoids. *Mol. Psychiatr.* **2002**, *7*, 234–235. [CrossRef] [PubMed]
56. Freund, T.F.; Katona, I.; Piomelli, D. Role of endogenous cannabinoids in synaptic signaling. *Physiol. Rev.* **2003**, *83*, 1017–1066. [CrossRef]
57. Pertwee, R.G.; Ross, R.A. Cannabinoid receptors and their ligands. *Prostaglandins Leukot. Essent. Fat. Acids* **2002**, *66*, 101–121. [CrossRef] [PubMed]
58. Szabo, B.; Schlicker, E. Effects of cannabinoids on neurotransmission. *Handb. Exp. Pharm.* **2005**, 327–365.
59. Russo, E.B.; Burnett, A.; Hall, B.; Parker, K.K. Agonistic properties of cannabidiol at 5-HT1a receptors. *Neurochem. Res.* **2005**, *30*, 1037–1043. [CrossRef]
60. Yang, K.-H.; Galadari, S.; Isaev, D.; Petroianu, G.; Shippenberg, T.S.; Oz, M. The nonpsychoactive cannabinoid cannabidiol inhibits 5-hydroxytryptamine3A receptor-mediated currents in Xenopus laevis oocytes. *J. Pharm. Exp.* **2010**, *333*, 547–554. [CrossRef]
61. Rock, E.; Bolognini, D.; Limebeer, C.; Cascio, M.; Anavi-Goffer, S.; Fletcher, P.; Mechoulam, R.; Pertwee, R.G.; Parker, L.A. Cannabidiol, a non-psychotropic component of cannabis, attenuates vomiting and nausea-like behaviour via indirect agonism of 5-HT1A somatodendritic autoreceptors in the dorsal raphe nucleus. *Br. J. Pharm.* **2012**, *165*, 2620–2634. [CrossRef]
62. Gonca, E.; Darıcı, F. The effect of cannabidiol on ischemia/reperfusion-induced ventricular arrhythmias: The role of adenosine A1 receptors. *J. Cardiovasc. Pharm.* **2015**, *20*, 76–83. [CrossRef]
63. Izzo, A.A.; Borrelli, F.; Capasso, R.; Di Marzo, V.; Mechoulam, R. Non-psychotropic plant cannabinoids: New therapeutic opportunities from an ancient herb. *Trends Pharm. Sci.* **2009**, *30*, 515–527. [CrossRef] [PubMed]
64. Scuderi, C.; Steardo, L.; Esposito, G. Cannabidiol promotes amyloid precursor protein ubiquitination and reduction of β amyloid expression in SHSY5YAPP+ cells through PPARγ involvement. *Phytother. Res.* **2014**, *28*, 1007–1013. [CrossRef] [PubMed]
65. Bakas, T.; van Nieuwenhuijzen, P.S.; Devenish, S.O.; McGregor, I.S.; Arnold, J.C.; Chebib, M. The direct actions of cannabidiol and 2-arachidonoyl glycerol at GABAA receptors. *Pharm. Res.* **2017**, *119*, 358–370. [CrossRef] [PubMed]

66. Newmeyer, M.N.; Swortwood, M.J.; Barnes, A.J.; Abulseoud, O.A.; Scheidweiler, K.B.; Huestis, M.A. Free and Glucuronide Whole Blood Cannabinoids' Pharmacokinetics after Controlled Smoked, Vaporized, and Oral Cannabis Administration in Frequent and Occasional Cannabis Users: Identification of Recent Cannabis Intake. *Clin. Chem.* **2016**, *62*, 1579–1592. [CrossRef]
67. Mechoulam, R.; Parker, L.A.; Gallily, R. Cannabidiol: An overview of some pharmacological aspects. *J. Clin. Pharm.* **2002**, *42*, 11s–19s. [CrossRef]
68. Grotenhermen, F. Pharmacokinetics and pharmacodynamics of cannabinoids. *Clin. Pharm.* **2003**, *42*, 327–360. [CrossRef] [PubMed]
69. Fasinu, P.S.; Phillips, S.; ElSohly, M.A.; Walker, L.A. Current Status and Prospects for Cannabidiol Preparations as New Therapeutic Agents. *Pharmacotherapy* **2016**, *36*, 781–796. [CrossRef] [PubMed]
70. Huestis, M.A. Pharmacokinetics and metabolism of the plant cannabinoids, delta9-tetrahydrocannabinol, cannabidiol and cannabinol. *Handb. Exp. Pharm.* **2005**, 657–690. [CrossRef]
71. Gaston, T.E.; Friedman, D. Pharmacology of cannabinoids in the treatment of epilepsy. *Epilepsy Behav.* **2017**, *70*, 313–318. [CrossRef]
72. Bergamaschi, M.M.; Queiroz, R.H.; Zuardi, A.W.; Crippa, J.A. Safety and side effects of cannabidiol, a Cannabis sativa constituent. *Curr. Drug Saf.* **2011**, *6*, 237–249. [CrossRef]
73. Zendulka, O.; Dovrtelova, G.; Noskova, K.; Turjap, M.; Sulcova, A.; Hanus, L.; Jurica, J. Cannabinoids and Cytochrome P450 Interactions. *Curr. Drug Metab.* **2016**, *17*, 206–226. [CrossRef]
74. Jiang, R.R.; Yamaori, S.; Okamoto, Y.; Yamamoto, I.; Watanabe, K. Cannabidiol Is a Potent Inhibitor of the Catalytic Activity of Cytochrome P450 2C19. *Drug Metab. Pharm.* **2013**, *28*, 332–338. [CrossRef] [PubMed]
75. Premoli, M.; Aria, F.; Bonini, S.A.; Maccarinelli, G.; Gianoncelli, A.; Pina, S.D.; Tambaro, S.; Memo, M.; Mastinu, A. Cannabidiol: Recent advances and new insights for neuropsychiatric disorders treatment. *Life Sci.* **2019**, *224*, 120–127. [CrossRef] [PubMed]
76. Dal Ben, D.; Lambertucci, C.; Buccioni, M.; Marti Navia, A.; Marucci, G.; Spinaci, A.; Volpini, R. Non-Nucleoside Agonists of the Adenosine Receptors: An Overview. *Pharmaceuticals* **2019**, *12*, 150. [CrossRef]
77. Kamp, T.J.; Hell, J.W. Regulation of cardiac L-type calcium channels by protein kinase A and protein kinase C. *Circ. Res.* **2000**, *87*, 1095–1102. [CrossRef] [PubMed]
78. Borea, P.A.; Gessi, S.; Merighi, S.; Vincenzi, F.; Varani, K. Pharmacology of Adenosine Receptors: The State of the Art. *Physiol. Rev.* **2018**, *98*, 1591–1625. [CrossRef]
79. Jacobson, K.A.; Gao, Z.G. Adenosine receptors as therapeutic targets. *Nat. Rev. Drug Discov.* **2006**, *5*, 247–264. [CrossRef] [PubMed]
80. Chen, J.F.; Eltzschig, H.K.; Fredholm, B.B. Adenosine receptors as drug targets-what are the challenges? *Nat. Rev. Drug Discov.* **2013**, *12*, 265–286. [CrossRef]
81. Carrier, E.J.; Auchampach, J.A.; Hillard, C.J. Inhibition of an equilibrative nucleoside transporter by cannabidiol: A mechanism of cannabinoid immunosuppression. *Proc. Natl. Acad. Sci. USA* **2006**, *103*, 7895–7900. [CrossRef]
82. Magen, I.; Avraham, Y.; Ackerman, Z.; Vorobiev, L.; Mechoulam, R.; Berry, E.M. Cannabidiol ameliorates cognitive and motor impairments in mice with bile duct ligation. *J. Hepatol.* **2009**, *51*, 528–534. [CrossRef] [PubMed]
83. Magen, I.; Avraham, Y.; Ackerman, Z.; Vorobiev, L.; Mechoulam, R.; Berry, E.M. Cannabidiol ameliorates cognitive and motor impairments in bile-duct ligated mice via 5-HT1A receptor activation. *Br. J. Pharm.* **2010**, *159*, 950–957. [CrossRef] [PubMed]
84. Martin-Moreno, A.M.; Reigada, D.; Ramirez, B.G.; Mechoulam, R.; Innamorato, N.; Cuadrado, A.; de Ceballos, M.L. Cannabidiol and other cannabinoids reduce microglial activation in vitro and in vivo: Relevance to Alzheimer's disease. *Mol. Pharm.* **2011**, *79*, 964–973. [CrossRef]
85. Raymond, J.R.; Mukhin, Y.V.; Gettys, T.W.; Garnovskaya, M.N. The recombinant 5-HT1A receptor: G protein coupling and signalling pathways. *Br. J. Pharm.* **1999**, *127*, 1751–1764. [CrossRef]
86. Bevilaqua, L.; Ardenghi, P.; Schroder, N.; Bromberg, E.; Schmitz, P.K.; Schaeffer, E.; Quevedo, J.; Bianchin, M.; Walz, R.; Medina, J.H.; et al. Drugs acting upon the cyclic adenosine monophosphate protein kinase A signalling pathway modulate memory consolidation when given late after training into rat hippocampus but not amygdala. *Behav. Pharm.* **1997**, *8*, 331–338. [CrossRef] [PubMed]
87. Mishima, K.; Hayakawa, K.; Abe, K.; Ikeda, T.; Egashira, N.; Iwasaki, K.; Fujiwara, M. Cannabidiol prevents cerebral infarction via a serotonergic 5-hydroxytryptamine1A receptor-dependent mechanism. *Stroke* **2005**, *36*, 1077–1082. [CrossRef]

88. Gomes, F.V.; Del Bel, E.A.; Guimaraes, F.S. Cannabidiol attenuates catalepsy induced by distinct pharmacological mechanisms via 5-HT1A receptor activation in mice. *Prog. Neuro-Psychopharmacol. Biol. Psychiatry* **2013**, *46*, 43–47. [CrossRef]
89. Sonego, A.B.; Gomes, F.V.; Del Bel, E.A.; Guimaraes, F.S. Cannabidiol attenuates haloperidol-induced catalepsy and c-Fos protein expression in the dorsolateral striatum via 5-HT1A receptors in mice. *Behav. Brain Res.* **2016**, *309*, 22–28. [CrossRef]
90. Pelz, M.C.; Schoolcraft, K.D.; Larson, C.; Spring, M.G.; Lopez, H.H. Assessing the role of serotonergic receptors in cannabidiol's anticonvulsant efficacy. *Epilepsy Behav.* **2017**, *73*, 111–118. [CrossRef] [PubMed]
91. De Gregorio, D.; McLaughlin, R.J.; Posa, L.; Ochoa-Sanchez, R.; Enns, J.; Lopez-Canul, M.; Aboud, M.; Maione, S.; Comai, S.; Gobbi, G. Cannabidiol modulates serotonergic transmission and reverses both allodynia and anxiety-like behavior in a model of neuropathic pain. *Pain* **2019**, *160*, 136–150. [CrossRef] [PubMed]
92. Avraham, Y.; Grigoriadis, N.; Poutahidis, T.; Vorobiev, L.; Magen, I.; Ilan, Y.; Mechoulam, R.; Berry, E. Cannabidiol improves brain and liver function in a fulminant hepatic failure-induced model of hepatic encephalopathy in mice. *Br. J. Pharm.* **2011**, *162*, 1650–1658. [CrossRef]
93. Marichal-Cancino, B.A.; Fajardo-Valdez, A.; Ruiz-Contreras, A.E.; Mendez-Diaz, M.; Prospero-Garcia, O. Advances in the Physiology of GPR55 in the Central Nervous System. *Curr. Neuropharmacol.* **2017**, *15*, 771–778. [CrossRef]
94. Kremshofer, J.; Siwetz, M.; Berghold, V.M.; Lang, I.; Huppertz, B.; Gauster, M. A role for GPR55 in human placental venous endothelial cells. *Histochem. Cell Biol.* **2015**, *144*, 49–58. [CrossRef]
95. Sylantyev, S.; Jensen, T.P.; Ross, R.A.; Rusakov, D.A. Cannabinoid- and lysophosphatidylinositol-sensitive receptor GPR55 boosts neurotransmitter release at central synapses. *Proc. Natl. Acad. Sci. USA* **2013**, *110*, 5193–5198. [CrossRef] [PubMed]
96. Ryberg, E.; Larsson, N.; Sjögren, S.; Hjorth, S.; Hermansson, N.O.; Leonova, J.; Elebring, T.; Nilsson, K.; Drmota, T.; Greasley, P. The orphan receptor GPR55 is a novel cannabinoid receptor. *Br. J. Pharm.* **2007**, *152*, 1092–1101. [CrossRef] [PubMed]
97. Jones, N.A.; Hill, A.J.; Smith, I.; Bevan, S.A.; Williams, C.M.; Whalley, B.J.; Stephens, G.J. Cannabidiol displays antiepileptiform and antiseizure properties in vitro and in vivo. *J. Pharm. Exp.* **2010**, *332*, 569–577. [CrossRef]
98. Kaplan, J.S.; Stella, N.; Catterall, W.A.; Westenbroek, R.E. Cannabidiol attenuates seizures and social deficits in a mouse model of Dravet syndrome. *Proc. Natl. Acad. Sci. USA* **2017**, *114*, 11229–11234. [CrossRef]
99. Celorrio, M.; Rojo-Bustamante, E.; Fernandez-Suarez, D.; Saez, E.; Estella-Hermoso de Mendoza, A.; Muller, C.E.; Ramirez, M.J.; Oyarzabal, J.; Franco, R.; Aymerich, M.S. GPR55: A therapeutic target for Parkinson's disease? *Neuropharmacology* **2017**, *125*, 319–332. [CrossRef]
100. Gonzalez-Garcia, C.; Torres, I.M.; Garcia-Hernandez, R.; Campos-Ruiz, L.; Esparragoza, L.R.; Coronado, M.J.; Grande, A.G.; Garcia-Merino, A.; Sanchez Lopez, A.J. Mechanisms of action of cannabidiol in adoptively transferred experimental autoimmune encephalomyelitis. *Exp. Neurol.* **2017**, *298*, 57–67. [CrossRef]
101. De Petrocellis, L.; Ligresti, A.; Moriello, A.S.; Allara, M.; Bisogno, T.; Petrosino, S.; Stott, C.G.; Di Marzo, V. Effects of cannabinoids and cannabinoid-enriched Cannabis extracts on TRP channels and endocannabinoid metabolic enzymes. *Br. J. Pharm.* **2011**, *163*, 1479–1494. [CrossRef]
102. Ross, R.A. Anandamide and vanilloid TRPV1 receptors. *Br J. Pharm.* **2003**, *140*, 790–801. [CrossRef] [PubMed]
103. Vyklicky, L.; Novakova-Tousova, K.; Benedikt, J.; Samad, A.; Touska, F.; Vlachova, V. Calcium-dependent desensitization of vanilloid receptor TRPV1: A mechanism possibly involved in analgesia induced by topical application of capsaicin. *Physiol. Res.* **2008**, *57*, 59–68.
104. Jara-Oseguera, A.; Simon, S.A.; Rosenbaum, T. TRPV1: On the road to pain relief. *Curr. Mol. Pharm.* **2008**, *1*, 255–269. [CrossRef]
105. Novakova-Tousova, K.; Vyklicky, L.; Susankova, K.; Benedikt, J.; Samad, A.; Teisinger, J.; Vlachova, V. Functional changes in the vanilloid receptor subtype 1 channel during and after acute desensitization. *Neuroscience* **2007**, *149*, 144–154. [CrossRef]
106. Planells-Cases, R.; Garcia-Sanz, N.; Morenilla-Palao, C.; Ferrer-Montiel, A. Functional aspects and mechanisms of TRPV1 involvement in neurogenic inflammation that leads to thermal hyperalgesia. *Pflug. Arch. Eur. J. Phys.* **2005**, *451*, 151–159. [CrossRef]
107. Srivastava, M.D.; Srivastava, B.I.; Brouhard, B. Delta9 tetrahydrocannabinol and cannabidiol alter cytokine production by human immune cells. *Immunopharmacology* **1998**, *40*, 179–185. [CrossRef]

108. Bisogno, T.; Hanus, L.; De Petrocellis, L.; Tchilibon, S.; Ponde, D.E.; Brandi, I.; Moriello, A.S.; Davis, J.B.; Mechoulam, R.; Di Marzo, V. Molecular targets for cannabidiol and its synthetic analogues: Effect on vanilloid VR1 receptors and on the cellular uptake and enzymatic hydrolysis of anandamide. *Br. J. Pharm.* **2001**, *134*, 845–852. [CrossRef]
109. Luo, H.L.; Rossi, E.; Saubamea, B.; Chasseigneaux, S.; Cochois, V.; Choublier, N.; Smirnova, M.; Glacial, F.; Perriere, N.; Bourdoulous, S.; et al. Cannabidiol Increases Proliferation, Migration, Tubulogenesis, and Integrity of Human Brain Endothelial Cells through TRPV2 Activation. *Mol. Pharm.* **2019**, *16*, 1312–1326. [CrossRef] [PubMed]
110. Nabissi, M.; Morelli, M.B.; Santoni, M.; Santoni, G. Triggering of the TRPV2 channel by cannabidiol sensitizes glioblastoma cells to cytotoxic chemotherapeutic agents. *Carcinogenesis* **2013**, *34*, 48–57. [CrossRef] [PubMed]
111. Esposito, G.; De Filippis, D.; Carnuccio, R.; Izzo, A.A.; Iuvone, T. The marijuana component cannabidiol inhibits β-amyloid-induced tau protein hyperphosphorylation through Wnt/β-catenin pathway rescue in PC12 cells. *J. Mol. Med.* **2006**, *84*, 253–258. [CrossRef]
112. Cheng, D.; Low, J.K.; Logge, W.; Garner, B.; Karl, T. Chronic cannabidiol treatment improves social and object recognition in double transgenic APP. (swe)/PS1a Delta E9 mice. *Psychopharmacology* **2014**, *231*, 3009–3017. [CrossRef] [PubMed]
113. Cai, W.; Yang, T.; Liu, H.; Han, L.J.; Zhang, K.; Hu, X.M.; Zhang, X.J.; Yin, K.J.; Gao, Y.Q.; Bennett, M.V.L.; et al. Peroxisome proliferator-activated receptor γ (PPARγ): A master gatekeeper in CNS injury and repair. *Prog. Neurobiol.* **2018**, *163*, 27–58. [CrossRef]
114. Surgucheva, I.; Surguchov, A. Γ-synuclein: Cell-type-specific promoter activity and binding to transcription factors. *J. Mol. Neurosci. Mn* **2008**, *35*, 267–271. [CrossRef] [PubMed]
115. Hou, Y.Z.; Moreau, F.; Chadee, K. PPAR γ is an E3 ligase that induces the degradation of NF-κB/p65. *Nat. Commun.* **2012**, *3*, 1–11. [CrossRef]
116. O'Sullivan, S.E. An update on PPAR activation by cannabinoids. *Br J. Pharm.* **2016**, *173*, 1899–1910. [CrossRef]
117. Hughes, B.; Herron, C.E. Cannabidiol Reverses Deficits in Hippocampal LTP in a Model of Alzheimer's Disease. *Neurochem. Res.* **2019**, *44*, 703–713. [CrossRef]
118. Aia, P.G.; Revuelta, G.J.; Cloud, L.J.; Factor, S.A. Tardive dyskinesia. *Curr. Treat. Options Neurol.* **2011**, *13*, 231–241. [CrossRef] [PubMed]
119. Sonego, A.B.; Prado, D.S.; Vale, G.T.; Sepulveda-Diaz, J.E.; Cunha, T.M.; Tirapelli, C.R.; Del Bel, E.A.; Raisman-Vozari, R.; Guimaraes, F.S. Cannabidiol prevents haloperidol-induced vacuos chewing movements and inflammatory changes in mice via PPARγ receptors. *Brain Behav. Immun.* **2018**, *74*, 241–251. [CrossRef]
120. Dos-Santos-Pereira, M.; da-Silva, C.A.; Guimaraes, F.S.; Del-Bel, E. Co-administration of cannabidiol and capsazepine reduces L-DOPA-induced dyskinesia in mice: Possible mechanism of action. *Neurobiol. Dis.* **2016**, *94*, 179–195. [CrossRef]
121. Hind, W.H.; England, T.J.; O'Sullivan, S.E. Cannabidiol protects an in vitro model of the blood-brain barrier from oxygen-glucose deprivation via PPARγ and 5-HT1A receptors. *Br J. Pharm.* **2016**, *173*, 815–825. [CrossRef]
122. Giacoppo, S.; Pollastro, F.; Grassi, G.; Bramanti, P.; Mazzon, E. Target regulation of PI3K/Akt/mTOR pathway by cannabidiol in treatment of experimental multiple sclerosis. *Fitoterapia* **2017**, *116*, 77–84. [CrossRef]
123. Sigel, E.; Steinmann, M.E. Structure, function, and modulation of GABA(A) receptors. *J. Biol. Chem.* **2012**, *287*, 40224–40231. [CrossRef] [PubMed]
124. Rabow, L.E.; Russek, S.J.; Farb, D.H. From ion currents to genomic analysis: Recent advances in GABAA receptor research. *Synapse* **1995**, *21*, 189–274. [CrossRef]
125. Cifelli, P.; Ruffolo, G.; De Felice, E.; Alfano, V.; van Vliet, E.A.; Aronica, E.; Palma, E. Phytocannabinoids in Neurological Diseases: Could They Restore a Physiological GABAergic Transmission? *Int. J. Mol. Sci.* **2020**, *21*, 723. [CrossRef] [PubMed]
126. Holley, S.M.; Galvan, L.; Kamdjou, T.; Dong, A.; Levine, M.S.; Cepeda, C. Major Contribution of Somatostatin-Expressing Interneurons and Cannabinoid Receptors to Increased GABA Synaptic Activity in the Striatum of Huntington's Disease Mice. *Front. Synaptic Neurosci.* **2019**, *11*, 14. [CrossRef]
127. Palma, E.; Ruffolo, G.; Cifelli, P.; Roseti, C.; Vliet, E.A.V.; Aronica, E. Modulation of GABAA Receptors in the Treatment of Epilepsy. *Curr. Pharm. Des.* **2017**, *23*, 5563–5568. [CrossRef]
128. Olsen, R.W.; Liang, J. Role of GABAA receptors in alcohol use disorders suggested by chronic intermittent ethanol (CIE) rodent model. *Mol. Brain* **2017**, *10*, 45. [CrossRef] [PubMed]

129. Bambico, F.R.; Li, Z.; Oliveira, C.; McNeill, S.; Diwan, M.; Raymond, R.; Nobrega, J.N. Rostrocaudal subregions of the ventral tegmental area are differentially impacted by chronic stress. *Psychopharmacol. (Berl.)* **2019**, *236*, 1917–1929. [CrossRef]
130. Gugliandolo, A.; Silvestro, S.; Chiricosta, L.; Pollastro, F.; Bramanti, P.; Mazzon, E. The Transcriptomic Analysis of NSC-34 Motor Neuron-Like Cells Reveals That Cannabigerol Influences Synaptic Pathways: A Comparative Study with Cannabidiol. *Life* **2020**, *10*, 227. [CrossRef]
131. Perucca, E. Cannabinoids in the Treatment of Epilepsy: Hard Evidence at Last? *J. Epilepsy Res.* **2017**, *7*, 61–76. [CrossRef]
132. Tang, R.R.; Fang, F. Trial of Cannabidiol for Drug-Resistant Seizures in the Dravet Syndrome. *N. Engl. J. Med.* **2017**, *377*, 699.
133. Devinsky, O.; Cross, J.H.; Laux, L.; Marsh, E.; Miller, I.; Nabbout, R.; Scheffer, I.E.; Thiele, E.A.; Wright, S.; Cannabidiol in Dravet Syndrome Study, G. Trial of Cannabidiol for Drug-Resistant Seizures in the Dravet Syndrome. *N. Engl. J. Med.* **2017**, *376*, 2011–2020. [CrossRef] [PubMed]
134. Ruffolo, G.; Cifelli, P.; Roseti, C.; Thom, M.; van Vliet, E.A.; Limatola, C.; Aronica, E.; Palma, E. A novel GABAergic dysfunction in human Dravet syndrome. *Epilepsia* **2018**, *59*, 2106–2117. [CrossRef] [PubMed]
135. FDA, U.S. FDA approves first drug comprised of an active ingredient derived from marijuana to treat rare, severe forms of epilepsy. *Silver Spring: Us Food Drug Adm.* **2018**. Available online: https://www.fda.gov/news-events/press-announcements/fda-approves-first-drug-comprised-active-ingredient-derived-marijuana-treat-rare-severe-forms (accessed on 16 December 2019).
136. Geffrey, A.L.; Pollack, S.F.; Bruno, P.L.; Thiele, E.A. Drug-drug interaction between clobazam and cannabidiol in children with refractory epilepsy. *Epilepsia* **2015**, *56*, 1246–1251. [CrossRef]
137. Hess, E.J.; Moody, K.A.; Geffrey, A.L.; Pollack, S.F.; Skirvin, L.A.; Bruno, P.L.; Paolini, J.L.; Thiele, E.A. Cannabidiol as a new treatment for drug-resistant epilepsy in tuberous sclerosis complex. *Epilepsia* **2016**, *57*, 1617–1624. [CrossRef]
138. Gaston, T.E.; Bebin, E.M.; Cutter, G.R.; Liu, Y.; Szaflarski, J.P.; Program, U.C. Interactions between cannabidiol and commonly used antiepileptic drugs. *Epilepsia* **2017**, *58*, 1586–1592. [CrossRef]
139. Nakajima, H. A pharmacological profile of clobazam (Mystan), a new antiepileptic drug. *Nihon Yakurigaku Zasshi. Folia Pharm. Jpn.* **2001**, *118*, 117–122. [CrossRef] [PubMed]
140. Sankar, R. GABA A receptor physiology and its relationship to the mechanism of action of the 1,5-benzodiazepine clobazam. *Cns. Drugs* **2012**, *26*, 229–244. [CrossRef] [PubMed]
141. Jensen, H.S.; Nichol, K.; Lee, D.; Ebert, B. Clobazam and its active metabolite N-desmethylclobazam display significantly greater affinities for α 2-versus α 1-GABA A–receptor complexes. *PLoS ONE* **2014**, *9*, e88456. [CrossRef]
142. Anderson, L.L.; Absalom, N.L.; Abelev, S.V.; Low, I.K.; Doohan, P.T.; Martin, L.J.; Chebib, M.; McGregor, I.S.; Arnold, J.C. Coadministered cannabidiol and clobazam: Preclinical evidence for both pharmacodynamic and pharmacokinetic interactions. *Epilepsia* **2019**, *60*, 2224–2234. [CrossRef]
143. Aso, E.; Andres-Benito, P.; Ferrer, I. Delineating the Efficacy of a Cannabis-Based Medicine at Advanced Stages of Dementia in a Murine Model. *J. Alzheimers Dis.* **2016**, *54*, 903–912. [CrossRef] [PubMed]

Sample Availability: Not available.

Publisher's Note: MDPI stays neutral with regard to jurisdictional claims in published maps and institutional affiliations.

© 2020 by the authors. Licensee MDPI, Basel, Switzerland. This article is an open access article distributed under the terms and conditions of the Creative Commons Attribution (CC BY) license (http://creativecommons.org/licenses/by/4.0/).

Review

Identification of Key Functional Motifs of Native Amelogenin Protein for Dental Enamel Remineralisation

Shama S. M. Dissanayake [1], Manikandan Ekambaram [2], Kai Chun Li [2], Paul W. R. Harris [1,3,4,*] and Margaret A. Brimble [1,3,4,*]

1. School of Chemical Sciences, 23 Symonds St, The University of Auckland, Auckland 1142, New Zealand; sdis008@aucklanduni.ac.nz
2. Paediatric Dentistry, Biomaterials, Faculty of Dentistry, The University of Otago, Dunedin 9016, New Zealand; mani.ekambaram@otago.ac.nz (M.E.); kc.li@otago.ac.nz (K.C.L.)
3. School of Biological Sciences, 3b Symonds St, The University of Auckland, Auckland 1142, New Zealand
4. Maurice Wilkins Centre for Molecular Biodiscovery, 3b Symonds St, The University of Auckland, Auckland 1142, New Zealand
* Correspondence: paul.harris@auckland.ac.nz (P.W.R.H.); m.brimble@auckland.ac.nz (M.A.B.); Tel.: +64-9-373-7599 (P.W.R.H. & M.A.B.); Fax: +64-9-373-7422 (P.W.R.H. & M.A.B.)

Academic Editors: Francesca Cardona, Camilla Parmeggiani and Camilla Matassini
Received: 24 August 2020; Accepted: 11 September 2020; Published: 14 September 2020

Abstract: Dental caries or tooth decay is a preventable and multifactorial disease that affects billions of people globally and is a particular concern in younger populations. This decay arises from acid demineralisation of tooth enamel resulting in mineral loss from the subsurface. The remineralisation of early enamel carious lesions could prevent the cavitation of teeth. The enamel protein amelogenin constitutes 90% of the total enamel matrix protein in teeth and plays a key role in the biomineralisation of tooth enamel. The physiological importance of amelogenin has led to the investigation of the possible development of amelogenin-derived biomimetics against dental caries. We herein review the literature on amelogenin, its primary and secondary structure, comparison to related species, and its' in vivo processing to bioactive peptide fragments. The key structural motifs of amelogenin that enable enamel remineralisation are discussed. The presence of several motifs in the amelogenin structure (such as polyproline, N- and C-terminal domains and C-terminal orientation) were shown to play a critical role in the formation of particle shape during remineralization. Understanding the function/structure relationships of amelogenin can aid in the rational design of synthetic polypeptides for biomineralisation, halting enamel loss and leading to improved therapies for tooth decay.

Keywords: dental caries; enamel remineralisation; hydroxyapatite; amelogenin; amelogenin-derived peptides; leucine-rich amelogenin peptides; tyrosine-rich amelogenin peptides

1. Introduction

Dental caries has a high prevalence of affecting permanent teeth of an estimated 2.3 billion people and primary teeth of more than 530 million children worldwide [1]. It is the result of manifestation of dental plaque on the tooth enamel. Plaque is a form of biofilm created by bacteria such as *Streptococcus mutans*, *Streptococcus sobrinus* [2] and *Lactobacilli* [3]. These cariogenic bacteria produce acid (formic-, acetic- and propionic acid) as a by-product during the metabolic process of fermentable carbohydrates, which leads to incipient caries or white spot lesions on tooth enamel [4], resulting in damaging the enamel, leading to formation of nanoscale pores [5].

Tooth enamel is highly mineralised and recognised as an acellular tooth tissue and is not capable of self-regeneration when damaged by injury or decay [6]. Ameloblasts secrete a group of unique enamel matrix proteins (EMPs) that consist of 90% of a protein called amelogenin, and 10% non-amelogenin proteins known as enamelins, ameloblastin and amelotin [7,8]. EMPs regulate the intricate matrix-like formation of hydroxyapatite (HA) crystals contributing to 70–80% of enamel weight within the human tooth. HA is a calcium phosphate mineral with the following chemical composition:

$Ca_{10}(PO_4)_6(OH)_2$ and a P/Ca molar ratio of 1:1.67 [9]. Each single cell unit is observed to be long, hexagonal, nanorod-like structures consisting of repeating units of morphological dimensions of ~60 nm in length, 20 nm in diameter and 2–5 nm thick. The collation of these single cell units forms bundles of prisms or rods that are 60–100 nm long [10]. Each rod is surrounded by a sheath formed from the non-amelogenin enamelin proteins, giving its characteristic matrix appearance (Figure 1).

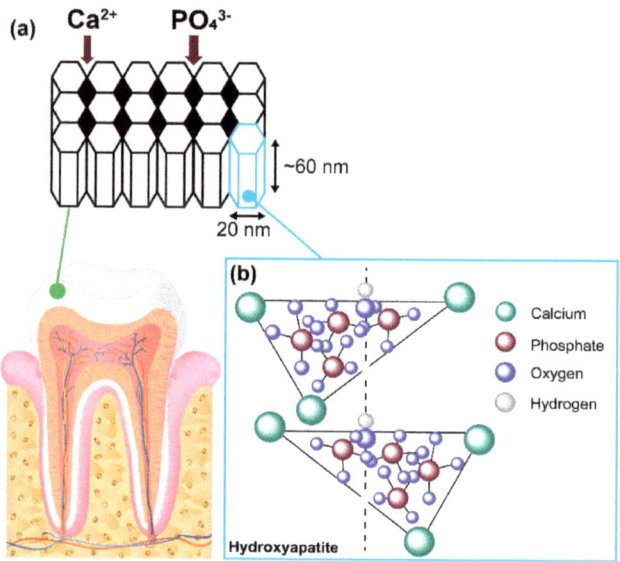

Figure 1. Schematic representation of enamel microstructure. (**a**) Arrangement of the hydroxyapatite crystals, the cavities shown by the black diamond shape, allows Ca^{2+} and PO_4^{3-} ions from saliva to pass through and assist in remineralization. (**b**) Molecular arrangement of individual unit cells making up the hydroxyapatite.

In between these rods, nanoscale pores are formed that enable the permeability of extracellular fluid containing important chemicals such as hydroxyl ions, chloride, carbonate and fluoride ions. Proteins such as statherin in saliva hold calcium and phosphate ions in a supersaturated form, making them readily available in a constant and sufficient supply to help the regeneration of HA [11]. Saliva also contains acidic and basic proline-rich proteins (PRPs), which provides a protective rind by attaching itself onto oral bacteria. The strength and rigidity of HA is mainly due to intricately woven crystals consisting of calcium, phosphate and oxygen.

1.1. Process of Microbial Attack on Dental Enamel

During the microbial invasion of the enamel, the nanoscale pores allow bacteria and acidic fluid to freely circulate through the channels within the matrix. Once the acid has reached a susceptible site, calcium and phosphate ions are dissolved into the surrounding extracellular environment from the hydroxyapatite (HA) crystal matrix, leading to the loss of calcium and phosphate ions, thus hindering the natural mineralisation process of enamel. This phenomenon is known as demineralisation, where the HA crystalline structure will decrease in size while the pores enlarge [12]. This combination of food fermentation and attack by cariogenic bacteria initiates the early manifestation of dental caries by beginning to demineralise the dental enamel [13].

1.2. Potential Approaches for Dental Remineralisation

Due to high prevalence of dental caries amongst any given population globally, the development of a potential therapeutic for dental caries is essential. Therefore, the use of organic scaffolds [14], dendrimers [15], chitosan [16], charged amino acids [17] and bioactive glasses [18] have been explored for remineralisation properties. Unfortunately, these approaches are yet to yield any commercial products. However, biomimetic in vitro strategies employing proteins which contribute to the EMP layer such as native amelogenin and leucine-rich amelogenin peptide (LRAP) have demonstrated the ability to bind hydroxyapatite crystals, which is crucial for remineralisation. The use of LRAP in vitro has demonstrated the de novo formation of HA [19]. This is of significant importance to researchers and dental clinicians to develop biomimetic peptides as potential treatments to initiate remineralisation. The critical role amelogenin protein plays in controlling enamel remineralisation has become evident over the last four decades [20].

2. Determination of Primary and Secondary Amelogenin Protein Structures amongst Different Species

Amelogenin protein first gained recognition in 1980 for its vital role in dental mineralisation during enamel matrix development [21]. These proteins self-assemble into ~17–18 nm globular supramolecular aggregates (nanospheres), microribbons and nanochains during enamel formation [22]. This self-assembling capability of amelogenin is known to be a key driving force in guiding the formation of HA crystals [23]. Therefore, dental clinicians and researchers have taken an interest in the ability of these proteins to effect enamel remineralisation. In the past four decades, several studies have been conducted to characterise the primary and secondary structure of the native amelogenin protein amongst various species to determine its active motifs [24]. Studies carried out to deduce the primary native amelogenin proteins demonstrated that amelogenins vary in sequence length (150–180 amino acids) depending on the origin of different species, human [25], bovine [26], porcine [27] and murine [28] (Figure 2). The primary structure of amelogenin protein is shown to be highly conserved (>80%) amongst different species consisting of three main domains: An N-terminal domain, a mid-section and the C-terminal domains [29]. Determination of the secondary structures would assist in understanding how these individual amelogenin proteins contribute to enamel crystal growth [30]. Techniques such as circular dichroism (CD), nuclear magnetic resonance (NMR) spectroscopy, isothermal titration calorimetry (ITC), selected area electron diffraction (SAED) and Fourier transform-infrared (FT-IR) have been used to fully characterise primary and secondary structure of native amelogenin proteins.

```
Human    → MGTWILFACLLGAAFAMPLPPHPGHPGYINFSYEVLTPLKWYQS-I   ··45¶
Porcine  → ---------------MPLPPHPGHPGYINFSYEVLTPLKWYQNMI    ··30¶
Murine   → MGTLILFACLLGAAFAMPLPPHPGSPGYINLSYEVLTPLKWYQSMI   ··46¶
Bovine   → MGTWILFACLLGAAFSMPLPPHPGHPGYINFSYEVLTPLKWYQSMI   ··46¶
¶
Human    ····RPPYPSYGYEPMGGWLHHQIIPVLSQQHPPTHTLQPHHHIPVVPAQ   ···91¶
Porcine  ····RHPYTSYGYEPMGGWLHHQIIPVVSQQTPQSHALQPHHHIPMVPAQ   ···76¶
Murine   ····RQPYPSYGYEPMGGWLHHQIIPVLSQQHPPSHTLQPHHHLPVVPAQ   ···92¶
Bovine   → RHPYPSYGYEPMGGWLHHQIIPVVSQQTPQNHALQPHHHIPMVPAQ     ···92¶
¶
Human    ····QPVIPQQPMMPVPGQHSMTPIQHHQPNLPPPA----QQPYQPQPVQ   ··133¶
Porcine  ····QPGIPQQPMMPLPGQHSMTPTQHHQPNLPLPA----QQPFQPQPVQ   ··118¶
Murine   ····QPVAPQQPMMPVPGHHSMTPTQHHQPNIPPSAQQPFQQPFQPQAIP   ··138¶
Bovine   ····QPVVPQQPMMPVPGQHSMTPTQHHQPNLPLPA----QQPPQPQSIQ   ··134¶
¶
Human    ····PQPH--------------------QPMQPQPPVHPMQPLPPQPPL   ··157¶
Porcine  ····PQPH--------------------QPLQPQSPMHPIQPLLPQPPL   ··143¶
Murine   ····PQSH--------------------QPMQPQSPLHPMQPLAPQPPL   ··163¶
Bovine   ····PQPHQPLQPHQPLQPMQPMQPLQPLQPLQPQPPVHPIQPLPPQPPL   ··180¶
¶
Human    ····PPMFPMQPLPPMLPDLTLEAWPSTDKTKREEVD →           ··180¶
Porcine  ····PPMFSMQSL---LPDLPLEAWPATDKTKREEVD →           ··173¶
Murine   ····PPLFSMQPLSPILPELPLEAWPATDKTKREEVD →           ··196¶
Bovine   ····PPIFPMQPLPPMLPDLPLEAWPATDKTKREEVD →           ··213¶
```

Figure 2. Alignment of native amelogenin protein from different species. The blue boxed areas indicate the N-terminal domain, the green boxed area indicates the C-terminal domain, underlined residues indicate the highly charged C-terminal tail, and red indicates the amino acid residues, which are non-homologous in comparison to the native human amelogenin protein. The (-) dash indicates the alignment gap.

2.1. Native Bovine Amelogenin Structure

The primary structure of native bovine amelogenin protein was reported by Takagi et al. [31] in 1984, and is seen to be the largest of the native amelogenins, with a molecular weight of 32 kDa consisting of 213 amino acid residues. This report led to the study of the secondary structure by

Renugopalakrishnan et al. [32] with the use of CD, FT-IR and 2D NMR [33] and the use of Raman spectroscopy by Zheng et al. [34].

The N-terminal domain is reported as constituting of β-turns at residues His^6-Tyr^{12} and Thr^{21}-Lys^{24}, Gln^{40}–Gly^{43}, Ile^{51}–Val^{54}, Thr^{58}–Asp^{61}, Ile^{70}–Val^{73}, Gln^{77}–Gln^{83}, Val^{88}–Glu^{91} and Gln^{93}–Leu^{96} and polyproline repeat units from Gln^{112} to His^{139} [32].

The presence of certain band values of frequencies between amide I depicting 1645–159 cm^{-1} and amide II depicting 1262–1300 cm^{-1} frequencies in FT-IR indicates an α-helix within the segment Gly^{44}-Ile^{50}, as deduced by Renugopalakrishnan et al. [32]. Having only seven amino acid residue segments contributing towards the α-helical structure is classed as a minimal contribution to native bovine amelogenin secondary structure.

The mid-section from Pro^{66} to Pro^{160} consists of a very high proportion of proline residues and is reported as forming the poly-L-proline type II (PPII) helical region. A series of β-sheet subdomains coexisting within the β-turns at the C-terminal end have also been described (Figure 3a) [32].

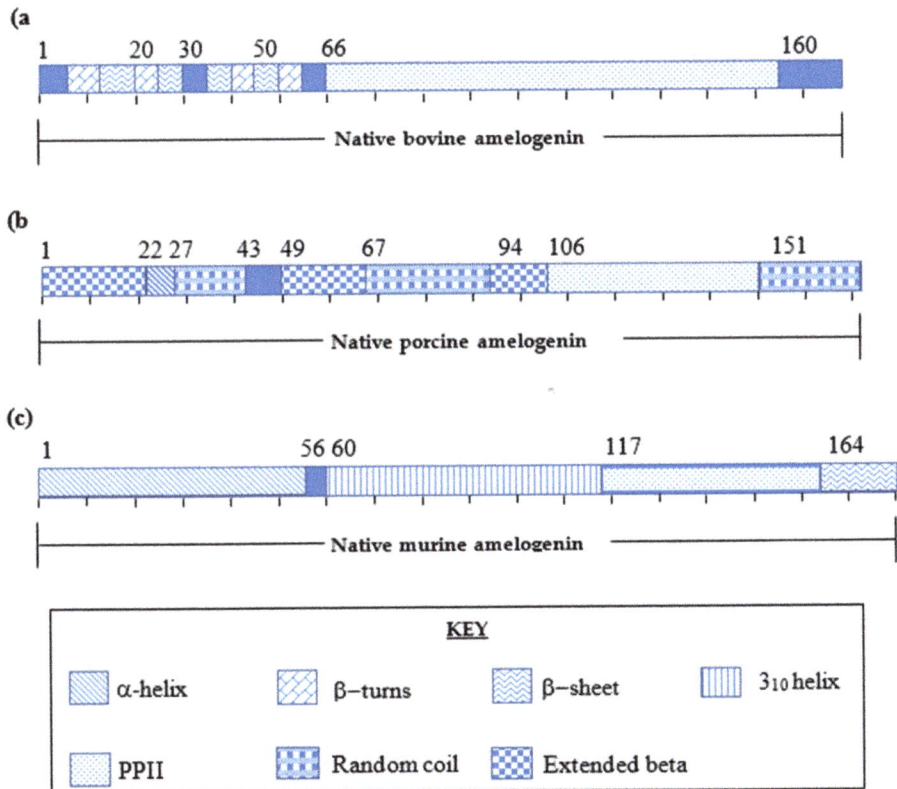

Figure 3. (a) Secondary structure of native bovine amelogenin determined by CD, FT-IR and 2D-NMR. (b) Secondary structure of native porcine amelogenin determined by VT-CD. (c) Secondary structure of native murine amelogenin determined by 3D-NMR.

2.2. Native Porcine Amelogenin Structure

The study of the primary and secondary structure of native porcine amelogenin was defined by Lakshminarayanan et al. [35] using FT-IR and CD. The native porcine amelogenin was demonstrated to be 173 amino acid residues with a molecular weight of 27kDa. [36] Similarities between both porcine and bovine native amelogenins are indicated by the presence of β-sheets within the N-terminal domain, high proline content and a prominent PPII domain within the midsection [37,38]. Native porcine amelogenin composition is observed to be highly similar to the bovine native amelogenin. The primary structure of porcine is slightly varied by the absence of Met^1–Ser^{16}, as seen in bovine, and a midsection polyproline segment from Gln^{139}–Leu^{159} is absent in comparison to bovine native sequence (Figure 2). In addition to secondary structure studies, the self-assembling capabilities were identified as being driven by hydrophobic interactions composed of β-sheets utilising variable temperature circular dichromism (VT-CD) [35] (Figure 3b).

2.3. Native Murine Amelogenin Protein

The secondary structure of the 196 amino acid residue native murine protein was determined using 3D-NMR techniques by Zhang et al. [30]. For NMR analysis, the native protein was split into three shorter peptide fragments and synthesised as three separate peptides (N-terminus; Amel-N AA 1–92, the mid-section; Amel-M AA 34-154 and the C-terminus; Amel-C AA 86-193) to facilitate characterisation. These synthetic fragments were mapped against the native murine amelogenin, which revealed four major functional domains contributing to the overall 3-D structure of the protein (Figure 3c) [30].

The N-terminal domain consists of four α-helical motifs (Ser^9-Val^{19}, Thr^{21}-Pro^{33}, Tyr^{39}-Trp^{45}, and Val^{53}-Gln^{56}). The mid-section is composed of both an elongated random coil motif with two 3$_{10}$ helices (Pro^{60}-Gln^{117}) and a PPII helical region (Pro^{118}-Leu^{165}). The PPII helices are known to exhibit a well-defined conformation, yet they are flexible, proving advantageous in mineral binding domains [30]. The hydrophobicity of amelogenin is attributed to a high proportion (50–60% of the protein sequence) of proline residues in the mid-section [7,35]. The C-terminal domain (Leu^{165}-Asn^{193}) contains a charged hydrophilic motif comprised of a 13 residue long peptide sequence rich in glutamic acid, arginine and lysine. The overall structure of the C-terminal domain is recognised to give rise to β-sheets and β-turns [30,39] (Figure 3c).

From these varied techniques used to determine both primary and secondary structures of native amelogenin amongst different species, a range of secondary structures motifs were reported, α-helical, polyproline, 3$_{10}$ helices, β-sheets, and β-spirals [30,33,38,40]. Of particular note is the presence of a highly conserved domain such as the PPII motif in the midsection amongst different species [41]. It is evident that the high proline content in the midsection contributes to these well-defined PPII helical conformations [40].

2.4. Contribution of Polyproline Motifs of Amelogenin towards Mineralisation

Intramolecular interactions of native amelogenin were determined using a combination of CD and NMR spectroscopic techniques. It is reported that the mobility and flexibility of the PPII helices in the midsection enable the interaction between α-helices and β-sheets commonly found within the N-terminus [30]. This flexibility enables the formation of elongated fibril like structures, allowing interactions with biological surfaces such as the HA layer. A recent study carried out by Jin et al. [40] showed the important contribution of the polyproline-rich motifs to matrix assembly and their vital role in biological mineralisation over a diverse range of different phyla [42]. Further studies by Lakshminarayanan et al. [20] using isothermal titration calorimetry (ITC) studies have reported that PPII motifs adopt a β-sheet secondary structure to facilitate self-assembly. This transformation is driven by the entropy gained via the hydrophobic amino acid residues, known as the hydrophobic effect which enables a conformation that facilitates mineral binding, resulting in HA crystal growth [43].

To assess these proposed effects of HA crystal growth a series of varying length polyproline repeats based on human amelognein were designed and synthesised by Jin et al. [40]. A polyproline motif unit from the latter part of the mid-section was selected (Pro^{12}–Pro^{153}) (Table 1). A series of proline repeat motifs—PXX12, PXX24 and PXX33, comprising 12, 24 and 33 amino acid residues, respectively—were prepared whereby proline is represented by P and X is substitutes for any amino acid residue [40]. Analysis of these peptides via atomic force microscopy (AFM) and dynamic light scattering (DLS) methods revealed that increasing the length of the polyproline motifs yielded longer

HA crystals. The 12 residue motif PXX12 exhibited shorter HA crystals that were 21.6 ± 6.5 nm in length, whereas the PXX24 exhibited 42.9 ± 8.5 nm long crystalline mineralisation and PXX33 exhibited crystal lengths of 102.1 ± 36.3 nm. It is noteworthy that native full-length murine amelogenin protein shows HA crystal formation with crystals of 102.1 ± 19.3 nm in length, similar in size to PXX33 [40].

Table 1. Series of polyproline repeating motifs derived from the tail end of the mid-section (P^{120} to L^{152} represented in blue) of native human amelogenin and substitution of glutamine for alanine in PX33 to give PQA peptide. The latter most peptide resulted in unsuitable crystals for remineralisation.

Midsection of Native Human Amelogenin	YPSYGYEPMGGWLHHQIIPVLSQQHPPTHTLQPHHHIPVVPAQQPVIPQ QPMMPVPGQHSMTPIQHHQPNLPPPAQQPYQPQPVQPQPHQP^{120}MQPQ PPVHPMQ$^{131}P^{132}$LPPQPPLPPMF$^{143}P^{144}$MQPLPPML^{152}
PXX12	P^{120}MQPQPPVHPMQ^{131}
PXX24	P^{120}MQPQPPVHPMQ$^{131}P^{132}$LPPQPPLPPMF^{143}
PXX33	P^{120}MQPQPPVHPMQ$^{131}P^{132}$LPPQPPLPPMF$^{143}P^{144}$MQPLPPML^{152}
PQA	PM<u>A</u>P<u>A</u>PPVHPM<u>A</u>PLPP<u>A</u>PPLPPMFPM<u>A</u>PLPPML

The glutamine residues are the second most prevalent amino acid in the PPII helix. Glutamine is recognised as the only other amino acid residue apart from proline to have favourable Chou–Fasman conformation parameters [44]. The Chou–Fasman method is used to predict elements of protein structure by assigning a certain value to each amino acid and then applying an algorithm to the assigned values [45]. From the Chou–Fasman conformational parameters, both glutamine and proline are observed to have higher values assigned for a PPII structure than for any other secondary structure. Glutamine is therefore the preferred amino acid residue together with proline for the preferential formation of PPII helices. To validate this hypothesis, the glutamine residues of PXX33 were substituted with alanine, and the resultant peptide coined PQA (Table 1). Comparison of HA crystal formation by PQA peptide or PXX33 peptide revealed far larger nanosphere structures with PQA peptide. These larger nanosphere structures formed by PQA peptide was seen to give rise to flake-like particles that are 18.6 +/− 3.2 nm in diameter. HA crystals can be formed with flake-like, sphere, needle-like or rod-shaped crystals, highly dependent on the particle size. For a successful formation of HA crystal layer, a certain density of 550 crystallites/μm^2 is crucial [46]. This is only accomplished by the rod-shaped crystals with a specific particle size (60–100 nm in length) [46]. Thus, a larger particle size, as formed by the PQA peptide, gives rise to flake-like crystals, which are unable to form crystalline mineral and, therefore, do not contribute to HA crystal formation. Thus, this confirms that the presence of Gln in the PX33 sequence is required to ensure correct HA crystal growth.

3. Splice Variants of Native Amelogenin from Different Species

Proteolytic degradation of the native amelogenin protein at the enamel maturation stage has been shown to involve two key proteases, matrix metalloproteinase (MMP-20) and kallikrein-related peptidase 4 (KLK-4). These two proteases are responsible for the complete degradation of native amelogenin, to yield two distinctive lower molecular weight amelogenin peptides of 5–6 kDa. Due to the amino acid composition of these peptides, they are designated as tyrosine-rich amelogenin peptides (TRAP) and leucine-rich amelogenin peptides (LRAP) [47].

Comparisons of amino acid sequence data for TRAP and LRAP peptides are reported as being conservative amongst bovine, human and porcine for the first 27 amino acid residues (N-terminus), and enrichment of tyrosine and leucine are seen only in the C-terminal regions of these 45 to 46 amino acid residue peptides [48,49]. TRAP peptides are formed by proteolytic cleavage and LRAP formed by proteolytic splicing of the native amelogenin by MMP-20 and KLK-4 (Figure 4a) [50].

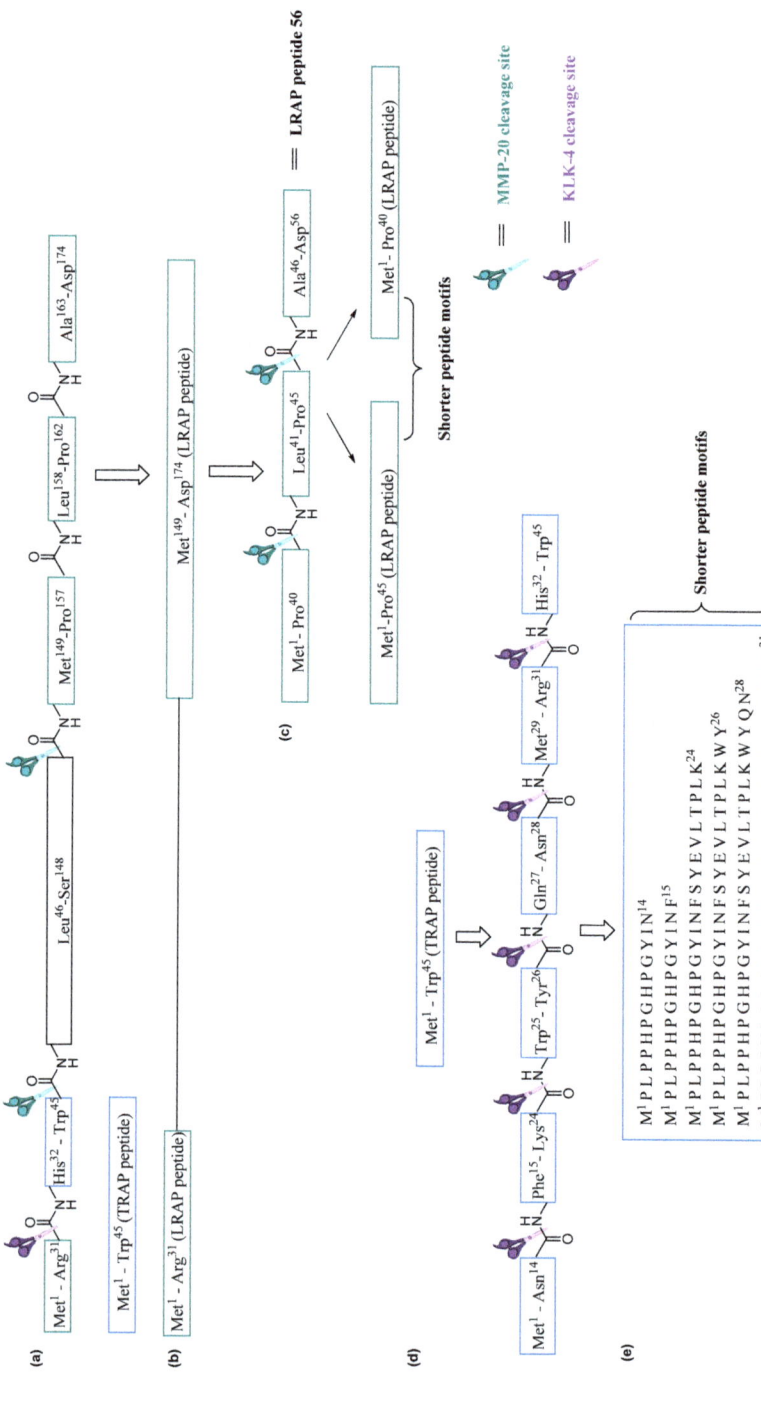

Figure 4. (**a**) Native porcine amelogenin protein indicating its major cleavage sites by KLK-4 and MMP-20 protease generating TRAP and LRAP peptide (**b**) cleaved N-terminal segment and C-terminal segment giving rise to LRAP (**c**) LRAP peptide is further cleaved by MMP-20 to generate shorter motifs of LRAP of 45 and 40 amino acid residues (**d**) TRAP peptide is further cleaved by KLK-4 protease to generate shorter motifs of TRAP peptides (**e**) Amino acid sequence of KLK-4 generated TRAP peptides.

The splicing of LRAP is directed by amelogenin promoter for cDNA encoding for LRAP and is beyond the scope of this review [50]. These spliced and degraded TRAP and LRAP peptides undergo further proteolytic cleavage by MMP-20 and KLK-4 to produce shorter fragments [51,52]. The TRAP peptide within the N-terminus of up to 44 to 45 amino acid residues and the LRAP segment span both the N- and C-terminal domains of up to 56 to 59 amino acid residues in length [53].

Yamakoshi et al. [36] has reported that the native 27 kDa porcine amelogenin undergoes splicing to form a major 25 kDa amelogenin that is secreted into the enamel consisting of 173 amino acid residues. The 25 kDa amelogenin is seen to be lacking the region from Lys^{19} to Gln^{35} found in 27 kDa amelogenin [36]. Further studies on these fragments led to the recognition of TRAP polypeptide embedded within the N-terminal domain of native amelogenin proteins [37].

Nagano et al. [54] reported the liquid chromatography mass spectrometry LC-MS analysis of MMP-20 cleavage of native porcine amelogenin protein at site Trp^{45}–Leu^{46} and at Ser^{148}–Met^{149} to generate two TRAP peptides, Met^1–Trp^{45} and Leu^{46}–Ser^{148}.[47] The Met^1–Trp^{45} (TRAP) peptide motif is further cleaved by KLK-4 at Asn^{14}, Phe^{15}, Lys^{24}, Tyr^{26}, Asn^{28}, and Arg^{31} to generate smaller fragments (Figure 4d,e). MMP-20 is also seen as the primary protease responsible for generating LRAP by cleaving native porcine amelogenin N-terminal amino acid residues (Met^1–Arg^{31}) and at C-terminal amino acid residues (Met^{149}–Asp^{174}) facilitating the splicing resulting in the formation of LRAP peptides [53]. Further cleavage at Pro^{157} or Pro^{162} generates shorter LRAP peptides (Figure 4b,c) [55].

3.1. Development of Shorter Peptides from Lrap Motif

As described in Figure 4, cleavage sites generating TRAP and LRAP are seen to be conserved amongst different species. The use of shorter motifs has demonstrated the ability to initiate remineralization; therefore, murine LRAP peptides have been further investigated by Mukherjee et al. [56] to design shorter amino-acid-containing peptides of 26 and 32 residues yielding peptide P26 and P32, respectively (Table 2). The primary aims of these shortened murine amelogenin-derived peptides were to study the potential capabilities of HA binding and remineralisation initiating capabilities.

The design of P26 was driven by retaining the last 12 amino acid residues of the C-terminus and 14 amino acid residues from the N-terminus (residues Met^1–Pro^4, and Ser^{16} and Asn^{25}), including the phosphorylation site at Ser^{16}. P32 was designed in a similar manner to P26, with the addition of two polyproline repeat motifs from the midsection. To determine the initiation of HA crystal growth peptide, P26 and P32 were incubated separately in artificial saliva for 2 days. It was observed that P26 contributed in rapid crystal overgrowth (ca. ≤ 100 nm width) and was characterised as bundles of needle-like crystals, perpendicular to the enamel surface. In comparison, P32 also initiated the growth of crystals (ca. ≤ 100nm width) and was seen to be parallel to the enamel surface. The extent of HA crystal formation and size induced by P26 and P32 was compared to the full-length native murine amelogenin, which showed crystal formation of ca. ≤ 100 nm in width similar to the crystals induced by both P26 and P32 peptide. However, it was P32, which showed preferential crystal growth on the same axis as the formation of the native enamel crystals [56]. Therefore from this study, it is noteworthy to highlight the importance of P32 as a shorter peptide used for enamel remineralisation.

3.2. The Importance of N-Terminal Domains of Amelogenin

It was determined that the N-terminal domain embedded within the TRAP segment provides the only available phosphorylation sites (serine or threonine) within the peptide. Phosphate and calcium ions are known to aid in enamel mineralisation, therefore exploiting any functionality within the sequence that enhances phosphorylation is of interest. It is postulated that incorporation of a phosphorylation site on the peptide sequence would facilitate calcium ion binding.

Le Norcy et al. [57] studied porcine-derived amelogenin with phosphorylation (+P) and non-phosphorylation (−P) of serine at position 16 of the TRAP segment (Figure 5) [58]. The aim was to induce the formation of HA crystals by mixing the two peptide analogues in separate solutions mimicking physiological concentrations found in the tooth enamel. Peptides (−P) (Figure 5a) and (+P) (Figure 5b) were used at concentrations of 2 mg/mL in a 2.5 mM calcium and 1.5 mM phosphate containing solution at pH 7.4 and 37 °C [57]. HA crystal formation was identified by SAED, FT-IR and transmission electron microscopy TEM. It was anticipated that the peptides would act as HA crystal forming directors with the aid of the calcium and phosphate ions. A decrease in calcium and phosphate in the solution due to calcium phosphate precipitation from HA crystal formation results in a pH change, and this can be used to determine the effectiveness of each peptide in HA crystal formation.

(a) (−P)

MPLPPHPGHPGYINFSer^{16}YEVLTPLKWYQNMIRHPSLLPDLP

(b) (+P)

MPLPPHPGHPGYINFSerPYEVLTPLKWYQNMIRHPSLLPDLP

Figure 5. (a) Non-phosphorylated sequence at Ser^{16} of porcine N-terminus (blue) in the presence of partial C-terminal motif (purple). (b) Phosphorylated sequence at Ser^{16} of porcine N-terminus (blue) in the presence of partial C-terminal motif (purple).

The pH of a control sample in the absence of peptide was recorded as being ~pH 7.2. In the presence of N-terminal non-phosphorylated peptide (−P), a significant decrease in pH was observed. Mineral formation from (−P) was analysed by FT-IR, indicating that the control sample and (−P) induced similar crystal particle size. In comparison, phosphorylated peptide has a slight pH decrease, and spherical nanoparticles of calcium phosphate were observed with smaller diameters (29.1 ± 6.8 nm) when compared to calcium phosphate particles observed in the control sample (84 ± 5.2 nm).

From this study, it was deduced that the N-terminal domain alone is capable of interacting with calcium phosphate minerals to produce crystalline structures. The diameter of the different particles produced indicates the extent of the crystal formation is far more apparent with the phosphorylated peptide. The particle sizes formed by the phosphorylated peptide (~29.1 nm) are similar to particle sizes formed during natural enamel mineralisation (~20 nm) [10,58]. Therefore, the importance of the N-terminal motif contributing towards HA crystal formation is highlighted. The phosphorylation status of the N-terminus was noted for its capability to contribute to assisting mineralisation. Recognition of these domains is also validated by in vitro phosphorylation studies carried out by Nagano et al. [54].

3.3. Importance of the C-Terminal Domain on Formation of Particle Shape

In addition to the N-terminal domain, the presence of the C-terminal domain is recognised as having remineralisation capabilities. Some literature reports little to no influence by the native amelogenin on HA crystal nucleation and growth [59,60], whereas some report that in the absence of the C-terminal domain in the native amelogenin a significant reduction in enamel remuneration is observed [61,62]. Therefore, over the past four decades various in vitro experiments such as mineralisation experiments, HA crystal growth experiments with native amelogenin and apatite binding experiments of recombinant and native amelogenins were carried out by Beniash et al. [7], Aoba et al. [61,63], Moradian-Oldak et al. [64], respectively, to study the functionality of the C-terminal domain.

Two separate studies by Kwak et al. [58] and Wiedemann-Bidlack et al. [65] further validates the importance of the 13 amino acid residue, highly charged C-terminal tail (WPATDKTKREEVD) [22]. These studies involved the comparison of C-terminal containing recombinant amelogenin from mouse (rM179) (Figure 6a) and porcine (rP172) (Figure 6b) with C-terminal absent recombinant amelogenin mouse (rM166) (Figure 6c) and porcine (rP148) (Figure 6d). These four peptides were incubated separately for 150 min in a solution of pH 7.2 at 37 °C, closely mimicking the conditions of the oral cavity during enamel remineralisation. TEM analysis of structures formed under these conditions indicated that the peptides with the C-terminal domain of rM179 were reported as forming elongated tightly connected assemblies that are made up of chain-like structures of 7 ± 1.3nm wide and 58 ±

27.0nm long. The rP172 is reported to form structures of similar size and shape with 7 ± 2.1 nm wide and 51 ± 26.2 nm long, which are also seen to be tightly connected [65]. Peptides lacking the C-terminal domain (rM166 and rP148) were observed as forming spheres of 9 ± 2.5 nm and 20 ± 4.4nm in diameter, respectively, which are seen to be loosely connected to one another. HA crystals formed during natural mineralisation via amelogenin is reported as forming long nanorod-like structures rather than spheres [10]. The observation that the C-terminal containing peptides enable the initiation of formation of similar shapes validates the importance of this motif and highlights the key role it plays in regulating the shape and organisation of HA crystals [7].

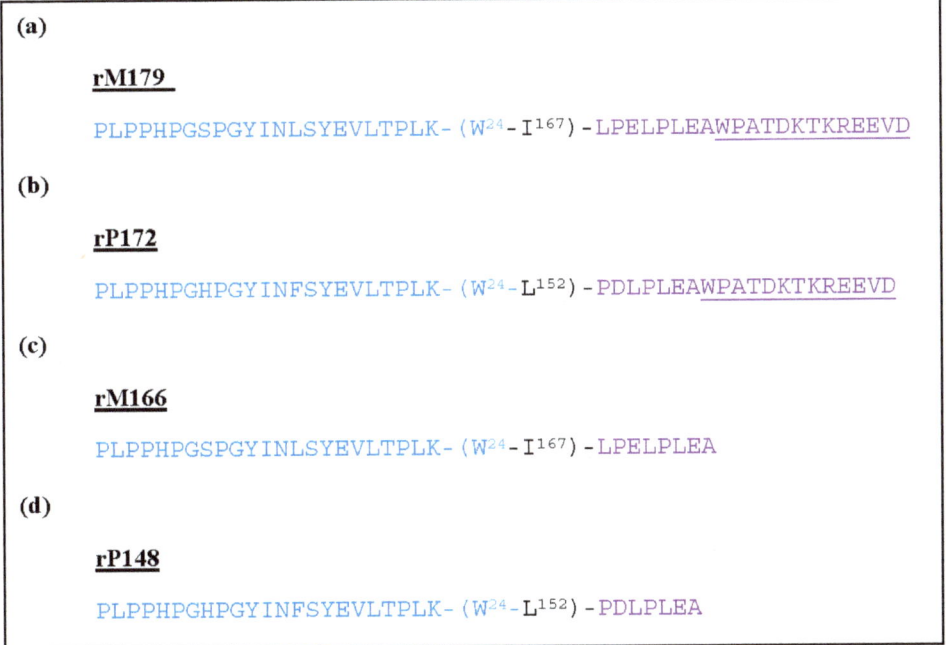

Figure 6. Comparison of N-terminal segment of mouse and porcine (blue) in the presence of the charged C-terminal motif (purple) with the C-terminal tail (purple and underlined) (**a**) is derived from mouse amelogenin containing the charged C-terminal tail and (**b**) porcine amelogenin containing the charged C-terminal tail. (**c,d**) retain the N-terminal region but without the charged portion of the C-terminal tail.

3.4. C-Terminal Orientation Studies for HA Crystal Growth

Solid state NMR studies carried out by Shaw et al. [53] established that the C-terminal domain orientates itself in a manner which enhances interactions with HA crystals by orientating itself in a favourable conformation [58]. The orientation can be attributed to the charged side chains of Lys and Asp containing a direct ionic interaction with extracellular calcium and phosphate, directing HA crystal formation [58]. To study these ionic interactions, rotational echo double resonance (REDOR) NMR was employed [66]. This technique provided site specific structural information of the ionic interactions between the peptide motif and the HA crystals. The REDOR technique involves isotopic labelling of either the side chains or backbone of the C-terminal domain to measure the distance between the labelled residues to any ^{31}P residues found on the surface of HA crystals. Images obtained by Le Norcy et al. [57] using TEM indicate small calcium phosphate nanoparticles that align and form needle-like structures in the presence of the C-terminal domain, which are favourable in forming nanorods and HA crystals [57]. In the absence of the C-terminal domain, flake-like structures were observed. Flake-like crystals do not have the ability to form the necessary nanoscale structures that aid HA crystal formation. The results from these experiments indicate that the C-terminal domain is orientated in a favourable conformation next to the HA crystal surface and it was concluded that the C-terminus plays an influential role in regulating the shape and organisation of HA crystal formation and growth during remineralization [67].

3.5. Calcium Binding Capabilities of Human LRAP (hLRAP)

The calcium-binding capabilities of human recombinant amelogenin (rH174) in comparison with the shorter LRAP version were studied by Le et al. [41] rH174 and human LRAP (hLRAP) were studied using isothermal titration calorimetry (ITC) techniques carried out in 4-(2-hydroxyethyl)-1-piperazineethanesulfonic acid (HEPES) buffer (10 mM) at pH 7.5 with a rLRAP concentration of 0.12 mM using a calcium chloride (5 mM) and a rH174 concentration of 0.064 mM with a calcium chloride concentration of 15 mM. The binding affinity of calcium to hLRAP and rH174 was determined by the number of calcium ions (N) bound to these molecules. ITC studies elucidated that hLRAP has a 6.4 higher calcium-binding affinity over the native rH174 peptide. Calcium binding is recognised as a key parameter responsible for HA crystal formation [41]. This in turn indicates the potential of employing a truncated version of full-length amelogenin as LRAP that contains both N-terminal and C-terminal motifs, can achieve the same outcome on HA crystal formation in dental mineralisation (Table 2) [68].

Table 2. Rationally designed peptides derived from native amelogenin protein to form potential mineral regeneration polypeptides.

Species	Name of Peptide	Peptide Sequence [a]	Ref
Porcine	Porcine LRAP 56 AA	MPLPPHPGHPGYINFSYEVLTPLKWYQNMIRHPSLLPDLPLEAWPATDKTKREEVD	[57,68]
	P45	MPLPEHPGHPGYINFSYEVLTPLKWYQNMIRHPSLLPDLPLEAWP	
	P40	MPLPPHPGHPGYINFSYEVLTPLKWYQNMIRHPSLLPDLP	
Murine	Murine LRAP 59AA	MPLPPHPGSPGYINLSYEVLTPLKWYQSMIRQPPLSPILPELPLEAWPATDKTKREEVD	[56]
	P32	MPLP———SYEVLTPLKWPVHPMQPS———————TDKTKREEVD	
	P26	MPLP————SYEVLTPLKWPS————————TDKTKREEVD	
Bovine	Bovine LRAP 59 AA	MPLPPHPGHPGYINFSYEVLTPLKWYQSMIRHPPLPPMLPDLPLEA WPATDKTKREEVD	[47]
	Bovine TRAP	MPLPPHPGHPGYINFSYEVLTPLKWYQSMIRHPYSPYGYEPMGGT	
Human	hLRAP 58 AA	MPLPPHPGHPGYINFSYEVLTPLKWYQS-IRPPPLPPMLPDLTLEAWPSTDKTKREEVD	[69]

[a] Blue represents the N-terminal domain, purple represents the C-terminal domain.

4. Successful Peptide Based Therapeutics

P11-4 is an N-terminally acetylated, C-terminal amidated 11 amino acid residue (Ac-QQRFEWEFEQQ-NH$_2$) peptide It was one of five peptides (P11-1, P11-2, P11-3, P11-4 and P11-5) initially designed and synthesised as a self-assembling peptide for bone regeneration in 1997 by Aggila et al. [70] To facilitate self-assembling capabilities, the peptide sequence is arranged with alternating hydrophobic and hydrophilic amino acid residues as seen in P11-4 [71]. Studies carried out by Aggila et al. [70] established that P11-4 forms a β-sheet, which is commonly seen among self-assembling peptides. The 3D matrix formed by self-assembling P11-4 peptide is known to have high affinity towards calcium ions which acts as a nucleator facilitating de novo HA crystal formation 14]. This capability gained recognition and was applied to the regeneration of demineralised enamel. P11-4 has since then been developed into a novel peptide based therapeutic for dental caries by inducing mineralisation. It was patented and marketed as Curodont^{TM} Repair and is now widely available for the treatment of dental caries [72].

5. Conclusions

Structural studies carried out over the past four decades on amelogenin proteins have elucidated the importance of the N- and C-terminal domains of amelogenin on HA crystal formation and their essential role in successful in vitro enamel remineralisation to date. Phosphate and calcium ions are known to aid in enamel remineralization; therefore, exploiting the N-terminal domain by also incorporating a unique site capable of calcium binding is of great interest. In addition, the presence of the highly charged C-terminal domain is also postulated to assist in remineralisation due to its ability to adopt a favourable conformation to facilitate the initiation of remineralisation.

An area of the amelogenin midsection consisting of polyproline containing motifs has been shown to attribute to HA crystal formation while the random coil portion also present in the midsection does not greatly contribute to self-assembling properties, nor contribute in a significant manner towards HA crystal formation.

The exemplary example, Curodont^{TM} Repair, is a self-assembling peptide (P11-4), currently used by dental clinicians to induce enamel remineralisation. The physicochemical properties of the amino acid residues constituting P11-4 have shown remarkable similarities to amelogenin-derived peptides reported in the literature. Therefore, in designing future peptide-based therapeutics for enamel remineralization, the following key characteristics should be considered:

1. **Self-assembling capabilities**—To ensure that a 3D-matrix-forming scaffold is provided that will facilitate HA crystal formation and maintain similarity to natural enamel formation by the native amelogenin protein.

2. **Calcium-binding motif**—The phosphorylation of serine at position 16 in the N-terminus can assist in increasing the ability to effectively bind calcium ions, crucial to the remineralisation process.

3. **C-terminal domain motif**—Regarded as an important domain to facilitate HA formation, as the loss of the C-terminal tail led to no HA crystal formation. This highly charged tail region orientates itself in a favourable manner, allowing for ionic interactions with the HA crystal surface and thereby facilitating HA crystal formation.

4. **Polyproline motifs**—To enable HA crystal formation, eleven or more polyproline repeats are required, as having less proline repeat units has been shown to hinder HA crystal formation.

5. **Glutamine residues**—Crucial to the composition of the polyproline motifs, as they preferentially interact with proline residues of the polyproline motifs forming PPII helixes.

This brief review of the development of effective peptides that promote HA regrowth is still in its infancy. However, the key structural motifs of amelogenin protein highlighted in this review could aid in developing shorter amelogenin-derived peptides for use in enamel remineralisation.

Author Contributions: M.E., P.W.R.H. and M.A.B conceptualized the review, S.S.M.D. and K.C.L. wrote the original draft, M.E., K.C.L., P.W.R.H., M.A.B. and S.S.M.D. all contributed to the editing, proofreading and final version. All authors have read and agreed to the published version of the manuscript.

Funding: This research received no external funding.

Conflicts of Interest: The authors declare no conflict of interest.

References

1. Disease, G.B.D.; Injury, I.; Prevalence, C. Global, regional, and national incidence, prevalence, and years lived with disability for 354 diseases and injuries for 195 countries and territories, 1990–2017: A systematic analysis for the Global Burden of Disease Study 2017. *Lancet* **2018**, *392*, 1789–1858.
2. Badet, C.; Thebaud, N.B. Ecology of lactobacilli in the oral cavity: A review of literature. *Open Microbiol. J.* **2008**, *2*, 38–48. [CrossRef] [PubMed]
3. Caufield, P.W.; Schon, C.N.; Saraithong, P.; Li, Y.; Argimon, S. Oral lactobacilli and dental caries: A model for niche adaptation in humans. *J. Dent. Res.* **2015**, *94*, 110S–118S. [CrossRef] [PubMed]
4. Chu, J.; Zhang, T.; He, K. Cariogenicity features of Streptococcus mutans in presence of rubusoside. *BMC Oral Health* **2016**, *16*, 54. [CrossRef] [PubMed]
5. Ahmadian, E.; Shahi, S.; Yazdani, J.; Maleki Dizaj, S.; Sharifi, S. Local treatment of the dental caries using nanomaterials. *Biomed. Pharmacother.* **2018**, *108*, 443–447. [CrossRef] [PubMed]
6. Hu, J.C.; Chun, Y.H.; Al Hazzazzi, T.; Simmer, J.P. Enamel formation and amelogenesis imperfecta. *Cellstissuesorgans* **2007**, *186*, 78–85. [CrossRef]
7. Beniash, E.; Simmer, J.P.; Margolis, H.C. The effect of recombinant mouse amelogenins on the formation and organization of hydroxyapatite crystals in vitro. *J. Struct. Biol.* **2005**, *149*, 182–190. [CrossRef]
8. Bartlett, J.D.; Ganss, B.; Goldberg, M.; Moradian-Oldak, J.; Paine, M.L.; Snead, M.L.; Wen, X.; White, S.N.; Zhou, Y.L. 3. Protein-protein interactions of the developing enamel matrix. *Curr. Top. Dev. Biol.* **2006**, *74*, 57–115.
9. Habibah, T.U.; Salisbury, H.G. Hydroxyapatite Dental Material. 2020; StatPearls [Internet]. Available online: https://www.ncbi.nlm.nih.gov/books/NBK513314/ (accessed on 14 September 2020).
10. Robinson, C.; Brookes, S.J.; Shore, R.C.; Kirkham, J. The developing enamel matrix: Nature and function. *Eur. J. Oral Sci.* **1998**, *106*, 282–291. [CrossRef]
11. Anderson, H.C. Matrix vesicles and calcification. *Curr. Rheumatol. Rep.* **2003**, *5*, 222–226. [CrossRef]
12. Simmer, J.P.; Hu, J.C. Dental enamel formation and its impact on clinical dentistry. *J. Dent. Educ.* **2001**, *65*, 896–905. [CrossRef] [PubMed]
13. Peterson, L.R.; Thomson, R.B., Jr. Use of the clinical microbiology laboratory for the diagnosis and management of infectious diseases related to the oral cavity. *Infect. Dis. Clin. N. Am.* **1999**, *13*, 775–795. [CrossRef]
14. Kirkham, J.; Firth, A.; Vernals, D.; Boden, N.; Robinson, C.; Shore, R.C.; Brookes, S.J.; Aggeli, A. Self-assembling peptide scaffolds promote enamel remineralization. *J. Dent. Res.* **2007**, *86*, 426–430. [CrossRef] [PubMed]
15. Wu, D.; Yang, J.; Li, J.; Chen, L.; Tang, B.; Chen, X.; Wu, W.; Li, J. Hydroxyapatite-anchored dendrimer for in situ remineralization of human tooth enamel. *Biomaterials* **2013**, *34*, 5036–5047. [CrossRef] [PubMed]
16. Cicciu, M.; Fiorillo, L.; Cervino, G. Chitosan use in dentistry: A systematic review of recent clinical studies. *Mar. Drugs* **2019**, *17*, 417. [CrossRef] [PubMed]
17. Wang, H.; Xiao, Z.; Yang, J.; Lu, D.; Kishen, A.; Li, Y.; Chen, Z.; Que, K.; Zhang, Q.; Deng, X.; et al. Oriented and Ordered Biomimetic Remineralization of the Surface of Demineralized Dental Enamel Using HAP@ACP Nanoparticles Guided by Glycine. *Sci. Rep.* **2017**, *7*, 40701. [CrossRef]
18. Dai, L.L.; Mei, M.L.; Chu, C.H.; Lo, E.C.M. Mechanisms of Bioactive Glass on Caries Management: A Review. *Materials* **2019**, *12*, 4183. [CrossRef]
19. Pandya, M.; Diekwisch, T.G.H. Enamel biomimetics-fiction or future of dentistry. *Int. J. Oral Sci.* **2019**, *11*, 8. [CrossRef]
20. Lakshminarayanan, R.; Fan, D.; Du, C.; Moradian-Oldak, J. The role of secondary structure in the entropically driven amelogenin self-assembly. *Biophys. J.* **2007**, *93*, 3664–3674. [CrossRef]
21. Termine, J.D.; Belcourt, A.B.; Christner, P.J.; Conn, K.M.; Nylen, M.U. Properties of dissociatively extracted fetal tooth matrix proteins. I. Principal molecular species in developing bovine enamel. *J. Biol. Chem.* **1980**, *255*, 9760–9768.
22. Fincham, A.G.; Moradian-Oldak, J.; Diekwisch, T.G.; Lyaruu, D.M.; Wright, J.T.; Bringas, P., Jr.; Slavkin, H.C. Evidence for amelogenin "nanospheres" as functional components of secretory-stage enamel matrix. *J. Struct. Biol.* **1995**, *115*, 50–59. [CrossRef] [PubMed]

23. Diekwisch, T.; David, S.; Bringas, P., Jr.; Santos, V.; Slavkin, H.C. Antisense inhibition of AMEL translation demonstrates supramolecular controls for enamel HAP crystal growth during embryonic mouse molar development. *Development* **1993**, *117*, 471–482. [PubMed]
24. Lu, J.X.; Burton, S.D.; Xu, Y.S.; Buchko, G.W.; Shaw, W.J. The flexible structure of the K24S28 region of Leucine-Rich Amelogenin Protein (LRAP) bound to apatites as a function of surface type, calcium, mutation, and ionic strength. *Front. Physiol.* **2014**, *5*, 254. [CrossRef] [PubMed]
25. Lau, E.C.; Mohandas, T.K.; Shapiro, L.J.; Slavkin, H.C.; Snead, M.L. Human and mouse amelogenin gene loci are on the sex chromosomes. *Genomics* **1989**, *4*, 162–168. [CrossRef]
26. Gibson, C.W.; Golub, E.E.; Abrams, W.R.; Shen, G.; Ding, W.; Rosenbloom, J. Bovine amelogenin message heterogeneity: Alternative splicing and Y-chromosomal gene transcription. *Biochemistry* **1992**, *31*, 8384–8388. [CrossRef]
27. Yamakoshi, Y. Porcine Amelogenin: Alternative Splicing, Proteolytic Processing, Protein-Protein Interactions, and Possible Functions. *J. Oral Biosci.* **2011**, *53*, 275–283. [CrossRef]
28. Buchko, G.W.; Shaw, W.J. Improved protocol to purify untagged amelogenin-Application to murine amelogenin containing the equivalent P70–>T point mutation observed in human amelogenesis imperfecta. *Protein Expr. Purif.* **2015**, *105*, 14–22. [CrossRef]
29. Paine, M.L.; White, S.N.; Luo, W.; Fong, H.; Sarikaya, M.; Snead, M.L. Regulated gene expression dictates enamel structure and tooth function. *Matrix. Biol.* **2001**, *20*, 273–292. [CrossRef]
30. Zhang, X.; Ramirez, B.E.; Liao, X.; Diekwisch, T.G. Amelogenin supramolecular assembly in nanospheres defined by a complex helix-coil-PPII helix 3D-structure. *PLoS ONE* **2011**, *6*, e24952. [CrossRef]
31. Takagi, T.; Suzuki, M.; Baba, T.; Minegishi, K.; Sasaki, S. Complete amino acid sequence of amelogenin in developing bovine enamel. *Biochem. Biophys. Res. Commun.* **1984**, *121*, 592–597. [CrossRef]
32. Renugopalakrishnan, V.; Strawich, E.S.; Horowitz, P.M.; Glimcher, M.J. Studies of the secondary structures of amelogenin from bovine tooth enamel. *Biochemistry* **1986**, *25*, 4879–4887. [CrossRef] [PubMed]
33. Renugopalakrishnan, V.; Prabhakaran, M.; Huang, S.G.; Balasubramaniam, A.; Strawich, E.; Glimcher, M.J. Secondary structure and limited three-dimensional structure of bovine amelogenin. *Connect. Tissue Res.* **1989**, *22*, 131–138. [CrossRef] [PubMed]
34. Zheng, S.; Tu, A.T.; Renugopalakrishnan, V.; Strawich, E.; Glimcher, M.J. A mixed beta-turn and beta-sheet structure for bovine tooth enamel amelogenin: Raman spectroscopic evidence. *Biopolymers* **1987**, *26*, 1809–1813. [CrossRef] [PubMed]
35. Lakshminarayanan, R.; Yoon, I.; Hegde, B.G.; Fan, D.; Du, C.; Moradian-Oldak, J. Analysis of secondary structure and self-assembly of amelogenin by variable temperature circular dichroism and isothermal titration calorimetry. *Proteins* **2009**, *76*, 560–569. [CrossRef] [PubMed]
36. Yamakoshi, Y.; Tanabe, T.; Fukae, M.; Shimizu, M. Porcine amelogenins. *Calcif. Tissue Int.* **1994**, *54*, 69–75. [CrossRef] [PubMed]
37. Goto, Y.; Kogure, E.; Takagi, T.; Aimoto, S.; Aoba, T. Molecular conformation of porcine amelogenin in solution: Three folding units at the N-terminal, central, and C-terminal regions. *J. Biochem.* **1993**, *113*, 55–60. [CrossRef]
38. Delak, K.; Harcup, C.; Lakshminarayanan, R.; Sun, Z.; Fan, Y.; Moradian-Oldak, J.; Evans, J.S. The tooth enamel protein, porcine amelogenin, is an intrinsically disordered protein with an extended molecular configuration in the monomeric form. *Biochemistry* **2009**, *48*, 2272–2281. [CrossRef]
39. Khan, F.; Li, W.; Habelitz, S. Biophysical characterization of synthetic amelogenin C-terminal peptides. *Eur. J. Oral Sci.* **2012**, *120*, 113–122. [CrossRef]
40. Jin, T.; Ito, Y.; Luan, X.; Dangaria, S.; Walker, C.; Allen, M.; Kulkarni, A.; Gibson, C.; Braatz, R.; Liao, X.; et al. Elongated polyproline motifs facilitate enamel evolution through matrix subunit compaction. *PLoS Biol.* **2009**, *7*, e1000262. [CrossRef]
41. Le, T.Q.; Gochin, M.; Featherstone, J.D.; Li, W.; DenBesten, P.K. Comparative calcium binding of leucine-rich amelogenin peptide and full-length amelogenin. *Eur. J. Oral Sci.* **2006**, *114*, 320–326. [CrossRef]
42. Zhang, B.; Xu, G.; Evans, J.S. Model peptide studies of sequence repeats derived from the intracrystalline biomineralization protein, SM50. II. Pro,Asn-rich tandem repeats. *Biopolymers* **2000**, *54*, 464–475. [CrossRef]
43. Gopinathan, G.; Jin, T.; Liu, M.; Li, S.; Atsawasuwan, P.; Galang, M.T.; Allen, M.; Luan, X.; Diekwisch, T.G. The expanded amelogenin polyproline region preferentially binds to apatite versus carbonate and promotes apatite crystal elongation. *Front. Physiol.* **2014**, *5*, 430. [CrossRef]

44. Williamson, M.P. The structure and function of proline-rich regions in proteins. *Biochem. J.* **1994**, *297*, 249–260. [CrossRef] [PubMed]
45. Chou, P.Y.; Fasman, G.D. Prediction of protein conformation. *Biochemistry* **1974**, *13*, 222–245. [CrossRef]
46. Abou Neel, E.A.; Aljabo, A.; Strange, A.; Ibrahim, S.; Coathup, M.; Young, A.M.; Bozec, L.; Mudera, V. Demineralization-remineralization dynamics in teeth and bone. *Int. J. Nanomed.* **2016**, *11*, 4743–4763. [CrossRef]
47. Fincham, A.G.; Belcourt, A.B.; Termine, J.D.; Butler, W.T.; Cothran, W.C. Dental enamel matrix: Sequences of two amelogenin polypeptides. *Biosci. Rep.* **1981**, *1*, 771–778. [CrossRef] [PubMed]
48. Fukae, M.; Tanabe, T.; Ijiri, H.; Shimizu, M. Studies on porcine enamel proteins: A possible original enamel protein. *Tsurumi Shigaku* **1980**, *6*, 87–94. [PubMed]
49. Fincham, A.G.; Belcourt, A.B.; Termine, J.D.; Butler, W.T.; Cothran, W.C. Amelogenins. Sequence homologies in enamel-matrix proteins from three mammalian species. *Biochem. J.* **1983**, *211*, 149–154. [CrossRef]
50. Gibson, C.W.; Golub, E.; Ding, W.D.; Shimokawa, H.; Young, M.; Termine, J.; Rosenbloom, J. Identification of the leucine-rich amelogenin peptide (LRAP) as the translation product of an alternatively spliced transcript. *Biochem. Biophys. Res. Commun.* **1991**, *174*, 1306–1312. [CrossRef]
51. Simmer, J.P.; Hu, Y.; Lertlam, R.; Yamakoshi, Y.; Hu, J.C. Hypomaturation enamel defects in Klk4 knockout/LacZ knockin mice. *J. Biol. Chem.* **2009**, *284*, 19110–19121. [CrossRef]
52. Moradian-Oldak, J.; Gharakhanian, N.; Jimenez, I. Limited proteolysis of amelogenin: Toward understanding the proteolytic processes in enamel extracellular matrix. *Connect. Tissue Res.* **2002**, *43*, 450–455. [CrossRef] [PubMed]
53. Shaw, W.J.; Campbell, A.A.; Paine, M.L.; Snead, M.L. The COOH terminus of the amelogenin, LRAP, is oriented next to the hydroxyapatite surface. *J. Biol. Chem.* **2004**, *279*, 40263–40266. [CrossRef] [PubMed]
54. Nagano, T.; Kakegawa, A.; Yamakoshi, Y.; Tsuchiya, S.; Hu, J.C.; Gomi, K.; Arai, T.; Bartlett, J.D.; Simmer, J.P. Mmp-20 and Klk4 cleavage site preferences for amelogenin sequences. *J. Dent. Res.* **2009**, *88*, 823–828. [CrossRef] [PubMed]
55. Bartlett, J.D.; Simmer, J.P.; Xue, J.; Margolis, H.C.; Moreno, E.C. Molecular cloning and mRNA tissue distribution of a novel matrix metalloproteinase isolated from porcine enamel organ. *Gene* **1996**, *183*, 123–128. [CrossRef]
56. Mukherjee, K.; Ruan, Q.; Nutt, S.; Tao, J.; De Yoreo, J.J.; Moradian-Oldak, J. Peptide-Based Bioinspired Approach to Regrowing Multilayered Aprismatic Enamel. *ACS Omega* **2018**, *3*, 2546–2557. [CrossRef]
57. Le Norcy, E.; Kwak, S.Y.; Wiedemann-Bidlack, F.B.; Beniash, E.; Yamakoshi, Y.; Simmer, J.P.; Margolis, H.C. Potential role of the amelogenin N-terminus in the regulation of calcium phosphate formation in vitro. *Cellstissuesorgans* **2011**, *194*, 188–193. [CrossRef]
58. Kwak, S.Y.; Wiedemann-Bidlack, F.B.; Beniash, E.; Yamakoshi, Y.; Simmer, J.P.; Litman, A.; Margolis, H.C. Role of 20-kDa amelogenin (P148) phosphorylation in calcium phosphate formation in vitro. *J. Biol. Chem.* **2009**, *284*, 18972–18979. [CrossRef]
59. Hunter, G.K.; Curtis, H.A.; Grynpas, M.D.; Simmer, J.P.; Fincham, A.G. Effects of recombinant amelogenin on hydroxyapatite formation in vitro. *Calcif. Tissue Int.* **1999**, *65*, 226–231. [CrossRef]
60. Moradian-Oldak, J.; Leung, W.; Fincham, A.G. Temperature and pH-dependent supramolecular self-assembly of amelogenin molecules: A dynamic light-scattering analysis. *J. Struct. Biol.* **1998**, *122*, 320–327. [CrossRef]
61. Aoba, T.; Fukae, M.; Tanabe, T.; Shimizu, M.; Moreno, E.C. Selective adsorption of porcine-amelogenins onto hydroxyapatite and their inhibitory activity on hydroxyapatite growth in supersaturated solutions. *Calcif. Tissue Int.* **1987**, *41*, 281–289. [CrossRef]
62. Doi, Y.; Eanes, E.D.; Shimokawa, H.; Termine, J.D. Inhibition of seeded growth of enamel apatite crystals by amelogenin and enamelin proteins in vitro. *J. Dent. Res.* **1984**, *63*, 98–105. [CrossRef] [PubMed]
63. Aoba, T.; Moreno, E.C.; Kresak, M.; Tanabe, T. Possible roles of partial sequences at N- and C-termini of amelogenin in protein-enamel mineral interaction. *J. Dent. Res.* **1989**, *68*, 1331–1336. [CrossRef] [PubMed]
64. Moradian-Oldak, J.; Bouropoulos, N.; Wang, L.; Gharakhanian, N. Analysis of self-assembly and apatite binding properties of amelogenin proteins lacking the hydrophilic C-terminal. *Matrix Biol. J. Int. Soc. Matrix Biol.* **2002**, *21*, 197–205. [CrossRef]
65. Wiedemann-Bidlack, F.B.; Beniash, E.; Yamakoshi, Y.; Simmer, J.P.; Margolis, H.C. pH triggered self-assembly of native and recombinant amelogenins under physiological pH and temperature in vitro. *J. Struct. Biol.* **2007**, *160*, 57–69. [CrossRef] [PubMed]

66. Gibson, J.M.; Raghunathan, V.; Popham, J.M.; Stayton, P.S.; Drobny, G.P. A REDOR NMR study of a phosphorylated statherin fragment bound to hydroxyapatite crystals. *J. Am. Chem. Soc.* **2005**, *127*, 9350–9351. [CrossRef]
67. Moradian-Oldak, J. Amelogenins: Assembly, processing and control of crystal morphology. *Matrix Biol. J. Int. Soc. Matrix Biol.* **2001**, *20*, 293–305. [CrossRef]
68. Le Norcy, E.; Kwak, S.Y.; Wiedemann-Bidlack, F.B.; Beniash, E.; Yamakoshi, Y.; Simmer, J.P.; Margolis, H.C. Leucine-rich amelogenin peptides regulate mineralization in vitro. *J. Dent. Res.* **2011**, *90*, 1091–1097. [CrossRef]
69. Ali, S.; Farooq, I. A Review of the Role of Amelogenin Protein in Enamel Formation and Novel Experimental Techniques to Study its Function. *Protein Pept. Lett.* **2019**, *26*, 880–886. [CrossRef]
70. Aggeli, A.; Bell, M.; Boden, N.; Keen, J.N.; Knowles, P.F.; McLeish, T.C.; Pitkeathly, M.; Radford, S.E. Responsive gels formed by the spontaneous self-assembly of peptides into polymeric beta-sheet tapes. *Nature* **1997**, *386*, 259–262. [CrossRef]
71. Zhang, S. Discovery and design of self-assembling peptides. *Interf. Focus* **2017**, *7*, 20170028. [CrossRef]
72. Doberdoli, D.; Bommer, C.; Begzati, A.; Haliti, F.; Heinzel-Gutenbrunner, M.; Juric, H. Randomized clinical trial investigating self-assembling peptide P11-4 for treatment of early occlusal caries. *Sci. Rep.* **2020**, *10*, 4195. [CrossRef] [PubMed]

© 2020 by the authors. Licensee MDPI, Basel, Switzerland. This article is an open access article distributed under the terms and conditions of the Creative Commons Attribution (CC BY) license (http://creativecommons.org/licenses/by/4.0/).

Review

New Frontiers on Human Safe Insecticides and Fungicides: An Opinion on Trehalase Inhibitors

Camilla Matassini [1,*], Camilla Parmeggiani [1,2] and Francesca Cardona [1,*]

[1] Dipartimento di Chimica "Ugo Schiff", Università degli Studi di Firenze, Via della Lastruccia 3-13, 50019 Sesto Fiorentino, Italy; camilla.parmeggiani@unifi.it
[2] European Laboratory for Non-linear Spectroscopy via Nello Carrara 1, 50019 Sesto Fiorentino, Italy
* Correspondence: camilla.matassini@unifi.it (C.M.); francesca.cardona@unifi.it (F.C.); Tel.: +39-055-4573536 (C.M.); +39-055-4573504 (F.C.)

Academic Editor: Baoan Song
Received: 18 June 2020; Accepted: 26 June 2020; Published: 1 July 2020

Abstract: In the era of green economy, trehalase inhibitors represent a valuable chance to develop non-toxic pesticides, being hydrophilic compounds that do not persist in the environment. The lesson on this topic that we learned from the past can be of great help in the research on new specific green pesticides. This review aims to describe the efforts made in the last 50 years in the evaluation of natural compounds and their analogues as trehalase inhibitors, in view of their potential use as insecticides and fungicides. Specifically, we analyzed trehalase inhibitors based on sugars and sugar mimics, focusing on those showing good inhibition properties towards insect trehalases. Despite their attractiveness as a target, up to now there are no trehalase inhibitors that have been developed as commercial insecticides. Although natural complex pseudo di- and trisaccharides were firstly studied to this aim, iminosugars look to be more promising, showing an excellent specificity profile towards insect trehalases. The results reported here represent an overview and a discussion of the best candidates which may lead to the development of an effective insecticide in the future.

Keywords: antibiotics; biochemical studies; iminosugars; inhibitors; insect trehalase; trehalose; in vivo studies; mammalian trehalase; natural compounds; selectivity

1. Introduction

Insecticides and fungicides have played a fundamental role in raising the quality of our lives, not only for crop protection in agriculture, but also to avoid the spreading of harmful pests causing lethal human diseases, such as malaria. In the past, the non-restricted use of highly dangerous insecticides such as dichlorodiphenyltrichloroethane (DDT) has provoked negative effects in the environment and to mammals. DDT is a very toxic and persistent organic compound, its chemical stability provokes a long range transport into the environment and its non-hydrophilicity causes bioaccumulation in the tissues of animals and human beings. For these reasons, the US banned the use of DDT in 1972. However, DDT is still very efficient against malaria and other diseases caused by insects. In September 2006, the World Health Organization (WHO) declared its support for the indoor use of DDT in African countries where malaria remains a major health problem, citing that the benefits of the pesticide outweigh the health and environmental risks. The development of non-toxic, environmentally friendly insecticides and fungicides for human health and for crop protection is of great interest, especially for less developed countries affected by pandemic and starvation. The identification of a new target which is specific for insects and does not affect humans is therefore of particular relevance.

Trehalose (**1**, Figure 1) is a peculiar non-reducing disaccharide featured by the presence of two glucose units linked through an α,α-1,1-glycosidic linkage. This disaccharide is present in a wide variety of organisms, including yeast, fungi, bacteria, insects, some invertebrates, and lower and higher plants. However, trehalose is not found in mammals. In the 1970s, trehalose was merely regarded as a storage form of glucose for energy and/or for cellular components structure [1]. Since then, knowledge about the various functions of this simple disaccharide greatly expanded, and it is now evident that trehalose is much more than a simple storage compound [2], although its exact function in many organisms is still under investigation. In yeast and plants, trehalose is a signaling compound able to regulate certain metabolic pathways [3]. In addition, trehalose can act as a chemical "chaperone" [4] by stabilizing proteins in their native structure and thus preventing cellular damage from inactivation or denaturation caused by stress conditions such as desiccations, dehydration, heat, cold, and damage by oxygen radicals. When unicellular organisms are exposed to stress, they adapt by synthesizing huge amounts of trehalose, which contributes to retain cellular integrity by stabilizing protein structures in many different ways [5]. The equilibrium between trehalose storage and degradation needs to be finely tuned in response to different cellular states. In this regard, an important enzyme in trehalose metabolism is α-trehalase (EC 3.2.1.28), which has a regulatory role in controlling the levels of trehalose in cells by lowering its concentration once the stress is alleviated. Trehalose is indeed the main sugar circulating in the blood or hemolymph of most insects [6]. α-Trehalase (EC 3.2.1.28) is an inverting glycosidase [7] that promotes the conversion of trehalose into two molecules of glucose, which is vital for insect flight and essential for larvae resistance to stress factors [8].

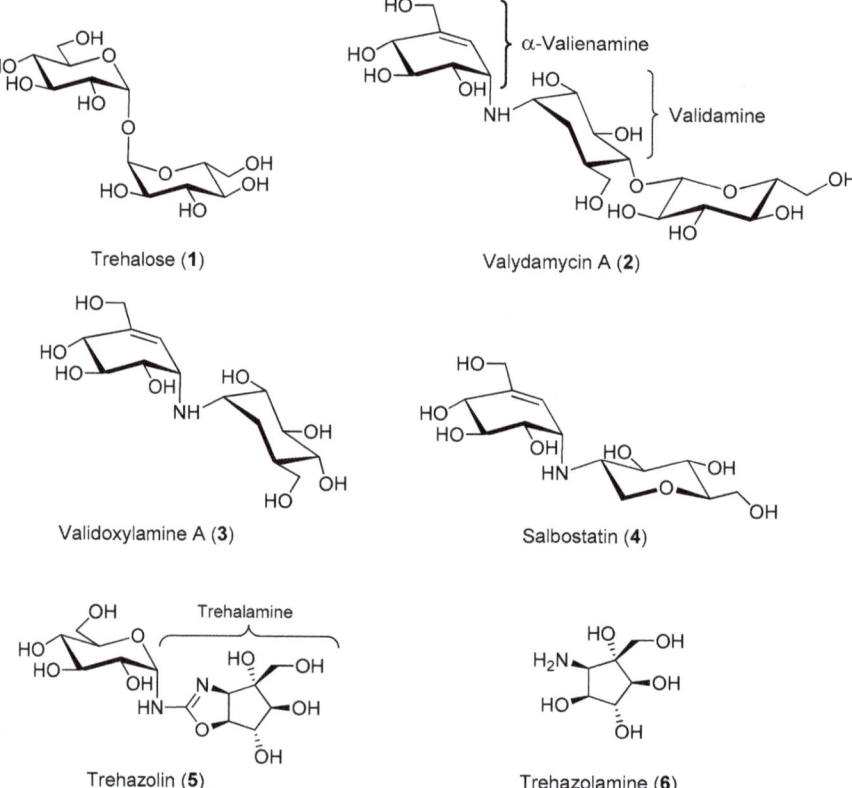

Figure 1. Structures of trehalose (**1**), the natural substrate of trehalase, and of some natural occurring carbohydrate- and carbocyclic-based trehalase inhibitors.

The enzyme α-trehalase is also found in mammals both in the kidney brush border membranes [9] and in the intestinal villae membranes [10]. While its function in the kidney is not clear, the intestinal enzyme has the role of occasionally hydrolyzing ingested trehalose. Intolerance to fungi has been correlated with a deficit or a defect of the intestinal trehalase in mammals [2]. Trehalose is also an important component in fungal spores, where its hydrolysis plays a major role during early germination presumably serving as a source of glucose for energy [11].

Due to the biological relevance of trehalose and trehalose processing enzymes in pathological and physiological states, in particular for the important role of trehalose-derived glucose in larvae survival and development, trehalase inhibitors have been regarded in recent years as an interesting target for the identification of novel insecticides and fungicides. Therefore, in the last 50 years, several natural trehalose analogues and synthetic trehalose mimetics have been studied as potential fungicides and antibiotics.

Nowadays, the search for green pesticides is an urgent issue and therefore, due to the presence of trehalase also in mammals, specificity towards the insect enzyme has become crucial for the development of drugs that are in principle safe for plants and mammals [12]. In addition, trehalase inhibitors are considered as valuable tools for studying the molecular physiology of trehalase function and sugar metabolism in insects [13].

In the past, several potent trehalase inhibitors were isolated from natural sources. These compounds include pseudodisaccharide structures and their glycosyl derivatives containing a sugar or a polyhydroxylated carbocycle (such as validamycins, validoxylamines, salbostatin and trehazolin, see Section 2). More recently, natural iminosugars were also reported as strong trehalase inhibitors (see Section 3). Iminosugars are nitrogenated glycomimetics with a nitrogen replacing the endocyclic oxygen of sugars, and widely known as glycosidase inhibitors [14].

This review aims to describe the efforts made in the last 50 years in the evaluation of natural compounds and their analogues as trehalase inhibitors, in view of their potential use as insecticides and fungicides. However, since isolation and purification of such compounds from natural source is very difficult and expensive, some of these structures have never been obtained in a chemically pure form. Therefore, considerable efforts have been devoted to their total synthesis and that of related analogues. In this regard, the syntheses of these compounds are highly challenging, due to the presence of several contiguous carbon stereocenters, which limited the number of synthetic approaches described, but in the same way resulted in the description of very elegant total syntheses. This review collects the efforts made by the researchers in the evaluation of carbohydrate-, carbocyclic-, and iminosugar-based compounds as trehalases inhibitors of different origins.

2. Carbohydrate- and Carbocyclic-Based Inhibitors

Between the 1970s and the 1980s, the pseudo-oligosaccharides complex of validamycins, a family of naturally occurring antibiotics, was isolated from *Streptomyces hygroscopicus subsp. limoneus* and characterized [15–17]. Validamycin A, the major component of the complex, rapidly emerged for its activity in controlling rice sheath blight caused by the phytopathogenic fungus *Rizochtonia solani* (*R. solani*) [15].

Validamycin A (2, Figure 1) possesses a unique pseudo-trisaccharide structure composed of an unsaturated carbasugar (α-valienamine), an amino carbasugar (validamine) and a β-D-glucose residue. The hydrolysis of the glycosidic bond affords the pseudo-disaccharide validoxylamine A (3, Figure 1), which showed an outstanding activity as trehalase inhibitor. Indeed, while validamycin A inhibited trehalase from *R. solani* with an IC_{50} = 72 μM, validoxylamine A resulted in a much more potent competitive inhibitor with a K_i = 1.9 nM (IC_{50} = 140 nM) (Table 1) [18]. For both 2 and 3 no significant activity was exhibited against cellulase, pectinases, α-amylase, and α- and β-glucosidases, showing an important selectivity for the trehalase enzyme. Interestingly, studies conducted to determine the uptake of these compounds into the mycelia of *R. solani* demonstrated that 2 was more readily taken up into the cells than its aglycone 3. In addition, 2 significantly suppressed the degradation of intracellular trehalose at very low concentration (0.1 μg/mL) when incubated in mycelia of *R. solani*, thus showing remarkable in vivo activity as a trehalase inhibitor and explaining the origin of the antifungal properties. Taken together, these data suggest that validamycin A is efficiently transported into the mycelia and probably hydrolyzed therein by a β-glucosidase yielding validoxylamine A, which possesses a greater inhibitory activity [18]. The inhibitory activity of 2, 3 and some structural analogues was tested on various trehalases from porcine intestine, rat intestine, rabbit kidney, baker's yeast, *Mycobacterium smegmantis*, and *Spodoptera litura* (*S. litura*) insect [19]. Once more, **3** emerged as the best competitive inhibitor with K_i values from 0.31 µM (rat intestine) to 0.27 nM (baker's yeast). The screening of validoxylamine A analogues demonstrated that the double bond in validoxylamine A is not essential to the activity as the configuration of the hydroxyl and hydroxymethyl groups is. Later reports confirmed that **3**, resembling trehalose in structure, showed potent and specific inhibitory activity towards trehalases in various organisms, from microorganisms to mammals [20–22] and in most cases the inhibition mechanism by validoxylamine A was competitive, with the exception of a non-competitive slow binding process reported in porcine kidney trehalase [22]. It is well known that insect hemolymph contains a high concentration of trehalose which is used as an energy source in various tissues. Administration of the potent trehalase inhibitor **3** to insects was thought to block energy metabolism leading to abnormal physiological function. Therefore, in the last 20 years, the effects of validoxylamine A on insects was investigated. Some studies reported only the inhibitory activity towards the trehalase enzyme extracted and purified from insects, such as the case of termites trehalase for which **3** showed to be a one order of magnitude more potent inhibitor than **2** (K_i = 3.2 µM vs. K_i = 402 µM) [23]. Other studies reported both the in vitro and in vivo inhibitory activity of validoxylamine A on insect trehalases, as well as its insecticidal activity. For example, the injection of validoxylamine A and its analogues to *S. litura* larvae showed that validoxylamine A was both the best inhibitor towards the *S. litura* trehalase (K_i = 43 nM, Table 1) and the most potent insecticide, providing 100% mortality at a dose of 10 µg/last instar larva [20]. Lethal activity exhibited by **3** in the common silkworm (*Bombyx mori*, *B. mori*) [24,25], in *Mamestra brassicae* (*M. brassicae*) [26,27] and in the American cockroach (*Periplaneta americana*, *P. americana*) [28,29] was also reported.

In the early 1990s the trehalase inhibitor salbostatin (**4**, Figure 1) was isolated from *Streptomyces albus* ATCC21838 and its full characterization revealed a pseudo-disaccharide structure, which can be considered a validoxylamine A analogue with a 1-deoxyglucosamine residue attached to the α-valienamine (Figure 1) [30]. A convenient total synthesis of salbostatin **4**, reported by Ogawa and coworkers, allowed screening of the inhibitory activity of **4** towards different trehalases [31]. Unfortunately, **4** showed a stronger inhibitory activity towards trehalase from porcine kidney (K_i = 0.18 µM) [30] than from silkworm (IC$_{50}$ = 8.3 µM) [31] and this hampered its use as a non-toxic insecticide (Table 1).

Isolated in 1991 from a culture broth of *Micromonospora* strain SANK 62390 [32,33], trehazolin (**5**, Figure 1) is a pseudo-disaccharide consisting of an α-D-glucopyranose moiety bonded to a unique aminocyclopentitol (trehazolamine **6**, Figure 1) through a fused 2-aminooxazoline ring. Although **5** can be considered a glucosylamine derivative, formally obtained by the reaction of trehalamine and D-glucose, this synthetic approach has never appeared in the literature. All reported syntheses employed thiourea derivative **7** (Figure 2) as precursor, which can be in turn obtained by diversely combining two subunits bearing an isothiocyanate moiety and an amino group, respectively. The synthesis of such fragments from different sugars and the subsequent strategies to access trehazolin (**5**) and analogues have been exhaustively reviewed by El Ashry and El Nemr in 2011 [34]. Compound **5** rapidly emerged as a slow, tight-binding inhibitor of silkworm trehalase [35], showing a remarkable selectivity with respect to other glycosidases (e.g., α- and β-glycosidases, maltase, isomaltase, sucrase, amyloglucosidase, etc.) and a reversible, competitive type of inhibition (K_i = 10 nM [35], IC$_{50}$ = 49 nM [36], Table 1). The presence of the sugar moiety was proven essential for the inhibitory activity, being trehalamine much less potent in inhibiting silkworm trehalase (IC$_{50}$ = 1 mM vs. IC$_{50}$ = 52 nM) [35]. Compound **5** is a potent inhibitor of porcine trehalase as attested by two independent studies that reported IC$_{50}$ = 16 nM [35] and IC$_{50}$ = 19 nM (purified pig kidney trehalase) [21], respectively (Table 1). In order to investigate the structure–activity relationships with regard to the stereochemistry of the aminocyclitol moiety and that of the anomeric position, as well as the role played by hydroxyl functions, several analogues of **5** (epimers, β-anomer, deoxygenated aminocyclitol moiety congeners) were synthesized and tested.

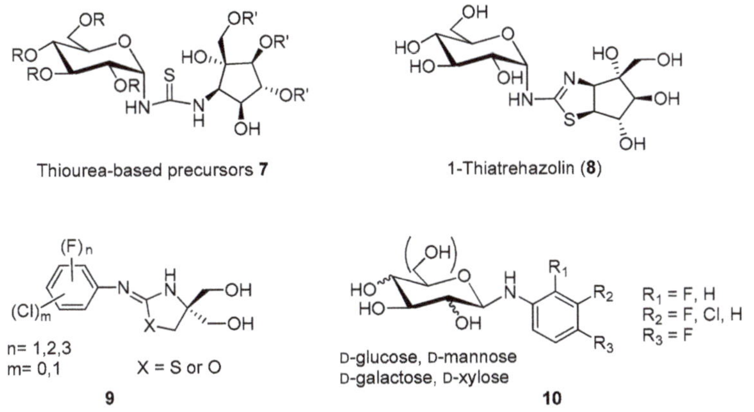

Figure 2. Structures of synthetic carbohydrate-, carbocyclic-, and heterocyclic-based trehalase inhibitors.

Different studies confirmed that any structural modification at the pyranose or cyclopentitol moieties appreciably decreases **5** inhibitory potency, suggesting that its close resemblance to the substrate α,α-trehalose is responsible for the high and selective trehalase inhibitory activity [36–38]. This assumption was also extended to the modification of the oxazolidine ring obtained by replacing the oxygen atom with a sulfur atom. Indeed, Chiara and coworkers reported a highly stereoselective and efficient synthesis of 1-thiatreazolin **8** (Figure 2), which was proven to be a nanomolar, slow, tight-binding inhibitor of porcine trehalase, even though it is less potent than trehazolin itself (K_i = 30.4 nM vs. K_i = 2.1 nM, Table 1) [39]. However, 1-thiatrehazolin **8** is more stable than trehazolin **5** and for this reason in 2007 it was co-crystallized with Tre37A trehalase from *Escherichia coli* (*E. coli*), in the first report on the three-dimensional structure of a trehalase [40]. Davies and coworkers determined the enzyme structure in complex with validoxylamine A (**3**) and 1-thiatreazolin (**8**). Availability of these structures definitely proved the hypothesis earlier emerged from kinetic studies performed with porcine kidney trehalase in the presence of two types of competitive inhibitors [41], which suggested that the active centre of the enzyme may comprise two subsites, one for catalysis and one for recognition, acting separately on each glucose unit of trehalose. In particular, the complexes of Tre37A with **3** and **8** revealed the interactions in the −1 ("catalytic") and +1 ("leaving-group") subsites: the valienamine of **3** and the cyclopentane ring of **5** are placed in the −1 subsite, even though with different positions, while in the +1 subsite the pseudosugar ring of **3** and the glucose moiety of **8** lie in a nearly identical position, which promotes favorable hydrogen bonding with the same residues. In addition, the complexes of Tre37A with **3** and **8** strongly implicate Asp312 and Glu469 as the catalytic acid and base, confirming that the catalysis by trehalases occurs with inversion of the anomeric configuration [40].

In 1995, Nakajima and coworkers reported for the first time the pesticidal activity (namely, antifungal and insecticidal) of trehazolin **5** [42]. First, they investigated the protective and curative activities of **5** against plant pathogenic fungi and obtained remarkable results with the infection of rice seedlings caused by *R. solani*, which was completely inhibited by spraying a 100 ppm **5** solution one day after inoculation. This curative effect was ascribed to the activity of **5** as potent inhibitor of *R. solani* trehalase (IC_{50} = 66 nM, Table 1). It is worth noting that validamycin A was more effective in suppressing the growth of *R. solani* (curative effect at 7.5 ppm concentration) probably due to a more efficient uptake of this antibiotic into the mycelia of *R. solani* [18]. Due to the well-known inhibitory activity of **5** towards trehalase from silkworm (*B. mori*), the same authors investigated the insecticidal activity of this compound in *B. mori* larvae. Interestingly, 50 μg and 100 μg doses of **5** were insecticidal (7/10 and 10/10 injected larvae were killed, respectively), while no toxicity towards mice (intravenous administration at a dose of 100 mg/kg) was observed [42].

Table 1. Inhibition of insect trehalases from silkworm (*B. mori*) and tobacco cutworm (*S. litura*), fungi trehalase from *R. solani*, porcine kidney and porcine intestine trehalases by compounds **2, 3, 4, 5** and **8**.

Compound	Silkworm (*B. mori*), IC_{50} (K_i)	Tobacco Cutworm (*S. litura*), IC_{50} (K_i)	*R. solani*, IC_{50} (K_i)	Porcine Kidney, IC_{50} (K_i)	Porcine Intestine, IC_{50} (K_i)
2	n.d.	370 nM [19]	72 µM [18]	250 µM [21]	420 nM [19]
3 [a]	n.d.	48 nM [19] (43 nM) [20]	140 nM (1.9 nM) [18]	2.4 nM [21]	14 nM [19]
4	8.3 µM [31]	n.d.	n.d.	(0.18 µM) [30]	n.d.
5 [b,c]	49 nM [36] 52 nM [35] (10 nM) [35]	n.d.	66 nM [42]	15.5 nM [39] [d] (2.1 nM) [39]	n.d.
8 [a]	n.d.	n.d.	n.d.	83.0 nM [39] (30.4 nM) [39]	n.d.

[a] Compounds **3** and **8** are potent inhibitors of Tre37A enzyme from *E. coli* with K_i values of 10 nM and 9 nM, respectively [40]; [b] IC_{50} values of 5.5 nM and 3.7 nM were also reported towards silkworm and porcine trehalase, respectively [32]; [c] **5** also inhibits locust flight muscle trehalase with K_i = 10 nM (in extracts) and K_i = 8 nM (on purified enzyme) [13]; [d] IC_{50} = 16 nM [35] and IC_{50} = 19 nM [21] were also reported for natural **5**; n.d.: not determined.

With the aim of clarifying the mechanism of toxicity exerted by trehalase inhibitors in insects, the effect of **5** on trehalase activity of locust flight muscles was investigated by Wegener and coworkers [13]. In vitro experiments showed that **5** acts as a competitive, tight binding inhibitor of locust flight muscle trehalase both when tested in extracts and on the purified enzyme (K_i = 10 nM vs. K_i = 8 nM) (Table 1). In vivo experiments revealed that **5** differentiates between an 'overt' and a 'latent' trehalase, the latter being catalytically inactive in vitro and probably derived from a trehalase form that is protected from inhibition by **5** in the intact insect. In addition, the insecticidal activity of **5** was proven: 50 µg injected into locusts completely and selectively blocked the overt form of muscle trehalase and killed 50% of the insects within 24 h. This study also demonstrated that **5** caused dramatic hypoglycemia: indeed, after injection of a 10 µg dose, glucose levels decreased by over 90% in 24 h. More interestingly, feeding glucose to the locusts fully neutralized the effect of a potentially lethal dose of **5**, indicating that hypertrehaloseanemia is not acutely toxic while lack of glucose causes organ failure, and that sufficient hemolymph glucose can be only generated from trehalose by trehalase. This and other similar studies [43] perfectly show that trehalase inhibitors are valuable tools for studying the molecular physiology of trehalase function and sugar metabolism in insects.

Aiming at developing artificial synthetic inhibitors of trehalase with simpler structures than the natural compounds **3** and **5**, Qian and co-workers prepared a series of fluorine-containing arylaminooxazo(thiazo)lidines **9** [44] and fluorinated *N*-arylglycosylamines **10** [45], respectively (Figure 2). Although compounds **9** were moderate porcine trehalase inhibitors (best inhibitor: IC_{50} = 31.6 µM), they were expected to increase the hydrophobicity and the penetrating ability in vivo. Indeed, some compounds of the series showed remarkable larvicidal activity and inhibition action on insect flight, when tested on fruit fly *Drosophila melanogaster* at 200 ppm concentration. Conversely, compounds **10**, designed to improve the antifungal activity in vivo with respect to validoxylamine A (**3**), failed in accomplishing this task toward *R. solani*, showing 133-fold lower antifungal activity as compared to validamycin A **2**.

3. Iminosugar-Based Inhibitors

Natural iminosugars are classified into five structural classes: polyhydroxylated pyrrolidines, piperidines, indolizidines, pyrrolizidines, and nortropanes [14]. For clarity, the discussion on the trehalase inhibitory activity of iminosugars has been divided based on their monocyclic or bicyclic structure.

3.1. Monocyclic Iminosugars and Their Derivatives

Although the inhibitory activity of iminosugars towards glycosidases was well documented from the 1970s, only in the early 1980s was the first evidence of the trehalase inhibition by monocyclic iminosugars emerged and the relationship between structure and selectivity towards trehalase over other glycosidases investigated.

The polyhydroxy piperidine family was tested first. Nojirimycin (**11**, Figure 3), known to be a more potent inhibitor of β-glucosidases than deoxynojirimycin (DNJ, **12**) [46,47] showed no inhibition towards insect trehalase [48], while DNJ (isolated from *Streptomyces lavendulae*) was reported to inhibit trehalase from *Chaetomium aureum* and from rabbit [49]. However, in 1994 Asano and coworkers reported that **12** strongly inhibited also α-glucosidases (IC$_{50}$ = 0.3–0.4 µM) as well as porcine kidney trehalase (IC$_{50}$ = 41 µM) [50], demonstrating its scarce selectivity. A better selectivity was observed for the D-mannose analogue of DNJ, deoxymannonojirimycin (DMJ, **13**), which showed an IC$_{50}$ = 55 µM towards insect trehalase [48] and no inhibition towards mammalian trehalase (Table 2) [50]. Finally, the α-homonojirimycin-7-*O*-β-D-glucopyranoside (MDL 25637, **14**, Figure 3), whose synthesis was reported by Liu and coworkers in 1989 [51], showed an outstanding inhibitory activity towards porcine kidney trehalase (K_i = 3 nM) [22], suggesting that the presence of a glucose unit strongly enhances the affinity with the enzyme active site.

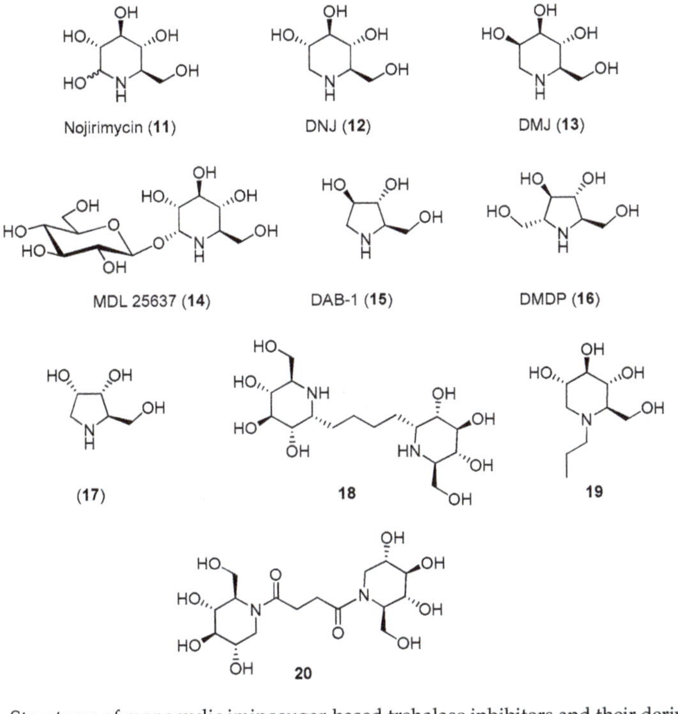

Figure 3. Structures of monocyclic iminosugar-based trehalase inhibitors and their derivatives.

Regarding pyrrolidine iminosugars, 1,4-dideoxy-1,4-imino-D-arabinitol (DAB-1, **15**) was a good inhibitor of mammalian trehalase (IC$_{50}$ = 4.8 µM towards porcine kidney trehalase). However, it was scarcely selective, showing good inhibitory activity also towards α-glucosidases (IC$_{50}$ = 5.8–100 µM) and α-mannosidases (IC$_{50}$ = 46–110 µM) [50]. In addition, its analogue DMDP (**16**), inhibited bacteria and insect trehalases (IC$_{50}$ = 0.35 µM towards *Corynebacterium* sp. and IC$_{50}$ = 10 µM towards *Plutella xylostella*) but did not inhibit porcine kidney trehalase (no inhibition at 1 mM) (Table 2) [52].

The relevant role played by the iminosugar moiety in trehalase inhibition was demonstrated by Bini and coworkers: a series of pseudo-disaccharide structures was synthesized by cross-metathesis (CM) reactions between suitably functionalized piperidine (namely nojirimycin **11**) or pyrrolidine (namely DAB-1 **15** and 1,4-dideoxy-1,4-imino-D-ribitol **17**, Figure 3) iminosugars and preliminary assays toward commercial porcine trehalase were reported [53]. While very small inhibition was observed with pyrrolidine-based dimers, a good inhibition was detected with nojirimycin dimer **18**

(Figure 3), which showed $K_i = 44$ µM. Even more interestingly, reference glucose dimer did not show any inhibition at a concentration of 1 mM, highlighting that the nitrogen atom is essential for inhibition.

More recently, the inhibition of a membrane-bound trehalase from the larvae of the midge *Chironomus riparius* (*C. riparius*) was investigated with the aim of identifying differences between midge trehalase and mammalian trehalase [54]. The membrane-bound isoform of trehalase (trehalase-2) is believed to be the latent (inactive) form, more abundant in larvae, which is activated at the beginning of the prepupal period, thus representing a very interesting target. Forcella and coworkers reported for the first time the isolation and purification of the trehalase-2 isoform from *C. riparius* and demonstrated that this enzyme is highly specific for trehalose, with an affinity about 5-fold higher when compared to the mammalian trehalase. DNJ showed a competitive inhibition both towards *C. riparius* and porcine kidney trehalase (with $IC_{50} = 2.83$ µM and $IC_{50} = 5.96$ µM, respectively, Table 2), but there was significant dissimilarity in the kinetic behavior between the two enzymes. All these data, supporting the idea that the catalytic sites of trehalases from porcine kidney and insects have different recognition requirements, prompted the design of new DNJ and DAB-1 derivatives, aimed at enhancing the inhibitor specificity.

For example, the introduction of a propyl chain on the endocyclic nitrogen of DNJ (**19**, Figure 3) caused a slight decrease of activity with respect to DNJ ($IC_{50} = 9.7$ µM vs. $IC_{50} = 2.8$ µM towards *C. riparius* trehalase) but imparted a 10-fold selectivity towards the insect trehalase ($IC_{50} = 109$ µM vs. $IC_{50} = 5.96$ µM towards porcine kidney trehalase) (Table 2) [55]. A good degree of selectivity (3–7 times) was obtained also with *N*-bridged DNJ dimers, in which the two iminosugar units are linked through 2-, 3-, or 4-carbon atoms amidic spacers. In particular, dimer **20** (Figure 3) was 7-fold more active towards insect trehalase and thus more specific than DNJ (only twice as active on insect trehalase) (Table 2) [56]. More recently, a series of DNJ derivatives presenting thiolated or unsaturated *N*-alkyl chains of various lengths were synthesized [57]. Although they potently inhibited insect and mammalian trehalases with inhibition in the low micromolar range (IC_{50} values ranging from 0.27 µM to 6.94 µM), no selectivity toward insect trehalases was observed.

Table 2. Inhibition of *C. Riparius* and porcine kidney trehalases by compounds **12, 13, 15, 16, 19,** and **20**.

Compound	*C. riparius*, IC_{50} (K_i)	Porcine Kidney, IC_{50} (K_i)	Selectivity [a]
12	2.83 µM [54] (1.39 µM) [54]	5.96 µM [54] [b] (2.98 µM) [54]	
13	55 µM [48]	n.i. [50]	18.2
15	19 µM [54] (9.3 µM) [54]	4.8 µM [50] [c] (5.3 µM) [54]	-
16	10 µM [52] [d]	n.i. [52]	100
19	9.7 µM [55]	109 µM [55]	11.2
20	11 µM [56]	76 µM [56]	6.9

[a] Selectivity is the ratio between the IC_{50} value against porcine kidney trehalase and the IC_{50} value against *C. riparius* trehalase; [b] $IC_{50} = 41$ µM [50] and $IC_{50} = 43$ µM [21] have been previously reported; [c] $IC_{50} = 10.6$ µM was also reported [54]; [d] IC_{50} reported towards insect trehalase from *Plutella xylostella* instead of *C. riparius*; n.i.: no inhibition at 1 mM.

Surprisingly, only one report appeared on the insecticidal activity of monocyclic iminosugars. In 2007, Hirayama and coworkers, inspired by the fact that mulberry latex contains very high concentrations of DAB-1 and DNJ as a defense against herbivory insects [58], added these inhibitors to the diet of *Samia ricini* (*S. ricini*) larvae and observed a significant increase in the hemolymph trehalose concentrations (trehalose concentration was approximately 50% higher than that in larvae fed on the standard diet) [59]. This effect, although interesting, is lower than the one observed in various other insect species with validoxylamine A (**3**), which is a strictly trehalase-specific inhibitor.

3.2. Bicyclic Iminosugars and Their Glycosyl Derivatives

The tropane alkaloids are bridged bicyclic amines combining a five-membered ring (pyrrolidine) and a six membered-ring (piperidine), with the nitrogen atom on the bridge. The tropane structure is also found in important medicinal alkaloids such as cocaine and atropine. These alkaloids are named calystegines and, due to their structural similarities with monocyclic iminosugars (such as DNJ, **12**), behave as glycosidase inhibitors.

In 1996, Asano and coworkers isolated several tropane alkaloids from the extracts of *Scopolia japonica*, a plant of the Solanaceae family [60]. A careful structural determination allowed the identification of a new compound, which was named as calystegine B$_4$ (**23**, Figure 4) in addition to other known calystegines. Interestingly, the calystegine A$_3$ (**21**), B$_2$ (**22**), and B$_4$ (**23**) (Figure 4) showed a remarkable and similar activity toward rat intestinal trehalase (IC$_{50}$ values of 12.0, 9.0, and 9.8 µM, respectively). This indicated that the presence of a hydroxy group and the stereochemistry at C-4 of calystegines had no significant effect on inhibitory activity towards rat small intestinal trehalase. The three calystegines—**21**, **22**, and **23**—were also active towards pig kidney trehalase, with IC$_{50}$ values of 13.0, 10.0, and 4.8 µM, respectively. These compounds were also tested towards non-mammalian trehalases of different origins. Calystegines B$_2$ (**22**) and B$_4$ (**23**) had very weak activities towards the enzyme from the pathogenic fungus *R. solani* (IC$_{50}$ values of 700 and 540 µM, respectively), and none of the calystegines exhibited any appreciable inhibition toward baker's yeast trehalase. More interestingly, calystegines **21**, **22**, and **23** exhibited good inhibitory activities toward the last instar larvae midgut trehalases form *B. mori* and *S. litura* (IC$_{50}$ in a 19–50 µM range). Among the three calystegines shown in Figure 4, calystegine B$_4$ (**23**) was more effective towards pig kidney trehalase than towards rat intestinal trehalase (IC$_{50}$ = 4.8 µM vs. IC$_{50}$ = 9.8 µM) and showed competitive inhibition with a K_i value of 1.2 µM. There are very few total syntheses of calystegine B$_4$, which is due to the structural complexity of this compound [61–63].

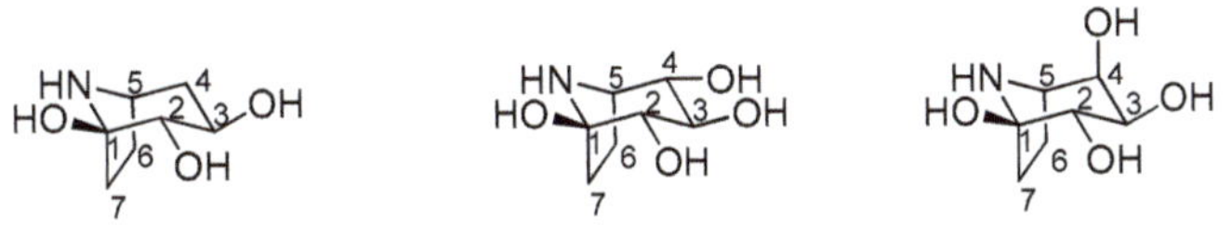

Calystegine A$_3$ (**21**) Calystegine B$_2$ (**22**) Calystegine B$_4$ (**23**)

Figure 4. Structures of calystegines A$_3$, B$_2$, and B$_4$.

Pyrrolizidines are bicyclic iminosugars bearing two fused pyrrolidine rings and a bridgehead nitrogen atom. Alexine (**24**, Figure 5) was isolated in 1988 from the pods of the legume *Alexa leiopetala* [64], and, differently from the previously known and broad class of pyrrolizidine alkaloids bearing a carbon substituent at C-1 [65,66], represented the first example of a pyrrolizidine alkaloid with a carbon substituent at C-3. Casuarine (**25**, Figure 5), a highly oxygenated pyrrolizidine, was isolated from the bark of *Casuarina equisetifolia* L. (Casuarinaceae) and from the leaves of *Eugenia jambolana Lam.* (Myrtaceae) [67], two plants widely employed in phytomedicine for their therapeutic action [68–71] together with its 6-*O*-α-D-glucopyranoside (**26**) [72]. During a study aimed at the structural elucidation of australine and related alkaloids, Asano and coworkers found that alexine (**24**) was a moderate inhibitor of porcine kidney trehalase (IC$_{50}$ = 55 µM) [73]. In the same work, casuarine (**25**) was reported to inhibit this enzyme quite strongly (IC$_{50}$ = 12 µM), and its 6-*O*-α-D-glucopyranoside (**26**) was an even more potent inhibitor. The glucoside **26**, indeed, inhibited porcine kidney trehalase with an IC$_{50}$ = 340 nM and in a competitive manner, with a K_i value of 18 nM. However, the compound isolated from natural source showed a purity of about 60%, thus, presumably, the pure compound was expected to have a lower K_i. Other examples of glycosyl iminosugars were later isolated from natural sources, indicating that their occurrence in nature is not so uncommon [74].

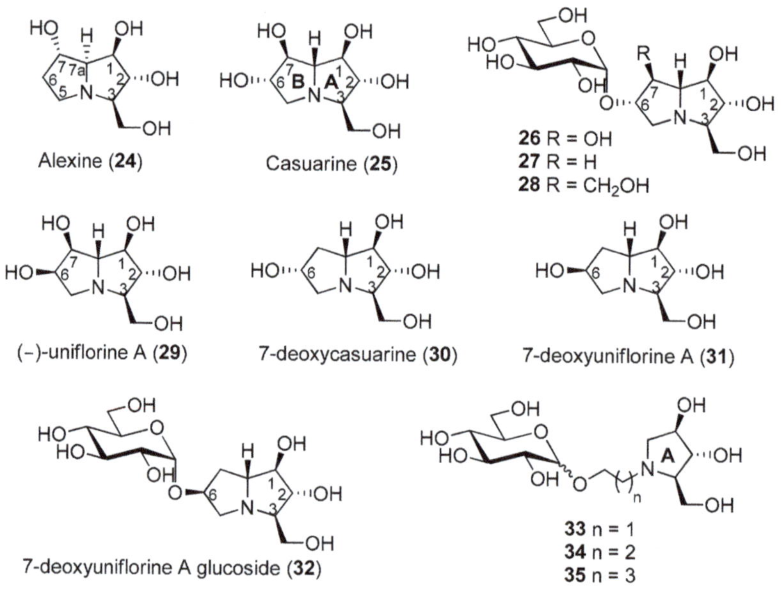

Figure 5. Structures of bicyclic or glycosylated iminosugar-based trehalase inhibitors.

The total synthesis of these pyrrolizidine alkaloids is quite challenging, since they bear several contiguous and well defined stereocenters. The first asymmetric synthesis of (+)-alexine (**24**) was accomplished by Yoda and co-workers in 2000 through synthetic elaboration of a key lactam derived from 2,3,5-tri-*O*-benzyl-β-arabinofuranose [75]. Since then, several other total syntheses of (+)-alexine (**24**) were reported [76–79].

Regarding casuarine (**25**), its first total synthesis dates back to 1999 and was reported by Denmark and co-workers, who employed a tandem [4 + 2]/[3 + 2] nitroalkene cycloaddition as the key step of the strategy, which involved a nitroalkene, a chiral vinyl ether and a vinyl silane as starting materials [80,81]. Later on, (+)-casuarine (**25**) was synthesized by other research groups starting from carbohydrate derivatives [82–84]. Goti and coworkers developed a synthetic strategy aimed at the total synthesis of (+)-casuarine (**25**), based on a stereoselective carbohydrate-based nitrone cycloaddition with a proper alkene [85]. The key step of the synthesis involved a Tamao–Fleming oxidation for the correct installation of the OH group at C-7. The same strategy also allowed the first total synthesis of its 6-*O*-α-glucoside (**26**), through a selective α-glucosylation of a glucosyl donor with a pyrrolizidine acceptor bearing a non-protected OH group at C-6. The use of different alkenes allowed the synthesis of other casuarine derivatives and their biological evaluation towards a wide range of glycosidases [86]. The total synthesis of **26** allowed to obtain this compound for the first time in a chemically pure form. Two novel casuarine-6-*O*-α-D-glucoside analogues **27** and **28** (Figure 5) were also synthesized by Cardona and coworkers, by means of a similar selective α-glucosylation of a glucosyl donor with pyrrolizidine acceptors bearing no substitution or the CH_2OH substitution at C-7, respectively [87]. The activity of this set of compounds was tested towards trehalases of different origins, namely porcine kidney, *E. coli* Tre37A [40] and *C. riparius* trehalases (Table 3) [54].

Table 3. Inhibition of porcine kidney, Tre37A and *C. riparius* trehalases by compounds **25**, **26**, **27**, and **28**.

Compound	Porcine Kidney, K_i	*E. coli* Tre37A, K_i	*C. riparius*, K_i
25	12 µM [73] [a]	17 µM [85]	0.12 µM [54]
26	11 nM [87]	12 nM [87]	0.66 nM [87]
27	138 nM [87]	86 nM [87]	22 nM [87]
28	>10 µM [87]	2.8 µM [87]	157 nM [87]

[a] Expressed as IC_{50}.

It was found that (+)-casuarine (**25**) was a strong inhibitor of both *E. coli* Tre37A and *C. riparius* trehalases (K_i = 17 µM and 0.12 µM, respectively). However, its 6-*O*-α-D-glucoside **26** was around 1000-fold more active (K_i = 12 nM and 0.66 nM, respectively), thus proving to be one of the most potent trehalase inhibitors described to date. The potency of glycosides **26**, **27**, and **28** towards the three trehalases reported in Table 3 showed a similar trend, demonstrating that the substitution at the C-7 position of the pyrrolizidine portion had a significant effect on the inhibition: the OH group at C-7 (in **26**) was preferential to a hydrogen atom (in **27**), and both were more potent than a CH_2OH group (as in **28**). However, in terms of specificity of inhibition, compounds **27** and **28** were more interesting, since they showed to be more selective towards the insect *C. riparius* trehalase over the mammalian enzyme. In particular, glucoside **27** displayed 6.3-fold selectivity, while glucoside **28** was more than 60-fold selective towards the *C. riparius* over the porcine kidney trehalase [87].

Tre37A enzyme has a buried cavity with two subsites. The −1 subsite is the catalytic site and it is specific for the substrate, while the +1 subsite is specific for the glucoside leaving group [40]. X-ray analysis of compound **26** in complex with the enzyme Tre37A was collected to 1.9 Å resolution [85] and revealed that the casuarine portion in **26** was bound to the −1 subsite of Tre37A, while the glucose moiety of **26** was placed in the +1 leaving group subsite. As the binding of casuarine (**25**) leaves free one of the two subsites of Tre37A, fewer interactions can form, and this rationalizes the lower affinity of 'monosaccharide-like' inhibitors with respect to 'disaccharide-like' ones. Interestingly, a sulfate group was co-crystallized together with **26** and formed hydrogen bonds with the OH at C-6 position of the sugar portion. More or less the same situation was found for the complex of **28** with Tre37A, which was collected to 2.1 Å resolution [87]. The pyrrolizidine portion was found in the -1 catalytic site with the CH_2OH group forming key hydrogen bonds interaction, and again a sulfate group was co-crystallized and formed hydrogen bonds with near residues.

(−)-Uniflorine A (6-*epi*-casuarine, **29**, Figure 5) was isolated in 2000 from the leaves of the tree *Eugenia uniflora* L. (Myrtaceae). This plant, widely distributed in South America, was used to prepare infusions used in folk medicine in antidiabetic preparations [88]. In 2004, the group of Pyne and co-workers proved, through the total synthesis of putative (−)-Uniflorine A, that the originally proposed structure (a polyhydroxylated indolizidine alkaloid) was not correct, thus demonstrating again the importance of total synthesis [89]. The structural reassignment was accomplished through the synthesis of its enantiomer (+)-uniflorine A [90]. Goti and coworkers accomplished the first total synthesis of **29** exploiting a similar strategy that allowed also the synthesis of (+)-casuarine (**25**) [91]. More recently, (−)-Uniflorine A was synthesized by SmI_2-mediated radical cross-coupling [92].

As shown in Table 4, **29** and its 7-deoxy analogue (7-deoxyuniflorine A, **31**, Figure 5) showed remarkable inhibitory properties against *C. riparius* trehalase (IC_{50} = 177 and 175 nM, respectively) in spite of not bearing an additional glucosyl moiety at C-6. More importantly, they both showed excellent selectivity (>5000) towards the insect trehalase. This finding suggested that the stereochemistry at position 6 was crucial to impart such selectivity and that the presence of an additional glucosyl moiety at C-6 was not essential. Compound **30**, with the opposite stereochemistry at C-6, was active only in the low micromolar range (IC_{50} = 1.22 µM) toward the insect enzyme, while the glucoside derivative **27** was more active (IC_{50} = 44 nM) but less selective (Table 4) [93]. However, preliminary tests in vivo of compound **31** on *S. littoralis* larvae failed.

Table 4. Inhibition (IC_{50}) of porcine kidney and *C. riparius* trehalases by compounds **29, 30, 31, 27, 32, 33, 34,** nd **35**.

Compound	Porcine Kidney, IC_{50}	*C. riparius*, IC_{50}	Selectivity [a]
29	>1 mM [93]	177 ± 18 nM [93]	>5649
30	20.6 ± 2.2 µM [93]	1.22 ± 0.08 µM [93]	17
31	>1 mM [93]	175 ± 12 nM [93]	>5714
27	479 ± 45 nM [93]	44 ± 1.0 nM [93]	10
32	190.60 ± 34.14 µM [94]	29.49 ± 7.26 µM [94]	6
33 α,β	7.67 ± 3.91 µM [94]	2.30 ± 0.13 µM [94]	3
33 α	27.64 ± 5.35 µM [94]	9.36 ± 1.49 µM [94]	3
33 β	5.84 ± 0.26 µM [94]	0.784 ± 0.059 µM [94]	7
34 α,β	n.d.	>1 mM	-
35 α,β	n.d.	>1 mM	-

[a] Selectivity is the ratio between the IC_{50} value against porcine kidney trehalase and the IC_{50} value against *C. riparius* trehalase; n.d.: not determined.

With the aim of synthesizing a more active glucoside derivative, Matassini and Cardona embarked with the total synthesis of glucoside **32**, which bears the same stereochemistry at C-6 of **29** and **31** [94]. Quite disappointingly, this compound showed activity only in the micromolar range (IC_{50} = 29.49 µM towards *C. riparius*), thus showing that a pyrrolizidine with this configuration is not able to place the glucosyl moiety within the enzyme cavity with favorable orientations. In order to reduce the overall number of synthetic steps, they also synthesized a series of simple and more flexible disaccharide mimetics **33–35** (Figure 5) with a pyrrolidine nucleus instead of the pyrrolizidine core, different spacers connecting the sugar and the iminosugar portion, and both configurations at the glucosyl moiety. Among the compounds synthesized, only compound **33 β** was active in the low micromolar range (IC_{50} = 0.784 µM towards *C. riparius*, Table 4) and showed good selectivity towards the insect enzyme.

4. Conclusions

The development of non-toxic environmentally friendly insecticides and fungicides for human health and for crop protection is of particular relevance, especially for less developed countries. Trehalase (EC3.2.1.28) has been identified in this regard, as a new target which is specific for insects and does not affect humans. Trehalase hydrolyses trehalose to two glucose units, a process which is essential to the life functions of several organisms, in particular fungi and insects, but does not affect vertebrates, who do not depend on the hydrolysis of this sugar. Thus, at least in principle, trehalase inhibitors were excellent candidates as new 'green and safe insecticides' to be used for protecting humans from diseases brought by insects and for crop protection.

Despite their attractiveness as a target, up to now there are no trehalase inhibitors that have been developed as commercial insecticides. This could be ascribed to a non-favorable plant uptake due to the high hydrophilicity of these compounds. Indeed, validamycin A, the only trehalase inhibitor that has been commercialized from the late 1960s (by Takeda Chemical Industries) for its curative effect toward the pathogen *R. solani*, was shown to behave as a pro-drug, being able to form the active pseudo-disaccharide trehalase inhibitor validoxylamine A in situ.

However, carbohydrate and carbocyclic based inhibitors with a pseudo-disaccharide structure (such as validoxylamine A, **3**) are very potent trehalase inhibitors but they are often not specific towards insects over the mammalian enzymes, and therefore cannot be considered as non-toxic. An effective alternative seems to be offered by the natural pseudo-disaccharide trehazolin (**5**), which shows remarkable insecticidal activity in larvae from different species, although it is not toxic towards mice. However, trehazolin strongly inhibits porcine kidney trehalase, the general model for the mammalian enzyme, thus suggesting that a deeper comprehension of the inhibition profile is necessary to develop safer inhibitors.

Indeed, when designing an inhibitor, the specificity is a very significant issue to be considered: the design of an effective insecticide which is able to target a given trehalase would require to be safe for plants and mammals, and if possible, also for insects which are of benefit in nature. In this regard, some iminosugar-based inhibitors showed an excellent specificity profile towards insect *C. riparius* trehalase. In particular, glucosyl derivatives of pyrrolizidine iminosugars showed excellent inhibitory activities toward this insect enzyme but, more importantly, it came out that the presence of an additional glucosyl moiety is not essential for selectively targeting the insect trehalase. In our opinion, the simplest pyrrolizidines **29** and **31** showing the highest selectivity towards the insect trehalases are among the most attractive hit compounds that deserve further studies in this direction. Although preliminary in vivo experiments did not give the desired effect, we deem that further studies are needed, in particular aiming at the transformation of the hit-compounds into pro-drugs with more favorable physico-chemical properties that will improve plant-uptake.

Funding: We thank MIUR-Italy ("Progetto Dipartimenti di Eccellenza 2018–2022" for the funds allocated to the Department of Chemistry "Ugo Schiff"), and for a fellowship to C.M.

Conflicts of Interest: The authors declare no conflict of interest.

References

1. Elbein, A.D. The metabolism of α,α-trehalose. *Adv. Carbohydr. Chem. Biochem.* **1974**, *30*, 227–256. [CrossRef] [PubMed]
2. Elbein, A.D.; Pan, Y.T.; Pastuszak, I.; Carroll, D. New insights on trehalose: A multifunctional molecule. *Glycobiology* **2003**, *13*, 17R–27R. [CrossRef] [PubMed]
3. Lunn, J.E.; Delorge, I.; Figueroa, C.M.; Van Dijck, P.; Stitt, M. Trehalose metabolism in plants. *Plant J.* **2014**, *79*, 544–567. [CrossRef] [PubMed]
4. Crowe, J.H. Trehalose as a "chemical chaperone": Fact and fantasy. In *Molecular Aspects of the Stress Response: Chaperones, Membranes and Networks*; Csermely, P., Vígh, L., Eds.; Springer: New York, NY, USA, 2007; Volume 594, pp. 143–158, ISBN 978-0-387-39975-1. [CrossRef]
5. Jain, N.K.; Roy, I. Effect of trehalose on protein structure. *Protein Sci.* **2009**, *18*, 24–36. [CrossRef] [PubMed]
6. Thompson, S.N. Trehalose—The insect "blood" sugar. In *Advances in Insect Physiology*; Simpson, S.J., Ed.; Elsevier Ltd.: Oxford, UK, 2003; Volume 31, pp. 205–285, ISBN 0-12-024231-1. [CrossRef]
7. Defaye, J.; Driguez, H.; Henrissat, B.; Bar-Guilloux, E. Stereochemistry of the hydrolysis of α,α-trehalose by trehalase, determined by using a labelled substrate. *Carbohydr. Res.* **1983**, *124*, 265–273. [CrossRef]
8. Becker, A.; Schlöder, P.; Steele, J.E.; Wegener, G. The regulation of trehalose metabolism in insects. *Experentia* **1996**, *52*, 433–439. [CrossRef] [PubMed]
9. Yoneyama, Y.; Lever, J.E. Apical trehalase expression associated with cell patterning after inducer treatment of LLC-PK1 monolayers. *J. Cell Physiol.* **1987**, *131*, 330–341. [CrossRef]
10. Dahlqvist, A. Assay of intestinal disaccharides. *Anal. Biochem.* **1968**, *22*, 99–107. [CrossRef]
11. Thevelein, J.M. Regulation of trehalose metabolism in fungi. *Microbiol. Rev.* **1984**, *48*, 42–59. [CrossRef]
12. Bini, D.; Cardona, F.; Gabrielli, L.; Russo, L.; Cipolla, L. Trehalose mimetics as inhibitors of trehalose processing enzymes. In *Carbohydrate Chemistry Chemical and Biological Approaches*; Rauter, A.M., Lidhorst, T., Eds.; The Royal Society of Chemistry: Cambridge, UK, 2012; Volume 37, pp. 259–302, ISBN 978-1-84973-154-6. [CrossRef]
13. Wegener, G.; Tschiedel, V.; Schlöder, P.; Ando, O. The toxic and lethal effects of the trehalase inhibitor trehazolin in locusts are caused by hypoglycaemia. *J. Exp. Biol.* **2003**, *206*, 1233–1240. [CrossRef]
14. Compain, P.; Martin, O.R. *Iminosugars: From Synthesis to Therapeutic Applications*; John Wiley & Sons Ltd.: Chichester, West Sussex, UK, 2007; ISBN 978-0-470-03391-3. [CrossRef]
15. Iwasa, T.; Higashide, E.; Yamamoto, H.; Shibata, M. Studies on validamycins, new antibiotics. II. Production and biological properties of validamycins A and B. *J. Antibiot.* **1971**, *24*, 107–113. [CrossRef] [PubMed]
16. Horii, S.; Kameda, Y.; Kawahara, K. Studies on validamycins, new antibiotics. VIII. Isolation and characterization of validamycins C, D, E and F. *J. Antibiot.* **1972**, *25*, 48–53. [CrossRef]
17. Kameda, Y.; Asano, N.; Yamaguchi, T.; Matsui, K.; Horii, S.; Fukase, H. Validamycin G and validoxylamine G, new members of the validamycins. *J. Antibiot.* **1986**, *39*, 1491–1494. [CrossRef] [PubMed]
18. Asano, N.; Yamaguchi, T.; Kameda, Y.; Matsui, K. Effect of Validamycins on Glycohydrolases of *Rhizoctonia Solani*. *J. Antibiot.* **1987**, *40*, 526–532. [CrossRef]
19. Kameda, Y.; Asano, N.; Yamaguchi, T.; Matsui, K. Validoxylamines as trehalase inhibitors. *J. Antibiot.* **1987**, *40*, 563–565. [CrossRef] [PubMed]

20. Asano, N.; Takeuchi, M.; Kameda, Y.; Matsui, K.; Kono, Y. Trehalase inhibitors, validoxylamine A and related compounds as insecticides. *J. Antibiot.* **1990**, *43*, 722–726. [CrossRef] [PubMed]
21. Kyosseva, S.V.; Kyossev, Z.N.; Elbein, A.D. Inhibitors of pig kidney trehalase. *Arch. Biochem. Biophys.* **1995**, *316*, 821–826. [CrossRef] [PubMed]
22. Salleh, H.M.; Honek, J.F. Time-dependent inhibition of porcine kidney trehalase by aminosugars. *FEBS Lett.* **1990**, *262*, 359–362. [CrossRef]
23. Jin, L.-Q.; Zheng, Y.-G. Inhibitory effects of validamycin compounds on the termites trehalase. *Pest. Biochem. Physiol.* **2009**, *95*, 28–32. [CrossRef]
24. Takeda, S.; Kono, Y.; Kameda, Y. Induction of non-diapause eggs in Bombyx mori by a trehalase inhibitor. *Entomol. Exp. Appl.* **1988**, *46*, 291–294. [CrossRef]
25. Kono, Y.; Takeda, S.; Kameda, Y.; Takahashi, M.; Matsushita, K.; Nishina, M.; Hori, E. Lethal activity of trehalase inhibitor, validoxylamine A, and its influence on the blood sugar level in Bombyx mori (Lepidoptera: Bombycidae). *Appl. Entomol. Zool.* **1993**, *28*, 379–386. [CrossRef]
26. Kono, Y.; Takeda, S.; Kameda, Y. Lethal activity of a trehalase inhibitor, validoxylamine A, against Mamestra brassicae and Spodoptera litura. *J. Pestic. Sci.* **1994**, *19*, 39–42. [CrossRef]
27. Kono, Y.; Takeda, S.; Kameda, Y.; Takahashi, M.; Matsushita, K.; Nishina, M.; Hori, E. NMR analysis of the effect of validoxylamine A, a trehalase inhibitor, on the larvae of the cabbage armyworm. *J. Pestic. Sci.* **1995**, *20*, 83–91. [CrossRef]
28. Kono, Y.; Takahashi, M.; Matsushita, K.; Nishina, M.; Kameda, Y.; Hori, E. Inhibition of flight in *Periplaneta americana* (Linn.) by a trehalase inhibitor, validoxylamine A. *J. Insect Physiol.* **1994**, *40*, 455–461. [CrossRef]
29. Kono, Y.; Takahashi, M.; Mihara, M.; Matsushita, K.; Nishina, M.; Kameda, Y. Effect of a Trehalase Inhibitor, Validoxylamine A, on Oocyte Development and Ootheca Formation in Periplaneta americana (Blattodea, Blattidae). *Appl. Entomol. Zool.* **1997**, *32*, 293–301. [CrossRef]
30. Vértesy, L.; Fehlhaber, H.-W.; Schulz, A. The Trehalase Inhibitor Salbostatin, a Novel Metabolite from *Streptomyces albus*, ATCC21838. *Angew. Chem. Int. Ed. Engl.* **1994**, *33*, 1844–1846. [CrossRef]
31. Yamagishi, T.; Uchida, C.; Ogawa, S. Total Synthesis of Trehalase Inhibitor Salbostatin. *Chem. Eur. J.* **1995**, *9*, 634–636. [CrossRef]
32. Ando, O.; Satake, H.; Itoi, K.; Sato, A.; Nakajima, M.; Takahashi, S.; Haruyama, H.; Ohkuma, Y.; Kinoshita, T.; Enokita, R. Trehazolin, a new trehalase inhibitor. *J. Antibiot.* **1991**, *44*, 1165–1168. [CrossRef]
33. Ando, O.; Nakajima, M.; Hamano, K.; Itoi, K.; Takahashi, S.; Takamatsu, A.; Sato, A.; Enokita, R.; Okazaki, T.; Haruyama, H.; et al. Isolation on trehalamine, the aglycon of trehazolin, from microbial broths and characterization of trehazolin related compounds. *J. Antibiot.* **1993**, *46*, 1116–1125. [CrossRef]
34. El Nemr, A.; El Ashry, E.H. Chapter 3—Potential trehalase inhibitors: Syntheses of trehazolin and its analogues. In *Advances in Carbohydrate Chemistry and Biochemistry*; Horton, D., Ed.; Elsevier: Amsterdam, The Netherlands, 2011; Volume 65, pp. 45–114. [CrossRef]
35. Ando, O.; Nakajima, M.; Kifune, M.; Fang, H.; Tanzawa, K. Trehazolin, a slow, tight-binding inhibitor of silkworm trehalase. *Biochim. Biophys. Acta* **1995**, *1244*, 295–302. [CrossRef]
36. Uchida, C.; Yamagashi, T.; Kitahashi, H.; Iwaisaki, Y.; Ogawa, S. Further chemical modification of trehalase inhibitor trehazolin: Structure and inhibitory-activity relationship of the inhibitor. *Bioorg. Med. Chem.* **1995**, *3*, 1605–1624. [CrossRef]
37. Berecibar, A.; Grandjean, A.; Siriwardena, A. Synthesis and Biological Activity of Natural Aminocyclopentitol Glycosidase Inhibitors: Mannostatins, Trehazolin, Allosamidins, and Their Analogues. *Chem. Rev.* **1999**, *99*, 779–844. [CrossRef] [PubMed]
38. Kobayashi, Y. Chemistry and biology of trehazolins. *Carbohydr. Res.* **1999**, *315*, 3–15. [CrossRef]
39. Chiara, J.L.; Storch de Gracia, I.; García, Á.; Bastida, Á.; Bobo, S.; Martín-Ortega, M.D. Synthesis, inhibition properties, and theoretical study of the new nanomolar trehalase inhibitor 1-thiatrehazolin: Towards a structural understanding of trehazolin inhibition. *ChemBioChem* **2005**, *6*, 186–191. [CrossRef] [PubMed]
40. Gibson, R.P.; Gloster, T.M.; Roberts, S.; Warren, R.A.J.; Storch de Gracia, I.; García, Á.; Chiara, J.L.; Davies, G.J. Molecular Basis for Trehalase Inhibition Revealed by the Structure of Trehalase in Complex with Potent Inhibitors. *Angew. Chem. Int. Ed.* **2007**, *46*, 4115–4119. [CrossRef] [PubMed]
41. Asano, N.; Kato, A.; Matsui, K. Two Subsites on the Active Center of Pig Kidney Trehalase. *Eur. J. Biochem.* **1996**, *240*, 692–698. [CrossRef] [PubMed]

42. Ando, O.; Kifune, M.; Nakajima, M. Effects of Trehazolin, a Potent Trehalase Inhibitor, on *Bombyx mori* and Plant Pathogenic Fungi. *Biosci. Biotechnol. Biochem.* **1995**, *59*, 711–712. [CrossRef]
43. Wegener, G.; Macho, C.; Schlöder, P.; Kamp, G.; Ando, O. Long-term effects of the trehalase inhibitor trehazolin on trehalase activity in locust flight muscles. *J. Exp. Biol.* **2010**, *213*, 3852–3857. [CrossRef]
44. Qian, X.; Liu, Z.; Li, Z.; Li, Z.; Song, G. Synthesis and Quantitative Structure-Activity Relationships of Fluorine-Containing 4,4-Dihydroxylmethyl-2-aryliminooxazo(thiazo)lidines as Trehalase inhibitors. *J. Agric. Food Chem.* **2001**, *49*, 5279–5284. [CrossRef]
45. Qian, X.; Li, Z.; Liu, Z.; Song, G.; Li, Z. Syntheses and activities as trehalase inhibitors of N-arylglycosylamines derived from fluorinated anilines. *Carbohydr. Res.* **2001**, *336*, 79–82. [CrossRef]
46. Niwa, T.; Inouye, S.; Tsuruoka, T.; Koaze, Y.; Niida, T. "Nojirimycin" as a potent inhibitor of glucosidase. *Agric. Biol. Chem.* **1970**, *34*, 966–968. [CrossRef]
47. Schmidt, D.D.; Frommer, W.; Muller, L.; Trusheit, E. Glucosidase-inhibitoren aus Bazillen. *Naturwissenschaften* **1979**, *66*, 584–585. [CrossRef]
48. Evans, S.V.; Fellows, L.E.; Bell, E.A. Glucosidase and trehalase inhibition by 1,5-dideoxy-1,5-imino-D-mannitol, a cyclic amino alditol from *Lonchocarpus sericeus*. *Phytochemistry* **1983**, *22*, 768–770. [CrossRef]
49. Murao, S.; Miyata, S. Isolation and characterization of a new trehalase inhibitor, S-GI. *Agric. Biol. Chem.* **1980**, *44*, 219–221. [CrossRef]
50. Asano, N.; Oseki, K.; Kizu, H.; Matsui, K. Nitrogen-in-the-ring pyranoses and furanoses: Structural basis of inhibition of mammalian glycosidases. *J. Med. Chem.* **1994**, *37*, 3701–3706. [CrossRef] [PubMed]
51. Anzeveno, P.B.; Creemer, L.J.; Daniel, L.K.; King, C.H.R.; Liu, P.S. A facile, practical synthesis of 2,6-dideoxy-2,6-imino-7-O-β-D-glucopyranosyl-D-glycero-L-gulo-heptitol (MDL 25,637). *J. Org. Chem.* **1989**, *54*, 2539–2542. [CrossRef]
52. Watanabe, S.; Kato, H.; Nagayama, K.; Abe, H. Isolation of 2R,SR-Dihydroxymethyl-3R,4R-dihydroxypyrrolidine (DMDP) from the Fermentation broth of *Streptomyces* sp. KSC-S791. *Biosci. Biotechnol. Biochem.* **1995**, *59*, 936–937. [CrossRef]
53. Bini, D.; Forcella, M.; Cipolla, L.; Fusi, P.; Matassini, C.; Cardona, F. Synthesis of Novel Iminosugar-Based Trehalase Inhibitors by Cross-Metathesis Reactions. *Eur. J. Org. Chem.* **2011**, 3995–4000. [CrossRef]
54. Forcella, M.; Cardona, F.; Goti, A.; Parmeggiani, C.; Cipolla, L.; Gregori, M.; Schirone, R.; Fusi, P.; Parenti, P. A membrane-bound trehalase from *Chironomus riparius* larvae: Purification and sensitivity to inhibition. *Glycobiology* **2010**, *20*, 1186–1195. [CrossRef]
55. Bini, D.; Cardona, F.; Forcella, M.; Parmeggiani, C.; Parenti, P.; Nicotra, F.; Cipolla, L. Synthesis and biological evaluation of nojirimycin- and pyrrolidine-based trehalase inhibitors. *Beilstein J. Org. Chem.* **2012**, *8*, 514–521. [CrossRef]
56. Cipolla, L.; Sgambato, A.; Forcella, M.; Fusi, P.; Parenti, P.; Cardona, F.; Bini, D. N-Bridged 1-deoxynojirimycin dimers as selective insect trehalase inhibitors. *Carbohydr. Res.* **2014**, *389*, 46–49. [CrossRef] [PubMed]
57. Cendret, V.; Legigan, T.; Mingot, A.; Thibaudeau, S.; Adachi, I.; Forcella, M.; Parenti, P.; Bertrand, J.; Becq, F.; Norez, C.; et al. Synthetic deoxynojirimycin derivatives bearing a thiolated, fluorinated or unsaturated N-alkyl chain: Identification of potent α-glucosidase and trehalase inhibitors as well as F508del-CFTR correctors. *Org. Biomol. Chem.* **2015**, *13*, 10734–10744. [CrossRef] [PubMed]
58. Konno, K.; Ono, H.; Nakamura, M.; Tateishi, K.; Hirayama, C.; Tamura, Y.; Hattori, M.; Koyama, A.; Kohno, K. Mulberry latex rich in anti-diabetic sugar-mimic alkaloids forces dieting on caterpillars. *Proc. Natl. Acad. Sci. USA* **2006**, *103*, 1337–1341. [CrossRef]
59. Hirayama, C.; Konno, K.; Wasano, N.; Nakamura, M. Differential effects of sugar-mimic alkaloids in mulberry latex on sugar metabolism and disaccharidases of Eri and domesticated silkworms: Enzymatic adaptation of *Bombyx mori* to mulberry defense. *Insect Biochem. Mol. Biol.* **2007**, *37*, 1348–1358. [CrossRef] [PubMed]
60. Asano, N.; Kato, A.; Kizu, H.; Matsui, K.; Watson, A.A.; Nash, R.J. Calystegine B_4, a novel trehalase inhibitor from *Scopolia japonica*. *Carbohydr. Res.* **1996**, *293*, 195–204. [CrossRef]
61. Skaaanderup, P.R.; Madsen, R. Short syntheses of enantiopure calystegines B_2, B_3, and B_4. *Chem. Commun.* **2001**, 1106–1107. [CrossRef]
62. Skaaanderup, P.R.; Madsen, R. A Short Synthetic Route to the Calystegine Alkaloids. *J. Org. Chem.* **2003**, *68*, 2115–2122. [CrossRef]
63. Moosophon, P.; Baird, M.C.; Kanokmedhakul, S.; Pyne, S.G. Total Synthesis of Calystegine B_4. *Eur. J. Org. Chem.* **2010**, 3337–3344. [CrossRef]

64. Nash, R.J.; Fellows, L.E.; Dring, J.V.; Fleet, G.W.J.; Derome, A.E.; Hamor, T.A.; Scofield, A.M.; Watkin, D.J. Isolation from alexia leiopetala and X-ray crystal structure of alexine, (1r,2r,3r,7s,8s)-3-hydroxymethyl-1,2,7-trihydroxypyrrolizidine[(2r,3r,4r,5s,6s)-2-hydroxymethyl-1-azabicyclo [3.3.0]octan-3,4,6-triol], a unique pyrrolizidine alkaloid. *Tetrahedron Lett.* **1988**, *29*, 2487–2490. [CrossRef]
65. Wróbel, J.T. Chapter 7 Pyrrolizidine Alkaloids. In *The Alkaloids: Chemistry and Pharmacology*; Brossi, A., Ed.; Elsevier: Amsterdam, The Netherlands, 1985; Volume 26, pp. 327–385, ISBN 9780080865508.
66. Robins, D.J. Chapter 1 Biosynthesis of Pyrrolizidine and Quinolizidine Alkaloids. In *The Alkaloids: Chemistry and Pharmacology*; Cordell, G.A., Ed.; Elsevier: Amsterdam, The Netherlands, 1995; Volume 46, pp. 1–61, ISBN 978-0-12-469546-7.
67. Nash, R.J.; Thomas, P.I.; Waigh, R.D.; Fleet, G.W.J.; Wormald, M.R.; Lilley, P.M.d.Q.; Watkin, D.J. Casuarine: A very highly oxygenated pyrrolizidine alkaloid. *Tetrahedron Lett.* **1994**, *35*, 7849–7852. [CrossRef]
68. Chopra, R.N.; Nayar, S.L.; Chopra, I.C. *Glossary of Indian Medicinal Plants*; Council of Scientific and Industrial Research: New Delhi, India, 1956; ISBN 8172360487.
69. Nair, R.B.; Santhakumari, G. Anti–Diabetic Activity of the Seed Kernel of Syzygium Cumini Linn. *Anc. Sci. Life* **1986**, *6*, 80–84. [PubMed]
70. Grover, J.K.; Yadav, S.; Vats, V. Medicinal plants of India with anti-diabetic potential. *J. Ethnopharmacol.* **2002**, *81*, 81–100. [CrossRef]
71. Mentreddy, S.R. Medicinal plant species with potential antidiabetic properties. *J. Sci. Food. Agric.* **2007**, *87*, 743–750. [CrossRef]
72. Wormald, M.R.; Nash, R.J.; Watson, A.A.; Bhadoria, B.K.; Langford, R.; Sims, M.; Fleet, G.W.J. Casuarine-6-α-D-glucoside from Casuarina equisetifolia and Eugenia jambolana. *Carbohydr. Lett.* **1996**, *2*, 169–174.
73. Kato, A.; Kano, E.; Adachi, I.; Molyneux, R.J.; Watson, A.A.; Nash, R.J.; Fleet, G.W.J.; Wormald, M.R.; Kizu, H.; Ikeda, K.; et al. Australine and related alkaloids: Easy structural confirmation by ^{13}C NMR spectral data and biological activities. *Tetrahedron Asymmetry* **2003**, *14*, 325–331. [CrossRef]
74. Asano, N.; Yamauchi, T.; Kagamifuchi, K.; Shimizu, N.; Takahashi, S.; Takatsuka, H.; Ikeda, K.; Kizu, H.; Chuakul, W.; Kettawan, A.; et al. Iminosugar-Producing Thai Medicinal Plants. *J. Nat. Prod.* **2005**, *68*, 1238–1242. [CrossRef]
75. Yoda, H.; Katoh, H.; Takabe, K. Asymmetric total synthesis of natural pyrrolizidine alkaloid, (+)-alexine. *Tetrahedron Lett.* **2000**, *41*, 7661–7665. [CrossRef]
76. Dressel, M.; Restorp, P.; Somfai, P. Total Synthesis of (+)-Alexine by Utilizing a Highly Stereoselective [3 + 2] Annulation Reaction of an N-Tosyl-α-Amino Aldehyde and a 1,3-Bis(silyl)propene. *Chem. Eur. J.* **2008**, *14*, 3072–3077. [CrossRef]
77. Takahashi, M.; Maehara, T.; Sengoku, T.; Fujita, N.; Takabe, K.; Yoda, H. New asymmetric strategy for the total synthesis of naturally occurring (+)-alexine and (−)-7-*epi*-alexine. *Tetrahedron* **2008**, *64*, 5254–5261. [CrossRef]
78. Yu, L.; Somfai, P. Enantioselective synthesis of *anti*-3-alkenyl-2- amido-3-hydroxy esters: Application to the total synthesis of (+)-alexine. *RSC Adv.* **2019**, *9*, 2799–2802. [CrossRef]
79. Myeong, I.-S.; Jung, C.; Ham, W.-H. Total Syntheses of (−)-7-*epi*-Alexine and (+)-Alexine Using Stereoselective Allylation. *Synthesis* **2019**, *51*, 3471–3476. [CrossRef]
80. Denmark, S.E.; Hurd, A.R. Synthesis of (+)-Casuarine. *Org. Lett.* **1999**, *1*, 1311–1314. [CrossRef] [PubMed]
81. Denmark, S.E.; Hurd, A.R. Synthesis of (+)-Casuarine. *J. Org. Chem.* **2000**, *65*, 2875–2886. [CrossRef] [PubMed]
82. Izquierdo, I.; Plaz, M.T.; Tamayo, J.A. Polyhydroxylated pyrrolizidines. Part 6: A new and concise stereoselective synthesis of (+)-casuarine and its 6,7-di*epi* isomer, from DMDP. *Tetrahedron* **2005**, *61*, 6527–6533. [CrossRef]
83. Ritthiwigrom, T.; Willis, A.C.; Pyne, S.G. Total Synthesis of Uniflorine A, Casuarine, Australine, 3-*epi*-Australine, and 3,7-Di-*epi*-australine from a Common Precursor. *J. Org. Chem.* **2010**, *75*, 815–824. [CrossRef]
84. Parmeggiani, C.; Cardona, F.; Giusti, L.; Reissig, H.-U.; Goti, A. Stereocomplementary Routes to Hydroxylated Nitrogen Heterocycles: Total Syntheses of Casuarine, Australine, and 7-epi-Australine. *Chem. Eur. J.* **2013**, *19*, 10595–10604. [CrossRef]
85. Cardona, F.; Parmeggiani, C.; Faggi, E.; Bonaccini, C.; Gratteri, P.; Sim, L.; Gloster, T.M.; Roberts, S.; Davies, G.J.; Rose, D.R.; et al. Total Synthesis of Casuarine and Its 6-*O*-α-Glucoside: Complementary

Inhibition towards Glycoside Hydrolases of the GH31 and GH37 Families. *Chem. Eur. J.* **2009**, *15*, 1627–1636. [CrossRef]
86. Bonaccini, C.; Chioccioli, M.; Parmeggiani, C.; Cardona, F.; Lo Re, D.; Soldaini, G.; Vogel, P.; Bello, C.; Goti, A.; Gratteri, P. Synthesis, Biological Evaluation and Docking Studies of Casuarine Analogues: Effects of Structural Modifications at Ring B on Inhibitory Activity Towards Glucoamylase. *Eur. J. Org. Chem.* **2010**, 5574–5585. [CrossRef]
87. Cardona, F.; Goti, A.; Parmeggiani, C.; Parenti, P.; Forcella, M.; Fusi, P.; Cipolla, L.; Roberts, S.M.; Davies, G.J.; Gloster, T.M. Casuarina-6-O-α-D-glucoside and its analogues are tight binding inhibitors of insect and bacterial trehalases. *Chem. Commun.* **2010**, *46*, 2629–2631. [CrossRef]
88. Matsumura, T.; Kasai, M.; Hayashi, T.; Arisawa, M.; Momose, Y.; Arai, I.; Amagaya, S.; Komatsu, Y. a-glucosidase Inhibitors from Paraguayan natural medicine, Ñangapiry, the leaves of *Eugenia uniflora*. *Pharm. Biol.* **2000**, *38*, 302–307. [CrossRef]
89. Davis, A.S.; Pyne, S.G.; Skelton, B.W.; White, A.H. Synthesis of Putative Uniflorine A. *J. Org. Chem.* **2004**, *69*, 3139–3143. [CrossRef] [PubMed]
90. Ritthiwigrom, T.; Pyne, S.G. Synthesis of (+)-Uniflorine A: A Structural Reassignment and a Configurational Assignment. *Org. Lett.* **2008**, *10*, 2769–2771. [CrossRef]
91. Parmeggiani, C.; Martella, D.; Cardona, F.; Goti, A. Total Synthesis of (−)-Uniflorine A. *J. Nat. Prod.* **2009**, *72*, 2058–2060. [CrossRef]
92. Liu, X.-K.; Qiu, S.; Xiang, Y.-G.; Ruan, Y.-P.; Zheng, X.; Huang, P.-Q. SmI$_2$-Mediated Radical Cross-Couplings of α-Hydroxylated Aza-hemiacetals and N,S-Acetals with α,β-Unsaturated Compounds: Asymmetric Synthesis of (+)-Hyacinthacine A$_2$, (−)-Uniflorine A, and (+)-7-*epi*-Casuarine. *J. Org. Chem.* **2011**, *76*, 4952–4963. [CrossRef] [PubMed]
93. D'Adamio, G.; Sgambato, A.; Forcella, M.; Caccia, S.; Parmeggiani, C.; Casartelli, M.; Parenti, P.; Bini, D.; Cipolla, L.; Fusi, P.; et al. New Synthesis and biological evaluation of uniflorine A derivatives: Towards specific insect trehalase inhibitors. *Org. Biomol. Chem.* **2015**, *13*, 886–892. [CrossRef] [PubMed]
94. D'Adamio, G.; Forcella, M.; Fusi, P.; Parenti, P.; Matassini, C.; Ferhati, X.; Vanni, C.; Cardona, F. Probing the Influence of Linker Length and Flexibility in the Design and Synthesis of New Trehalase Inhibitors. *Molecules* **2018**, *23*, 436. [CrossRef] [PubMed]

© 2020 by the authors. Licensee MDPI, Basel, Switzerland. This article is an open access article distributed under the terms and conditions of the Creative Commons Attribution (CC BY) license (http://creativecommons.org/licenses/by/4.0/).

MDPI
St. Alban-Anlage 66
4052 Basel
Switzerland
Tel. +41 61 683 77 34
Fax +41 61 302 89 18
www.mdpi.com

Molecules Editorial Office
E-mail: molecules@mdpi.com
www.mdpi.com/journal/molecules

www.ingramcontent.com/pod-product-compliance
Lightning Source LLC
LaVergne TN
LVHW070145100526
838202LV00015B/1892